HANDBOOK OF THE
Psychology of Science

Gregory J. Feist, PhD, is an associate professor of Psychology in Personality and Adult Development and director of the Experimental Graduate Program in Psychology at San Jose State University. He is widely published in the psychology of creativity, the psychology of science, and the development of scientific talent.

Michael E. Gorman, PhD, is a professor at the University of Virginia, where he has taught such classes as Scientific and Technological Thinking and Social Implications of Nanotechnology. From 2000 to 2003, he was chair of the Division of Technology, Culture, and Communication. He is also an associate editor of *Journal of Psychology of Science and Technology.*

HANDBOOK OF THE
Psychology of Science

Gregory J. Feist, PhD

Michael E. Gorman, PhD

Editors

SPRINGER PUBLISHING COMPANY
NEW YORK

Springer Publishing Company, LLC
11 West 42nd Street
New York, NY 10036
www.springerpub.com

Acquisitions Editor: Nancy Hale
Associate Editor: Kathryn Corasaniti
Composition: Newgen Imaging

ISBN: 978-0-8261-0623-0
E-book ISBN: 978-0-8261-0624-7

12 13 14 15/ 5 4 3 2 1

The author and the publisher of this Work have made every effort to use sources believed to be reliable to provide information that is accurate and compatible with the standards generally accepted at the time of publication. The author and publisher shall not be liable for any special, consequential, or exemplary damages resulting, in whole or in part, from the readers' use of, or reliance on, the information contained in this book. The publisher has no responsibility for the persistence or accuracy of URLs for external or third-party Internet Websites referred to in this publication and does not guarantee that any content on such Websites is, or will remain, accurate or appropriate.

Library of Congress Cataloging-in-Publication Data

Handbook of the psychology of science / editors, Gregory J. Feist, PhD & Michael E. Gorman, PhD.
   pages cm
Includes index.
 ISBN 978-0-8261-0623-0 — ISBN 978-0-8261-0624-7 (ebk.) (print)
 1. Science—Psychological aspects. 2. Science and psychology. I. Feist, Gregory J. II. Gorman, Michael E., 1952-
Q175.H2646 2013
 501'.9—dc23   2012033698

Special discounts on bulk quantities of our books are available to corporations, professional associations, pharmaceutical companies, health care organizations, and other qualifying groups.

If you are interested in a custom book, including chapters from more than one of our titles, we can provide that service as well.

For details, please contact:
Special Sales Department, Springer Publishing Company, LLC
11 West 42nd Street, 15th Floor, New York, NY 10036-8002s
Phone: 877-687-7476 or 212-431-4370; Fax: 212-941-7842
Email: sales@springerpub.com

Printed in the United States of America by Gasch Printing.

*I dedicate this book to my lovely wife, Erika,
and two vibrant sons Jerry and Evan.*

— GJF

Contents

Contributors

E. J. Capaldi, PhD
Department of Psychological
  Sciences
Purdue University
West Lafayette, Indiana

Steve Croker, PhD
Department of Psychology
Illinois State University
Normal, Illinios

Gregory J. Feist, PhD
Associate Professor of Psychology
Director, MA Program for
  Psychological Science
San Jose State University
San Jose, CA

Steve Fuller, PhD
Department of Sociology
University of Warwick
Coventry, United Kingdom

Michael E. Gorman, PhD
Professor
University of Virginia
Charlottesville, Virginia

Sven Hemlin, PhD
Department of Psychology
University of Gothenburg
Gothenburg, Sweden

Jamie Jirout, PhD
Department of Psychology
Temple University
Philadelphia, Pennsylvania

Joanne E. Kane, PhD
Department of Psychology
Princeton University
Princeton, New Jersey

David Klahr, PhD
Department of Psychology
Carnegie Mellon University
Pittsburgh, Pennsylvania

Barbara Koslowski, EdD
Department of Human
  Development
Cornell University
Ithaca, New York

Neelam Kumar, PhD
National Institute of Science and
  Technology Development
New Delhi, India

Anton E. Lawson, PhD
School of Life Sciences
Arizona State University
Tempe, Arizona

Sofia Liberman, PhD
Facultad de Psicologiá
Universidad Nacional Autónoma
   de México
Mexico City, Mexico

Bryan Matlen, PhD
Department of Psychology
Carnegie Mellon University
Pittsburgh, Pennsylvania

Anne Moyer, PhD
Department of Psychology
Stony Brook University
Stony Brook, New York

Lisa Olsson, PhD
Department of Psychology
University of Gothenburg
Gothenburg, Sweden

Robert W. Proctor, PhD
Department of Psychological
   Sciences
Purdue University
West Lafayette, Indiana

William McKinley Runyan, PhD
School of Social Welfare
Institute of Personality and Social
   Research
University of California
Berkeley, California

Anna Dorothea Schulze, PhD
Institute of Psychology
Humboldt Universität zu Berlin
Berlin, Germany

Christian D. Schunn, PhD
Learning Research and
   Development Center
University of Pittsburgh
Pittsburgh, Pennsylvania

Verena Seuffert, MA
Institute of Psychology
Humboldt Universität zu Berlin
Berlin, Germany

Dean Keith Simonton, PhD
Department of Psychology
University of California
Davis, California

J. Gregory Trafton, PhD
HCI Laboratory
Naval Research Laboratory
Washington, DC

Ryan D. Tweney, PhD
Department of Psychology
Bowling Green State University
Bowling Green, Ohio

Gregory D. Webster, PhD
Department of Psychology
University of Florida
Gainesville, Florida

Kurt Bernardo Wolf, PhD
Instituto de Ciencias Físicas
Universidad Nacional Autónoma de
   México
Cuernavaca, México

Corinne Zimmerman, PhD
Department of Psychology
Illinois State University
Normal, Illinios

Preface

Science and technology are the defining features of the modern world. It is not surprising that scholars in history, philosophy, and sociology have long turned their attention to the products and processes of science. The psychology of science is the newest addition to the studies of science, taking its place alongside philosophy, history, and sociology of science. Indeed the first organized society, the International Society for the Psychology of Science and Technology (ISPST), was established only in 2006 and the first peer-reviewed journal (*Journal of the Psychology of Science and Technology*) was launched 2 years later. As Feist (2006) wrote in *The Psychology of Science and the Origins of the Scientific Mind*, new fields become established once they progress through a known sequence of events; first, lone voices argue for or make discoveries before others; second, groups of individuals coagulate around the originators and identify with the movement; and third, societies, journals, and graduate programs form the infrastructure. Crucial to a field maintaining its infrastructure is having an avenue for young scholars to identify with the field and choose it as their career path. University courses and graduate programs need to provide that foundation. Although much progress has been made in the psychology of science—a new society, regular conferences, and peer-reviewed journals—courses and university programs are still lacking or in the early stages of development.

This is where a handbook becomes especially important. Indeed, handbooks are offered once a field has matured to the point that a definitive state-of-the-art source is needed to highlight the current knowledge on a given topic. We are happy to say that the psychology of science has finally reached the point in its development where a handbook is necessary. As the one and only handbook on the topic, the *Handbook of the Psychology of Science* will be the definitive source for students and scholars interested in how psychological forces shape and mold scientific thought and behavior.

This volume differs in some ways from a typical handbook. Like other handbooks, the chapters are designed to give state-of-the-art overviews and reviews of current research areas in the various topics in the psychology of science. The chapters, however, are different in that they also are meant to provoke future research. The field is still young, in its formative stages of development, and there is much to be done. These chapters implicitly and explicitly offer ideas for the next steps. In addition, in the conclusion section, the editors gather these trends and make additional suggestions about what need to be done if the field is to survive and strive well into the future.

We have organized the handbook into six major sections:

- Introduction and History
- Foundational Psychologies of Science
- Development and Theory Change
- Special Topics
- Applied Psychologies of Science
- Past and Future of Psychology of Science

In the first section, we provide both an introduction to the field and offer a historical account for how the psychology of science has been an integral part of the history of psychology. In the second section, we offer an overview of the foundations of a mature psychology of science. Here, leading figures overview and summarize each of the major areas of psychology—developmental, cognitive, personality, social—in the context of how research and theory have been applied to understand the nature of scientific thought and behavior.

How scientific thinking and concepts are learned and change from infancy, adolescence, and adulthood are the focus of the third section of the book—development and theory change. The chapters in this section combine developmental and cognitive approaches to show the categorical similarities and differences in scientific concepts and thinking among children, adolescents, adults, and scientists. Being curious about the world and having ideas of how the physical, biological, and social worlds operate are natural products of the human mind, but special training and strategies are required to actually think like a scientist.

The fourth section is special topics and highlights the breadth and depth of a psychological perspective in the studies of science, from creativity and genius, gender, conflict and cooperation, to postmodernism, and psychobiography. These are all important problems for a fully developed psychology of science:

- Why do some individuals consistently make scientific discoveries and breakthroughs, and what role does "genius" play in the process?
- Why are relatively few women in science and technology careers?
- How do conflict and cooperation help or hinder scientific progress?
- Is scientific knowledge merely a social construction or something that approaches real explanation of the world?
- How does a scientist's life history affect his or her science and scientific thinking?

The fifth section of the book—applied psychologies of science—offers findings and ideas that can be put to use by educators, policy makers, and administrators of science. The invention of the airplane and telephone offer insight into how mental models are often critical in solving difficult technical problems. In addition, leadership and organizational structure are important factors in whether science groups and labs are innovative or not. Other chapters in this section detail the psychological factors (motivation, interest, resistance) of being a participant in science and how these factors influence the results of the study. Errors, biases, and uncertainty all play roles in science and scientific discovery, and only by being explicit about how and when they occur do we have a chance to control and lessen their effects.

The last theme and section of the book provides a look back on the trends in the literature, as well as looks forward to the future health of the field. This section provides a quantitative analysis of how the psychology of science is growing in influence and, in particular, the effect that starting a society and journal have on the field's development. Finally, we conclude with discussion and suggestions for how to continue to grow and build a healthy psychology of science.

In summary, the chapters in the handbook bring together the best in psychological explorations on the nature of scientific thought and behavior, and offer a unique glimpse into the scientific mind. We hope these chapters stimulate your own thinking and understanding of scientific thought and behavior and encourage further forays into two of the most important forces in the modern world, science and technology.

HANDBOOK OF THE
Psychology of Science

Introduction and History

CHAPTER 1

Introduction: Another Brick in the Wall

Gregory J. Feist and Michael E. Gorman

Science and technology are driving forces of modern society. In fact, they may be *the* defining features of the modern world. Science began by providing explanations and predictions for phenomena we encounter in everyday life on the planet, including the motions of familiar objects in the sky. Science now reaches into realms that test our ability to comprehend. How can an electron behave like both a wave and a particle? What are the dark matter and energy that comprise over 95% of the universe? New technologies have helped make these and other discoveries possible, and have also transformed the way human beings communicate: consider the progression from mail to telegraphy to telephony to the Internet. Advances in biology and medicine mean that age 60, at least in some parts of the world, is no longer old.

Science is a transformational force and rightly deserves to be the object of intense study by scholars both in and out of science. For both theoretical and practical reasons, philosophers, historians, and sociologists have examined the nature of scientific thought, behavior, and institutions for decades. In fact, in the case of history and philosophy of science, these disciplines began around the turn of the 20th century with the establishment of formal societies, conferences, and journals (Feist, 2006b). But, the psychology of science was a missing brick in the wall of science studies until the mid 2000s. The psychology of science has shown significant signs of growth within the last decade, with the formation of a society and conferences, and the handbook you hold in your hands (or, increasingly, see on your computer screen) is evidence of that. The goal of this volume, in fact, is to bring together in one place all of the cutting edge work psychologists have done, and are doing, to understand the processes by which scientific discoveries and technological inventions are created.

Before we get ahead of ourselves, we must define the psychology of science. If psychology is the study of human thought and behavior (Feist & Rosenberg, 2010), then the psychology of science is simply the study of scientific thought and behavior both narrowly and broadly defined (Feist, 2006b). Narrowly defined, science refers to the thoughts and behaviors of professional scientists, technologists, inventors, and mathematicians. Broadly defined, science refers

3

to the thought and behavior of any person or people (present or past) of any age (from infants to the elderly) engaged in theory construction, learning scientific or mathematical concepts, model building, hypothesis testing, scientific reasoning, problem finding or solving, or creating or working on technology. In short, science narrowly defined is explicit science, whereas science broadly defined is implicit science (Feist, 2006b). The psychology of science examines both explicit and implicit forms of scientific thought and behavior.

Indeed, psychology can offer not only a psychological perspective to science studies, but also a methodological and theoretical one. Compared to other studies of science—with the exception of history—psychology of science is the only discipline to focus on the development of the individual scientist in the context of his or her social environment and group. Moreover, psychology of science is unique in that it focuses on influences such as intelligence, motivation, personality, and development over the lifespan of scientific interest, thought, ability, and achievement (Feist, 2006b; Gholson, Shadish, Neimeyer, & Houts, 1989; Proctor & Capaldi, 2012; Shadish & Fuller, 1994; Simonton, 2009).

Stepping back, let's examine briefly where the psychology of science is in the context of a new field's establishment. Building on Mullins (1973), Feist (2006b) argued that fields become established by moving through three distinct phases: isolation, identification, and institutionalization. Philosophy and history of science became institutionalized and established around 1900, and the sociology of science by around 1970. Psychology of science hovered between isolation and identification until the 2000s. In this introductory chapter, we briefly review and summarize some of this history and then overview the contributions you will find in this book.

A BRIEF HISTORY OF THE PSYCHOLOGY OF SCIENCE

The history of the psychology of science has been a struggle for existence. The very first inklings of its development extend back into the second half of the 19th century. The French historian, Alphonse de Candolle, published a book in 1873 titled *Histoire des Sciences et des Savants depuis deux Siècles*. Candolle's book analyzed the history of 200 scientific men who lived between the late 1600s and 1800s. Not only did the book inspire Francis Galton to publish what is often regarded as the first work in the psychology of science, *English Men of Science: Their Nature and Nurture*, but it also touched off one of the earliest examples of the now classic "nature–nurture" debate in psychology (Fancher, 1983). Indeed, Candolle's 1873 book was a response to Galton's earlier work, *Hereditary Genius*, and was an attempt to show how environmental forces more than genetics, created scientific eminence (Fancher, 1983). Because it took more of a cultural/sociological perspective on the development of scientific eminence, Candolle's work might more appropriately belong to the sociology of science than the psychology of science.

Galton's *English Men of Science* was the first clear empirical work on the psychology of scientists. He surveyed scientists about their personal histories, attitudes, and beliefs about their interests in science. Here are Galton's own words:

> The intent of this book is to supply what may be termed a Natural History of the English Men of Science of the present day. It will describe their earliest antecedents, including their hereditary influences, the inborn qualities of their mind and body, the causes that first induced them to pursue science, the education they received and their opinions on its merits. (Galton, 1874/1895, p. 1)

Unfortunately, there was very little follow-up to Candolle's and Galton's work for nearly 50 years. Some philosophers occasionally touched on psychological factors in scientific thought and creativity. For example, in the mid 1930s, a French philosopher, Gaston Bachelard, took a psychoanalytic approach to understanding the scientific mind in *The Scientific Mind* (1934) and *The Formation of the Scientific Mind* (1938). Moreover, Bachelard's concept of "epistemological rupture" apparently influenced Kuhn's thinking and, more specifically, Kuhn's idea of "paradigm shift" (Bachelard, n.d.). A few years later an American, Stevens (1939), wrote an article titled "Psychology and the Science of Science," in which he was the first to coin the phrase "psychology of science." The Stevens article, therefore, is an important turn of events for the origins of a psychology of science. The field now had a name, if not much else.

Between the 1940s and 1970s, however, Eastern bloc countries were more active in the psychology of science than Western countries. Choynowski (1948a, 1948b), for example, was president of the circle for the science of science and he founded the Institute for the Science of Sciences in 1948 in Cracow, Poland. The institute consisted of departments for all the major areas of science studies, including history, sociology, pedagogy/education, and psychology. The department of psychology of science was described as "The study of different kinds of scholars, the psychology of research work, the hygiene of intellectual work, the changes of specialty by well-known scientists, the correlation between scientific interests and abilities, etc." (Choynowski, 1948b, p. 4). His main work on the psychology of science and other studies of science was a monograph titled *Life of Science* (Choynowski, 1948a; cf. Walentynowicz, 1975).

A similar movement was begun by Mikhail Grigoryevich Yaroshevsky (1915–2000) in Russia in 1960 (Volodarskaya, 2005). He guided the "Department of Psychology of Scientific Creative Works" at the Institute of the History of Science and Technology at the Russian Academy of Science. Yaroshevsky's group defined the psychology of science as "the study of the psychological peculiarity of people whose activity produces new knowledge; it is also the study of the psychological regularity of the development and functioning of scientific collectives" (Volodarskaya, 2005, p. 478).

We should point out, however, that these Eastern bloc movements had absolutely no impact on psychologists in the West and, in fact, were only discovered on obscure websites in 2010. What impact they had on Eastern bloc psychologists is also unknown. What is known is that no societies, conferences, or journals stemmed from them, and psychology of science remained in its fetal stage of development.

During the 1950s, there was an upswing in research and theory on the psychology of creativity, including scientific creativity. Especially post-Sputnik, anything that helped foster an interest in science was encouraged and relatively well funded. Noteworthy among the efforts from the 1950s and early 1960s were Anne Roe's *The Making of a Scientist* (1952), Bernice Eiduson's *Scientists: Their Psychological World* (1962), and Calvin Taylor and Frank Barron's *Scientific Creativity*. Also in the early 1950s, Terman (1954) published an important monograph on the psychological qualities of scientists. By the mid-1960s, Maslow (1966) published the very first book with psychology of science in the title.

Between the late 1960s and mid-1980s, however, there was very little systematic work done by psychologists on scientific thought, reasoning, or behavior in Western countries. Starting in the mid-1960s, however, one university was about to generate a very productive and impressive string of scholars who would go on to conduct many ground-breaking studies in the psychology of science. That school was Carnegie-Mellon, and its founding scholar was Herb Simon. Simon published one of the first explicit pieces in the psychology of science, a chapter on scientific discovery and the psychology of scientific problem solving (Simon, 1966). Just as importantly for the psychology of science, Simon produced a graduate student, David Klahr, who himself would go on to produce a string of one of the more impressive lists of graduate students and postdoctorial fellows in the psychology of science for the 1990s and 2000s: Kevin Dunbar, Jeffrey Shrager, Amy Masnick, and Chris Schunn. Indeed, we could dub this Simon lineage the "Carnegie-Mellon School."

In the early 1980s, Tweney, Doherty, and Mynatt (1981) edited a seminal work on the cognitive psychology of science, *On Scientific Thinking*. Indeed by the mid-to-late 1980s, the field really stood at the precipice of being a full-fledged field, or so it appeared. The 1986 conference held in Memphis and organized by William Shadish, Barry Gholson, Robert Neimeyer, and Arthur Houts was a good beginning. Indeed, the books that followed from that conference (Gholson et al., 1989; Simonton, 1988) were important beginnings in moving toward an established discipline. However, no society, regular conferences, or journal sprouted up afterward. In the 1990s, Ron Westrum at Eastern Michigan University started a newsletter ("Social Psychology of Science Newsletter"), but even that lasted but a few years. These works, however, did inspire Feist, while he was working on his dissertation on scientific creativity at University of California, Berkeley, and he would later argue that psychology of science should be an independent discipline of study.

Starting in the 1990s, a new and exciting avenue in the psychology of science was opened up by developmentalists. In fact, developmental work in the psychology of science owes its existence to the seminal research and theory of Jean Piaget on the development of children's understanding of the physical, biological, and numeric worlds (Inhelder & Piaget, 1958; Piaget, 1952). Starting in the late 1980s, Deanna Kuhn and her students blazed the trail in examining how scientific reasoning develops in children and adolescents (Kuhn, 1989, 1993; Kuhn Amsel, & O'Loughlin, 1988; Kuhn & Pearsall, 2000). Others soon followed (developmentalists) (e.g., Brewer & Samarapungavan, 1991; Fay, Klahr, & Dunbar, 1990; Klaczynski & Narasimham, 1998; Koslowski, 1996). One work of primary importance in this movement was the publication of *The Scientist in the Crib: Minds, Brains and How Children Learn* by Gopnik, Meltzhoff, and Kuhl (1999). One general argument they put forth in the book is the relationship between children and scientists; it is best to think of scientists as big children rather than children as little scientists.

Starting with his participation in the Memphis meetings, Dean Simonton continued to publish many important works in the 1990s and 2000s on the psychology of science, the most influential of which was his book *Scientific Genius: A Psychology of Science* (1988; see also Simonton, 1989, 1991, 1992, 1995, 1999, 2000, 2004, 2008, 2009). Most of these works focused on the cognitive, personality, and motivational forces behind scientific creativity and scientific genius.

After completing his dissertation in 1991 at the University of California–Berkeley on the psychology of science, Feist published a few papers from that program of research (Feist, 1993, 1994, 1997). The most interesting finding was that blind raters (from tape recordings) perceived the most eminent scientists as arrogant and hostile (Feist, 1993). From this work, he became acquainted with Mike Gorman and his work in the psychology of science. Initially, Feist organized a symposium at the American Psychological Association in Washington, DC, in August of 1992 titled "The Foundations for a Psychology of Science." The presenters were central figures in the early days of psychology of science: Ryan Tweney, Dean Simonton, Steve Fuller, Mike Gorman, and Greg Feist. The APA meeting signified the first collaboration between Feist and Gorman and was the beginning of their joint efforts to launch a new discipline. That same year, Gorman (1992) published his book *Simulating Science: Heuristics, Mental Models, and Technoscientific Thinking*.

A few years later, in 1995, Feist and Gorman organized a symposium on the psychology of science at the "Society for the Social Studies of Science" (4S) annual conference and from there they collaborated on the first review article of the published literature (Feist & Gorman, 1998). That article, in turn, inspired Feist to work up a book-length review of the literature, which he began working on in 2001. The book, *The Psychology of Science and the Origins of the Scientific Mind* (Feist, 2006b), made the argument that a field does exist and there is exciting literatures in the developmental, cognitive, personality, and social psychologies of science. In the book and a subsequent article, he lamented the fact that the discipline still had no society, no conferences, and no journal (Feist, 2006a, 2006b). Little did we know at the time, however, that 2 years later, all three would exist.

Sure enough, soon after the book appeared, Sofia Liberman and Javier Rayas organized the second conference on the psychology of science in September of 2006 in Zacatecas, Mexico. It was a very small, but very productive conference. Attendees at this conference went on to establish the International Society for the Psychology of Science and Technology (ISPST) and elected Gregory Feist as its first president. The society, too, started off small with only about 40 members in its first year, 2007. By 2008, the society, in collaboration with Springer Publishing Company, published the first peer-reviewed journal, the *Journal of Psychology of Science and Technology* (*JPST*).[1] Feist (2008) editorialized in the inaugural issue of *JPST* that the "psychology of science has finally arrived" (p. 1). Indeed, conferences are now biennial, with the second one taking place in Berlin, Germany, in 2008; one in

Berkeley, California, in 2010; and most recently in Pittsburgh, Pennsylvania, in 2012. The society is still small, but if the Berlin, Berkeley, and Pittsburgh conferences are any indication, the field is vibrant and continues to produce exciting ideas and programs of research.

OVERVIEW OF THE CURRENT VOLUME

The psychology of science building is being built, wall-by-wall, brick-by-brick. The wall, however, is not a divider but rather a connector to other rooms in the studies of science. Let us now look a little more closely at the bricks being offered in each of the six sections of the book.

Historical Perspective

As a psychologically informed sociologist, Fuller is in a unique position to provide a rich context on the historical, sociological, and philosophical roots of psychology of science. Fuller argues that the history of experimental psychology began as a psychology of the scientist. The introspectionism of William James and company was a prime example of how science involved the self-conscious psychologist. The move outward and away from internal subjective processes began with behaviorism and continued with the cognitive revolution of the 1950s and 1960s. But by the middle of the last century, two branches of the psychology of science were reforming: one by Edward Tolman and the second by Abraham Maslow. Tolman's work of the 1930s and 1940s, further explicated by his student, Donald Campbell, focused on mental and psychological epistemologies, especially in their rigorous experimental forms. The Maslow branch veered toward so-called psychohistorical studies that mix hermeneutical and experimental methods to arrive at conclusions about the motivational structure of "exemplary" scientists.

Foundational Psychologies of Science

Zimmerman and Croker review research on the development of scientific inquiry, focusing on studies in which children and adults take part in all phases of scientific inquiry, including the generation of hypotheses and experiments to test them. The authors focus on systems where participants have to consider and manipulate multiple variables to determine their effect on an outcome. These studies can be conducted in physical or virtual environments; an example of the former is comparing heavier and lighter weights to see which falls faster, and an example of the latter is a computer microworld testing the effect of multiple variables on the speed of racing cars.

A particular focus of the authors is the development of inquiry skills. The development of effective inquiry involves the complementary growth of both domain-specific content knowledge and metacognition. Most of the studies reviewed in this article focus on how participants deal with anomalies. An anomaly occurs when a result contradicts a prior belief. Noticing an anomaly requires a certain amount of domain knowledge. Hypotheses developed from prior knowledge limit the search space; in the absence of such knowledge, it is hard to generate effective experiments. Strong prior beliefs make it more difficult to generate and test hypotheses that could falsify those beliefs.

Both children and adults create hypotheses about causal relations, but studies show that adults are better at designing experiments that can distinguish between hypotheses, and also better at following a control-of-variables strategy (CVS). Children are also more likely than adults to jump to conclusions after one or a few experiments. Children are less metacognitively aware of the limits of their memory and therefore keep more sporadic results of experiments than adults. The authors suggest that improvement in children's metacognitive skills requires instruction and practice, not just learning from experience, but more research should be done on this question.

Tweney advocates a cognitive–historical approach to understanding science. He cites the way in which a cognitive–historical approach can reveal shifts in scientific representations. Research by Hanne Andersen showed that, before Kepler, angular position was the key attribute of an object in its celestial orbit; after Kepler, the shape and center of the orbit were the key attributes, as embodied in his laws, shifting focus from the orb, an object concept, to the orbit, an event concept.

The cognitive–historical method helps not only with understanding shifts in scientific concepts but also with the step-by-step details of scientific problem. Frederic Holmes's detailed historical account of the discovery of the ornithine cycle was used by Kulkarni and Simons to build a computational simulation of Krebs's discovery that captured the process of going from one experiment to another, including what happened when there was a surprise. Nersessian has focused on the way Maxwell deployed a rich variety of mental models in his work. Tweney and Gooding have graphs and maps of Michael Faraday's scientific processes, using his diary as data. Gooding even replicated some of Faraday's experiments. These historical cases can and should be complemented by real-time studies of living scientists (as Nersessian is now doing). The advantage of the historical cases is we know in hindsight that these discoveries are important. The disadvantage is we need the equivalent of a kind of control group: studies of failed research programs.

Feist makes a case for the importance of scientific personality, that is, personality traits that consistently affect scientific thought, behavior, and achievement. Feist reviews the last few decades of research on personality and scientific interest, thought, and creative achievement, and quantifies and categorizes these effects. He codifies these with an updated model of the scientific personality. More specifically, Feist explicates how genetic, epigenetic, neuropsychological, cognitive, social, motivational–affective, and clinical personality traits shape scientific thought and behavior. In addition, he summarizes the research on how the personalities of scientists affect their scientific theories. He concludes the chapter with discussion on the directional influence of personality on scientific thinking and achievement.

Liberman and Wolf provide a social–psychological perspective and begin their chapter by reviewing the literature on scientific communication before presenting their own model. They distinguish between inner and outer cycles of scientific communication. Inner cycles are the informal communications among a scientific workgroup. Outer cycles are the published record of this work. Contacts are established at scientific meetings and these contacts may grow into scientific work groups and invisible colleges of colleagues who keep abreast of, and comment on, each others' evolving work prior to publication.

Scientific coauthorship is an indicator of collaboration. Coauthorship can be granted for generating the research idea, choosing the method, selecting or developing the instruments, making measurements, and/or interpreting the results. Publication norms vary from field to field, including order of authorship. Liberman and Wolf looked at the number of bonds, or author pairings, across mathematics, physics, biotechnology, and anthropologists, and found significant differences by field. Recently, they are finding more articles, with as many as 100 coauthors from multiple institutions, and the idea that these authors have bonded or even work in a group is a stretch. A division of labor among experts is more likely.

Development and Theory Change

Koslowski focuses on practical scientific reasoning, primarily about causation. Formal, logical models of reasoning do not account for all, or even most, of what scientists and engineers do. She covers abductive inference, or inference to the best explanation, using evolution as an example. She points out that confirmation is not always a bias—it can be an effective reasoning strategy in certain situations, for example, when there is a possibility of error in the data. Koslowski would agree with Klahr's argument that children can do scientific reasoning. In Koslowski's case, it is practical, abductive reasoning, involving consideration of alternate explanations for phenomena and what data might distinguish between explanations. The problems may be simpler and the children do not always have the knowledge to properly assess the alternatives, but according to Koslowski, the reasoning process is the same. She reviews the developmental literature on this and other issues, such as whether children can distinguish between theory and explanation (yes, according to Koslowski). She also describes how scientists actually deal with anomalies, using research on scientific laboratories by Dunbar; lone anomalies are dismissed as errors, but anomalies that occur in series are taken seriously and provoke new hypotheses and research. Group reasoning is an important theme in Koslowski's chapter, and she extends it to the kinds of expertise that facilitate exchanges of knowledge and methods among laboratories.

Lawson's chapter begins by describing the roles of induction, deduction, and abduction in scientific discovery. He uses Galileo's discovery of three bright lights next to Jupiter as an example. Galileo first made the abductive inference that these were stars: if these are bright points of light in the sky and other, similar bright points are stars, then these lights are stars. But these stars were in a straight line going out from Jupiter. So, Galileo abductively inferred another hypothesis that these were moons centered on Jupiter. He deduced that if the fixed-line pattern was repeated on subsequent nights, these bright lights were moons, not stars. He used further observations to induce that he was correct.

Lawson argues that the brain is hardwired to use the same hypothetico-deductive reasoning as Galileo, but more rapidly. If a hunter–gatherer hears a faint sound nearby, looks toward it and sees what might be stripes, then the salient hypothesis that is a tiger and the response is fight or flight, depending on the circumstances. According to Lawson, this process is hypothetico-deductive: input sound and stripes, hypothesis tiger—but in this case, the hunter–gatherer does not stick around for confirmation or disconfirmation! To prove that the brain is hypothetico-deductive, Lawson demonstrates that a neural net analysis could account for Galileo's discovery. (Presumably, this is an abductive inference—if a neural net works like the brain and the neural net models Galileo's discovery process, then we know the neural basis for Galileo's discovery.)

Lawson sketches the implications for reasoning biases, a list of which are provided in a table. He advocates a constructivist approach to education, in which children generate and test their own hypotheses, on the grounds that this approach is best for developing presynaptic and postsynaptic connections on the brain. (Klahr and colleagues reach a different conclusion about constructivist approaches in their chapter in this volume.)

Klahr, Matlen, and Jirout focus on the development of scientific thinking. They use the Herbert Simon problem space framework in which heuristics are used to reduce the size of the search space; their chapter contains a list of common heuristics that can be used on both scientific and nonscientific problems, establishing their point that there is nothing distinct about scientific reasoning. Children think about scientific problems in ways appropriate to their developmental level. For example, children can engage in analogical reasoning about scientific concepts but need much more scaffolding from a teacher to be as effective as an adult at seeing where the analogy works and where it breaks down. The authors of this chapter argue against pure guided discovery learning; teacher scaffolding is essential. Learning about scientific methods and concepts is a lifelong pursuit—even for expert scientists.

Special Topics

Simonton focuses on the origins of creative genius in science. Simonton borrows from Kant to create a definition of genius: it produces work that is original, exemplary, and cannot be derived from existing rules. Note the similarity to the patent office's requirements for an invention: It must be novel (original), useful (exemplary), and nonobvious (not derivable from existing rules). Simonton assesses scientific genius using four methods:

- Historiometric evidence, based on the number and extent of references to a scientist in histories, encyclopedias, and biographical dictionaries, a method employed by Galton and his student Cattell
- Identification of geniuses done by experts in a domain
- Journal citations
- Scientific awards, such as the Nobel Prize

Identifying geniuses requires multiple methods, especially given the danger of confounding genius with the Matthew effect. Surveys can help spot scholars who have high integrative complexity.

What factors lead to genius? Genetics plays a role, as do family factors, like birth order; eminent scientists tend to be first born, but revolutionary scientists are more likely to be born last. Mentoring is also important; Nobel laureates often worked under mentors who were themselves Nobel laureates. Scientific productivity over the lifespan generally follows an inverted "U," peaking in a person's 30s or 40s, depending on the expertise domain: some fields require much longer to complete successful research than others. Geniuses tend to defy this pattern by making high-impact contributions both earlier and later in their careers than most scientists. Simonton concludes by suggesting more research on whether geniuses have unique cognitive processes, why there is an apparent bias toward male geniuses in science and engineering, and how particular times in history and in the life of a field make it easier to become a genius.

Kumar's chapter provides a review of the literature on the psychology of gender and relates this literature to science. She also details the way in which women were excluded from science: the British Royal Society did not admit women until 1945. She also covers feminist perspectives on scientific methodology and epistemology and refers to recent feminist work on technology. Kumar next discusses the reasons more women do not adopt careers in STEM (science, technology, engineering, and mathematics) today, despite all the efforts to encourage more gender balance. The situation is better in some other countries than the United States, especially Latin America—Kumar provides detailed statistics, based on National Science Foundation reports. Women tend to be clustered in the lower academic ranks.

This gender stratification creates a research opportunity for psychologists of science, and psychological literature on this topic is growing. Explanations for the gender stratification include the following:

- Mathematical ability appears to be equal for young men and women, though there may be more variance for the men. But there is sex stereotyping: women are socialized to believe they are not as good at mathematics.
- Women may simply be less motivated to undertake careers in science and engineering and more oriented toward people-oriented professions and family life. This preference can, of course, result partly from stereotyped perceptions of roles for women.

Once women become scientists, they tend to publish less than males pretenure, but more after. According to Rossiter, there is a "Matilda Effect," a gendered equivalent of the Matthew Effect in which women's accomplishments in STEM areas are often credited to men. Although there are hundreds of studies on gender and science, Kumar finds fewer studies of gender and technology, creating an opportunity for future research.

Schulze applies organizational psychology to conflicts and cooperation in science and technology. Scientists use a hypothetico-deductive format to present their research, which makes it sound like disputes could be resolved simply through logic and evidence. But heated controversies are common in science, especially during the genesis and gradual acceptance of innovative ideas. Minority dissent is an important catalyst of innovation (for more on minority influence in science, see Rosenwein, 1994). Schulze's longitudinal research on biologists, physicists, and computer scientists shows that effective conflict resolution strategies depend on the stage of the research. Her results sound like they would fit Kuhn's model of science: During the normal science phase when a research program is being implemented, dissent over fundamentals is not productive; rising dissent signals the beginning of a new research program, which initially encounters significant resistance.

Schulze details strategies for overcoming the resistance, including submarine research carried on beneath management's radar, like the skunkworks that led to the Sidewinder missile (Westrum & Wilcox, 1989).

Schulze's chapter concludes with a case study of over 200 life science researchers focusing on conflict and cooperation. Scientists in the life sciences know they are in a competitive field where they will have to cooperate to innovate. The competition is particularly acute in applied areas where profits are at stake. Psychology of science could help develop mechanisms for cooperation in a competitive environment.

Capaldi and Proctor highlight one of the issues that has made it hard for psychologists of science to discuss their work with some of the sociologists and anthropologists who take a constructivist, postmodern approach to science studies. The authors argue that postmodernism assumes reality is little else but an agreement among members of a culture. The authors argue against postmodern approaches to psychology, particularly Kenneth Gergen, who argues that experiments simply produce the conclusions the researcher wanted and therefore other, more reflexive approaches would be better. Kuhn is often cited as an advocate of this kind of constructivist, relativist approach, but, in fact, Kuhn was a realist.

The "reality is a cultural construction" view puts sociology of science into a central position among fields of inquiry, because these sociologists might be the best diviners of what constitutes a relative truth for a culture. The authors argue that psychologists of science should ally with those philosophers of science who are realists, and continue the empirical study of science. They note a recent turn by sociologists toward empirical, quantitative studies of expertise. (Leading sociologist of science, Harry Collins, and several of his expert researchers participated in the ISPST conference in Berkeley in 2010 and stayed for an extra day to work with psychologists.) The authors also discuss and critique the inclusion of a Society for Qualitative Inquiry into APA's Division 5. The authors are careful to distinguish qualitative methods from qualitative inquiry. The former are essential tools in psychology of science (witness the use of such methods in multiple chapters in this volume). The latter reject the objective standards of science.

William McKinley Runyan has been a pioneer in the field of psychobiography—analyzing people's lives in the context of their biographical life history and gleaning psychological insight from these biographical details. In his chapter for this handbook, he continues this line of investigation chapter by applying it to the lives of individual scientists, theoreticians, and philosophers, namely Bertrand Russell, Ludwig Wittgenstein, Sigmund

Freud, Henry Murray, Karen Horney, B. F. Skinner, Paul Meehl, and Michel Foucault. In so doing, Runyan demonstrates how personal and experiential forces shape the science not only of "soft" scientists (e.g., psychoanalysts) but also the more rigorous and tough-minded ones as well (behaviorists and psychometricians). Runyan's chapter makes clear that the psychology of science is quite varied in its conceptual and methodological approaches, and the field can benefit greatly from adding biographical methods to its experimental and correlational ones.

Applied Psychologies of Science and Technology

Gorman's chapter on invention includes a cognitive–historical analysis of the discovery of the telephone, showing the potential for this method to be applied to technological innovation as well as scientific discovery. Gorman uses a narrative structure to organize his chapter, so we learn a bit about his research process, which relied on collaborating with W. Bernard Carlson, a historian of technology. The Wright brothers have also been studied by cognitive scientists. Johnson-Laird used mental models to account for the fact that the brothers were far ahead of their contemporaries in their design. Gary Bradshaw, in contrast, uses a dual-space search framework and attributes their success to alternating between design and function spaces. Gorman also describes briefly Nersessian's current work in bioengineering laboratories, looking at model systems that are hybrids of mental models and physical systems. There is far less work on invention than on science.

Hemlin and Olsson describe factors that influence group creativity in science and focus on leadership as a major influence. Why groups? Because more and more scientific research is done by groups or teams, which often span different disciplines. The authors go through research on variables that could potentially relate to creativity, including:

- A group climate of openness and respect tends to promote creativity
- Group size is not related to creativity
- Strong connections to social networks outside of the group tend to increase creativity
- Leaders who promote creativity have high-quality relationships with those who work with them and collaborate as much as possible

The chapter concludes with two studies of Swedish biomedical research groups that measured creativity by the number of high-quality publications. The first study focused on leader–member interactions, using a standard scale. The second study used critical incident interviews on a smaller sample of the leaders and members from the first study. The two studies concluded that leader's who were able to provide expertise and coordinate research contributed to creativity. The chapter includes a good discussion of the measures and methods used to assess creativity in research and development (R&D) teams. The authors also provide suggestions for future research.

Moyer discusses the psychology of human research participation, beginning with Rosenthal's groundbreaking work on experimenter bias. Researcher expectations can be communicated to participants implicitly, and influence the results of studies. Moyer devotes her chapter to ways of correcting for this bias, including:

- Study how undergraduates, the usual participants in many psychological experiments, understand their research experience
- Expand the subject pool, which encounters resistance from the public and from patients. One method is financial incentives, and there is research on what constitutes the right amount to have the desired impact
- Reluctance of participants to be randomly assigned to control conditions—especially when the treatment condition may have some benefit. Participant attrition is another problem. Both may be solved by putting participants in more of a coinvestigator role (Epstein, 1996, discusses what AIDS activists did to avoid being in the control group and how they became coinvestigators, improving the research design).

Moyer advocates making the participant role an object of study for psychology of science.

Kane and Webster consider errors and biases in scientific judgment and decision making, focusing on psychological research. Their list of biases include the following:

- The Hawthorne Effect—observing people affects their performance
- Experimenter expectancies can affect data collection and interpretation, if double-blind controls are not used. Participants can also infer experimenter expectancies, creating a response bias.
- HARKing, which is hypothesizing after the results are known. This strategy is often used to ensure publication: get the result, then generate a hypothesis and a prediction and put them at the beginning of the article.
- Confirmation bias, demonstrated by Blondlot and his refusal to give up his N-ray theory despite failures to replicate his discovery. These replications illustrate the way in which another researcher, by trying to confirm a pattern, can end up disconfirming it.
- A variety of biases in review and publication practices, including a preference for publishing articles that fit with the existing paradigm within a field.

The authors suggest mechanisms for combating these biases in science, demonstrating the value psychology of science can have for scientific practice. Even just being aware of potential sources of bias makes the scientist more likely to anticipate and correct for them.

Schunn and Trafton apply psychology of science for reducing uncertainty in data analysis. They begin by providing a taxonomy of types of uncertainty and how scientists cope with them, based on years of observation and analysis of scientists in action. As an example of a scientific domain, they use cognitive neuroscience, in particular, functional magnetic resonance imaging (fMRI). As an example of a more applied domain, they use weather forecasting. Types of uncertainty include the following:

- Physics uncertainty, or uncertainty in the raw information from the system. In fMRI, the way the signal moves through the skull can introduce uncertainty. In weather forecasting, smoke, pollution, and precipitation can interfere with measurements of cloud height.

- Computational uncertainty, caused by the elaborate computational procedures used to convert measurements into the data analyzed by a scientist. One subtype is the uncertainties caused by aggregating measurements. In fMRI, this can remove small areas of activation in favor of larger areas of sustained activation. In weather forecasting, important microclimates can be disguised under larger patterns.
- Visualization uncertainty, or the uncertainty introduced by putting data into the sorts of visual forms that are particularly appropriate for human pattern recognition. One problem is multiple visualizations shown at the same time. The advantage is that one way of visualizing does not dominate; the disadvantage is that the options may stem from different assumptions. In the case of weather forecasting, they can be based on different models. In the case of fMRI, visualizations based on time or activating condition can disagree.
- Cognitive uncertainties include information overload, which can be induced by the different types of visualization above: processing different visualizations with different assumptions may simply overload the cognitive system, especially if a decision based on the data is time-critical.

Each of these types of uncertainty is divided into subtypes for a more detailed description. The authors also suggest ways scientists can identify and deal with uncertainties.

One way psychologists of science can identify uncertainties in the scientists they observe is by noting gestures. The authors video-recorded cognitive neuroscientists and engineers working on the Mars Rover to see what gestures they used when they encountered uncertainty. The fMRI technology may be an additional method psychologists of science can use in the future.

Past and Future of Psychology of Science

Webster's chapter is a good complement to Capaldi and Proctor's. Webster gives an overview of three metasciences—philosophy, history, and sociology of science— and contrasts their development with psychology of science. Webster builds on Feist's proposal of three stages in a field's development, namely isolation, identification, and institutionalization, and attempts to quantify where to put the psychology on this scale of development. Webster uses hit counts on Google Scholar to compare the relative impacts of establishing a regular meeting and establishing a journal. Establishment of regular conferences had no statistically significant effect on hits, but establishing a journal did. The *Journal of Psychology of Science and Technology* appeared in 2008, too early for its impact to be analyzed using Webster's methodology, but psychology of science has grown since 1950 when the first hits appeared in the records of Google Scholar. He concludes that psychology of science is somewhere between the identification and institutionalization stages. The *Journal of Psychology of Science and Technology* has lapsed;

if psychology of science is to become institutionalized, the journal needs to be revived. Webster includes a detailed discussion of his scientometric methodology, including its limitations, and gives advice on how it can be used for other psychology of science issues, like the Matthew Effect, where a few individuals dominate the citations in any scientific area, including the metasciences.

In the concluding chapter, Gorman and Feist synthesize and collate the themes and variations that exist in this volume and ask the question, "What can we do to ensure the next generation wants to carry the tradition forward and that scholars start to identify themselves as psychologists of science?" Any field, young or old, needs to excite the interest of young scholars, and in this sense, the psychology of science is no different from any other discipline. Indeed it stands to gain from learning how the other studies of science—philosophy, history, and psychology—made this transition. In order to assure the next generation inherits an active discipline, current psychologists of science need to do three things: teach exciting courses, carry out and publish exciting research, and reward student excellence.

The psychology of science is a rich repository of methods, findings, and insights into the nature and nurture of scientific thought and behavior. In fact, it offers a model for psychology as to how a single topic can unify all of the disjointed subdisciplines of a parent discipline (Staats, 1999). Just as important, the psychology of science also offers its methods, findings, and insights to the other studies of science—in an attempt to build a viable and vibrant "trading zone" (Collins, Evans, & Gorman, 2010). Psychology is no longer is a missing brick in the wall of studies of science, but rather is a cornerstone in its foundation.

NOTE

1. The journal is no longer published by Springer Publishing Company. The ISPST is currently working on converting the journal to an online, open-access, peer-reviewed journal.

REFERENCES

Brewer, W. F., & Samarapungavan, A. (1991). Child theories vs. scientific theories: Differences in reasoning or differences in knowledge? In R. R. Hoffman & D. S. Palermo (Eds.), *Cognition and the symbolic processes: Applied and ecological perspectives* (pp. 209–232). Hillsdale, NJ: Erlbaum.

Choynowski, M. (1948a). Life of science. *Synthese, 6*, 248–251.

Choynowski, M. (1948b, May 31). *Institute for the Science of Science.* Report to the United Nations Educational Scientific and Cultural Organization: Natural Sciences Section Occasional Paper No. 11. Retrieved May 15, 2010, from http://unesdoc.unesco.org/images/0015/001540/154044eb.pdf

Collins, H., Evans, R., & Gorman, M. E. (2010). Trading zones and interactional expertise. In M. E. Gorman (Ed.), *Trading zones and interactional expertise: Creating new kinds of collaboration* (pp. 7–23). Cambridge, MA: MIT Press.

Eiduson, B. T. (1962). *Scientists: Their psychological world*. New York, NY: Basic Books.

Epstein, S. (1996). *Impure science: AIDS, activism, and the politics of knowledge*. Berkeley, CA: University of California Press.

Fancher, R. E. (1983). Alphonse de Candolle, Francis Galton, and the early history of the nature-nurture controversy. *Journal of the History of the Behavioral Sciences, 19,* 341–352.

Fay, A. L., Klahr, D., & Dunbar, K. (1990). Are there developmental milestones in scientific reasoning? In *Proceedings of the twelfth annual conference of cognitive science society,* Conference Proceedings. MIT, July 1990.

Feist, G. J. (1993). A structural model of scientific eminence. *Psychological Science, 4,* 366–371.

Feist, G. J. (1994). Personality and working style predictors of integrative complexity: A study of scientists' thinking about research and teaching. *Journal of Personality and Social Psychology, 67,* 474–484.

Feist, G. J. (1997). Quantity, impact, and depth of research as influences on scientific eminence: Is quantity most important? *Creativity Research Journal, 10,* 325–335.

Feist, G. J. (2006a). The past and future of the psychology of science. *Review of General Psychology, 10,* 92–97.

Feist, G. J. (2006b). *The psychology of science and origins of the scientific mind*. New Haven, CT: Yale University Press.

Feist, G. J. (2008). The psychology of science has arrived. *Journal of Psychology of Science and Technology, 1,* 2–5.

Feist, G. J., & Gorman, M. E. (1998). Psychology of science: Review and integration of a nascent discipline. *Review of General Psychology, 2,* 3–47.

Feist, G. J., & Rosenberg, E. (2010). *Psychology: Making connections*. New York, NY: McGraw-Hill.

Galton, F. (1874/1895). *English men of science: Their nature and nurture*. New York, NY: Appleton.

Gholson, B., Shadish, W. R., Neimeyer, R. A., & Houts, A. C. (Eds.). (1989). *The psychology of science: Contributions to metascience*. Cambridge, UK: Cambridge University Press.

Gopnik, A., Meltzoff, A. N., & Kuhl, P. K. (1999). *The scientist in the crib: Minds, brains, and how children learn*. New York, NY: William Morrow.

Gorman, M. E. (1992). *Simulating science: Heuristics, mental models and technoscientific thinking*. Bloomington, IN: Indiana University Press.

Inhelder, B., & Piaget, J. (1958). *The growth of logical thinking from childhood to adolescence* (Trans. A. Parsons & S. Milgram). New York, NY: Basic Books.

Klaczynski, P., & Narasimham, G. (1998). Development of scientific reasoning biases: Cognitive versus ego-protective explanations. *Developmental Psychology, 34,* 175–187.

Koslowski, B. (1996). *Theory and evidence: The development of scientific reasoning*. Cambridge, MA: MIT Press.

Kuhn, D. (1989). Children and adults as intuitive scientists. *Psychological Review, 964,* 674–689.

Kuhn, D. (1993). Connecting scientific and informal reasoning. *Merrill-Palmer Quarterly, 39,* 74–103.

Kuhn, D., & Pearsall, S. (2000). Developmental origins of scientific thinking. *Journal of Cognition and Development, 1,* 113–129.

Kuhn, D., Amsel, E., & O'Loughlin, M. (1988). *The development of scientific thinking skills*. Orlando, FL: Academic.

Maslow, A. (1966). *The psychology of science*. New York, NY: Harper and Row.

Mullins, N. (1973). *Theories and theory groups in contemporary American sociology*. New York, NY: Harper & Row.

Piaget, J. (1952). *The child's concept of number.* New York, NY: Norton.

Proctor, R. W., & Capaldi, E. J. (Eds.). (2012). *Psychology of science: Implicit and explicit processes.* New York, NY: Oxford University Press.

Roe, A. (1952). *The making of a scientist.* New York, NY: Dodd, Mead.

Rosenwein, R. (1994). Social influence in science: Agreement and dissent in achieving scientific consensus. In W. R. Shadish & S. Fuller (Eds.), *The social psychology of science* (pp. 262–285). New York, NY: Guilford Press.

Shadish, W. R., & Fuller, S. (Eds.). (1994). *Social psychology of science.* New York, NY: Guilford Press.

Simonton, D. K. (1988). *Scientific genius: A psychology of science.* Cambridge, UK: Cambridge University Press.

Simonton, D. K. (1991). Career landmarks in science: Individual differences and interdisciplinary contrasts. *Developmental Psychology, 27,* 119–130.

Simonton, D. K. (1992). The social context of career success and course for 2,026 scientists and inventors. *Personality and Social Psychology Bulletin, 18,* 452–463.

Simonton, D. K. (1995). Behavioral laws in histories of psychology: Psychological science, metascience, and the psychology of science. *Psychological Inquiry, 6,* 89–114.

Simonton, D. K. (1999). *Origins of genius.* New York, NY: Oxford University Press.

Simonton, D. K. (2000). Methodological and theoretical orientation and the long-term disciplinary impact of 54 eminent psychologists. *Review of General Psychology, 4,* 13–21.

Simonton, D. K. (2004). *Creativity in science.* Cambridge, UK: Cambridge University Press.

Simonton, D. K. (2008). Scientific talent, training, and performance: Intellect, personality, and genetic endowment. *Review of General Psychology, 12*(1), 28–46. Doi: 10.1037/1089–2680.12.1.28

Simonton, D. K. (2009). Applying the psychology of science to the science of psychology: Can psychologists use psychological science to enhance psychology as a science? *Perspectives on Psychological Science, 4,* 2–4.

Staats, A. (1991). Unified positivism and unification psychology. *American Psychologist, 46,* 899–912.

Stevens, S. S. (1939). Psychology and the science of science. *Psychological Bulletin, 36,* 221–263.

Taylor, C. W., & Barron, F. (1963). *Scientific creativity: Its recognition and development.* New York, NY: John Wiley.

Terman, L. M. (1954). Scientists and nonscientists in a group of 800 men. *Psychological Monographs, 68,* Whole No. 378.

Tweney, R. D., Doherty, M. E., & Mynatt, C. R. (Eds.). (1981). *On scientific thinking.* New York, NY: Columbia University Press.

Walentynowicz, B. (1975). The science of science in Poland: Present state and prospects of development. *Social Studies of Science, 5,* 213–222.

Westrum, R., & Wilcox, H. A. (1989, Fall). Sidewinder. *Invention & Technology,* 57–63.

CHAPTER 2

History of the Psychology of Science

Steve Fuller

Considering the formative role that the "personal equation" in astronomical observation played in the emergence of psychophysics, experimental psychology itself arguably began as the psychology of the scientist. The idea of the scientist as self-conscious psychologist was most explicitly endorsed by the Würzburg School, based on introspection, to considerable effect in both psychology and philosophy, but was fatally undermined in the 1920s by behavioristically inspired critiques of the unreliability of self-reports. The rise of a nonintrospective experimental cognitive psychology in the 1960s aimed to study scientific reasoning in both lay and expert populations, but the results here pointed to the profoundly alien character of the "scientific method" even among working scientists. It turns out that already in the 1930s, the self-styled "purposive behaviorist" E. C. Tolman called for a field named "psychology of science" in the science–reformist spirit of the logical positivist movement. The chief legacy of this initiative was the work of Tolman's student, the great methodologist, Donald Campbell, who figured prominently at the three main conferences expressly dedicated to launching a "psychology of science" (Fuller, De Mey, Shinn, & Woolgar, 1989; Gholson, Shadish, Neimeyer, & Houts, 1989; Shadish & Fuller, 1994), prior to the establishment of the International Society for the Psychology of Science and Technology in 2006. However, the first book called "psychology of science" was authored by Abraham Maslow, a Neo-Freudian who championed a "humanist" psychology that diagnosed the alien character of the scientific method in terms of scientists' psychodynamically underdeveloped sense of their being-in-the-world. In this respect, Maslow saw the problem not in terms of scientists (and lay people) failing to adhere to the scientific method, but that the modes of Kuhnian "normal science" that passed for adherence to such a method did not enable scientists to become fully "self-actualized" persons. Although Tolman was generally regarded as the most liberal of behaviorists, not least by Maslow, the Tolman brand of psychology of science has veered toward experimental studies aimed at enhancing various senses of "validity," whereas the Maslow brand has veered toward so-called psychohistorical studies that mix hermeneutical and experimental methods to arrive at conclusions about the motivational structure of "exemplary" scientists.

This chapter proceeds in six parts, which present the argument as follows: (a) When introspection was the dominant method in psychology in the late 19th century, it was often self-applied, as scientists were thought to operate in a heightened state of self-consciousness. However, experimental findings generally did not confirm this assumption. On the contrary, scientists themselves did not seem to be especially "scientific" in their reasoning. (b) In that case, "psychology of science" might be identified as a field dedicated to improving the conduct of science by instilling in scientists a greater awareness of both their own thought processes and those of their subjects. This was the position advanced by E. C. Tolman in the 1930s. (c) One of Tolman's students, Donald Campbell, operationalized this vision in ways that profoundly influenced social–scientific methodology in the second half of the 20th century, generally resulting in a more sophisticated and fallibilist conception of scientific validity. (d) An alternative vision of the psychology of science was advanced by Abraham Maslow in the 1960s, and was less concerned with maximizing truth over error than with realizing human potential as such. For Maslow, science's cognitive shortcomings reflected stunted motivational development, an important symptom of which was the strong separation of science from religion as forms of experience. (e) Psychohistorical inquiry of the sort endorsed by Maslow has been used to try to recover some of science's lost motivational structure, with religion on offer as one—but by no means the only—animating force behind the 17th-century Scientific Revolution, whereby the West came to acquire its distinctive modern worldview. (f) However, Maslow's version of the psychology of science has not prevailed, having to endure skepticism about its psychoanalytic provenance and unrealistic assumptions about the normal distribution of scientific creativity, most of which happens to be inhibited by the constitution of scientific institutions and society in general. In contrast, much of today's psychology of science presumes a natural scarcity of scientific talent and ambivalence toward the survival value of the pursuit of science for the human species.

IN THE BEGINNING, THE SCIENTIST WAS A SELF-CONSCIOUS PSYCHOLOGIST

It is often not appreciated that experimental psychology itself began as the psychology of the scientist. This certainly applies to the pivotal role that intersubjective differences in astronomical observation—the so-called personal equation—played in laying the foundations of psychophysics in the mid-19th century (Boring, 1929/1950, p. 31). It was assumed that science's special epistemic powers resulted from scientists' heightened self-consciousness about human cognition in general. This assumption was taken in two different directions in the early 20th century. The first, championed by Dewey (1910), amounted to advising teachers that children should be exposed to the "scientific method" (i.e., hypothesis-testing) in order to learn how to think.

The second, relevant to the subsequent development of psychology as a discipline, was that scientists may be expected to be more articulate about what nonscientists always already do whenever they think. This view, shared by the physicists Max Planck and Albert Einstein, justified the Würzburg School's controversial application of the introspective method to themselves and each other, very often using complex philosophical problems as stimuli (Kusch, 1999). These early steps in what is now called "cognitive psychology" were followed up in the 1920s and 1930s by the Gestalt psychologists Otto Selz and Karl Duncker, who moved from their Würzburg roots in clearly separating the experimenter and subject roles. They studied scientists' notebooks to construct general testable accounts of problem-solving (Petersen, 1984; Wettersten, 1985). Moreover, the Gestalt thought leader Wertheimer (1945) probed the directive nature of Einstein's own discovery process in a quarter-century correspondence with the revolutionary physicist. This work in turn inspired the field of "protocol analysis" and other techniques to infer thought processes from verbal behavior that are still used by a wide range of psychologists today (Ericsson & Simon, 1985).

In terms of the philosophy of science, this original version of the psychology of science treated in a "realist" fashion what psychologists themselves might normally treat as "methodology," a general set of principles designed to amplify truth-tending and minimize false-tending inferences in the scientific community (Campbell, 1988). The implied difference here is between taking scientists' self-accounts of their reasoning as recapitulating their actual thought processes (psychology), and as rationally reconstructing those thought processes for purposes of peer evaluation (methodology). To be sure, the treatment of methodology as literal "rules for the direction of the mind" recalls the original 17th-century sense of "scientific method," promoted by Francis Bacon and René Descartes as extending the Protestant Reformation's call for the faithful to become more personally accountable to the deity (Yates, 1966). This psychologically realist sense of the scientific method was formalized (and secularized) in the 18th and 19th centuries as a quest for "the logic of discovery," paradigmatically the train of thought that culminated in Newton's grand synthesis (Laudan, 1981, Chap. 11). However, by the start of the 20th century, the failure of such philosophers as the pragmatist Charles Sanders Peirce to provide a compelling account of a distinct mental process that at once genuinely advances our knowledge and is reliably self-correcting—what Peirce had called "abduction"—resulted in the more psychologically antirealist or constructivist stance toward the scientific method that persists in philosophy to this day (Chap. 14). Thus, methodological principles such as the logical positivists' verifiability, Popper's falsifiability, and probability-based formulae such as Bayes's Theorem, have been primarily proposed as ex post tools for evaluating hypotheses, not for hypothesis generation. This helps to explain, and perhaps even justify, the radically reconstructed accounts of the history of science in which philosophers have indulged such methods to make their case for one or another (e.g., Lakatos, 1981).

Something of the original Würzburg and Gestalt sensibility returned to the psychology of science that emerged, now as an independent field, in the wake of the so-called cognitive revolution in psychology that began in the

1950s and peaked in the 1970s (Baars, 1986). However, this work, typically indebted to the experimental paradigms developed at University College London by Peter Wason and Philip Johnson-Laird in the 1960s, shifted the focus of inquiry from the introspective reports of scientists to the verbal behavior of nonscientists confronted with scientifically structured tasks (Wason & Johnson-Laird, 1972). The first work to self-consciously identify itself in terms of this lineage was Tweney, Doherty, and Mynatt's (1981), which showcased research from Bowling Green State University. Its epistemological legacy remains conflicted to this day. On the one hand, the very idea that subjects might be tested for, say, their propensity to look for evidence that disconfirms a hypothesis suggested that they could and should do such a thing in everyday life. Here, the new cognitive psychologists of science appeared to be one with the Würzburgers and Gestaltists. On the other hand, most of the work done under this paradigm showed how distinct canonical forms of scientific reasoning are from the ways of ordinary cognition. While one might hope that all this shows is that scientific reasoning requires specialized training, in fact, scientists themselves turned out to be no better at, say, falsifying their hypotheses than nonscientists. Similar points had been long observed and theorized by Paul Meehl in the context of the unreliability of clinical judgment. At first, Meehl (1967) understood this problem as limited to psychologists' tendencies to overestimate intuition and individual differences, at least vis-à-vis what he had presumed to be the superior—broadly Popperian—practice of physicists. However, Meehl's student, Faust (1984), subsequently incorporated the emerging literature in the new cognitive psychology, drawing generally skeptical conclusions about human cognitive capacities, both lay and expert, when compared to, say, actuarial tables and computerized expert systems capable of ranging over large and varied data sets.

Reflecting on this research with hindsight, one might conclude that the new cognitive psychology of science never quite escaped the shadow cast by behaviorism on mid-20th-century experimental psychology. In particular, it could never quite decide whether it wanted to treat mental representations of the scientific method (either by the experimenter or the subject) as revisable models (i.e., prototypes for some presumed cognitive process) or testable hypotheses (i.e., accounts for predicting behavioral regularities). In the next section, we explore the source of this ambivalence as alternative responses to Bertrand Russell's critique of the introspective method, but some recent researchers have tried to turn this ambivalence into a virtue (e.g., Gorman, 1992). In any case, by holding subjects accountable to one or another version of the scientific method, psychologists appeared to have unwittingly revealed a deep sense of human irrationality intrinsic to the cognitive process, and not simply the product of motivated interference, say, from conscious self-interest or an unconscious defense mechanism.

In the past quarter century, there have been two general responses to this development, one social constructivist and the other, more strictly, social psychological. The first aims for no less than a deconstruction of the very idea of the scientific self-consciousness, according to which, independent behavioral corroboration being absent, any scientific self-reporting should be presumed to be entirely the product of rationalization, better explained by the norms of self-accounting at the time the scientist is writing than any generally identifiable psychological processes (Brannigan, 1981; Fuller, 1989/1993). Arguably, this sort of critique served to undermine the episte-mological basis for the most systematic attempt at founding a "social psy-chology of science" (Shadish & Fuller, 1994). Turner (1994), a social theorist active in the science studies disciplines, observed that the repeated finding of discrepancies between psychologists' normative expectations of what scientists should do and what scientists actually do, whether studied in experimental or field settings, placed psychologists of science in the tricky rhetorical position of arguing that scientists do not know how to do science properly, a conclusion that would then rebound on the epistemic authority of the psychologists of science themselves, just as it had on followers of Karl Mannheim in the sociology of knowledge earlier in the 20th century. An honorable attempt to carry on as a practicing psychologist reflexively aware of the various levels at which this problem arises—from personal encounters with subjects to formal peer-review processes—may be found in the corpus of Mahoney (1976/2004).

The second response to this apparent irrationality of scientists was already foreshadowed in the first major systematic presentation of the cogni-tive limits research: It is to argue that all of the formally fallacious forms of inference observed in the laboratory must serve some sort of "adaptive" func-tion for the human organism (Tversky & Kahneman, 1974). Here, "adaptive" may be understood in a purely social–psychological sense, such as referring to a body of "lay knowledge," roughly an articulated version of "common sense," that offers a coherent world-view in normal settings. However, such knowledge tends to be culturally sensitive and does not handle extreme cases especially well. Some self-styled "experimental philosophers" have tried to redefine the task of philosophy as the systematic probing of those limits (Knobe & Nichols, 2008), while some social psychologists have examined how those limits are negotiated and perhaps even transcended by scien-tists who, after all, are lay people for all but their specific area of expertise (Kruglanski, 1989, Chap. 10). Since the advent of evolutionary psychology in the 1980s, it has become customary to contrast the *adaptive* character of our comprehensively flawed normal modes of reasoning—typically dignified under the rubric of "heuristics"—with the *exaptive* character of methodologi-cally sound scientific reasoning, which casts it as the "co-opted by-products of adaptations" (Feist, 2006, p. 217). In short, science is a humanly significant, unintended consequence of the overall evolutionary process.

To be sure, this response has done nothing to eliminate the apparent strangeness of scientific modes of thinking. However, it has inspired research into the difficulties that children face in mastering scientific concepts in ways that have served to undermine the plausibility of Jean Piaget's long influential "genetic epistemology," which had portrayed the maturing child as recapitulating in its own cognitive development the increasingly counter-intuitive character of the history of science (Feist, 2006, Chap. 3). Perhaps a more straightforward normative response to dealing with the alien character of scientific reasoning has been to reassert the need for strong, methodologi-cally driven scientific institutions to counteract our default cognitive liabili-ties (Fuller, 1989/1993). However, scientists who studied the psychological and sociological evidence for the "unnatural nature of science" tended to interpret this charge narrowly to involve requiring scientists to stick to what is "evidence-based" and not consider "value-laden" issues—thereby leaving the weighting of evidence to public debate and policymakers (Wolpert, 1992). A relevant precedent for thinking about the matter is the U.S. Constitution, which is subject to various "separations of powers" and "checks and balances" to ensure that people's inevitable conflicts of interest and short-sightedness do not undermine their collective capacity to serve the public good (Fuller, 2000a). Popper held a similar view but did not believe that "facts" and "values" could be separated so neatly, as scientists were no different from others in overestimating their own views and underestimating those of their opponents. Indeed, science," for Popper, is the organization of those liabilities in aid of producing epistemic virtue, via "the method of conjectures and refutations," as imperfectly regulated by the peer-review process (Jarvie, 2001).

THE NEED FOR PSYCHOLOGY OF SCIENCE TO IMPROVE THE CONDUCT OF SCIENCE: THE CASE OF E. C. TOLMAN

The crucial step taken in the research design of the new cognitive psychology of science from its Würzburg precursors was to study scientific thought processes in a manner alienated from the scientist as a psychological agent. After Kant, the teleological character of science had been underwritten by a conception of thinking as *judgment* that stressed its self-generated, self-directed, discriminating, and determining character. (As we shall see, William James understood this very well.) The paradigm case of judgment was the assertion of the truth or falsity of a proposition, what logicians still call, after the psychophysicist Gustav Fechner's student Hermann Lotze, "the assignment of a truth-value" (Fuller, 2009). The Würzburg School had attempted to found experimental psychology on the study of judgment in its most primitive form, namely, between two sensations (of weight), in which subjects detected an "attitude" or "determining tendency" associated with task but no corresponding image (Kusch, 1999, Chap. 1). For them, the fundamental difference between "scientific" and "lay" thought resided in neither the content nor the processes of thought, but in the subject's degree of self-consciousness (Gigerenzer & Murray, 1987, Chap. 5). In this respect, Popper (1935/1957), though rightly seen as more methodologist than psychologist of science, was true to his own Würzburg roots—Karl Bühler was his doctoral supervisor in Vienna—when he proposed that a mark of a truly revolutionary scientist such as Einstein was his capacity to conduct in his own mind stringent tests of a pet theory (i.e., "thought-experiments") for purposes of reaching a critical judgment (Wettersten, 1985). This is in sharp contrast with the nemesis of Popper's later career, Thomas Kuhn, who held that even great scientists lacked the psychological wherewithal for such potentially paradigm-shifting feats (Fuller, 2000b, Chap. 6; Fuller, 2003).

However, Russell's (1927) popular work, *An Outline of Philosophy*, launched a two-pronged attack on the very idea of scientifically studying mental contents, based on the then vanguard fields of symbolic logic and behaviorist psychology, which together effectively eliminated introspection as a method in scientific psychology. The first call for a dedicated field of "psychology of science" emerged as a subtle response to Russell's critique. The key figure was Naess (1965), eventually the *magus* of Norwegian philosophy, who was one of two graduate students (the other being W. V. O. Quine) present at the original meetings of the logical positivists' Vienna Circle. Ness heard E. C. Tolman promote the need for such a field in an address given to the 1936 International Congress for Unified Science in Copenhagen (Smith, 1986, Chap. 5). Naess had been visiting Tolman, founder of Berkeley's psychology department, who had studied with Ralph Barton Perry, the successor to William James at Harvard's philosophy-cum-psychology department. Tolman's self-styled "purposive behaviorism" involved endowing the full range of organisms he studied—from rats and apes to humans—with "cognitive maps" that supposedly inform their responses to experimental tasks. Here, Tolman was following up the Jamesian insight that each individual has its own spontaneous way of organizing the data it receives prior to any experimenter-driven protocols or stimuli.

Originally, James (1884) was arguing for a more self-critical sense of introspection, which he then developed into a general account of "conceptual schemes," his influential expression for how the mind processes data as objects that then function as a means to the individual's ends (James, 1890/1983, Chap. 12). Features of the environment that fail to be captured in such patently functionalist terms—to mix two Jamesian metaphors—may escape through the mesh of the scheme, eventually only to blindside the individual altogether. However, James was clear that conceptual schemes do not change "naturally," but only as a result of individuals having to accommodate these previously ignored features in order to realize their goals. In this respect, James treated conceptual schemes as rigid but also replaceable, assuming that the organism has a strong sense of purpose. Although James was widely interpreted as retreating from idealism and opening the door to behaviorism, in fact he wanted to do almost the exact opposite—namely, to extend the locus of intentionality from the thinking head to the entire organism operating in the world (Bordogna, 2008).

True to the original spirit of James, Tolman (1932) held that the "psychology of science" should help the scientist become psychologically closer to the object of investigation—including not least, as Tolman himself stressed, by coming to understand how a rat thinks of its maze-running task (Maslow, 1966, p. 112). Despite being criticized by fellow behaviorists for conflating hypothetical constructs with real psychological states, Tolman insisted that the capacity to think like the experimental subject (whatever the species) provides a crucial source of clues for operationalizing hidden psychic processes that govern behavior, what Tolman called "intervening variables." This approach resonated with the Gestalt psychologists, who distinguished themselves from both introspectionists and behaviorists by denying that the subject necessarily frames the experimental task as the experimenter does, such that the only behavior worth examining would be that which bears directly on the task (Ash, 1988, Chap. 5). Perhaps because the most systematic Gestaltist, Koehler (1938), shared a medical background with James (and Freud), he recognized the importance of such "unobtrusive measures" as blood pressure, rate of breathing, and other incidental behaviors unrelated to the experiment's official task that nevertheless may provide clues to the subject's wider existential horizon (or "baseline"), which in turn may serve to amplify or subvert the experimenter's own assumptions about the subject. These measures provided evidence for the intervening variables that Tolman sought, a point that was not lost on his most distinguished student, Donald Campbell, who worked directly with the Gestaltist who migrated from the Vienna Circle to Berkeley, Egon Brunswik (Campbell, 1988, Chap. 13; cf. Brunswik, 1952).

Russell's critique of introspection was still more strongly felt in the radical behaviorism that would prevail at Harvard itself in the middle third of the 20th century, courtesy of W. V. O. Quine in philosophy and B. F. Skinner in psychology (Baars, 1986, p. 62). A key mediator in this transition was the psychophysicist E. G. Boring, the founding head of Harvard's university's free-standing psychology department, who underwent what he described as a "Damascene" conversion from the introspective methods that he had learned at Cornell University from its leading U.S. proponent, E. B. Titchener. Generalizing from the phenomenon of time delay associated with the transmission of both electrochemical impulses in the nerves (Helmholtz) and light rays in the heavens (Einstein), Boring came to believe by 1930 that any act of outer or inner perception amounted to a bifurcation of the self into two relatively independent parts, one largely a stranger to the other (Boring, 1955). He saw the perceiver-part as captive to a "*Zeitgeist*," the totality of the scientist's cultural heritage that functioned as a cognitive unconscious that inhibited his ability to see the perceived-part in all its significant novelty. (Intuitively, the *Zeitgeist* might be seen as the time lag that results from taking one extended moment in history as a stationary frame of reference.) Here, Boring specifically referred to Joseph Priestley's failure to abandon phlogiston in favor of the oxygen-based chemistry promoted by Antoine Lavoisier, an example he credited to Conant (1950), the then Harvard president who had torn asunder William James's original unification of psychology and philosophy, so as to render psychology a purely experimental unit. Conant's teaching assistant, Kuhn (1962/1970), would later infer from the Priestley–Lavoisier example a more restricted, discipline-based sense of *Zeitgeist*, the "paradigm."

Together, the two responses to Russell's *Outline*—purposive and radical behaviorism—generated the two main problematics of the philosophy of science during the final third of the 20th century: what after Quine (1960) is called the "indeterminacy of translation" and after Kuhn (1962/1970) the "incommensurability of paradigms" (Fuller, 1988, Chap. 5). While philosophers have tended to define these matters in strictly logical terms (i.e., in terms of whether truth is preserved as the same data are processed by different conceptual schemes or theoretical languages), the psychological implications were more vivid. Both Quine and Kuhn suggested that any piece of evidence might be regarded in any number of logically self-consistent ways, such that its most "natural interpretation" would depend largely on the personal history of the interpreter in question. Arguably, philosophy has failed to make progress as a field precisely because its history has been subject to the simultaneous propagation of several qualitatively different "world hypotheses," each with its own way of configuring what might appear, from a "common sense" standpoint, as the "same evidence." This point was associated with Tolman's Harvard classmate and chair of Berkeley's philosophy department, Pepper (1942), who originally used it to counter the logical positivist claim of theory-neutral data. However, one of Pepper's key hires in the late 1950s would go on to show how science has historically solved the problem that Pepper had identified in philosophy—namely, that science is governed by a succession of distinct paradigms, each dominant for a given period (Kuhn, 1962/1970). Thus, Kuhn gave an overriding role to disciplinary socialization, which reorients the scientist's world-view to such an extent as to make it difficult, if not impossible, for her to alter her working assumptions later in life, even in the face of cognitively better alternatives. The exceptions to this rule, originally stressed by Max Planck, were those scientists who have made relatively little existential investment—say, by virtue of being young practitioners or cross-disciplinary interlopers (De Mey, 1982, Chap. 6). It follows that a radically new science may require radically new scientists—though *pace* Planck it would be a mistake to interpret the relevant sense of "newness" primarily in terms of age (Simonton, 1988).

IF TRUTH IS TOO HARD, THEN HOW ABOUT VALIDITY? DONALD CAMPBELL AND THE ROUTE FROM TOLMAN

The centrality of judgment as a mental process that enabled scientists to be seen as self-conscious psychologists presupposed that science is a goal-directed activity with a clear objective, the ultimately comprehensive account of reality, what physicists continue to dignify after Einstein as the "Grand Unified Theory of Everything" but philosophers know more simply as "The Truth." Recalling the late 19th century, when physicists were routinely claiming to be on the verge of such an account, Max Planck characterized scientific theories as *Weltbilder* ("world constructions" or "world pictures"). Indeed, a secular legacy of Newton's own crypto-millenarian aspirations is that at the end of every century in the modern era, science has been seen as near the realization of a very grand *Weltbild* through its own initiative that would provide the basis for transcending humanity's intellectual if not physical burdens (Horgan, 1996; Passmore, 1970, Chap. 10). In the heyday of this conception—roughly, the quarter century on either side of 1800—Condorcet, Comte, and Hegel exemplified the guiding intuition here—that the history of science systematically records humanity's collective self-reflection, unreflective versions of which occur in the ordinary thought processes of individual humans.

However, the relativity and quantum revolutions in physics that Planck helped to usher into the early 20th century caused the *Weltbild* approach to science to retreat from first-order characterizations of scientific practice to second-order metaphysical interpretations that remain relevant even after a scientific revolution. Thus, Holton (1973), a student of Percy Bridgman, the Harvard professor of thermodynamics who wanted to reduce all scientific concepts to logical and instrumental "operations," relocated *Weltbilder* to the heuristic value provided by perennial philosophical questions—such as whether matter is continuous or discontinuous—to the conduct of scientific inquiry. These *themata* function as auxiliary constraints or desiderata that guide the empirical elaboration of scientific theories but are never conclusively proven by them. Unsurprisingly, the *Weltbild* conception of science was pursued most actively not by professional scientists but by philosophers, especially as a pragmatic interpretation of "convergent scientific realism," ranging from Peirce to Wilfrid Sellars and Nicholas Rescher (Laudan, 1981, Chap. 14). However, all of these figures stressed not only the increasing distance that the scientific world-view takes one from commonsense but also the diminishing epistemic returns of scientific effort as one continues down that path (e.g., Rescher, 1998). Moreover, once Kuhn (1962/1970) installed a proper Darwinian model of scientific evolution, whereby inquiry endlessly exfoliates into ever greater specialization and never to converge in some summative epistemic achievement, even the idea of *themata* as regulative ideals of science started to lose its appeal.

However, it would be a mistake to conclude that the loss of unified knowledge and ultimate truth as explicit aims of science in the second half of the 20th century undermined the idea of validity in scientific inquiry altogether. Here, Donald Campbell played an especially important role in translating the epistemic aspirations of the scientific method into operating procedures for fallible inquirers. Most notably, Campbell and Stanley (1963) introduced the "quasi-experimental" research design in response to demoralized education researchers who despaired of ever identifying a "crucial experiment" that could decide between rival hypotheses, each of which could claim some empirical support in explaining a common phenomenon. Campbell and Stanley concluded that such despair reflected an oversimplified view of the research situation, which demanded a more complex understanding of research design that enabled rivals to demonstrate the different senses and degrees in which all of their claims might be true. But, opening up the research situation in this fashion legitimized the loosening of the various controls associated with the experimental method, which in turn reinvited versions of the traditional problems of naturalistic observation. It was these problems that Campbell and Stanley attempted to address as "threats to validity."

Most of the 12 threats to validity enumerated in Campbell and Stanley (1963) involve insensitivity to the complex sociology of the research situation. For example, research may be invalidated because the researcher fails to take into account the interaction effects between researcher and subject, the subject's own response to the research situation over time, the subject's belonging to categories that remain formally unacknowledged in the research situation but may be relevant to the research outcome, salient differences between sets of old and new subjects, as well as old and new research situations, when attempting to reproduce an outcome, and so on. The "quasi-experimental" research design that Campbell and Stanley advocated as a strategy to avoid, mitigate, and/or compensate for such potential invalidations can be understood as a kind of "forensic sociology of science" that implicitly concedes that social life is not normally organized—either in terms of the constitution of individuals or the structures governing their interaction—to facilitate generalizable research. In effect, methodologically sound social research involves an uphill struggle against the society in which it is located, and which its "valid" interventions might improve.

This rather heroic albeit influential premise has come under increasing pressure as insights from the sociology of science have been more explicitly brought to bear on research design. Presented as a reflexive application of the scientific method to itself, the resulting analysis can be quite skeptical (e.g., Brannigan, 2004). Underlying this is ambiguity about how the aims of research might be undermined in its conduct. This may happen because a hypothesis fails to be tested in the manner that the researcher intended—a situation not to be confused with the researcher's failure to obtain the result she had expected. Much is presupposed in this distinction. Research is not a linear extension of common sense or everyday observation, but rather requires a prior theory or paradigm that yields an appropriate hypothesis, on the basis of which the researcher selects relevant variables that are then operationalized and manipulated in an environment of the researcher's creation and control. Two matters of validity arise: first, the reliability of the outcomes, vis-à-vis other experiments of similar design; second, the generalizability of the outcomes to the population that the experiment purports to model. The clarity of this distinction is due to the statistician Fisher (1925), who recommended the random assignment of subjects to groups of the researcher's choosing, given the variables that need to be operationalized to test her hypothesis.

Two features of Fisher's original context explain the *prima facie* clarity but long-term difficulties of his approach (Ziliak & McCloskey, 2008). First, Fisher developed his paradigm while working in an agricultural research station, which meant that the things subject to "random assignment" were seeds or soils, not human beings. Second, the station was operated by the British Government and so not for profit. While the former feature pertains to the concerns originally raised by Campbell and Stanley (1963), to which Fisher was perhaps understandably oblivious, the latter suggests a more general challenge to statistical significance as a meaningful outcome of hypothesis testing. Fisher's experiments on the relative efficacy of genetic and environmental factors on agricultural output were conducted in a setting where matters of utility and cost did not figure prominently. A state interested in acquiring a comprehensive understanding of what is likely to make a difference to food production presumed that every hypothesis was equally worthy of study. Thus, "statistical significance" came to be defined as the likelihood of an experimental outcome, given a particular hypothesis, which is tested by seeing whether the outcome would have been the same even if the hypothesis were false (i.e., the "null hypothesis"). In contrast, had Fisher been more explicitly concerned with utility and cost, he might have treated hypothesis testing as a species of normal rational decision making. In that case, the validity question would be posed as follows: Given the available evidence, what is the cost of accepting a particular hypothesis, vis-à-vis potentially better hypotheses, that might require further testing? Indeed, Ziliak and McCloskey (2008) observe that this was the stance taken by W. S. Gosset, a student of Karl Pearson who was lab director for Guinness breweries in the early 20th century.

The opportunity costs of not exploring alternative hypotheses raised in the economic critique of Fisher threaten to blur his distinction between internal and external validity by forcing the researcher always to calibrate the knowledge gained from a particular design in terms of the goals it is meant to serve. For their part, psychologists have tried to address the issue by following Campbell's Berkeley teacher (and Bühler's student) Brunswik (1952) and distinguishing *ecological* and *functional* validity, the latter being "external validity" in the strict sense of extending the laboratory to the larger world rather than trying to capture the prelaboratory world (Fuller, 1989/1993, Part 4). The distinction addresses exactly what should be reproducible in a piece of research: Ecological validity pertains to an interest in reproducing causes and functional validity effects. Very few experiments meet the standard of ecological validity but many may achieve functional validity, if they simulate in the target environment the combination of factors that produced the outcome in the original experiment, even if that means altering the target environment's default settings (cf. Shadish & Fuller, 1994, Chap. 1).

The key epistemological point made by dividing external validity this way is that experiments may provide valid guides to policy intervention without necessarily capturing anything historically valid about the underlying causal relations. Instead, they would test the limits of a hypothetical model of the world that has been already proven within specific lab-based parameters. Indeed, Meehl (1967) appealed to this use of experimentation to explain the scientific superiority of physics to psychology, a field that by his lights too often relied on research designs that aimed simply to mimic rather than genuinely test (and hence potentially extend) what the researcher already thought he knew about the target environment. But as Hedges (1987) subsequently observed, ignoring the ecological dimension of research validity tends to be accompanied by an omission of outlying data that may be crucial to identifying relevant contextual differences that point to other, perhaps even countervailing factors at work to the ones hypothesized. This point bears on the skepticism generated by proposed "evolutionary" explanations of the experiments conducted in the 1960s and 1970s by Stanley Milgram and Philip Zimbardo that demonstrated the relative ease with which subjects submit to authority and torture their fellows (Brannigan, 2004). However, one may grant the lack of ecological validity in such experiments, while acknowledging that they suggest how a different combination of factors might reach an equivalent outcome, say, in the context of wartime intelligence gathering. The success of such an extension of the original research paradigm would demonstrate its functional validity.

PSYCHOLOGY OF SCIENCE FOR UNTAPPING HUMAN POTENTIAL: THE CASE OF ABRAHAM MASLOW

If E. C. Tolman's original call for a "psychology of science" in the 1930s was made in a broadly positivistic spirit of ongoing improvements to the conduct of science, the first book actually to bear the title "psychology of science," published 30 years later, was concerned mainly with the very motivation for doing science, which suited a time when a science-based "Cold War" placed humanity at the brink of self-destruction. The book's author, Maslow (1966), was a thought leader in humanistic psychology, a school that aimed to provide an account of mental health that complemented Freud's psychodynamic theory of neurosis. Although Maslow is nowadays remembered for a cluster of concepts—"self-actualization," "hierarchy of needs," "peak experiences," and so on—that continue to inform the positive psychology movement, this legacy was very much informed by Maslow's own training as a scientist (originally as a primate behaviorist and then as a devotee of Alfred Adler's brand of individual psychology) and his encounters with natural and social scientists whom he came to regard as paradigm cases of self-actualized human beings (Albert Einstein, Max Wertheimer, Ruth Benedict).When the core thesis of Maslow (1966) was first delivered as the annual invited lecture to the John Dewey Society for the Study of Education and Culture, Maslow's words were clearly received as a challenge for those who considered themselves "scientists" to determine how much of their personal experience might enhance the epistemic power of their research (Maslow, 1966, Foreword).

Underlying this challenge was Maslow's broad commitment to science as "hypothesis testing" but understood in a particular way, namely, that hypotheses are artificial constructions that approximate but are not substitute for reality. At the start of the book, Maslow cites the influence of a Brandeis University colleague, the psychohistorian Manuel (1963), who wrote of Isaac Newton's secret interest in sacred history as reflecting an awareness of a deity perpetually dissatisfied with his own creation, whose "laws of nature" were an ideal that regularly needed to be enforced against the recalcitrant ways of matter. Far from reading divine order directly from the heavens, Newton had to postulate God's periodic interventions to square his mathematics with the astronomical data. As this sensibility was secularized in the modern era, the biblically fallen character of nature came to be internalized as humanity's sense of its own subordination to nature, a viewpoint that easily shaded (e.g., in Nietzsche) into nihilism, potentially undermining the motivation to study nature in Newton's systematic fashion. To be sure, Maslow was not the first to discover the problem of motivating a commitment to science. At the end of the 19th century, no less than "Darwin's bulldog," Huxley (1893) turned Herbert Spencer's confidence on its head by arguing that instead of providing a scientific basis for ethics, evolutionary theory—with its radically demystified ("species egalitarian") view of humanity—could subvert any belief in future scientific, let alone ethical progress. After all, if Darwin is correct, extinction is the only long-term guaranteed outcome to any collective project in which a species might engage (Fuller, 2010, Chap. 6).

Maslow took this general existential anxiety to have direct consequences for actual scientific practice. In a world where people no longer believe that they have been created "in the image and likeness of God," a Newton-like capacity to abstract heroically from an array of data points—so necessary for success in modern science—may reflect a psychological estrangement from concrete human existence, which in turn may lead to the application of methods that can be justified on theoretical but not phenomenological grounds. Thus, the fact that Neo-Darwinism says that humans, genetically speaking, differ very little from other animals does not mean that study-ing humans according to methods normally used to study apes or rats will yield equally interesting results. More specifically, Maslow diagnosed radi-cal behaviorists' lack of openness to differences in the individual histories of their human subjects in terms of a "fear of knowing" that failed to alleviate the behaviorists' own anxiety about the ultimate truth of their own hypoth-eses, but simply kept that anxiety in check—so to speak, Cold War-style. Maslow dubbed this defense mechanism *safety science*, in contrast to the psy-chologically healthier *growth science*. He drew a clear distinction between the safety scientist's need for a method that achieves attestable results and the growth scientist's greater tolerance for uncertainty and error—only the latter of which provides the relevant psychic environment for creative effort.

Interestingly, Maslow explicitly cast the two scientific world-views as glosses on Kuhn's (1962/1970) normal and revolutionary science. However, he went against the grain of Kuhn's own argument by associating the lat-ter, not the former, with the fully self-actualized growth scientist. Indeed, Maslow saw the safety scientist as subject to arrested development and regarded the relative rarity of revolutionaries in science as symptomatic of massive untapped human potential. Focusing on just its cognitive implica-tions, Maslow's safety–growth science distinction would seem to drive a wedge between "expertise" in the strict sense (i.e., knowledge that results from trained experiences of a certain sort) and a more authentic scientific existence, whose comprehension of reality extends beyond glorified pattern recognition. Here, it is worth recalling that an important sense of "para-digm" in Kuhn (1962/1970) is the kind of pattern recognition associated with the normal scientist's tendency to solve problems by finding textbook exem-plars for his or her solutions. Kuhn had learned of this approach from the self-styled "new look" to the study of thinking then championed at Harvard by Bruner, Goodnow, and Austin (1956). However, it involved just the sort of self-imposed cognitive boundedness that was anathema to Maslow, which by the 1960s had come to be popularized in advertising as "subliminal per-ception." Accordingly, one would overcome fear of the new (experimental stimulus, scientific finding, consumer good, etc.) by unconscious assimila-tion to already familiar things that had satisfied corresponding needs in the past. In this respect, one might see Maslow as providing an implicit method-ological critique of the new cognitive psychology of science.

In the near half century since the publication of Maslow (1966), support for the presence of the two scientific worldviews have come through the con-cept of the "self-regulatory focus," which social psychologists claim struc-tures a person's cognitive motivation to enable one to remain in equilibrium with his or her environment. In this context, safety science corresponds to a "prevention" focus and growth science to a "promotion" focus (Higgins, 1997). The distinction turns on what people take to be the greater evil: a harm that could have been avoided or a good that could have been realized. From the standpoint of statistical inference, the contrast is easily captured in terms of living one's life so as to err on the side of either missed opportunities or false alarms. In terms of the broader cultural context of science, one might distinguish, respectively, *precautionary* and *proactionary* uses of science, the former stressing the damage potentially done from not correctly anticipating the consequences of a science-led intervention, the latter the benefits (including learning from mistakes) potentially reaped from taking a chance on just such an intervention, regardless of its consequences. These perspectives are routinely played out in public debates about the impact of new technology on the future of the planet (Fuller, 2010, Chap. 1).

Although Maslow anchored the safety–growth science distinction in Kuhn (1962/1970), his principal source for the psychology of the growth scientist was, perhaps surprisingly, the philosophical chemist Michael Polanyi, again deployed in ways that stressed the differences from Kuhn. Nowadays, Polanyi (1957) is remembered as Kuhn's great precursor on the role of "tacit knowledge" as the aspect of scientists' socialization that enables them to judge intuitively what is normal and deviant practice in their fields. However, Polanyi (1961/1963) had criticized an early version of Kuhn (1962/1970) for suggesting that successful revolutionary science never occurred deliberately but only as an unintended consequence of the self-implosion of normal science. In contrast, Polanyi believed that science differed from "mere" technical expertise in its drive to test the limits of its fundamental assumptions. In this respect, Polanyi sounds very much like Karl Popper—except that Polanyi parted from Popper in stressing the scientific establishment's authority as the final arbiter of any revolutionary proposals a particular scientist might make. Here, Polanyi was probably reflecting on his own experience as a chemist, whose research failed to make headway with its nonmathematical, visually based approach just when his field was being absorbed into the decidedly nonvisual, mathematically driven paradigm of quantum mechanics (Nye, 2011). Unlike Popper, who continued to contest the correct interpretation of quantum mechanics and evolutionary theory into his old age, Polanyi conceded that, in some definitive sense, his field's research frontier had moved on, and so he migrated from a chair in chemistry at the University of Manchester to one in "social studies," where he forged an equally creative second career; indeed, the one for which he is now more widely known.

Interestingly, Maslow appeared to turn a blind eye to Polanyi's social conformism, which was based on an extended analogy between the scientific community and a monastic order, while agreeing with him that the difference between science and religion was "merely" institutional, a distinction that was drawn nowadays to the detriment of both. Indeed, one of the most jarring features of Maslow (1966) for today's secular reader is the ease with which the religious nature of science is maintained. Indeed, his concluding chapter calls for the "resacralization of science" (cf. Fuller, 2006, Chap. 5). However, Maslow does not identify religion with a church or dedicated clergy, but with the basic need to understand the ground of one's being. In tune with his times, he explicates this need in terms of complementary attitudes represented by the Eastern ("Taoistic") and Western ("Judaic") world religions: on the one hand, a receptivity to the structure of reality; on the other, a refusal to be satisfied with first appearances. What distinguishes religion and science in practice is simply the means by which they pursue these general regulative principles of inquiry, which then has implications for what one means by "success," "validity," and "progress." In the next section, we consider a now forgotten dispute in the "psychohistorical" vein that Maslow encouraged that illustrates how to argue about the contribution of religion to the motivational structure of modern science.

PSYCHOLOGY OF SCIENCE IN SEARCH OF THE MOTIVATION FOR SCIENCE: IS IT IN RELIGION?

Merton (1938/1970) and Feuer (1963), two philosophically minded sociologists and contemporaries at Harvard in the 1930s, arrived at radically different views about the affective foundations of the 17th-century Scientific Revolution that were enacted in a fraught encounter at the 1956 meeting of the Eastern Sociological Association in New York (Cushman & Rodden, 1997). Following Max Weber, Merton argued that Puritan asceticism fostered the necessary self-discipline, now divested from the overbearing authority of the Church of Rome, to enable a comprehensive investigation of nature through experimentation and close observation. The crucial premise—sometimes hidden in Merton's own argument—was that a literal reading of the Genesis-based belief of humans as created *in imago dei* suggested that "we," understood as a species rather than as particular individuals, likely to succeed in coming to grasp God's plan. Thus, Calvin's Elect was secularized as—to recall a slogan of Newton's that Merton later refashioned for his own purposes—the "lucky giants" on whose shoulders others great and small subsequently stand. This premise justified the long, arduous, and often seemingly pointless work of day-to-day research, central to what Kuhn (1962/1970) later characterized as "normal science."

Feuer countered this with an updated version of a view popularized by Cornell University founder White (1896) in the wake of the controversies surrounding Darwin's theory of evolution—namely, that the Scientific Revolution marked an end to the "Dark Ages," a fundamental break from religious attitudes altogether. In particular, Christian routines of self-denial were replaced by a hedonistic openness to a material world that we fully inhabit, a sensibility that Feuer interpreted as a shift in humanity's existential horizons from pessimism to optimism. To be sure, Merton had not denied science's latent optimism. However, he saw its role as sustaining collective effort toward a goal, the beneficiaries of which would most likely be not a given generation of scientists, but their glorious successors who finally fathom the divine plan. In contrast, the optimism that Feuer attributed to the Scientific Revolution was ultimately self-consuming, namely, the capacity, unfettered by religious orthodoxy, to ameliorate the human condition in one's own lifetime by inventions and discoveries that minimize ambient pain, while increasing personal convenience. Both Merton and Feuer could easily cite the founder of the scientific method, Francis Bacon, for their own purposes. But, where Merton clearly saw the Scientific Revolution as culminating in Newton and fostering an ideology of progress that secularized the Christian salvation narrative, Feuer's scientific exemplar from that period is Spinoza, whose spirit has informed the increasingly prominent participation of Jews in science over the last 350 years (most notably Einstein), where "Judaism" stands for the most earthbound and naturalistic of the Abrahamic faiths.

How to resolve this disagreement? First, there is no denying that the Renaissance revival of Epicurean philosophy—its atomic view of nature, its hedonistic conception of all animal life (including humans), and its skepticism about any transparent sense of cosmic design—helped to launch and propel the Scientific Revolution. Disagreement arises only over the spirit of Epicureanism's adoption: Did it destroy and replace the Christian world-view, or was it simply assimilated, leaving most of the Christian assumptions intact? That is *Feuer v. Merton* in a nutshell. Clearly, there is a hermeneutic problem here. Natural philosophers of the period rarely stressed points of conflict between Epicureanism and Christianity in their own thinking. That was left for opponents who accused them of "atheism," the sacrilege of which largely rested on its perceived psychological aberration (Febvre, 1947/1982). But difficulties in imagining sincere and sane atheists must be offset against a decline in substantive references to divine agency in nature—though that may be explainable as an expedient to avoid intractable doctrinal disputes. Merton takes this general line, namely, that Epicureanism was incorporated into a self-disciplining Christianity that subsequently privatized and then secularized matters of faith. In return, Christianity tamed the overriding role of chance in the Epicurean worldview, which had inclined it toward fatalism. The legacy came in the form of probability theory, a field where theologians played a formative role, as they attempted to resolve the uncertainties involved in deciding to believe in God, as well as explain how even stochastic processes in nature appear to operate within discernible limits. It is from these roots that modern subjective and objective conceptions of probability, respectively, derive (Hacking, 1975).

For his part, Feuer pointed to a freer, antischolastic style of discourse that had taken hold in the 16th and 17th centuries that he associated with the unfettering of the human imagination from the yoke of dogma. In addition, Feuer attempted to portray Merton as outright denying an emotional basis to the Scientific Revolution. But to be fair, Merton was simply operating with a different account of the cognitive work of emotions, one indebted to the Italian political economist and early sociologist Vilfredo Pareto, for whom all ideologies (including scientific ones) are sublimated versions of fundamental affective ties to the world (Shapin, 1988). Feuer presupposed a more phenomenological understanding of emotion that led him to doubt the reliability of official professions of faith made by early Royal Society members as indicators of the lives they actually led, given background information about the times—and sometimes the members themselves. What Feuer quite happily explained as hypocrisy on the part of these so-called Puritans, Merton understood in terms of a subtle, perhaps only semi-conscious, retooling of religious beliefs for scientific purposes that over the next two centuries became an unwitting vehicle of secularization. Of course, it is easy to imagine truth on both sides of the argument, but the main question of interest to the psychology of science is which one better explains the character of science, as it has actually come to be practiced. Here, Merton ultimately has the upper hand, but the contest is subtle.

On Feuer's side, the Scientific Revolution was striking in its antipathy to the sort of disciplinary specialization that Kuhn (1962/1970) and others have seen as the hallmark of scientific progress. Indeed, many of the 17th-century revolutionaries would regard such specialization as a sterile scholastic holdover. To be sure, the period featured many biblically inspired "trees of knowledge"—not least from Francis Bacon—that rationalized various divisions of cognitive labor. But, as if to anticipate the "modular" approach to the mind favored by today's evolutionary psychologists, Bacon organized the pursuit of knowledge to match the organization of our brains, so that we could learn quickly about the world so as to lead optimally adapted lives together. This meant that disciplines had to correspond to innate mental functions, which the artificially defined disciplines of the scholastic curriculum had distorted. Thus, Epicurean sympathizer Denis Diderot was convinced that the future history of science would consist of incorporating more embodied, praxis-based forms of knowledge (i.e., the chemical, biological, and social sciences) to supplement what he believed was Newton's overemphasis on our abstract, formal capacities that were exhaustively elaborated in the mechanical world view. Thus, when designing the Enlightenment's great editorial project, *L'Encyclopédie*, Diderot gave arts and crafts unprecedented visibility as forms of knowledge—but understood *à la* Bacon as extensions of memory, not *à la* Newton as applications of physics (Darnton, 1984, Chap. 5).

However, what ultimately gives Mertonian asceticism the upper hand are three distinctive features of the "research" orientation to modern scientific inquiry, namely, its *collective, trans-generational,* and *indefinitely extended* character. The sociologist Collins (1998) has drawn attention to the emotional energy that sustains intellectual networks, such self-generated and largely self-maintained enthusiasm, which seems to favor Feuer's hedonic approach. But, Collins also recognizes that more must be involved to distinguish science from, say, philosophical schools or, for that matter, religious sects. After all, many of the metaphysical ideas and empirical findings of modern science were also present in various intellectual social formations in ancient Greece, India, and China. Yet, by our lights, there seemed to be relatively little appetite for improving and adding to those insights to produce a cumulative epistemic legacy. If anything, such societies, while superficially tolerant of diverse perspectives, inhibited their development beyond the level of a concrete project, a hobby or other self-consuming activity, especially if it might result in self-perpetuating disciplines that could lay claim to producing reliable higher order knowledge of greater social relevance than that provided by established authorities (Fuller, 2010, Chap. 2). Thus, these societies, however advanced, lacked a sense of the scientist as a distinct kind of person whose primary social identity might derive from affiliations with like-minded people living in other times and places—and hence, potentially subversive of the local power structure. This distinctive pattern of affiliation marks the unique social role of the scientist, popularized by Max Weber in the 20th century in terms of "science as a vocation" (Ben-David, 1971).

Indicative of the establishment of such an identity is the gradual introduction of academic journals affiliated not to particular universities or even national scientific bodies, but to internationally recognized scientific disciplines in the 19th and 20th centuries. This greatly facilitated the conversion of scholarship to research by introducing criteria of assessment that went beyond the local entertainment value of a piece of intellectual work. One way to envisage this transformation is as an early—and, to be sure, slow and imperfect—form of *broadcasting*, whereby a dispersed audience is shaped to receive the same information on a regular basis, to which they are meant to respond in a similarly stylized fashion. This served to generate an autonomous collective consciousness of science and a clear sense of who was ahead and behind the "cutting edge" of research. However, there remained the question of how to maintain the intergenerational pursuit of a research trajectory, such that one can be sure that unsolved problems in what has become a clearly defined field of research are not simply forgotten but carried forward by the next generation. Enter the *textbook*.

National textbooks that aspired to global intellectual reach were a pedagogical innovation adumbrated by Napoleon, but fostered by Bismarck. They have served as vehicles for recruiting successive generations of similarly oriented researchers in fields that Kuhn would recognize as possessing a "paradigm" that conducts "normal science" (Kaiser, 2005). A relatively unnoticed consequence of the increased reliance on textbooks has been that people with diverse backgrounds and skills are regularly integrated into a common intellectual project, not least through the textbook's integration of textual, visual, and other multimedia representations. In this way, the ancient and medieval separation of "head" (i.e., book-learning) from "hand" (i.e., craft-learning) is finally broken down: Text is no longer solely chasing text (as in humanistic scholarship, even today), and the skills needed for research design need not be captive to esoteric rites of apprenticeship. In this respect, the textbook has been most responsible for converting the research mentality into a vehicle for both the democratization and collectivization of knowledge production.

THE TURN AWAY FROM MASLOW AND WHERE WE ARE NOW: HOW SCIENTIFIC CREATIVITY BECAME SCARCE

The current wave of work in the psychology of science that originated in the 1980s has not been especially kind to the research agenda Maslow (1966) originally outlined. In the intervening period, psychoanalysis has fallen from grace as a source of authoritative scientific knowledge of psychodynamic processes. Indeed, one touchstone volume in the field opens by observing that "fortunately" psychology of science was no longer dominated by Freudian psychohistories (Gholson et al., 1989, p. 10). However, it should be quickly added that the genre of psychological biographies of significant scientists was largely the innovation of E. G. Boring, who was quite critical of psychoanalysis, especially after having undergone it himself in the 1930s (Runyan, 2006). In any case, psychoanalytically inspired accounts of psychology of science suffered from the stigma of seeming to stress idiosyncratic features of a scientist's life in ways that were both ultimately unverifiable and ungeneralizable. While this criticism may have applied to some work done under the rubric of "psychohistory," it is unfairly applied to Maslow (1966), which repeatedly stressed the need for an eclectic mix of methods that allowed for the scaling up of the phenomenological to the objective. Thus, Maslow himself developed tests for creativity and endorsed the controversial electrode-based attempts by psychoanalyst Lawrence Kubie to find neural correlates for creativity-relevant psychodynamic processes such as free association (Winter, 2011, Chap. 4).

Beyond the issue of psychoanalysis, a more fundamental difference between Maslow (1966) and today's psychology of science may lie in the conceptualization of scientific creativity. Rather unlike Freud, but in line with such Neo-Freudians as Adler and Herbert Marcuse, Maslow believed that required explanation is not what makes particular individuals creative but, on the contrary, what prevents more, if not all, people from being creative. Thus, neurosis is associated with the inhibition, not the promotion, of creativity (Kubie, 1967). This position was developed quite explicitly against the more popular view—on which Freud appeared to confer his authority—that creative individuals possess a unique genius that reflects their tortured souls. In contrast, Maslow would diagnose what prevents people from realizing their full creative potential and then advise how they might change their relationships with other people and their environment. Philosophically speaking, this "emancipatory" approach to creativity, which continues in the positive psychology movement, was informed by the then recent English translation of the early "humanist" writings of Karl Marx, which argued that all socially relevant forms of inequality are not merely "unjust" in some abstract political sense, but reflect the arrested development of the human species. Maslow (1966) held that this point applied no less to the uneven distribution of creativity in the population.

However, more recent psychological research has tended to take the uneven distribution of creativity as given, rather than a problem as such. To be sure, some of the elitism inherent in the old "tortured genius" view has been mitigated by a more explicitly social–psychological approach to creativity. For example, drawing on some of the earliest cognitive limitations work, Mitroff (1974) identified, in the case of the Apollo moon scientists, three styles of inquiry that displayed complementary strengths and weaknesses that worked together best in certain proportions, in certain research environments. Mitroff, who went on to found the field of "crisis management," thought about the fostering of creativity from a systems design perspective that was indebted to C. West Churchman, founder of the journal *Philosophy of Science*. Perhaps the heyday for this approach appears in the first handbook of the science and technology studies, whose review essay on "psychology of science" (Fisch, 1977) focuses on the recruitment and retention of people with the right traits. Indeed, personality inventory tests targeting the nexus between "creativity" and "achievement" of the sort championed in McClelland (1962) formed the basis of the review. This version of the psychology of science continues to flourish in human resource management studies of research and development (R&D) recruitment that, given our neo-liberal times, are not any longer so clearly focused on retention.

More recent social–psychological studies present a mixed verdict on the approach to scientific creativity taken in Maslow (1966). While creativity clearly needs to be self-motivated and may be outright undermined by external incentives, nevertheless "effective creativity" is increasingly defined in terms of the judgment of presumed peer communities, rather than (per Maslow) a test administered by the psychologist that is designed to control for the inevitable conservative bias of such communities (e.g., Amabile, 1996). A more fundamental challenge to the Maslow world view, especially its assumption of a great untapped store of human potential, is the broadly "historiometric" approach that descends from Francis Galton's search for "hereditary genius," which presumes that society contains a scarce amount of human creativity, which needs to be genetically conserved and cultivated, lest it be dissipated through default patterns of mating and breeding. Galton saw his "eugenics" as an updating of Plato's own anticipation of human resource management, but where recruitment and retention of creative individuals occurs at quite an early age. This idea attracted many in the technocratic left in the early 20th century, including the Fabian founders of the London School of Economics, who envisaged a eugenically inspired "social biology" as the foundational social science (Renwick, 2011). Moreover, Galton's program was well received by both Marxists and positivists in the 1920s (Otto Neurath was Galton's German translator), until falling into disrepute with the rise of Nazism. But in recent decades, armed with a wider and better array of data, improved statistical methods, and formal recognition (and systematic measurement) of the increasing dimensions of "big science" (Price, 1963), historiometry has been given a new lease on life, one pursued most vigorously for the last four decades by Simonton (1988).

A good sense of the strongly anti-Maslovian cast of this entire line of research may be seen in Cole and Cole (1973), a statistically based study of "social stratification in science." Cole and Cole counterpose two hypotheses concerning the distribution of talent in science: one, the so-called Ortega Hypothesis, which argues that every scientist, however mediocre, contributes to the storehouse of knowledge from which "geniuses" draw together into a synthetic whole; the other, inspired by the authors' mentor Robert Merton, proposes that scientific talent travels in specific lineages (i.e., schools, jobs, achievements, rewards), implying that most actual scientific research is in principle eliminable noise. They clearly plump for the latter hypothesis, though neither hypothesis presumes the existence of a massive untapped scientific potential. Into the breach stepped Simonton (1988) with an equally anti-Maslovian resolution. His "chance-configuration" theory, based on Campbell's (1988) "selective retention" model of evolutionary epistemology, agreed with Cole and Cole on the elite nature of scientific talent but argued that it needs to be more carefully disaggregated from actual track record, which may not do the scientist's talent justice and could inhibit the ability of others in the future from benefiting from it. Thus, scientific talent should be assessed somewhat independently of lineage, so as to catch these talented "outliers" (Simonton, 1988, p. 97).

The turn against Maslow may boil down to a shift in attitudes to human biological evolution over the past half century. Campbell, Simonton, as well as Feist (2006) treat the Neo-Darwinian account of natural selection as a background-material constraint on human evolution that in turn shapes their sense of what the psychology of science can be. But, given that in strictly Darwinian terms the species-adaptive character of science is far from secure (i.e., it is not clear that more scientific creativity would increase the chances of human survival), it is perhaps not surprising for them to conclude that science is relatively unrepresentative of human psychology in general. In contrast, Maslow's main source for biological inspiration was the founder of general systems theory, Bertalanffy (1950), who started with the assumption that since our native biological equipment deprives us of a natural habitat, our species identity is tied to continuing to beat the odds against our long-term survival by turning a hostile nature into an anthropocentric life-world—or to put it in Dawkins's (1982) more contemporary terms—the species mark of the human is the will and the competence to extend its phenotype indefinitely. Whether von Bertalanffy's proactionary view of the human condition trumps Darwin's more precautionary one in the long term will be of interest, of course, not only to students of scientific creativity, but also to the rest of humanity that is bound to feel the effects of such creativity in the future.

REFERENCES

Amabile, T. (1996). *Creativity in context*. Boulder, CO: Westview.

Ash, M. (1988). *Gestalt psychology in German culture, 1890–1967*. Cambridge, UK: Cambridge University Press.

Baars, B. (1986). *The cognitive revolution in psychology*. New York, NY: Guilford Press.

Ben-David, J. (1971). *The scientist's role in society*. Englewood Cliffs, NJ: Prentice-Hall.

Bertalanffy, L. V. (1950). An outline of general system theory. *British Journal for the Philosophy of Science, 1*, 134–165.

Bordogna, F. (2008). *William James at the boundaries: Philosophy, science, and the geography of knowledge*. Chicago, IL: University of Chicago Press.

Boring, E. G. (1929/1950). *A history of experimental psychology* (2nd ed.). New York, NY: Appleton-Century Crofts.

Boring, E. G. (1955). Dual role of the zeitgeist in scientific creativity. *Scientific Monthly, 80*, 101–106.

Brannigan, A. (1981).*The social basis of scientific discoveries*. Cambridge, UK: Cambridge University Press.

Brannigan, A. (2004). *The rise and fall of social psychology: The use and misuse of the experimental method*. New York, NY: Aldine de Gruyter.

Bruner, J., Goodnow, J., & Austin, G. (1956). *A study of thinking*. New York, NY: John Wiley.

Brunswik, E. (1952). The conceptual framework of psychology. In *International encyclopedia of unified science* (Vol. 1, Chap. 10). Chicago, IL: University of Chicago Press.

Campbell, D. T. (1988). *Methodology and epistemology for social science*. Chicago, IL: University of Chicago Press.

Campbell, D. T., & Stanley, J. C. (1963). *Experimental and quasi-experimental designs for research*. Chicago, IL: Rand McNally.

Cole, J., & Cole, S. (1973). *The social stratification of science*. Chicago, IL: University of Chicago Press.

Collins, R. (1998). *The sociology of philosophies: A global theory of intellectual change*. Cambridge, MA: Harvard University Press.

Conant, J.B. (1950). The overthrow of the Phlogiston theory: The chemical revolution of 1775–1789. In *Harvard case studies in experimental science, case 2*. Cambridge, MA: Harvard University Press.

Cushman, T., & Rodden, J. (1997, Winter). Sociology and the intellectual life: An interview with Lewis S. Feuer. *The American Sociologist, 28*(4), 56–89.

Darnton, R. (1984). *The great cat massacre and other episodes in French cultural history*. New York, NY: Vintage.

Dawkins, R. (1982). *The extended phenotype*. Oxford, UK: Oxford University Press.

De Mey, M. (1982). *The cognitive paradigm*. Dordrecht, NL: Kluwer.

Dewey, J. (1910). *How we think.* Lexington, MA: DC Heath.

Ericsson, K. A., & Simon, H. (1985). *Protocol analysis: Verbal reports as data.* Hillsdale, NJ: Lawrence Erlbaum.

Faust, D. (1984). *The limits of scientific reasoning.* Minneapolis, MN: University of Minnesota Press.

Febvre, L. (1947/1982). *The problem of unbelief in the sixteenth century.* Cambridge, MA: Harvard University Press.

Feist, G. (2006). *The psychology of science and the origins of the scientific mind.* New Haven, CT: Yale University Press.

Feuer, L. (1963). *The scientific intellectual: The psychological and sociological origins of modern science.* New York, NY: Basic Books.

Fisch, R. (1977). Science, technology and society: A cross-disciplinary perspective. In I. Spiegel-Rösing & D. deSolla Price (Eds.), *Psychology of science* (pp. 277–318). London: Sage.

Fisher, R. (1925). *Statistical methods for research workers.* Edinburgh: Oliver and Boyd.

Fuller, S. (1988). *Social epistemology.* Bloomington, IN: Indiana University Press.

Fuller, S. (1989/1993). *Philosophy of science and its discontents* (2nd edn.). New York, NY: Guilford Press.

Fuller, S. (2000a). *The governance of science.* Milton Keynes, UK: Open University Press.

Fuller, S. (2000b). *Thomas Kuhn: A philosophical history for our times.* Chicago, IL: University of Chicago Press.

Fuller, S. (2003). *Kuhn vs. Popper: The struggle for the soul of science.* Cambridge, UK: Cambridge University Press.

Fuller, S. (2006). *The philosophy of science and technology studies.* London: Routledge.

Fuller, S. (2009). The genealogy of judgement: Towards a deep history of academic freedom. *British Journal of Educational Studies, 57,* 164–177.

Fuller, S. (2010). *Science: The art of living.* Durham, UK: Acumen.

Fuller, S., De Mey, M., Shinn, T, & Woolgar, S., (Eds.). (1989). *The cognitive turn: Psychological and sociological perspectives on science.* Dordrecht, NL: Kluwer.

Gholson, B., Shadish, W., Neimeyer, R., & Houts, A., (Eds.). (1989). *Psychology of science: Contributions to metascience.* Cambridge, UK: Cambridge University Press.

Gigerenzer, G., & Murray, D. (1987). *Cognition as intuitive statistics.* Hillsdale, NJ: Lawrence Erlbaum.

Gorman, M. E. (1992). *Simulating science: Heuristics, mental models and technoscientific thinking.* Bloomington, IN: Indiana University Press.

Hacking, I. (1975). *The emergence of probability.* Cambridge, UK: Cambridge University Press.

Hedges, L. (1987). How hard is hard science, how soft is soft science? *American Psychologist, 42,* 443–455.

Higgins, E. T. (1997). Beyond pleasure and pain. *American Psychologist, 52*(12), 1280–1300.

Holton, G. (1973). *The thematic origins of scientific thought.* Cambridge, MA: Harvard University Press.

Horgan, J. (1996). *The end of science.* Reading, MA: Addison-Wesley.

Huxley, T. H. (1893). *Evolution and ethics* (Romanes lecture). Retrieved from http://aleph0.clarku.edu/huxley/comm/OxfMag/Romanes93.html

James, W. (1884). On some omissions of introspective psychology. *Mind, 9,* 1–26.

James, W. (1890/1983). *The principles of psychology.* Cambridge, MA: Harvard University Press.

Jarvie, I. (2001). *The republic of science.* Amsterdam: Rodopi.

Kaiser, D. (Ed.). (2005). *Pedagogy and the practice of science.* Cambridge, MA: MIT Press.

Knobe, J., & Nichols, S. (Eds.). (2008). *Experimental philosophy.* Oxford: Oxford University Press.

Koehler, W. (1938). *The place of value in a world of facts.* New York, NY: Liveright.

Kruglanski, A. (1989). *Lay epistemics and human knowledge.* New York, NY: Plenum Press.

Kubie, L. (1967). *Neurotic distortion of the creative process.* New York, NY: Farrar, Straus and Giroux.

Kuhn, T. S. (1962/1970). *The structure of scientific revolutions* (2nd ed.). Chicago, IL: University of Chicago Press.

Kusch, M. (1999) *Psychological knowledge: A social history and philosophy.* London: Routledge.

Lakatos, I. (1981). History of science and its rational reconstructions. In I. Hacking (ed.), *Scientific revolutions.* Oxford: Oxford University Press, pp. 107–127.

Laudan, L. (1981). *Science and hypothesis.* Dordrecht, NL: Kluwer.

Mahoney, M. J. (1976/2004). *Scientist as subject: The psychological imperative* (2nd ed.). Clinton Corners, NY: Percheron Press.

Manuel, F. (1963). *Isaac Newton, historian.* Cambridge, MA: Harvard University Press.

Maslow, A. (1966). *The psychology of science.* New York, NY: Harper and Row.

McClelland, D. (1962). On the psychodynamics of creative physical scientists'. In H. Gruber, G. Terrel, & M. Wertheimer (Eds.), *Contemporary approaches to creative thinking* (pp. 141–174). New York, NY: Atherton.

Meehl, P. (1967). Theory-testing in psychology and physics: A methodological paradox. *Philosophy of Science, 34*, 103–115.

Merton, R. (1938/1970). *Science, technology and society in seventeenth century England.* New York, NY: Harper and Row.

Mitroff, I. (1974). *The subjective side of science.* Amsterdam: Elsevier.

Naess, A. (1965). Science as behavior. In B. Wolman and E. Nagel (Eds.), *Scientific psychology* (pp. 50–67). New York, NY: Basic Books.

Nye, M. J. (2011). *Michael Polanyi and his generation: Origins of the social construction of science.* Chicago, IL: University of Chicago Press.

Passmore, J. (1970). *The perfectibility of man.* London: Duckworth.

Pepper, S. (1942). *World hypotheses: A study in evidence.* Berkeley, CA: University of California Press.

Petersen, A. (1984). The role of problems and problem solving in Popper's early work on psychology. *Philosophy of the Social Sciences, 14*, 239–250.

Polanyi, M. (1957). *Personal knowledge.* Chicago, IL: University of Chicago Press.

Polanyi, M. (1961/1963). Commentary by Michael Polanyi. In A. Crombie (Ed.), *Scientific change* (pp. 375–380). London: Heinemann.

Popper, K. (1935/1957). *The logic of scientific discovery.* London: Routledge and Kegan Paul.

Price, D. de S. (1963). *Little science, big science.* New York, NY: Columbia University Press.

Quine, W. V. O. (1960). *Word and object.* Cambridge, MA: MIT Press.

Renwick, C. (2011). *British sociology's lost biological roots: A history of futures past.* London: Palgrave Macmillan.

Rescher, N. (1998). *The limits of science.* Pittsburgh, PA: University of Pittsburgh Press.

Runyan, W. M. (2006). Psychobiography and the psychology of science. *Review of General Psychology, 10*(2), 147–162.

Russell, B. (1927). *An outline of philosophy.* London, UK: George Allen and Unwin.

Shadish, W., & Fuller, S. (Eds.). (1994). *Social psychology of science.* New York, NY: Guilford Press.

Shapin, S. (1988). Understanding the Merton thesis. *Isis, 79*, 594–605.

Simonton, D. K. (1988). *Scientific genius: A psychology of science.* Cambridge, UK: Cambridge University Press.

Smith, L. (1986). *Behaviorism and logical positivism: A reassessment of the alliance.* Palo Alto, CA: Stanford University Press.

Tolman, E. C. (1932). *Purposive behavior in animals and men.* New York, NY: Appleton-Century Crofts.

Turner, S. (1994). Making scientific knowledge a social psychological problem. In W. R. Shadish & S. Fuller (Eds.), *The social psychology of science: The psychological turn* (pp. 345–351). New York, NY: Guilford.

Tversky, A., & Kahneman, D. (1974). Judgment under uncertainty: Heuristics and biases. *Science, 185*, 1124–1131.

Tweney, R., Doherty, M., & Mynatt, C. (Eds.). (1981). *On scientific thinking.* New York, NY: Columbia University Press.

Wason, P., & Johnson-Laird, P. (1972). *Psychology of reasoning: Structure and content.* Cambridge, MA: Harvard University Press.

Wertheimer, M. (1945). *Productive thinking.* New York, NY: Harper and Row.

Wettersten, J. (1985). The road through Würzburg, Vienna and Göttingen. *Philosophy of the Social Sciences, 15,* 487–505.

White, A. D. (1896). *A history of the warfare of science with theology in Christendom.* New York, NY: D. Appleton.

Winter, A. (2011). *Memory: Fragments of a modern history.* Chicago, IL: University of Chicago Press.

Wolpert, L. (1992). *The unnatural nature of science.* London, UK: Faber and Faber.

Yates, F. (1966). *The art of memory.* London, UK: Routledge and Kegan Paul.

Ziliak, S. T., & McCloskey, D. N. (2008). *The cult of statistical significance: How the standard error costs us jobs, justice, and lives.* Ann Arbor, MI: University of Michigan Press.

Foundational Psychologies of Science

CHAPTER 3

Learning Science Through Inquiry

Corinne Zimmerman and Steve Croker

Science is a complex human activity that results in the creation of new knowledge and, as such, has been studied by psychologists who are interested in how children and adults think, reason, and come to understand the natural and social world (Feist, 2006). In initial research efforts to understand the development of scientific thinking, researchers focused on whether domain-general thinking skills *or* conceptual knowledge within specific domains better accounted for how people learn science (see Zimmerman, 2000, for a review). Current research efforts, in contrast, support the idea that inquiry skills *and* relevant domain knowledge "bootstrap" one another, such that there is an interdependent relationship that underlies the development of scientific thinking. In the cyclical process of investigation and inference that supports the discovery of relationships among variables, knowledge about a domain can help bolster strategies, while at the same time, sound investigation strategies support valid inferences, thereby resulting in increases in knowledge and understanding.

WHAT IS SCIENTIFIC THINKING?

Scientific thinking is an umbrella term that encompasses the reasoning and problem-solving skills involved in generating, testing, and revising hypotheses or theories, and in the case of fully developed skills, reflecting on the process of knowledge acquisition and change. The focus of this chapter is on the inquiry skills that result in knowledge change, and how this developing knowledge influences the development of more sophisticated inquiry skills. Concurrently, as metacognitive and metastrategic knowledge develop, children and adolescents gain a greater understanding of the nature of inquiry and the skills used in the inquiry process. Thus, the entire process is iterative and cyclical and, depending on the time scale examined, could involve some or all of the components of scientific inquiry, such as designing experiments, evaluating evidence, and making inferences in the service of forming and/or revising theories about the phenomenon under investigation.

As developmental psychologists of science, we are interested in the factors that influence the origins and growth of inquiry skills and scientific concepts across the lifespan, from the child and adolescent through to the scientifically literate adult, as well as in the practicing scientist (e.g., Dunbar, 1995; Koerber, Sodian, Kropf, Mayer, & Schwippert, 2011). Young children demonstrate many of the requisite skills needed to engage in scientific thinking, and yet there are also conditions under which adults do not show full proficiency. The goal of our chapter is to focus on the skills and knowledge that are brought to bear on inquiry tasks, the ways in which these skills and knowledge facilitate or hinder scientific investigation, and how inquiry skills and conceptual knowledge bootstrap one another. To do this, we focus on studies of *self-directed inquiry*. In such studies, children or adults engage in first-hand investigations in which they conduct experiments to discover and confirm the causal relations in a *multivariable system*. A multivariable system is either a physical task or a computer simulation that involves manipulation of variables to examine the effect on an outcome variable.[1] Physical tasks involve hands-on manipulation, such as the ramps task (e.g., Masnick & Klahr, 2003), mixing chemicals (Kuhn & Ho, 1980), and the canal task (e.g., Gleason & Schauble, 2000). Computer simulations of virtual environments have been created, in domains such as electric circuits (Schauble, Glaser, Raghavan, & Reiner, 1992), genetics (Echevarria, 2003; Okada & Simon, 1997), earthquakes (Azmitia & Crowley, 2001; Kuhn, Pease, & Wirkala, 2009), flooding risk (e.g., Keselman, 2003), human memory (Schunn & Anderson, 1999), visual search (Métrailler, Reijnen, Kneser, & Opwis, 2008), as well as social science problems such as factors that affect TV enjoyment (e.g., Kuhn, Garcia-Mila, Zohar, & Andersen, 1995), CD catalog sales (e.g., Dean & Kuhn, 2007), athletic performance (e.g., Lazonder, Wilhelm, & Van Lieburg, 2009), or shoe store sales (e.g., Lazonder, Hagemans, & de Jong, 2010). Some studies are conducted over a single session, others may involve multiple sessions over the course of many weeks. The latter are referred to as microgenetic studies; the idea behind this method is to provide a density of observations to track the process of change in knowledge, skills, and metacognitive competencies, as learners engage in extended inquiry.

FRAMEWORKS FOR SCIENTIFIC INQUIRY: THE COGNITIVE AND THE METACOGNITIVE

With respect to the frameworks that guide this line of work, two are highly influential. The work of Klahr and his colleagues focuses on the cognitive and procedural aspects of the endeavor, whereas the work of Kuhn and her colleagues emphasizes evidence evaluation and question generation, with a particular focus on the metacognitive and metastrategic control of the process of coordinating evidence with theory.

The Cognitive Components

Klahr and Dunbar's (1988) Scientific Discovery as Dual Search (SDDS) model captures the complexity and the cyclical nature of the process of scientific discovery and includes both inquiry skills and conceptual change (see Klahr, 2000, for a detailed discussion). SDDS is an extension of a classic model of problem solving from the field of cognitive science (Simon & Lea, 1974) and explains how people carry out problem solving in varied science contexts, from simulated inquiry to professional scientific practice. There are three major cognitive components in the SDDS model: searching for hypotheses, searching for experiments, and evaluating evidence.

Individuals begin inquiry tasks with some existing or intuitive ideas, or perhaps no ideas at all, about how particular variables influence an outcome. Given some set of possible variables (i.e., independent variables) and asked to determine their effect on an outcome (i.e., the dependent variable), participants negotiate the process by coordinating search in the set of possible hypotheses, and the set of possible experiments. Experiments are conducted to determine the truth status of the current hypothesis, or to decide among a set of competing hypotheses. Experiments may also be conducted to generate enough data to be able to propose a hypothesis (as might be the case when one comes to the inquiry task with little or no prior knowledge). Evidence is then evaluated so that inferences can be made whether a hypothesis is correct or incorrect (or, in some cases, that the evidence generated is inconclusive). Depending on the complexity of the task, the number of variables, and the amount of time on task, these processes may be repeated several times as an individual negotiates search in the hypothesis and experiment spaces and makes inferences based on the evaluation of self-generated evidence. Factors such as task domain and the perceived goal of the task can also influence how these cognitive processes are deployed.

The Metacognitive Components

Kuhn (1989, 2002, 2005) has argued that the defining feature of scientific thinking is the set of cognitive and metacognitive skills involved in differentiating and coordinating theory and evidence. In particular, metacognitive awareness is what differentiates more from less sophisticated scientific thinking. Kuhn argues that the effective coordination of theory and evidence depends on three metacognitive abilities: (a) the ability to encode and represent evidence and theory separately, so that relations between them can be recognized; (b) the ability to treat theories as independent objects of thought (i.e., rather than a representation of "the way things are"); and (c) the ability to recognize that theories can be false, setting aside the acceptance of a theory so evidence can be assessed to determine the truth or falsity of the theory. These metacognitive abilities are necessary precursors to sophisticated scientific thinking, and represent one of the ways in which children, adults, and professional scientists differ. Kuhn suggests that problems involving relating evidence to theory arise because the acquisition of metacognitive skills is a gradual process. As a result, early strategies for coordinating theory and evidence are replaced with better ones, but there is not a stage-like change from using an older strategy to a newer one.

Kuhn argues that inquiry is one of the fundamental thinking skills (along with argumentation) that need to be developed via education (see Kuhn, 2005, for a detailed discussion). Children are inclined to notice and respond to causal events in the environment; even infants and young children have been shown to have rudimentary understanding of cause and effect (Baillargeon, 2004; Gopnik, Sobel, Schulz, & Glymour, 2001; for a review see Koslowski & Masnick, 2010). Inquiry goes to the next level, showing that there is something to find out about the world. This fundamental first step, getting individuals to understand there is something that can be found out, is crucial. Then, in order for inquiry to go beyond demonstrating the correctness of one's existing beliefs (which, as will be discussed below, is common for younger children), it is necessary for meta-level competencies to be developed and practiced. With metacognitive control over the processes involved, learners can change what they believe based on evidence and, in doing so, are aware not only that they are changing a belief, but also know *why* they are changing a belief. Thus, sophisticated inquiry involves inductive causal inference skills and an awareness that evidence and theory belong to different epistemological categories.

THE ACQUISITION OF KNOWLEDGE AND SKILLS DURING INQUIRY IN MULTIVARIABLE SYSTEMS

Inquiry tasks, whether hands-on or simulated laboratories, afford the opportunity for individuals to learn new concepts and to improve investigation and inference skills. We take a broad view of learning, and include both conceptual and procedural knowledge and skills. The types of knowledge that learners come away with after interacting with an inquiry task can vary from fairly simple to complex. For some systems, knowledge change may simply represent the successful discovery of the causal/noncausal status of all variables in the multivariable system (e.g., Reid, Zhang, & Chen, 2003). In other cases, learners may come away with a more sophisticated understanding of, for example, genetics (e.g., Echevarria, 2003), electricity (e.g., Schauble, Glaser, et al., 1991), or visual perception (e.g., Métrailler et al., 2008).

With respect to investigation skills, a foundational skill that supports valid inferences is the control-of-variables strategy (CVS). Through the course of inquiry, individuals may discover the logic behind designing controlled experiments. CVS is emphasized in this line of work, as it is a domain-general skill that is fundamental to experimental inquiry. Regardless of whether one is trying to determine which factors influence earthquake risk, the speed of a ball on a ramp, the period of a pendulum, or the inhibitory mechanisms of gene control, the multivariable systems that are explored require an understanding of the logic behind a controlled experiment. An experiment can be coded as either controlled (CVS) or *confounded* (in which learners manipulate more than one variable at once). The size of the experiment space can be calculated for multivariable systems (e.g., the canal task with varying boat characteristics and canal depths can produce 24 unique combinations of variables) and so the percentage of experiment space searched (including repetitions) can be measured.

To assess developing inference skills, researchers record the number of correct and incorrect *causal* ("inclusion"), *noncausal* ("exclusion"), or *indeterminate* inferences, as well as whether inferences are *valid* (i.e., based on sufficient evidence and a controlled design) or *invalid*. Justifications for inferences can be coded as either evidence-based or theory-based. The number of valid inferences is a common performance indicator because such inferences involve (a) the design of an unconfounded experiment, (b) the correct interpretation of evidence, and (c) a conclusion that is correctly justified.

WHAT WE KNOW ABOUT LEARNING SCIENCE THROUGH INQUIRY

The Effects of Prior Knowledge on Scientific Inquiry

A consideration of how knowledge changes through inquiry must necessarily include a discussion of prior or existing knowledge. Inquiry studies imitate the scientific discovery process in that participants start with an initial set of beliefs or knowledge, but these beliefs or knowledge are changed as a function of the evidence generated via observation and investigation. Numerous findings speak to the change in knowledge that occurs as a result of investigation and inferences drawn from different types of evidence (e.g., anomalous or surprising evidence, or evidence from physical vs. social domains) that are evaluated and interpreted in the context of existing knowledge and beliefs. In some cases, children or adults may encounter an inquiry task in an entirely novel domain. In other cases, individuals come to the task with some level of existing conceptual knowledge of the task domain. In both cases, people are typically willing to propose hypotheses about the causal status of relevant variables.

There are various ways in which prior beliefs (or those developed on the spot) can affect the process of generating and testing hypotheses (e.g., Croker & Buchanan, 2011), evaluating evidence (e.g., Koerber, Sodian, Thoermer, & Nett, 2005), and changing—or failing to change—beliefs. Learning through inquiry involves reconciling prior beliefs with new evidence that either confirms or disconfirms existing beliefs. In some cases, holding prior beliefs can facilitate inquiry by helping learners to generate plausible hypotheses. In other cases, prior beliefs may inhibit reasoning by preventing the learner from exploring aspects of the problem that may be crucial to understanding the relationships among variables.

If learners have little domain knowledge, they tend to use less sophisticated investigation and inference strategies (discussed in more detail below) and are not able to execute these strategies effectively. Domain knowledge, therefore, affects strategy use and, conversely, strategy knowledge helps with the acquisition of domain-specific knowledge (Alexander & Judy, 1988). More general epistemological beliefs about the nature of knowledge and processes of knowing can also have a positive effect on performance of inquiry tasks, and increased scientific knowledge may, in turn, lead to the development of sophisticated epistemological beliefs (Zeineddin & Abd-El-Khalick, 2010).

Because the bootstrapping hypothesis implies that there is concurrent development of conceptual knowledge and reasoning skills, several factors related to prior knowledge are important for learning via inquiry, including characteristics such as (a) the amount of prior knowledge, (b) the "direction" of causal knowledge (i.e., if a variable is initially believed to be causal or not), (c) the knowledge domain, and (d) the plausibility of the relationships between variables with respect to prior knowledge, which is influenced by the strength of prior knowledge.

Amount of Prior Knowledge

In order to discover the causal relationships in multivariable systems, learners must form hypotheses about the role of several variables on the outcome measure. The prior beliefs that are brought to a scientific inquiry task influence the *choice* of which hypotheses to test, including which hypotheses are tested *first*, which ones are tested *repeatedly*, and which receive the most *time and attention* (e.g., Echevarria, 2003; Klahr, Fay, & Dunbar, 1993; Penner & Klahr, 1996; Schauble, 1990, 1996; Zimmerman, Raghavan, & Sartoris, 2003). Children and adults are more likely to begin the discovery process by attending to variables that are *believed to be causal* (e.g., Kanari & Millar, 2004; Klahr et al., 2007; Schauble, 1990, 1996).

Prior beliefs that are inconsistent with the causal relationships in an inquiry task can *hinder inquiry*. When this happens, children and adults often cling to their prior beliefs rather than changing them to be in-line with the evidence (Chinn & Brewer, 1993, 1998; Chinn & Malhotra, 2002). In some cases, researchers may deliberately choose variables to exploit known misconceptions, some of which are robust and continue to adulthood (e.g., Renken & Nunez, 2009). "Surprising results" are an impetus for conceptual change in real science as discussed by Kuhn (1962). The effect of "anomalous" evidence that violates expectation has been examined empirically. For example, Penner and Klahr (1996) used a task for which there are rich prior beliefs: Most children believe heavy objects sink faster than light objects. For steel objects, sink times for heavy and light objects are very similar. Only 8 of 30 participants directly compared the heavy and light steel objects, and all noted that the similar sinking time was unexpected. The process of knowledge change was not straightforward. For example, some students suggested that the size of the smaller steel ball offset the fact that it weighed less because it was able move through the water as fast as the larger, heavier steel ball. Others tried to update their knowledge by concluding that both weight and shape make a difference. There was an attempt to reconcile the evidence with prior belief by appealing to causal mechanisms, alternate causes, or enabling conditions. However, by the end of the study, all participants had revised their knowledge to reflect the evidence that weight alone does not determine the rate at which objects sink.

It is important to note that the children in the Penner and Klahr study did in fact *notice* the surprising finding that heavy objects do not sink faster. For the finding to be "surprising" it had to be noticed, and therefore these participants did not ignore or misrepresent the data. They tried to make sense of the surprising finding by acting as theorists who conjecture about the causal mechanisms or boundary conditions (e.g., shape) to account for the unexpected result. Chinn and Malhotra (2002) found that the process of observation (or "noticing") was important for conceptual change (see also Kloos & Van Orden, 2005). Chinn and Malhotra examined children's responses to data from experiments to determine if there are particular cognitive processes that interfere with conceptual change in response to evidence that is inconsistent with current belief. Experiments from physical science domains were selected in which the outcomes produced either ambiguous or unambiguous data, and for which the findings are considered counterintuitive for most children. For example, most children assume that a heavy object falls faster than a light object. When two objects are dropped simultaneously, there is some ambiguity because it is difficult to observe both objects. An example of counterintuitive but unambiguous evidence is the reaction temperature of baking soda added to vinegar. Children believe that fizzing causes an increase or no change in temperature. Thermometers unambiguously show a temperature drop of about 4°C.

When examining such anomalous evidence, difficulties may occur at one of four cognitive processes: observation, interpretation, generalization, or retention. Prior belief may influence what is "observed," especially in the case of ambiguous data. At the interpretation stage, the resulting conclusion will be based on what was (or was not) observed (e.g., a child may or may not perceive two objects landing simultaneously). At the level of generalization, an individual may accept, for example, that these particular heavy and light objects fell at the same rate, but that it may not hold for other situations or objects. Prior beliefs may re-emerge even when conceptual change occurs, so retention failure could prevent long-term belief change.

Chinn and Malhotra (2002) concluded that children could change their beliefs based on unexpected evidence, but only when they are capable of making the correct observations. Difficulty in making observations was the main cognitive process responsible for impeding conceptual change (i.e., rather than interpretation, generalization, or retention). Certain interventions involving an explanation of what scientists expected to happen and why were very effective in mediating conceptual change, when encountering counterintuitive evidence. With particular scaffolds, children made observations independent of theory, and changed their beliefs based on observed evidence. For example, the initial belief that a thermometer placed inside a sweater would display a higher temperature than a thermometer outside a sweater was revised after seeing evidence that disconfirmed this belief and hearing an explanation that the temperature would be the same unless there was something warm inside the sweater.

Echevarria (2003) examined seventh-graders' reactions to anomalous data in the domain of genetics to see if unexpected evidence facilitated knowledge construction. In general, the number of hypotheses generated, tests conducted, and explanations generated were a function of students' ability to notice, and take seriously, an anomalous finding. Most learners (80%) developed some explanation for the pattern of anomalous data. When learners encountered surprising findings, they were more likely to generate hypotheses and experiments to test these potential explanations, but those without sufficient domain knowledge did not "notice" anomalies. At the beginning of the study, most learners were unable to explain how parents with the same dominant form of a trait could produce offspring with the recessive form of the trait, but by the end of the study, most of the participants gave logical explanations for this anomalous finding, demonstrating that the inquiry task led to changes in their knowledge of genetics.

Although prior knowledge can hinder performance, many studies have demonstrated the *facilitative effect* of prior knowledge on strategy use in inquiry-learning tasks (Hmelo, Nagarajan, & Roger, 2000; Lavoie & Good, 1988; Schauble, Glaser, et al., 1991). Wilhelm and Beishuizen (2003), for example, found better performance in a concrete inductive reasoning task than an abstract one. Following from this study, Lazonder, Wilhelm, and Hagemans (2008) gave undergraduates two tasks to complete. One was a concrete task in which participants had high prior knowledge. The students had to determine the effects of four variables on the amount of time it takes an athlete to run 10 km. Three factors were familiar (i.e., smoking, nutrition, and training frequency) and the fourth was a "mystery factor" called Xelam. For the second task, participants had no prior knowledge of any of the variables or what the relationships among them might be (i.e., determine the influence of four shapes on a numerical score). Participants adopted a theory-driven approach on the concrete task, but a data-driven approach on the abstract task, and were more successful at determining the influence of the factors on the outcome in the concrete task. But what influences success? Is it knowledge of the likely ways in which factors can affect the outcome, or just familiarity with the factors?

Lazonder et al. (2009) followed up on Lazonder et al.'s (2008) study by including an intermediate task in which the factors (musical instrument, cutlery, flower, clothing) were familiar to the participants, and their effects on the outcome (a numerical score) could not be predicted. Performance on this intermediate task was comparable to performance on the abstract task, suggesting that familiarity with the variables alone is of little benefit. In order to enhance performance on inquiry tasks, learners need to have some understanding of the relationships between the variables. In the intermediate task, participants were trying to find meaningful relationships between the causal variables and the outcome. They would guess that the outcome score referred to something concrete, such as weight or money, and try to construct explanations of why a variable would have an effect on this outcome. As the relationships were arbitrary, this approach could interfere with task success.

Lazonder et al. (2010) designed a task in which participants would be familiar with the variables, but would not have any prior knowledge about the way in which the causal variables might affect the outcome. Some of the participants were given extra information before and during the task specifying the effects of the variables on the outcome; these learners performed better on the task than those who were not given this information, even though no explanation was given for *why* there was an effect.

The prior knowledge that learners bring to an inquiry task can be used to constrain the hypothesis space. The learner can start proposing informed hypotheses that are used to guide experimentation. In contrast, if learners have little or no prior knowledge, they may start by exploring the experiment space, constructing experiments without hypotheses, and then using the outcomes of these experiments to constrain search of the hypothesis space. Klahr and Dunbar (1988) first observed strategy differences between *theorists* and *experimenters* in adults. Theorists tend to generate hypotheses and then select experiments to test the predictions of the hypotheses. Experimenters tend to make data-driven discoveries, by generating data and finding the hypothesis that best summarizes or explains that data. Dunbar and Klahr (1989) and Schauble (1990) also found that children conformed to the description of either theorists or experimenters. In Penner and Klahr's (1996) study, children conducted experiments to determine how the shape, size, material, and weight of an object influence sinking times. Students' approaches to the task could be classified as either "prediction orientation" (i.e., a theorist; e.g., "I believe that weight makes a difference") or a "hypothesis orientation" (i.e., an experimenter; e.g., "I wonder if …"). Ten-year-olds were more likely to take a prediction (or demonstration) approach, whereas 14-year-olds were more likely to explicitly test a hypothesis about an attribute without a strong belief or need to demonstrate that belief. At around the age of 12, students may begin to transition from using experiments to demonstrate a belief to using experiments to investigate the truth status of beliefs.

Zimmerman et al. (2003) were able to classify sixth-graders as either theorists ("theory-modifying" or "theory-preserving") or experimenters (or "theory-generating") in their approach to experimenting with three variables that did or did not influence a balance apparatus. Theorists approached the task by explicitly stating and testing their theories about how the apparatus worked, using a combination of controlled tests and free-form exploration of the apparatus. Theory-modifying students evaluated evidence and, when based on controlled comparisons, were able to revise their theories based on the evidence they generated, which led to discovery of the torque rule (i.e., that the force acting on each arm of a balance scale is the product of mass and distance from the fulcrum). In contrast, theory-preserving students distorted or interpreted evidence as consistent with theory. Experimenters did not state theories in advance. Instead, they conducted controlled comparisons to determine the effects of each variable and derived the torque rule (i.e., they *generated* the theory based on evidence). Because they only interacted with the apparatus for one session, students only made progress if they did not have strong prior beliefs that they set out to demonstrate. Given more time on task, it is possible that the theorists would have eventually discovered the causal status of all three variables.

Direction of Prior Belief

In exploring multivariable systems, there are both causal and noncausal variables. It is the learner's job to investigate which variables make a difference. Inevitably, some of these variables will be inconsistent with prior belief. In some cases, learners will generate evidence inconsistent with a *causal* belief. In other cases, a variable that was initially deemed to be *non-causal* will be found to have an effect. Several studies have shown that it is generally more difficult to integrate evidence that disconfirms a prior causal theory, than evidence that disconfirms a prior noncausal theory. The former case involves restructuring a belief system, while the latter involves incorporating a newly discovered causal relation (Holland, Holyoak, Nisbett, & Thagard, 1986; Koslowski, 1996). For example, many people hold robust ideas that the weight of an object makes a difference in the period of a pendulum, and that heavy objects fall (and sink) faster than light objects. When confronted with evidence that disconfirms those beliefs, learners may struggle with how to reconcile belief with newly generated evidence. In contrast, many children do not believe that string length is causal in the case of pendulums or that the size of an object affects how fast it sinks. When experimental evidence shows that these variables do make a difference, they are more likely to accept the evidence as veridical—they are less likely to distort or misinterpret evidence in such cases. For example, Kanari and Millar (2004) found that all participants were able to accept that string length has a causal effect on the period of a pendulum, but only half reached the correct conclusion that mass does not have an effect. The direction of belief also affects whether key experiments are conducted. As noted above, only 8 of Penner and Klahr's (1996) 30 participants actually carried out a direct comparison of heavy and light objects.

In Schauble's (1990) computer microworld, the problem space was set up such that the direction of causal factors was not always consistent with children's existing beliefs. Participants were asked to determine the effects of five factors on the speed of racing cars: engine size, wheel size, presence of muffler, presence of tailfin, and color. Initially, learners tended to believe that a large engine, large wheels, and the presence of a muffler increased speed. In reality, a large engine and medium-size wheels increased speed, presence or absence of a muffler and the color of the car were irrelevant, and the absence of a tailfin only increased speed when the engine was large. By the end of the study, most had come to the conclusion that the muffler was not a causal variable, but very few discovered the interaction between engine size and tailfin. Those who successfully learned about this interaction only did so by proposing an explanation for why these factors might affect speed. Similarly, in Schauble's (1996) study, in which participants had to investigate problems in hydrodynamics and hydrostatics, children and adults ignored parts of the experiment space that did not fit in with their existing knowledge about which factors should be causal.

Knowledge Domain

Research on inquiry has compared performance in physical domains (e.g., speed of cars) and social science domains (e.g., determining the factors that affect students' school achievement). Kuhn et al. (1995) found that performance in the social domains was inferior. In the social domain, both fourth-graders and adults made fewer valid inferences; they focused on factors believed to be causal and tended to ignore those believed to be noncausal.

When theories were generated to support factors believed to be causal, they were difficult to relinquish in the face of evidence. This was true whether the causal theories were previously held or formed on the basis of (often insufficient) evidence. Kuhn et al. (1995) suggested that participants had a richer and more varied array of existing theories in the social domains. Moreover, participants may have had some affective investment in their theories about school achievement and TV enjoyment, but not for their theories about the causal factors involved in the speed of boats or cars. Most of the adults, for example, had an initial belief that the weight of a boat would affect its speed. Many of these adults revised their beliefs to reflect an interaction between weight and size.

Alternatively, Kuhn and Pearsall (1998) proposed that the performance differences found between social and physical domains might be due to the fact that the outcomes were more understandable in the physical tasks. The fastest boat or fastest car is a more salient outcome than descriptors (from poor through excellent) representing "popularity" of a TV show or "achievement" in school. Operationally defining social science constructs requires additional conceptual understanding that may not be necessary for types of variables that are directly observable in physical tasks (e.g., time, temperature, distance). The time scales involved in social tasks may also be greater than in physical tasks. The social domain thus lacks "causal transparency" (Rozenblit & Keil, 2002) in that one can mentally animate the hypothesized effects of immediate and direct physical causes, whereas it is harder to construct a visual representation of how causes affect an outcome in the social domain. This finding has implications in that the specific domain of prior knowledge (e.g., social vs. physical) may be a factor in more or less proficient reasoning and conceptual development.

Strength of Prior Belief and Plausibility of Causal Relationships

We often use prior knowledge to consider the plausibility of a variable or a hypothesis to test. *Plausibility* is a general constraint with respect to belief formation and revision (Holland et al., 1986) and has been identified as a domain-general heuristic (Klahr et al., 1993) that can be used to guide the choice of which hypotheses and/or experiments to pursue. Klahr et al. provided third- and sixth-grade children and adults with hypotheses to test that were incorrect, but either plausible or implausible. For plausible hypotheses, children and adults tended to go about *demonstrating the correctness* of the hypothesis rather than setting up experiments to decide between rival hypotheses. Third-graders tended to propose a plausible hypothesis, but then ignored or forgot the initial implausible hypothesis, getting sidetracked in an attempt to demonstrate that the plausible hypothesis was correct. When provided with implausible hypotheses to test, adults and some sixth-graders proposed a plausible *rival hypothesis*, and set up an experiment that would discriminate between the two. For both children and adults, the ability to consider many alternative hypotheses was a factor contributing to success. One could argue that any hypothesis that is inconsistent with a prior belief could be considered "implausible."

Therefore, because learners tend to begin exploration of a causal system by focusing on variables consistent with prior beliefs, these are the variables that are considered to be plausibly related to the outcome. Concern for theoretical relationships is also evident by references to causal mechanisms. For example, in the domains of hydrostatics and hydrodynamics, learners referred to unobservable forces such as "currents," "resistance," "drag," and "aerodynamics" to help explain and make sense of the evidence (Schauble, 1996).

Thus, it is clear that adults and children alike conduct scientific investigations by considering explanations regarding why a cause, or set of causes, should have a particular effect. Other research supports the idea that learners seek out information about causal mechanisms and explanations when making causal attributions (e.g., Ahn & Bailenson, 1996; Ahn, Kalish, Medin, & Gelman, 1995; Brem & Rips, 2000). The need for a plausible causal mechanism can be seen as a useful heuristic. Koslowski (1996) argues that although there are many examples of covariation and correlation in the world, we tend to only take seriously those for which there is a causal mechanism. Considering causal mechanisms requires a knowledge base. Thus, the consideration of plausible mechanisms is a useful way of excluding artifactual causes and deciding whether a theory requires outright rejection or modification in the face of disconfirming evidence (Koslowski & Masnick, 2010). Legare (2012) suggests that outcomes that are inconsistent with prior beliefs can be used to assist children's reasoning, if learners are asked to generate explanations for the inconsistent data. She argues that having to explain a set of data requires the learner to generate and test hypotheses they may not otherwise have considered. Thus, generating ideas about possible causal mechanisms may open up the hypothesis space.

The Development of Investigation Skills

As discussed previously, there are a number of strategies for manipulating and isolating variables. Of these, the only one that results in an unconfounded design and is considered valid is the CVS (Chen & Klahr, 1999) also known as "vary one thing at a time" (VOTAT; Tschirgi, 1980). The "hold one thing at a time" (HOTAT) and "change all" (CA) strategies are considered invalid in this context, as they produce confounded comparisons that result in ambiguous findings. Experimentation can be conducted for two purposes: to test a hypothesis or to generate a pattern of findings with which to generate or induce a hypothesis to account for that pattern.

Klahr et al. (1993) identified several heuristics for successful discovery, two of which focused on investigation: design experiments that produce informative and interpretable results, and attend to one feature at a time. Adults were more likely than third- and sixth-grade children to restrict the search of possible experiments to those that were informative (Klahr et al., 1993). Schauble (1996) found that both children and adults started out by covering about 60% of the possible experiment space when first given either a hydrostatics or hydrodynamics task. When they were asked to switch to the second task domain, only adults' search of experiment space increased (to almost 80%). Over the course of 6 weeks, children and adults conducted approximately the same number of experiments, meaning that children were more likely to duplicate experiments. Thus, children's investigation was less informative relative to the adults, reducing their potential for learning about the two domains. Children's duplicate experiments focused on variables that were already well understood, whereas adults would move on to exploring variables they did not understand as well (Klahr et al., 1993; Schauble, 1996). This approach to experimentation is consistent with the idea that children may view experimentation as a way of demonstrating the correctness of their current beliefs.

The heuristic of attending to one feature at a time is also more difficult for children. For example, Schauble (1996) found that children used the CVS about a third of the time. In contrast, adults improved from 50% CVS usage on the first task to 63% on the second task. Children usually begin by designing confounded experiments (often as a means to produce a desired outcome), but with repeated practice, they began to use the CVS (e.g., Kuhn et al., 1995; Kuhn, Schauble, & Garcia-Mila, 1992; Schauble, 1990). A robust finding in microgenetic studies is the coexistence of valid and invalid strategies (e.g., Garcia-Mila & Andersen, 2007; Gleason & Schauble, 2000; Kuhn et al., 1992, 1995; Schauble, 1990; Siegler & Crowley, 1991; Siegler & Shipley, 1995) for both children and adults. Developmental transitions do not occur suddenly; participants do not progress from an inefficient or invalid strategy to a valid strategy without ever returning to the former. An individual may begin with invalid strategies, but even when the usefulness of the CVS is discovered, it is only slowly incorporated into an individual's set of strategies, as participants become dissatisfied with strategies that produce ambiguous evidence. Investigation and inference strategies often codevelop in microgenetic contexts because valid inferences require controlled designs.

The Development of Inference Skills

Recall that inferences made based on self-generated evidence are typically classified as causal (inclusion), noncausal (exclusion), indeterminate, or false inclusion, and can be either valid (i.e., supported by evidence) or invalid. Valid inferences of indeterminacy can be correctly "supported by" ambiguous evidence. False inclusion is by definition invalid but is of interest because children and adults often incorrectly (based on prior belief) infer that a variable is causal, when in reality it is not. *Valid inferences* are defined as causal or noncausal inferences that are based on controlled experiments and include observations for both levels of the target variable. As was seen with investigation strategies, participants may continue to use invalid inference strategies even after they have learned how to make valid inferences. Children and adults may persist in making inferences that are consistent with prior beliefs, based on a single instance of covariation (or noncovariation), or based on one level of the causal factor and one level of the outcome factor (e.g., Klahr et al., 1993; Kuhn et al., 1992, 1995; Schauble, 1990, 1996).

As mentioned previously, children tend to focus on *causal inferences* during their initial explorations of a causal system. Schauble (1990) found that fifth- and sixth-graders began by producing confounded experiments and relied on prior knowledge or expectations, and therefore were more likely to make incorrect causal inferences (i.e., false inclusions) during early efforts to discover the causal structure of a computerized microworld. In direct comparison, adults and children both focused on making causal inferences (about 75% of inferences), but adults made more valid inferences because they used controlled experiments. Schauble (1996) found improvement in children's valid inferences over the course of six sessions, starting at 25% but improving to almost 60%. In addition, adults were more likely to make exclusion inferences and inferences of indeterminacy than children (80% and 30%, respectively) (Schauble, 1996).

Kanari and Millar (2004) asked 10- to 14-year-olds to discover the factors that influence the period of a pendulum or the force needed to pull a box along a level surface. The participants struggled with exclusion inferences. Only half were able draw correct conclusions about factors that do not covary with outcome (e.g., the weight of a pendulum bob), and, in these cases, learners were more likely to either selectively record data, selectively attend to data, distort or "reinterpret" the data, or state that noncovariation experimental trials were "inconclusive." These findings are consistent with Kuhn, Amsel, and O'Loughlin's (1988) research on the evaluation of researcher-supplied evidence in which inference strategies that preserved prior beliefs were used. Similarly, Zimmerman et al. (2003) reported children who distorted or "reinterpreted" self-generated evidence to determine which factors influenced the tilt of a balance apparatus. Most of these children held prior beliefs that the vertical height of a weight should make a difference, but some were unable to reconcile this expectation with the data they collected with the balance apparatus. Most were able to reconcile the discrepancy between expectation and evidence by updating their understanding of the balance system and concluding that vertical distance was noncausal.

In microgenetic studies, however, children start making more exclusion inferences (that factors are not causal) and indeterminacy inferences (i.e., one cannot make a conclusive judgment about a confounded comparison), and do not focus solely on causal inferences (e.g., Keselman, 2003; Schauble, 1996). As the ability to distinguish between an informative and an uninformative experiment (e.g., by using CVS) improves, so does the ability to make valid inferences. Through repeated practice, invalid inferences drop in frequency. These performance improvements may indicate that students may be developing an awareness of the adequacy or inadequacy of their experimentation strategies for generating sufficient and interpretable evidence.

Children and adults also differ in generating *sufficient evidence* to support inferences. In contexts where it is possible, children often terminate their search early, believing that they have determined a solution to the problem (e.g., Dunbar & Klahr, 1989). In microgenetic contexts where children must continue their investigation, this is less likely because of the task requirements. Children are also more likely to refer to evidence that was salient, or most recently generated. Whereas children often jump to a conclusion after a single experiment, adults typically need to see the results of several experiments before making inferences.

Perceived Goal of Inquiry

As noted in previous sections, prior knowledge about a task domain can affect how learners deploy inference skills. The context of a task can also affect the approach that learners take. Children often attempt to demonstrate that a factor is causal rather than testing whether it is causal. A seminal study by Tschirgi (1980) examining the choice of investigation strategy (i.e., to vary, hold, or change all variables) used to test a given hypothesis showed that a participant's goal could affect the choice of experimentation strategy. For hypotheses involving a positive outcome, participants selected the invalid HOTAT strategy. For negative outcomes, in contrast, the valid CVS was selected. In inquiry contexts, this general pattern has been found by a number of researchers. For example, Kuhn and Phelps (1982) noted that some children approached the colorless-fluids task as though the goal was to produce the red color, rather than to identify which chemicals were responsible for the reaction. Carey, Evans, Honda, Jay, and Unger (1989) found that most seventh-graders believed that "a scientist 'tries it to see if it works'" (p. 520) and demonstrated a pragmatic concern for particular valued outcomes. They also behaved as though their goal was to reproduce a bubbling effect produced by mixing yeast, sugar, salt, flour, and warm water, rather than to understand which substance produced the phenomenon. Similarly, children exploring Schauble's (1990) microworld often behaved as if the goal was to produce the fastest car, rather than to determine the causal status of each of the variables. Kuhn and colleagues (1992, 1995) also reported that early in investigations, students tend to focus on desirable versus undesirable outcome.

Schauble, Klopfer, and Raghavan (1991) addressed the issue of goals directly by providing fifth- and sixth-grade children with an "engineering context" and a "science context." When the children worked as *scientists*, their goal was to determine which factors made a difference and which ones did not. As *engineers*, their goal was to produce a desired effect (i.e., the fastest boat in the canal task or the longest spring length in the springs task). In the science context, children worked more systematically by establishing the effect of each variable, alone and in combination. There was an effort to make both causal and noncausal inferences. In the engineering context, children selected highly contrastive combinations, and focused on factors believed to be causal, while overlooking factors believed or demonstrated to be noncausal. Typically, children took a "try-and-see" approach to experimentation when acting as engineers, but took a theory-driven approach when acting as scientists. It is not clear if these different approaches characterize an individual, or if they are invoked by task demand or implicit assumptions. It might be that an engineering approach makes most sense to children as their investigation and inference skills are still developing. Schauble, Klopfer, et al. (1991) found that children who received the engineering instructions first, followed by the scientist instructions, made the greatest improvements.

Metacognitive and Metastrategic Competence

Metamemory Strategies

In many of the inquiry tasks reviewed here, participants were provided with some type of external memory system, such as a data notebook or access to computer files for previous trials. Many studies demonstrate that both children and adults are not always aware of their memory limitations while engaged in scientific investigation tasks (e.g., Carey et al., 1989; Dunbar & Klahr, 1989; Garcia-Mila & Andersen, 2007; Gleason & Schauble, 2000; Siegler & Liebert, 1975; Trafton & Trickett, 2001). Recent studies corroborate the importance of an awareness of one's own memory limitations while engaged in scientific inquiry tasks, regardless of age. Children may differentially record the results of experiments, depending on familiarity or strength of prior beliefs. For example, 10- to 14-year-olds recorded more data points when experimenting with factors affecting the force produced by the weight and surface area of boxes, than when they were experimenting with pendulums (Kanari & Millar, 2004). Overall, it is a fairly robust finding that children are less likely than adults to record experimental designs and outcomes, or to review notes they do keep, despite task demands that clearly necessitate a reliance on external memory aids. Metamemory develops between the ages of 5 and 10, with development continuing through adolescence (Schneider, 2008), and so there may not be a particular age that memory and metamemory limitations are no longer a consideration for children and adolescents engaged in complex inquiry tasks.

Metacognitive Strategies

Metastrategic competence does not appear to routinely develop in the absence of instruction. Most of the foundational research on the development of learning through inquiry has not specifically measured this aspect of scientific thinking. Kuhn's theoretical work has drawn attention to the importance of meta-level competencies. Kuhn and her colleagues have incorporated the use of specific practice opportunities and prompts to help learners develop these types of competencies. For example, Kuhn, Black, Keselman, and Kaplan (2000) incorporated performance-level practice and metastrategic-level practice for sixth- to eighth-grade students. Performance-level exercise consisted of standard exploration of the task environment. Meta-level practice consisted of paper-and-pencil scenarios in which two individuals disagreed about the effect of a particular feature in a multivariable situation. Students then evaluated different strategies that could be used to resolve the disagreement. Such scenarios were provided twice a week during the course of 10 weeks. Although no performance differences were found between the two types of practice with respect to the number of valid inferences, there were more sizeable differences in measures of understanding the task objectives and strategies (i.e., metastrategic understanding).

Keselman (2003) compared performance-level exercise (the control group) with two practice conditions: one that included direct instruction and practice at making predictions, and one with prediction practice only.

Sixth-graders experimented with a multivariable system (an earthquake forecaster). Students in the two practice conditions showed better performance on a meta-level assessment relative to the control group. Only the direct instruction group showed an increase in use of evidence from multiple records and the ability to make correct inferences about noncausal variables.

Zohar and Peled (2008) explicitly taught metastrategic knowledge of the CVS to fifth-graders who were classified as either high- or low-academic achievers. Participants were given a computerized task in which they had to determine the effects of five variables on seed germination. After an initial investigation of the task, students in the control group were taught about seed germination, and students in the experimental group were given a metastrategic knowledge intervention over several sessions. The intervention consisted of describing CVS, discussing when it should be used, and discussing what features of a task indicate that CVS should be used. A second computerized task on potato growth was used to assess near transfer. A physical task in which participants had to determine which factors affect the distance a ball will roll was used to assess far transfer. The students who received the intervention showed gains on both the strategic and the metastrategic level. The latter was measured by asking participants to explain what they had done. These gains were still apparent on the near and far transfer tasks, when they were administered 3 months later. The group that showed the largest gains was the low-academic achievers. It is clear from these studies that although meta-level competencies may not develop routinely, they can certainly be learned via explicit instruction.

SUMMARY AND CONCLUSIONS

We have focused our review of learning science through inquiry on research in which participants engage in the full cycle of scientific investigation, including generating hypotheses, designing experiments to generate evidence, and then evaluating the newly generated evidence in light of current knowledge. Many other studies focus on specific components of the inquiry cycle and include other types of tasks in order to elucidate the processes that underlie scientific thinking (see Sodian & Bullock, 2008, for a collection of recent work). There are numerous studies that focus on the educational implications of this work, including lab studies (e.g., Strand-Cary & Klahr, 2008) and situated classroom work (Lehrer, Schauble, & Lucas, 2008). Moreover, many studies extend beyond the traditional focus on the *experiment* to understand the development of scientific thinking (for a discussion of this issue, see Lehrer, Schauble, & Petrosino, 2001).

As we have seen from our focused review, the results of the coordination of the various cognitive and metacognitive processes in inquiry tasks may or may not result in knowledge change. Scientific thinking involves a long developmental trajectory, and many of the investigation and inference skills take considerable practice before they are consolidated. Moreover, specific educational intervention is typically needed for inquiry skills to become metacognitive and for these skills to be deployed in a metastrategic way.

Some of the factors that have been found to influence the process of inquiry include the amount and strength of prior knowledge. In some cases, prior knowledge can inhibit sound inquiry by limiting the consideration of variables considered to be noncausal. At the same time, prior knowledge has a facilitatory effect in that it can be used to assess the plausibility of factors. In the case of implausible or unexpected surprising findings, explanations or causal mechanisms can be proposed, allowing a consideration of previously spurious correlations. With robust misconceptions (e.g., the role of weight in many physical science tasks), disconfirmation promotes the bootstrapping of scientific knowledge and scientific inquiry skills.

Although early precursors to scientific thinking are evident in early childhood (see Zimmerman, 2007, for a review), it takes time, practice, and—for metastrategic skills—instruction for the investigation and inquiry skills needed for scientific thinking to develop. As developmental psychologists of science, we take the view that scientific thinking can be studied across the lifespan and is not found only in practicing scientists. Scientific thinking represents a major achievement of human culture, with respect to the body of knowledge that has been accumulated and the sophisticated set of individual and collaborative inquiry skills that were required to give rise to that knowledge. Although there is a long trajectory to full competence, and even though the skills involved in sophisticated scientific thinking do not routinely develop in all adults, the roots of this cultural achievement start young, and are facilitated and scaffolded and bootstrapped via instruction, practice, and developing cognitive and metacognitive competencies.

NOTE

1. Triona and Klahr (2003) and Klahr, Triona, and Williams (2007) directly compared physical tasks and computer simulations and found no significant performance differences.

REFERENCES

Ahn, W., & Bailenson, J. (1996). Causal attribution as a search for underlying mechanisms: An explanation of the conjunction fallacy and the discounting principle. *Cognitive Psychology, 31,* 82–123.

Ahn, W., Kalish, C. W., Medin, D. L., & Gelman, S. A. (1995). The role of covariation versus mechanism information in causal attribution. *Cognition, 54,* 299–352.

Alexander, P. A., & Judy, J. E. (1988). The interaction of domain-specific and strategic knowledge in academic performance. *Review of Educational Research, 58,* 375–404.

Azmitia, M., & Crowley, K. (2001). The rhythms of scientific thinking: A study of collaboration in an earthquake microworld. In K. Crowley, C. D. Schunn, & T. Okada (Eds.), *Designing for science: Implications from everyday, classroom, and professional settings* (pp. 51–81). Mahwah, NJ: Lawrence Erlbaum.

Baillargeon, R. (2004). Infants' physical world. *Current Directions in Psychological Science,* *13,* 89–94.

Brem, S. K., & Rips, L. J. (2000). Explanation and evidence in informal argument. *Cognitive Science, 24,* 573–604.

Carey, S., Evans, R., Honda, M., Jay, E., & Unger, C. (1989). "An experiment is when you try it and see if it works": A study of grade 7 students' understanding of the construction of scientific knowledge. *International Journal of Science Education, 11,* 514–529.

Chen, Z., & Klahr, D. (1999). All other things being equal: Children's acquisition of the control of variables strategy. *Child Development, 70,* 1098–1120.

Chinn, C. A., & Brewer, W. F. (1993). The role of anomalous data in knowledge acquisition: A theoretical framework and implications for science instruction. *Review of Educational Research, 63,* 1–49.

Chinn, C. A., & Brewer, W. F. (1998). An empirical test of a taxonomy of responses to anomalous data in science. *Journal of Research in Science Teaching, 35,* 623–654.

Chinn, C. A., & Malhotra, B. A. (2002). Children's responses to anomalous scientific data: How is conceptual change impeded? *Journal of Educational Psychology, 94,* 327–343.

Croker, S., & Buchanan, H. (2011). Scientific reasoning in a real world context: The effect of prior belief and outcome on children's hypothesis testing strategies. *British Journal of Developmental Psychology, 29,* 409–424.

Dean, D., & Kuhn, D. (2007). Direct instruction vs. discovery: The long view. *Science Education, 91,* 384–397.

Dunbar, K. (1995). How scientists really reason: Scientific reasoning in real-world laboratories. In R. J. Sternberg & J. E. Davidson (Eds.), *The nature of insight* (pp. 365–395). Cambridge, MA: MIT Press.

Dunbar, K., & Klahr, D. (1989). Developmental differences in scientific discovery strategies. In D. Klahr & K. Kotovsky (Eds.), *Complex information processing: The impact of Herbert A. Simon* (pp. 109–143). Hillsdale, NJ: Lawrence Erlbaum.

Echevarria, M. (2003). Anomalies as a catalyst for middle school students' knowledge construction and scientific reasoning during science inquiry. *Journal of Educational Psychology, 95,* 357–374.

Feist, G. J. (2006). *The psychology of science and the origins of the scientific mind.* New Haven, CT: Yale University Press

Garcia-Mila, M., & Andersen, C. (2007). Developmental change in notetaking during scientific inquiry. *International Journal of Science Education, 29,* 1035–1058.

Gleason, M. E., & Schauble, L. (2000). Parents' assistance of their children's scientific reasoning. *Cognition and Instruction, 17,* 343–378.

Gopnik, A., Sobel, D. M., Schulz, L. E., & Glymour, C. (2001). Causal learning mechanisms in very young children: Two-, three-, and four-year-olds infer causal relations from patterns of variation and covariation. *Developmental Psychology, 37,* 620–629.

Hmelo, C. E., Nagarajan, A., & Roger, S. (2000). Effects of high and low prior knowledge on construction of a joint problem space. *Journal of Experimental Education, 69,* 36–56.

Holland, J. H., Holyoak, K. J., Nisbett, R. E., & Thagard, P. R. (1986). *Induction.* Cambridge, MA: The MIT Press.

Kanari, Z., & Millar, R. (2004). Reasoning from data: How students collect and interpret data in science investigations. *Journal of Research in Science Teaching, 41,* 748–769.

Keselman, A. (2003). Supporting inquiry learning by promoting normative understanding of multivariable causality. *Journal of Research in Science Teaching, 40,* 898–921.

Klahr, D. (2000). *Exploring science: The cognition and development of discovery processes.* Cambridge, MA: MIT Press.

Klahr, D., & Dunbar, K. (1988). Dual search space during scientific reasoning. *Cognitive Science, 12,* 1–48.

Klahr, D., Fay, A., & Dunbar, K. (1993). Heuristics for scientific experimentation: A developmental study. *Cognitive Psychology, 25,* 111–146.

Klahr, D., Triona, L. M., & Williams, C. (2007). Hands on what? The relative effectiveness of physical versus virtual materials in an engineering design project by middle school children. *Journal of Research in Science Teaching, 44,* 183–203.

Kloos, H., & Van Orden, G. C. (2005). Can a preschooler's mistaken belief benefit learning? *Swiss Journal of Psychology, 64*, 195–205.

Koerber, S., Sodian, B., Kropf, N., Mayer, D., & Schwippert, K. (2011). Die entwicklung des wissenschaftlichen denkens im grundschulalter: Theorieverständnis, experimentierstrategien, dateninterpretation [The development of scientific reasoning in elementary school age: Understanding theories, designing experiments, interpreting data]. *Zeitschrift fur Entwicklungspsychologie und Padagogische Psychologie, 43*, 16–21.

Koerber, S., Sodian, B., Thoermer, C., & Nett, U. (2005). Scientific reasoning in young children: Preschoolers' ability to evaluate covariation evidence. *Swiss Journal of Psychology, 64*, 141–152.

Koslowski, B. (1996). *Theory and evidence: The development of scientific reasoning.* Cambridge, MA: MIT Press.

Koslowski, B., & Masnick, A. (2010). Causal reasoning and explanation. In U. Goswami (Ed.), *The Wiley-Blackwell handbook of childhood cognitive development* (2nd ed., pp. 377–398). Oxford: Wiley-Blackwell.

Kuhn, D. (1989). Children and adults as intuitive scientists. *Psychological Review, 96*, 674–689.

Kuhn, D. (2002). What is scientific thinking and how does it develop? In U. Goswami (Ed.), *Blackwell handbook of childhood cognitive development* (pp. 371–393). Oxford: Blackwell Publishing.

Kuhn, D. (2005). *Education for thinking.* Cambridge, MA: Harvard University Press.

Kuhn, D., Amsel, E., & O'Loughlin, M. (1988). *The development of scientific thinking skills.* Orlando, FL: Academic Press.

Kuhn, D., Black, J., Keselman, A., & Kaplan, D. (2000). The development of cognitive skills to support inquiry learning. *Cognition and Instruction, 18*, 495–523.

Kuhn, D., Garcia-Mila, M., Zohar, A., & Andersen, C. (1995). Strategies of knowledge acquisition. *Monographs of the Society for Research in Child Development, 60*, 1–128.

Kuhn, D., & Ho, V. (1980). Self-directed activity and cognitive development. *Journal of Applied Developmental Psychology, 1*, 119–130.

Kuhn, D., & Pearsall, S. (1998). Relations between metastrategic knowledge and strategic performance. *Cognitive Development, 13*, 227–247.

Kuhn, D., Pease, M., & Wirkala, C. (2009). Coordinating the effects of multiple variables: A skill fundamental to scientific thinking. *Journal of Experimental Child Psychology, 103*, 268–284.

Kuhn, D., & Phelps, E. (1982). The development of problem-solving strategies. In H. Reese (Ed.), *Advances in child development and behavior* (Vol. 17, pp. 1–44). New York, NY: Academic Press.

Kuhn, D., Schauble, L., & Garcia-Mila, M. (1992). Cross-domain development of scientific reasoning. *Cognition & Instruction, 9*, 285–327.

Kuhn, T. S. (1962). *The structure of scientific revolutions.* Chicago, IL: University of Chicago Press.

Lavoie, D. R., & Good, R. (1988). The nature and use of prediction skills in a biological computer simulation. *Journal of Research in Science Teaching, 25*, 335–360.

Lazonder, A. W., Hagemans, M. G., & de Jong, T. (2010). Offering and discovering domain information in simulation-based inquiry learning. *Learning and Instruction, 20*, 511–520.

Lazonder, A. W., Wilhelm, P., & Hagemans, M. G. (2008). The influence of domain knowledge on strategy use during simulation-based inquiry learning. *Learning and Instruction, 18*, 580–592.

Lazonder, A. W., Wilhelm, P., & Van Lieburg, E. (2009). Unraveling the influence of domain knowledge during simulation-based inquiry learning. *Instructional Science, 37*, 437–451.

Legare, C. H. (2012). Exploring explanation: Explaining inconsistent information guides hypothesis-testing behavior in young children. *Child Development, 83*(1), 173–185.

Lehrer, R., Schauble, L., & Lucas, D. (2008). Supporting development of the epistemology of inquiry. *Cognitive Development, 23*, 512–529.

Lehrer, R., Schauble, L., & Petrosino, A. J. (2001). Reconsidering the role of experiment in science education. In K. Crowley, C. Schunn, & T. Okada (Eds.), *Designing for science: Implications from everyday, classroom, and professional settings* (pp. 251–277). Mahwah, NJ: Lawrence Erlbaum.

Masnick, A. M., & Klahr, D. (2003). Error matters: An initial exploration of elementary school children's understanding of experimental error. *Journal of Cognition and Development, 4*, 67–98.

Métrailler, Y. A., Reijnen, E., Kneser, C., & Opwis, K. (2008). Scientific problem solving in a virtual laboratory: A comparison between individuals and pairs. *Swiss Journal of Psychology, 67*, 71–83.

Okada, T., & Simon, H. A. (1997). Collaborative discovery in a scientific domain. *Cognitive Science, 21*, 109–146.

Penner, D. E., & Klahr, D. (1996). The interaction of domain-specific knowledge and domain-general discovery strategies: A study with sinking objects. *Child Development, 67*, 2709–2727.

Reid, D. J., Zhang, J., & Chen, Q. (2003). Supporting scientific discovery learning in a simulation environment. *Journal of Computer Assisted Learning, 19*, 9–20.

Renken, M. D., & Nunez, N. (2009). Evidence for improved conclusion accuracy after reading about rather than conducting a belief-inconsistent simple physics experiment. *Applied Cognitive Psychology, 24*, 792–811.

Rozenblit, L., & Keil, F. C. (2002). The misunderstood limits of folk science: An illusion of explanatory depth. *Cognitive Science, 26*, 521–562.

Schauble, L. (1990). Belief revision in children: The role of prior knowledge and strategies for generating evidence. *Journal of Experimental Child Psychology, 49*, 31–57.

Schauble, L. (1996). The development of scientific reasoning in knowledge-rich contexts. *Developmental Psychology, 32*, 102–119.

Schauble, L., Glaser, R., Raghavan, K., & Reiner, M. (1991). Causal models and experimentation strategies in scientific reasoning. *The Journal of the Learning Sciences, 1*, 201–238.

Schauble, L., Glaser, R., Raghavan, K., & Reiner, M. (1992). The integration of knowledge and experimentation strategies in understanding a physical system. *Applied Cognitive Psychology, 6*, 321–343.

Schauble, L., Klopfer, L. E., & Raghavan, K. (1991). Students' transition from an engineering model to a science model of experimentation. *Journal of Research in Science Teaching, 28*, 859–882.

Schneider, W. (2008). The development of metacognitive knowledge in children and adolescents: Major trends and implications for education. *Mind, Brain, and Education, 2*, 114–121.

Schunn, C. D., & Anderson, J. R. (1999). The generality/specificity of expertise in scientific reasoning. *Cognitive Science, 23*, 337–370.

Siegler, R. S., & Crowley, K. (1991). The microgenetic method: A direct means for studying cognitive development. *American Psychologist, 46*, 606–620.

Siegler, R. S., & Liebert, R. M. (1975). Acquisition of formal scientific reasoning by 10- and 13-year-olds: Designing a factorial experiment. *Developmental Psychology, 11*, 401–402.

Siegler, R. S., & Shipley, C. (1995). Variation, selection and cognitive change. In T. J. Simon & G. S. Halford (Eds.), *Developing cognitive competence: New approaches to process modeling* (pp. 31–76). Hillsdale, NJ: Erlbaum.

Simon, H. A., & Lea, G. (1974). Problem solving and rule induction: A unified view. In L. W. Gregg (Ed.), *Knowledge and cognition* (pp. 105–128). Hillsdale, NJ: Lawrence Erlbaum.

Sodian, B., & Bullock, M. (2008). Scientific reasoning—Where are we now? [Special issue]. *Cognitive Development, 23*(4), 472–487.

Strand-Cary, M., & Klahr, D. (2008). Developing elementary science skills: Instructional effectiveness and path independence. *Cognitive Development, 23*, 488–511.

Trafton, J. G., & Trickett, S. B. (2001). Note-taking for self-explanation and problem solving. *Human-Computer Interaction, 16*, 1–38.

Triona, L. M., & Klahr, D. (2003). Point and click or grab and heft: Comparing the influence of physical and virtual instructional materials on elementary school students' ability to design experiments. *Cognition and Instruction, 21*, 149–173.

Tschirgi, J. E. (1980). Sensible reasoning: A hypothesis about hypotheses. *Child Development, 51*, 1–10.

Wilhelm, P., & Beishuizen, J. J. (2003). Content effects in self-directed inductive learning. *Learning and Instruction, 13*, 381–402.

Zeineddin, A., & Abd-El-Khalick, F. (2010). Scientific reasoning and epistemological commitments: Coordination of theory and evidence among college science students. *Journal of Research in Science Teaching, 47*, 1064–1093.

Zimmerman, C. (2000). The development of scientific reasoning skills. *Developmental Review, 20*, 99–149.

Zimmerman, C. (2007). The development of scientific thinking skills in elementary and middle school. *Developmental Review, 27*(2), 172–223.

Zimmerman, C., Raghavan, K., & Sartoris, M. L. (2003). The impact of the MARS curriculum on students' ability to coordinate theory and evidence. *International Journal of Science Education, 25*, 1247–1271.

Zohar, A., & Peled, B. (2008). The effects of explicit teaching of metastrategic knowledge on low- and high-achieving students. *Learning and Instruction, 18*, 337–353.

CHAPTER 4

Cognitive–Historical Approaches to the Understanding of Science

Ryan D. Tweney

*T*o purloin a famous quote by Hermann Ebbinghaus (1908, p. 3), "Psychology (of Science) has a long past yet its real history is short." Inquiry into the nature of science, much of it explicitly psychological, dates back to Descartes, Locke, Hume, and many other philosophers (even Aristotle), all of whom devoted explicit attention to what we would now call the psychology of science, as have many working scientists over the years: Faraday, Mach, even Einstein. In the early years of psychology as a discipline, the topic was prominent; William James and Wilhelm Wundt sought to characterize scientific thought, an effort extended by several others. Wertheimer's study of Einstein's problem-solving methods, and Piaget's studies of the "child as scientist" and his genetic epistemology obviously also count as contributions to the psychology of science.

Over 30 years ago, my colleagues and I published a book (Tweney, Doherty, & Mynatt, 1981), which argued that what was known about human thinking could serve as a basis for the understanding of science. Even then, the body of work in cognitive psychology relevant to understanding scientific thinking was large and had reached a stage where its use as an interpretive tool was fruitful. In the ensuing decades, even more has become available. There is now a rich literature on the nature of inference (stemming partly from the pioneering work of Peter Wason, e.g., Wason & Johnson-Laird, 1972; Oaksford & Chater, 2010), on the nature of analogy (e.g., Clement, 2008; Gentner & Gentner, 1983), on the role of external representations (e.g., Norman, 1988; Trickett, Trafton, & Schunn, 2009), on the nature and development of expertise (e.g., Ericsson, Charness, Feltovich, & Hoffman, 2006), on the role of imagery and diagrammatic representation (e.g., Glasgow, Narayanan, & Chandrasekaran, 1995), and even a few on the use of mathematical representations (e.g., Lakoff & Núñez, 2000; Tweney, in press). Further, an increasing number of studies have used these resources to understand *historical* case studies in science.

The use of historical "case studies" (the reason for the quote marks will become evident) is ubiquitous in accounts of the psychology of science.

Examples drawn from history abound in present-day accounts from all of the science studies disciplines, but the examples are often sketchy and based on a small number of highly stylized episodes; Copernicus overturning the Ptolemaic universe, Galileo bucking church authority, Einstein's overthrow of Newtonian physics, and Darwin's evolutionary theory. Still, there have been, and continue to be, detailed approaches to the understanding of science that rely, not just on such "canned" examples, "nuggets of history," but instead upon the close examination of the actual practices—the thought, experiment, and social engagement—that actually characterized the history of science. In this chapter, I restrict attention to those studies that are truly *cognitive–historical,* that is, to those studies that have used cognitive frameworks for the understanding of historical episodes *in such a way* that the methodological niceties of both history and cognitive science are respected. Such studies have opened important new insights into the psychology of science and technology.

In the present chapter, I first consider some general methodological aspects of the cognitive–historical approach. I then present a brief sketch of the emergence of the approach with an explicit focus on studies reflecting both cognitive science and historical practices. Following these introductory sections, I describe in more detail three exemplars of the approach, chosen to reflect the variety of ways in which cognitive–historical methods can illuminate the nature of science: (a) Andersen, Barker, and Chen's (2006) use of Barsalou's theory of the embodiment of concepts to understand Kepler's reformulation of Copernican and Ptolemaic astronomy; (b) Nersessian's (2008) use of a cognitive theory of analogy to explicate James Clerk Maxwell's theory of electromagnetism; and, (c), studies that have utilized the very extensive laboratory diaries and notebooks of Michael Faraday.

SETTING THE STAGE

What is history and what is cognitive science that both may merge in a cognitive–historical psychology of science? In a common view, history is seen as dealing only with particular events, whereas science is concerned with generalities, with lawful regularities. On a closer look, however, this cannot be the entire difference, as many sciences (e.g., geology, astronomy) also deal with particular events. Perhaps the key difference is that history deals with unobservables, that is, with traces of events in the present that allow one to make inferences about the past. But this too cannot be the key difference, as science deals with a great many unobservables that must be inferred, as, for example, when the tracks of water droplets in a cloud chamber photograph are used to infer the existence of a moving subatomic particle.

Although a complete discussion would extend to several pages, let me propose that the aspect of historical analysis that is of most interest here is that history deals with the *meanings* of events and their antecedents and consequence. To state this differently, historical analysis concerns itself with what in cognitive science is often referred to as the *intentional stance* of human agents. Though this can include accounts of external, even physical, causes of those stances (disease, say, or social factors), the historian's goal is to explicate how human agents construed events, how they initiated and reacted to events, and the consequences for subsequent intentional stances. From this point of view, the analyses of historians and those of many cognitive scientists are not all that different. One major difference, of course, resides in the fact that, for cognitive scientists, discovering *how* an episode of thinking works is of paramount importance, and this directs attention to the close examination of details, closer than that usually used by historians. As we will see, that is one reason why meshing the two disciplines makes sense.

To illustrate his approach to problem solving, Herbert Simon (1967/1996) used the metaphor of an ant crossing a sandy beach. Though an ant is a relatively simple behavioral system, the geometry of its path across a pebbly beach is very complex. The complexity, however, is a result of the environment; the irregular size and placement of the pebbles are the reason for the complex path, not the ant. Simon argued for a similar consideration in human problem solving: the complexity of problem solving is in the symbolic environment of thought, not the thought itself that follows relatively simple heuristic principles. These general principles of problem solving can be used, according to Simon, to untangle the complex manifestations of specific problem-solving episodes carried out by a single individual (Newell & Simon, 1972). The approach was strikingly successful in such domains as chess playing and has been extended to claims about scientific discovery (e.g., Klahr, 2000; Langley, Simon, Bradshaw, & Zytkow, 1987; Kulkarni & Simon, 1990). In contrast to most psychological research, Simon and his colleagues tended to downplay the aggregative methods of experimental psychology in favor of computer simulation, often based on single-subject protocols of great complexity. The test of adequacy of such analyses resides in the close match of the simulations to the patterns of movement across the symbolic "landscape" of an actual subject episode. The research strategy necessitates a richness of detail that is reminiscent of, say, the specific accounts given by a geologist of the development across time of a landscape. Simon's analyses can be seen as "microhistorical" in that the course of problem solving is mapped across time and the "explanation" of an episode is based on inferences from a verbal protocol (Ericsson & Simon, 1993/1984) to the underlying inferred mental states and the operators that transform one state into another.

A prominent historian of science, Frederick Holmes, defended the use of "microstructural" analyses of scientific practices (Holmes, 1987; 1996) and produced accounts of scientific discovery that resembled the accounts of Simon and his colleagues. Indeed, many historians have centered in recent years on the understanding of the specific, day-by-day, experimental practices of scientists (see, e.g., the papers in Buchwald, 1995, and in Gooding, Pinch, & Schaffer, 1989). A classic example of how historical accounts can be approached via cognitive science was developed by one of Simon's students (Kulkarni & Simon, 1988; 1990). On the basis of a detailed historical account (Holmes, 1980) of the diaries and notebooks maintained by Hans Krebs during the discovery of the ornithine cycle (now known as the "Krebs Cycle"), Kulkarni constructed a simulation of the investigative pathways followed by Krebs. Kulkarni's simulation used heuristics that were both general in nature (e.g., means–ends analysis) and others that were specific to techniques known to have been used by Krebs. Letting the simulation conduct "experiments" by outputting an "experimental design" and getting the "results" as inputs, the program successfully recreated much of the fine detail of the original account. The simulation of Krebs's problem-solving activity was presented as an explanation, at a cognitive level, of the adequacy of the Newell and Simon approach for the understanding of real-world science. Krebs, like the subjects in Newell and Simon's (1972) research, could be seen as searching a *problem space* using experiments as *operators* on *mental states*.

Simon's approach to problem solving as problem-space search continues to be influential in many cognitive studies of scientific thinking (see the chapters by Schunn and by Klahr in this volume). Specific applications to other historical cases have also been conducted, first by Langley, Simon, Bradshaw, and Zytkow (1987), which used simulations to explore the way in which data from historically important scientific experimentation could be used to construct mathematical representations. Although this work was not based on detailed records of the actual processes followed by scientists, one of the authors later provided such detail in a problem-space analysis of the Wright Brothers' invention of the airplane (Bradshaw, 1992).

In spite of its successes, however, the problem-space approach has not sufficed to capture all aspects of the processes of scientific thinking. In particular, it cannot adequately capture the process by which representations are generated in the first place. Thus, Elke Kurz-Milcke studied the way in which mathematicians and scientists construct a representation (a differential equation) to solve a physical rate-flow problem (Kurz-Milcke, 1998; 2004). Kurz-Milcke's participants went through a lengthy process of struggling to find a representation that could then be solved using their expert knowledge. Their struggle could be loosely described as a search through "representation space." But for Kurz-Milcke's participants in the calculus problem, the real challenge was first to characterize the problem in some constructive fashion, in effect, to *find* a space to search!

In every case they did this by invoking one or another aspect of the intrinsic representational character of expressions in calculus and reworking their personal knowledge of calculus as they considered its use in the problem context. Thus, one of the participants, a physical chemist, seemed very "Newtonian" in her use of calculus; for her, a changing quantity was like a moving point, and her calculus was almost "fluxion-like." Another participant, a mathematician, was "Leibnizian"; his attempts to represent the problem relied upon Leibniz-like infinitesimals. And for a theoretical physicist, the problem was to search for functions that could be taken to a limit; he thus relied upon formulations of the calculus first worked out by Cauchy. For all three, the hard part was not developing the specific solution to the presented problem, but developing the representation of the problem in the first place. Note also that Kurz-Milcke's analysis depended upon her knowledge of the historical details of the development of calculus; an example of how cognitive–historical studies can enhance cognitive science itself.

The integration of cognitive science and history of science in the Kulkarni and Simon simulation and in Kurz-Milcke's study is an illustration of how much the two fields can enrich each other, and it also makes the point that such integration relies on the presence of rich detail about the processes by which science is actually carried out. It is easy (too easy, in fact) to search the past of science for examples that confirm any presupposition an investigator holds. Indeed, the literature of the psychology of science is filled with "nuggets of history" to make one or another point; short, textbook, examples to make a point arrived at by means other than historical analysis. Thus, Copernicus's rejection of the Ptolemaic system of astronomy has been cited frequently as an example of "insight," whereas Darwin's massive contributions to natural history in the years preceding his 1859 *Origin of Species* has been cited as an example of "perseverance," or an "inductive" approach to science. Such uses are not considered as cognitive–historical in the present chapter.

One feature common to both the Kulkarni and Simon and Kurz-Milcke examples is worth emphasizing: like the original Newell and Simon studies of problem solving, and like the usual history of science approach, the analysis is based on a *single-subject* model of research. That is, rather than concentrating on large aggregates across many subjects (the norm in most psychological research today), the focus is on singular examples (whether of episodes or of individuals) presented in rich and complex detail. How this can be made to work as a source of knowledge about the general laws that govern scientific practices will become clearer in the detailed accounts that follow later in the chapter. Here, I merely follow Lamiell's (2003) discussion of the difference between nomothetic and idiographic approaches in psychology: "There exists…a significant epistemic gap between statements about what is generally true in the sense of 'common to all'…and statements about what is generally true in the sense of 'on average'" (p. 183). Lamiell was speaking of personality psychology, but the remark applies to the psychology of science as well; you cannot "average" Faraday and Maxwell to find what is generally true about their approaches. Instead, to properly characterize their differing research styles (and to find what is common to both) requires a case-by-case analysis, a "theory of each person," and a subsequent comparison of the theories (see Tweney, 2009, 2011, 2012 for more on this specific comparison). In effect, cognitive–historical studies, like ethnographic studies, open the way to a *person-centered* account of scientific practice (Osbeck, Nersessian, Malone, & Newstetter, 2011).

Finally, note that cognitive–historical studies of science do more than merely apply psychological concepts and theories; they also serve as a means of enlarging those concepts and theories. Because the cognitive processes at play in "real" science are much more complex than those in the usual laboratory studies, they open new questions, pose issues that have not emerged in laboratory studies, and serve, in general, to enlarge the scope of cognitive science itself. Because science occurs in the midst of complex social, cultural, and material contexts, cognitive–historical studies offer a wider, more human-centered, perspective on cognition.

THE COGNITIVE–HISTORICAL APPROACH: FIRST STEPS

Thomas Kuhn's *The Structure of Scientific Revolutions* (1962/1970) can be seen as a seminal work for the cognitive–historical psychology of science, as it has been for the history and the philosophy of science (see, e.g., Bechtel, 1988). Kuhn represents the most important claim in our era that the actual conduct of scientific thinking might be understandable using methods derived from the social sciences in general and from psychology in particular. His emphasis was primarily on social factors, to be sure, but it is often forgotten that he placed almost as much emphasis on psychological mechanisms as on social factors. For Kuhn, it was just as plausible to seek cognitive explanations as social explanations. Only an accident of history, as it were, led most later commentators to focus more on the social aspects in Kuhn's writings.

For Kuhn, a particular scientist's acceptance or rejection of a point of view was not dependent on the objective evaluation of evidence, nor could it be reduced to a logical canon. Instead, Kuhn emphasized nonlogical mechanisms like social consensus formation, constraints imposed by already accepted formulations, puzzle solving (in the case of "normal science"), and perceptual and conceptual transformations (in the case of "revolutionary science."). In the process he opened a new window on all empirical studies of science. History was central in his work, of course, and he has had an enormous impact on the history of science, even though few historians explicitly accept his most famous notion, that of the "paradigm."

Kuhn drew upon some aspects of Gestalt psychology and the then emergent cognitive psychology as a means to understand change in scientific paradigms. Thus, the notion of a "Gestalt switch" (actually first used by Hanson, 1958, who reproduced the famous "young lady–old lady" reversible figure as an example) was likened by Kuhn to the sudden conceptual changes that underlie paradigm shifts in science (Kuhn 1962/1970, pp. 111–114). In discussing the role of anomalies in science, Kuhn also used the "New Look" perceptual researches of Bruner and Postman (1949), in which subjects were exposed for very short durations to playing cards that violated the usual appearance, for example, a black four of hearts or a red six of clubs. Subjects typically misreported the cards, "normalizing" them, as it were, and often stuck to the initial erroneous judgment, even as the exposure times were gradually lengthened; "Either as a metaphor or because it reflects the nature of the mind, that psychological experiment provides a wonderfully simple and cogent schema for the process of scientific discovery" (Kuhn 1962/1970, p. 64).

Deeply embedded within history as Kuhn was, however, his work was slow to lead to psychological research that utilized the data of history. The first explicitly Kuhnian work within the psychology of science was that of Marc De Mey (1982), whose book *The Cognitive Paradigm* took Kuhn's notion of paradigm as a central defining concept of scientific change. For De Mey, paradigms in the Kuhnian sense were central: "Paradigms are internal models, cognitive structures which give shape to the specific expectations that guide the research of 'normal' scientists" (1982, p. 89). De Mey argued that the cognitive correspondents of a paradigm were derivable from the concept of "frames" in artificial intelligence. De Mey used many historical examples in his book, although he did not use the fine detail, the microstructure of scientific practices, as a source for analysis.

Howard Gruber's cognitive and affective biography of Charles Darwin, *Darwin on Man* (1974), stands as the first cognitive–historical contribution.

Gruber presented a cogent "theory of the individual" based on an extension of Piaget's theory. Darwin's creative formulation of the theory of evolution by natural selection was presented as the outcome of tension between two broad ideas that haunted Darwin for decades: (a) a "conservation principle," in which Darwin saw tendencies to conserve the number of species and the number of individuals of one species, and (b) an "equilibration principle" in which he saw adaptive tendencies aimed at achieving a steady state. On Gruber's account, Darwin's creativity relied upon the "chancy interaction" of semi-independent systems of ideas, guided by his particular purposes. *Darwin on Man* had to be a biography to make this point, because only in the complexity of an entire life can we see such disparate, long-term, processes at work. Further, Gruber's careful analysis of the notebooks maintained by Darwin supported this view, showing how the practices of theory construction and inference moved across time to produce the eventual final theory.

For Gruber, creative science can be understood as the unfolding of a network of enterprise in which new ideas emerge as diverse elements mesh and combine. Gruber (1989) traced a complex network of interests and transactions with external "data" leading to the crucial insights of the theory of evolution by natural selection. To understand these processes, Gruber invoked Darwin's personal goals and motivations as well as his intellectual meanderings, arguing for a unity amid the diversity, a unity driven by the processes of equilibration.

Arthur Miller also used Piagetian ideas as a framework to explore the development of concepts in modern physics (Miller, 1984). In particular, Miller argued for a transition in physicists' use of images to represent concepts. Traditionally, images in physics were representations of the things about which physics was concerned; light, for example, was represented as a stream of particles. After 1900, however, Miller documented an increasing use of images as representations of the theoretical concepts of physics, rather than as representations of the things physics is about. Einstein's visualization was presented as a case in point: Einstein's world lines are not pictures of the world as such; they are instead pictures that visualize the relationships between time and space in the theory of relativity. Similarly, light quanta are not "pictures" of a physical world. Instead, light quanta are an *Anschauung,* an "intuition," roughly, of the dual nature of light as wave and as particle. The transition from images as pictures of things to images as pictures of theories was a fundamental change in the way physics was done: "The contrast between the simplicity of Einstein's mental images, their theoretical context and the dazzling theories that they spawned is so startling as to undercut available cognitive scientific models" (Miller, 1984, p. 248); by "available cognitive scientific models" he meant the traditional distinction between imagistic thinking and propositional thinking.

Miller accounted for the changed visual role of *Anschauungen* by using and extending Piaget's notion of the developmental emergence of abstract concepts out of more primitive perceptually rooted cognitions: "The dynamics of scientific progress is driven by the upward spiral of the assimilation/ accommodation process resulting in a hierarchical series of structures in which equilibration is achieved only in part" (Miller, 1989, p. 185). In the specific case of Einstein, Miller argued that such partial equilibration drove

Einstein to think in ways that proceeded beyond the usual stage of formal operational thinking to a realm that was in fact at a greater level of abstraction than the highest levels previously considered by Piaget.

Both Gruber and Miller produced works that were recognized within the history of science as positive contributions, but each also argued that consideration of historical cases makes a positive contribution to psychology as well. For Gruber, this constituted the point that creativity in science was a matter, not just of consistency with prior theory and evidence, but in the integration of the personal *experiences* of the individual (Gruber, 1974, esp. Ch. 12). For Miller, while extending Piaget's formal operational stage as the highest level of thinking, this required also a new perspective on the role of imagery. Rather than characterizing abstract mathematical operations at the formal stage (as had Piaget), Miller, following his careful analysis of the development of relativity theory in Einstein's thought, suggested that imagery played a central role at every stage—but imagery as a schema rather than as a picture (Miller, 1984, see esp. p. 288).

In the remainder of this chapter, I describe in more detail three recent cognitive–historical studies. Each highlights different aspects of the scientific process, and each draws upon different resources of cognitive science to explicate historical cases.

CONCEPTUAL CHANGE IN THE COPERNICAN REVOLUTION

Andersen, Barker, and Chen (2006) presented a cognitive formalization of some aspects of Kuhn's theory of scientific change, arguing that such an approach can revive Kuhn's utility for work in the history of science and in the philosophy of science. The formalization was based on recent work in cognitive science dealing with concepts and conceptual change in thinking. The larger goal, Kuhn aside, was to argue that "cognitive factors are ineliminable in reaching a historical understanding" (p. 98). The book describes the approach in some detail, applying it to three case histories of scientific change to show its power. Two of the cases (nineteenth century reclassifications of birds and the discovery of nuclear fission) are treated relatively briefly, and one, the Copernican Revolution, is examined in more depth.

All three authors are known for their substantial contributions to the history of science and for cognitive–historical accounts of particle physics (e.g., Andersen, 1996), the Copernican revolution (Barker, 1999), and the wave theory of light (Chen, 2000). In the 2006 book, they presented a unified approach based on conceptual change, developing a formal account that allowed them to make claims about the changes in astronomical theory, first from the earth-centered Ptolemaic system and Copernicus's development of the sun-centered system, then moving to Kepler's modification of that system. Kuhn had argued that in the course of a paradigm shift, the new theory was *incommensurable* with the old, that is, that aspects of the new theory could not be merely reinterpreted in terms of the old, but that the two were so radically different that even the very terms of the old system could not be expressed in terms of the new. Andersen et al. sought to provide a cognitive interpretation of this incommensurability by relying upon recent advances in the psychological understanding of concepts and conceptual systems.

Within psychology, the traditional "feature theory," that concepts consist of lists of defining attributes, was attacked during the early 1960s, first via Eleanor Rosch's "prototype theory," in which concepts are organized psychologically by prototypes that abstract away less important detail. Thus, a robin is a "better" bird than a penguin, because it is closer to the prototype. Later extensions by experimental psychologists were based on Wittgensteinian "family resemblances." Recently, one of these extensions, known as "frame theory" and exemplified in the work of the cognitive psychologist Barsalou (2003), has dominated discussion in cognitive science; this is the approach used by Andersen et al. to ground their formalization.

According to Barsalou, conceptual structures are organized by frames, that is, organized layers of nodes that include attributes at one level, with attribute values at a subordinate level. Internode relationships are included (e.g., a value of one attribute may constrain the values of another attribute), as well as levels of nodes that are not attributes. As an example of constraint, consider the example of the concept of "blueberry." A blueberry is a "nonred fruit," but one must then recognize that "blue" and "nonred" are both possible values of "color." "Blue" and "nonred" cross-conceptual boundaries, tying "blueberry" to "color" in a fashion that constrains the possible subordinate attribute values. There is a dynamic aspect to Barsalou's use of frames: what is activated in any single instance of a recalled concept will be affected by the context of its recall. Thus the tie to the color of blueberries might center on "nonredness" when the context involves discussion of the vitamin content of fruit, but might center on "blueness" in discussion of a graphic design for cereal boxes. Following Barsalou, Andersen et al. used a graphical way of describing frames, permitting an economy of description highlighting similarities and differences among multiple frame representations.

Before Copernicus, there were two natural kinds of physical objects, celestial objects and terrestrial objects, differing in that the former (but not the latter) were changeless and endowed with perfect circular motion. No such bifurcation was possible after Newton's time, however, because both natural kinds were now physical objects, a change that became possible following Kepler's modification of the Copernican system. In cognitive terms, later celestial objects were described by a frame in terms of (a) orbit center (for which the possible values were star, planet, other), (b) orbit shape (ellipse, hyperbola, other), and (c) other aspects such as distance, luminance, size, and so on. By contrast, the equivalent frame for Copernicus specified attributes such as path (with possible values of daily, proper, and retrograde), distance, luminance, and size. Note that there is no way to map the values of "orbit center" and "orbit shape" onto those of "path"; the structure of the frames is simply different. Thus the frames for physical objects were different in structure and hence were incommensurable. Kuhn's rather vague notion of "degree of incommensurability" became for Andersen et al. a set of questions about the nature of the structural difference: Is it among attribute values at a low level? Does it redistribute objects across category boundaries? Do the categories differ? At what level of the frame?

In applying this idea to the historical case, Andersen et al. noted that before Kepler, most astronomical theory concentrated on a celestial object's angular position, in accordance with the importance of the components of the path attribute of celestial objects (whether, in short, the path was a manifestation of daily, proper, or retrograde motion). After Kepler, angular position was no longer part of the frame; a Keplerian *orbit* (not an *orb*) was characterized by its shape and center, not its angular position in the sky. This generated incommensurability (and opened the way to Newton's synthesis) because it introduced new attributes *and* new values, rather than new attributes with the same set of values or new values for existing attributes. The frames were of different structure; it was not a matter of pruning a branch here or adding a branch there. Instead, the entire tree was different. According to Andersen et al., the *real* revolution comes with Kepler, not Copernicus, since the earth-centered system of Copernicus relied on exactly the same sorts of frames as the Ptolemaic system.

Kepler's concept of an orbit was an *event* concept, not an *object* concept; this also may mark a major incommensurability. Event concepts, so recent research in cognitive science suggests, are organized very differently from object concepts, as they embody values that can vary over time. This creates difficulties for the frame representation approach, because events seem to require multiple frames for each successive time interval. And, experimental evidence from the cognitive laboratory (partly due to Barsalou and his colleagues) suggests that different processes are involved when event concepts are memorized, retrieved, and communicated. For example, drawing on a different case study, Chen (2003) showed that John Herschel's partial understanding of the wave theory of light could be explained by the fact that Herschel had an object concept of waves, one that was incommensurable with the prevailing event concept of waves that emerged after the work of Fresnel and others. The difference mattered little in how the wave account of refraction and reflection were understood by Herschel, but his grasp of the theory of polarization, which required an event concept, was therefore never fully in tune with those of his contemporaries.

Andersen et al. have provided a clear example of how the analysis of a well-documented case study can be enhanced by a cognitive approach. Still, their analysis is based on broad aspects of the case, drawn from published results spread across many centuries.[1] The next example to be considered shows how the cognitive–historical approach can accommodate a finer-grained analysis within the thinking of a single scientist.

NERSESSIAN ON MAXWELL

Nancy Nersessian (1984) has dealt explicitly with the psychological nature of meaning change in science. Her work began with an extensive analysis of the notion of field in the theoretical physics of Faraday, Maxwell, Lorentz, and Einstein. Her later work showed how this analysis can be tied to specific cognitive developmental models of conceptual change, on the one hand,

and to cognitive conceptions of mental models on the other. In Nersessian (1992), she argued that the study of science has been unnecessarily limited by considering only inductive and deductive modes of reasoning; instead, a detailed analysis of "abstraction techniques" such as analogical reasoning, imagistic analysis, and thought experiments "are strongly implicated in the explanation of how existing conceptual structures play a role in constructing new, and sometimes radically different, structures" (Nersessian, 1992, p. 13). For her, meaning change is the cognitive particularization of the original questions raised by the notion of a paradigm shift.

To capture these issues in detail, Nersessian, who first coined the term "cognitive-historical analysis" (Nersessian, 1986) focused on analysis of the successive stages of Maxwell's development of electromagnetic field theory. There were four major steps in this development; three papers published between 1855 and 1864 and the landmark *Treatise on Electricity and Magnetism* (Maxwell, 1873). The *Treatise* presents the final form of Maxwell's theory, but it is a complex, many-layered work that still lacks full analysis by historians of science (but see Simpson, 2010, for a first large step in this direction). Most historians have focused on the three papers that preceded it (e.g., Siegel, 1991). In the first of the three papers, Maxwell developed an alternative way of construing Faraday's theory of electric and magnetic forces. Faraday's notion of "lines of force" was re-represented by Maxwell in mathematical form as "tubes of force," which allowed him to use the mathematics of fluid flow as an analogy for the action of electric and magnetic forces. In the second paper, Maxwell dropped this analogy, in favor of a mechanical analogy, one in which rotating "vortices" separated by ball-bearing-like rollers were used to analogize the dynamic relations between electricity and magnetism. In the third paper, this analogy was dropped. Instead of being presented as a "real" model of the electromagnetic field, the energy relationships of the mechanical model were used to analogize the interactions in the "ether," which was the medium of the field itself. This mode of using analogy was extended in the *Treatise*, which used only the energy relationships to characterize the ether and the field.

In *Faraday to Einstein*, Nersessian (1984) outlined the differing ways in which Maxwell used analogy in the three papers. In the first paper, analogy was used to develop "a spatial representation of the intensity and direction of the lines of force" (pp. 75, 76), in the second paper to show "*how* the lines, if they exist, could be produced, that is, by what sorts of forces" (p. 76). In the second paper, the mathematical representation led to a surprising result. If the electromagnetic field had the properties that Maxwell ascribed to it, then it should be the case that an electromagnetic disturbance in the field could propagate. Further, if such propagation did exist, Maxwell showed that its velocity would be equal to *the speed of light*. This was an astonishing result. Still, Maxwell insisted that he was presenting his account as provisional, as a hypothetical account. In the third paper, the strategy changed even more radically. To understand the field using energy relationships only (i.e., what Maxwell called "generalized dynamics"), one needs only minimal knowledge. Thus, for a system consisting of connected bodies (where the mode of connection can be, for example, gravitational relationships, or the dynamic relations among electric and magnetic phenomena), knowledge about velocities, momenta, and potential and kinetic energy suffices to specify the motion of the system as a whole. Maxwell showed that the electromagnetic field is such a system, a system defined over the space in and among the electric and magnetic bodies.

Nersessian's *Faraday to Einstein* (1984) centered on understanding the historical context of field theory; her analysis of Maxwell is preceded by an account of Faraday's field theory and followed by accounts of Lorentz's theory and Einstein's theory. In subsequent works, Nersessian developed a more explicit cognitive account of how Maxwell (in particular) developed his field theory. Two major ideas emerged from this work, the notion of "abstraction via generic modeling," and a characterization of the "model-based reasoning" that underlies scientific thinking (Nersessian, 2005; 2008). The first of these concepts elaborated the way in which Maxwell was able to generate analogies that could be used to further develop his theories. The key point is that analogies participate in a dynamic process by which the source of the analogy is successively fitted to correspond to the target in ways that capture the significant relationships in the target. For example, an isosceles triangle is specified by holding the length of two sides and the angles opposite them to be equal. By suppressing the equal sides and angles constraint, via generic abstraction, any triangle can be represented. By further relaxing the constraint that there be three sides, again via generic abstraction, any polygon can be represented. In the case of Maxwell, Nersessian showed that he developed his generalized dynamics using such a process—gradually relaxing constraints imposed by a specific mechanical model (the vortex wheels, say) until only the crucial energy relationships remained, which could then be used to characterize the electromagnetic field, in the absence of knowledge about its underlying causes. In the course of this work, Nersessian drew upon (and extended) many aspects of the methods and theories of cognitive science. For example, computer simulations were used to show how generic abstraction worked, and the structure-mapping theory of Gentner and her colleagues (e.g., Gentner & Bowdle, 2008) was extended to include the goal-directed character of the development of analogies by Maxwell and with a focus on the process by which representations are built.

Model-based reasoning, the second of the major ideas that emerged from Nersessian's work, rests on the claim that scientific thinking is largely a matter of the development of mental models of a richly varied sort; models that are involved in constructive and manipulative activities and that draw upon information in various formats, including linguistic and imagistic simulations, as well as external representations. This claim implicates many topics in cognitive science, of course, and the 2008 book, *Creating Scientific Concepts*, provides a full discussion of these. Extending more traditional cognitive views of mental models (e.g., Johnson-Laird, 1980), which centered on linguistic and propositional reasoning, Nersessian described a model-based reasoning process that is much richer, involving even the mental simulation of complex physical systems (see also Forbus, 1983). Thus, in the case of Maxwell, his mechanical vortex model invokes a range of physical and mathematical representations, and his papers show the way in which these are successively modified, partly via generic abstraction, partly by thought experiments, and partly by careful alignment with experimental results, drawn from both electric and magnetic experiments, but also from actual mechanical models.

In her 2008 book, Nersessian developed two case studies, Maxwell (as already noted) and an analysis of a single subject solving a difficult "spring problem" in physics, which provided a verbal protocol and many diagrams and videotaped gestural movements. The protocol was gathered by John Clement and subjected, along with many others, to very detailed analysis across many papers (summarized in Clement, 2008). Clement's analysis also rests on model construction activities, with special attention to the differing roles analogy played within the protocols. Nersessian's account is similar in that regard, but she also interleaved her analysis of the subject protocol with that of Maxwell, arguing that the consistency shown between the two cases was evidence for the generality of the processes found in both. Thus, the analysis of the real-time protocol and that of the record shown by Maxwell (based primarily on published articles) serves as a validation study of her cognitive–historical method.

In *Science as Psychology* (2011), Nersessian and her colleagues extended her cognitive–historical approach in a real-time study of scientific thinking in two bioengineering labs. Using primarily ethnographic approaches, a kind of "cognitive anthropology," Nersessian and her colleagues have developed a person-centered account of the work of teams of scientists, in which cognitive, personality, social, and cultural aspects are the focus (Osbeck, Nersessian, Malone, & Newstetter, 2011). Although not historical in the usual sense, this work draws upon the cognitive–historical approach in the way in which the time course of participant's actions is treated, even to include the "history" of the epistemic preparations used as the focus of the team efforts. As with her earlier work, the program of research shows the depth and richness that are possible with the cognitive–historical approach.

COGNITIVE ANALYSES OF MICHAEL FARADAY

Michael Faraday has been an important subject for cognitive–historical research, primarily because of the extensive diaries and notebooks that he kept detailing his research (Tweney, 1991). The sheer scope of his record keeping is amazing; the main "laboratory notebooks" (published in eight large volumes, Martin, 1932–1936) record tens of thousands of experiments and span 42 years of research. The records are detailed enough to allow analysis at a level approaching that of a real-time verbal protocol, and much of my own work has been based on such analyses (e.g., Tweney, 1985). When I began my work on Faraday's diaries, I was attempting to confirm (or disconfirm) hypotheses that we had found in laboratory studies of experimental subjects (e.g., Mynatt, Doherty, & Tweney, 1978). Thus, we had found a strong tendency for most subjects to become stuck in a kind of "confirmation bias" when testing a hypothesis. In one study, a student and I (Tweney & Hoffner, 1987) constructed a problem-behavior graph that traced the steps taken by Faraday following his 1831 discovery of electromagnetic induction. The graph revealed a striking feature of Faraday's experimentation, that he quickly abandoned a particular experimental approach when he got results that did not meet his expectations, instead turning to another method, or, quite often, picking up a different "thread" altogether. At first glance, this looked like Faraday was "ignoring disconfirmation," but a closer look revealed that this was not the case. Instead, Faraday was ignoring disconfirmation *at the early stages* of an inquiry. In all of these cases, he first pursued only confirmatory results and, only later, when he had satisfied himself that a given hypothetical claim was supportable by a number of confirmatory results, would he turn to deliberate attempts to disconfirm what he had found. Thus, we characterized his strategy as one involving a "confirm early, disconfirm late" heuristic, similar to that shown by some subjects in our laboratory studies. My work on Faraday was thus initially construed in a hypothetico-deductive framework, in which predictions were made based on earlier theoretical and empirical claims. Though the results of my early studies of Faraday were illuminating in this respect, it was also apparent that there was a great deal more going on than simply the deployment of a particular heuristic, however sophisticated.

In subsequent years, my work on Faraday became more historical: rather than using Faraday as a case study to test ideas from cognitive psychology, I began to take Faraday more completely as interesting in his own right, and I began to use cognitive theories, methods, and empirical results as an *interpretive framework* for understanding the history (Tweney, 1989). At the same time, I looked for cognitive insights *from* the history as a means of enlarging the scope of cognitive science itself (Ippolito & Tweney, Kurz-Milcke, 2004; Kurz-Milcke & Tweney, 1995; Tweney, 1992). In the early 1980s, I became aware of the work being done on Faraday by David Gooding at the University of Bath. Trained in the history and philosophy of science, Gooding had carried out closely detailed analyses of Faraday's mode of "making meaning," and there were interesting similarities to the cognitive approaches that I was using.

Like Nersessian, Gooding (1990) developed an account of the emergence of scientific concepts, which relied upon nonpropositional aspects of thought. Critical of the exclusively linguistic focus of analytically oriented philosophy of science, Gooding instead emphasized the role of a dynamic interaction between "hand, eye, and brain." Using a detailed reconstruction of Michael Faraday's experimentation, Gooding produced maps of the process by which vague construals were turned by Faraday into concrete communicable scientific concepts of a formal sort: "Making meaning is a process" (Gooding, 1990, p. 271). For both Nersessian and Gooding, a cognitive account of science is one that accounts for the emergence and transformation of meaning in scientific practice, and their method of analysis is an attempt to show that this can be done.[2]

Outwardly, Gooding's maps resemble a problem behavior graph, but they actually rely on different principles. Whereas a typical-problem behavior graph is chronologically oriented (time is represented, flowing from top to bottom and left to right), Gooding's maps are maps of conceptual relations and time is sometimes represented only incidentally. Many of the steps in his maps are inferred, that is, they are based on more than simply what is recorded in the diary or in publications, but instead rely upon *possible* paths linking one step to another. This was necessary because it is not always clear which of several paths Faraday actually followed, and it is possible to see multiple alternatives. It also allowed representation of the fact that Faraday often worked with more than one possibility in mind, simultaneously holding several possible interpretations, while exploring them serially.

Using these techniques, Gooding provided a richly detailed account of the microstructure of Faraday's 1821 experiments on the rotation of a current-carrying wire in a magnetic field (and the obverse, the rotation of a magnet in proximity to a current-carrying wire). Gooding showed that Faraday's mode of experimentation in this research was driven by more than linguistic or propositional moves; the active and dynamic imagistic aspects of the phenomena were centrally involved as well. This is the sense in which the meanings of the rotations were "made."

Because Gooding's maps resemble problem-behavior graphs, it might be thought that he was endorsing the problem-space search model that emerged from the original Newell and Simon (1972) approach. In fact, this is not quite the case. Gooding's work, like Nersessian's and like my own later work, has suggested that, while problem-space search is a valid model for some kinds of problem solving in science, it is less able to account for the construction of representations in the first place (Kurz-Milcke, 2004). At times, Faraday's experimentation does, in fact, resemble search through a problem space, as in my (1985) account of his work immediately following the discovery of electromagnetic induction in 1831 and in Kulkarni's account of Krebs. But in other cases, as in Gooding's account of Faraday's 1821 discovery of electric and magnetic rotations, the need for understanding centers on the way Faraday *constructed* a phenomenon in a representational fashion that allowed subsequent problem-space search. Such constructions depend upon different processes that cannot be reduced to heuristic search through a predefined space.

Search for a representation can be seen in Faraday's 1831 work on the acoustic vibration of fluid surfaces, the "crispations" that can be seen, for example, when a glass of water is struck by a spoon (Tweney, 1992). The observational challenge faced by Faraday was to slow down a rapid process, one that moved too quickly to be seen in detail. To achieve this, he first explored the limits of what the human eye could resolve temporally, conducting a series of studies of movement illusions. This gave him a specific quantifiable account of perceptually averaged motion, and the relation between such perceptions and the physical motions that produced them. He then studied the vibrating surfaces and "separated out" the eye's illusory contribution to the observed results. In effect, Faraday's representation of perceptual phenomena allowed him to construct a searchable problem space in the acoustic domain. After many experiments (which could be described as search of an "experiment space," see Klahr & Dunbar, 1988), he was able to construct a representation, a model, of the phenomena. But the model construction process was not a matter of problem-space search. Instead, the process by which he did this could be described in terms of Nersessian's generic abstraction process, and we were able to describe it (in a fashion similar to Nersessian's) as a movement from perceptions to perceptual rehearsal, to "inceptions" (i.e., to perception-like mental representations that stripped away certain aspects of the perceptions themselves) to, finally, a working mental model of the phenomena (Ippolito & Tweney, 1995).

An even more detailed picture of Faraday's exploratory experimentation became available following the discovery of a large set of experimental specimens prepared by Faraday in 1856 as part of his researches on the colors of gold. Gold interested him because of the huge variety of colors it manifests in different physical and chemical situations and because it is the only metal transparent in thin films (as in commercial gold leaf). Thus, gold could hold clues to the interaction of light and matter. The found specimens were mostly numbered microscope slides with thin deposits or films of gold and other materials; the numbering allowed correlation with the diary entries covering the experimentation. In the absence of the specimens, the diary alone had been puzzling to historians dealing with Faraday, but the specimens proved to be part of the diary—Faraday's "illustrations" as it were. Thus, the pathways used by Faraday in exploring this domain could finally be understood (Tweney, 2006). Further, by replicating some of his procedures, we were able to uncover various aspects of the experimental practices used by Faraday (Tweney, Mears, & Spitzmüller, 2005). This allowed us to characterize more fully the "hand–eye–brain" interrelatedness of Faraday's experimental and theoretical practices. As with Gooding's (1990) work, our research also makes clear the close association of cognitive–historical analysis with the current themes of situated cognition (e.g., Robbins & Aydede, 2009), distributed cognition (e.g., Giere, 2006, 2008), and embodied cognition (e.g., Barsalou, 2010).

Replication has been used by historians of science to good effect. For example, Heering (1994) replicated a famous experiment conducted by Coulomb in 1784 and 1785, a procedure that established the inverse square law of electrical attraction and repulsion. By carefully reconstructing Coulomb's apparatus and procedures, Heering showed that Coulomb could not have obtained the measurements he reported with the apparatus he described. While this left open the question of exactly how Coulomb did obtain his results (Heering suggested that perhaps he made a long series of experiments but reported only a "representative" or averaged result), the fact that Coulomb was working in the context of strong theoretical conceptions became clear in the analysis. In the absence of laboratory records, however (Heering had only Coulomb's published work as a source), it is not possible to know with certainty how the results were obtained.[3]

Within the cognitive–historical area as such, however, other replications have extended the power of the method. In particular, Elizabeth Cavicchi has carried out many cognitive–historical replications, primarily of electrical and magnetic research from the 19th century. Much of her work centered on Faraday (e.g., Cavicchi, 1997, 2003), often in the context of teaching courses in the history of science or in physics itself. In other work, she has replicated work done by Page and others on induction coils (Cavicchi, 2006a). In a particularly interesting comparative study (Cavicchi, 2006b), she replicated some of Piaget's experiments and compared them to replications of Faraday's experiments on acoustic vibrations, a comparison that suggested a strong similarity between the cognitive styles of experimentation of Faraday and those of Piaget; both employed an exploratory, "discursive," style of experiment in which successive experiments constituted a chain of inquiry, filled with uncertainty and wrong paths, as meaning was constructed. Cavicchi's work has emphasized the importance of cognitive tools in science, both material ones (Faraday's tuning forks, water bowls, and such; Piaget's toy blocks, napkins, and such) and cognitive ones (varying conditions, the playful exploration of combinations, and so on) as investigation proceeded.

CONCLUSIONS...AND THE FUTURE OF COGNITIVE–HISTORICAL STUDIES

In an influential review of research in the psychology of science, Klahr and Simon (1999) compared historical studies of science to laboratory studies of simulated science, direct observation, and computational simulations. In summarizing their comments (see esp. p. 530 in their review), Klahr and Simon suggested that historical studies possessed high "face validity" and were excellent as a source of "new phenomena," and hence well suited as exploratory studies. But they faulted historical studies for relative lack of "construct validity," by which they meant a lack of generalizability from one scientist to another.[4] Yet, in the light of the studies we have reviewed here, this seems too narrow a view. They are certainly correct about the face validity and value as a source of new ideas about scientific thinking—the historical cases reviewed here provide ample support for both claims. But to say that the studies lack generalizability is too strong. Thus, Nersessian's studies of Maxwell reveal aspects of the use of analogy that shed light on its uses in everyday thinking as well as the use of analogy by other scientists. The different types of analogy manifested in Maxwell's science are found in other domains by other people as well, as her comparisons to the subject from Clement's studies reveal. Further, the many studies of analogy carried out in laboratory studies of cognition have revealed a great deal about the processes that underlie analogy use in general, and these processes are the foundation for the *interpretive* use of analogy in understanding Maxwell. I argue that the true way to conceive the use of historical studies for cognitive science is the reverse of the view articulated by Klahr and Simon: the fact that cognitive psychological results can be used to interpret historical episodes in science is *confirmation* of the value of those laboratory results!

Earlier, I made the point that one cannot "average" Faraday and Maxwell to understand their way of thinking. Certainly there are commonalities, both in general methods used and in the specific content of their approaches (on this, see Tweney, 2009; in press). There are differences as well, and neither scientist can be understood if the differences are ignored. In this sense, cognitive–historical studies can help to restore the person to the center of thinking about scientific thinking. Thus, one of the major differences between Faraday and Maxwell resides in the fact that Faraday used no—absolutely no—mathematics in any of his research. Maxwell, by contrast, was a powerfully gifted mathematical thinker and mathematical representations pervade his work. This counts as a major stylistic difference, and the reasons for the difference have been traced to the very different backgrounds of the two, to Faraday's self-education in science and to his rather unorthodox religious views (Cantor, 1991; James, 2010), and to Maxwell's nurturance within a scientific family and his education in Edinburgh and Cambridge (Harman, 1998; Warwick, 2003). Still, there were commonalities, as Maxwell often emphasized, regarding Faraday as an "intuitive mathematician." In fact, Faraday used a kind of "dynamic geometry" (Gooding, 1992) to represent electric and magnetic fields, and Maxwell's mathematical representations can be seen as equivalent in many respects. Assessing these equivalences depends on knowing the specific characteristics of each scientist in terms that remain faithful to the uniqueness of each as a person (Tweney, 2009, 2012).

In recent years, the discipline of the history of science has taken something of a "social-cultural" turn. That is, there have been studies, not just of the internal movement of scientific ideas, but a broader contextualization of the history of science in which social, cultural, and even economic and political events have been examined for a richer understanding of the nature of science and scientific change. Thus, Smith and Wise (1989) presented a kind of extended biography of Lord Kelvin (William Thomson) in which his scientific and technological achievements were placed in the context of the industrial society of which he was a part and its specific manifestations both near to his base in Glasgow and extended internationally across the British Empire. Shapin (1994) described the way cultural notions of gentlemanly conduct affected the early development of the Royal Society. Caneva (2005) has argued for more consideration of the way in which scientific "discoveries" become constructed in the course of becoming *collective* products shaped by communities of scientists. Such studies suggest many possibilities for the integration of cognitive studies (as well as social-psychological studies and cultural-psychological studies) with the work done by historians.

Still, why use historical cases, given that studies of scientific thinking can be done by observation of ongoing science in "real time?" One obvious advantage is that truly great science can be studied; the accomplishments of a Faraday, a Kepler, and a Maxwell represent the highest accomplishments of science and understanding these is obviously valuable. In addition, as Gooding has shown, it opens perspectives on the multiple possible pathways of thinking. Furthermore, the study of historical cases allows a much more nuanced view of the social and cultural forces that shape science over both short and long durations. In this sense, the context of science becomes much easier to deal with, given the benefit of the kinds of hindsight available. Thus, one can assess the place of the science in question within the context of developments in the political, social, economic, and cultural aspects of the period in question, a task that is much harder with contemporaneous science. In this way, we can construct much richer understandings of the sources of an individual scientist's thinking, enriching also the *person-centered* character of the psychology of science (Osbeck et al., 2011).

The present chapter has illustrated the kinds of approaches that are possible with a combination of historical and cognitive methods. Drawing on the richness of historical studies of science and the richness of current cognitive science opens the way to an integrated account of "how science works." Like the psychology of science in general, the cognitive-historical study of science has, as well as a long past, a bright future.

ACKNOWLEDGMENTS

Thanks to Nancy Nersessian and Elizabeth Cavicchi for incisive comments on an earlier draft.

NOTES

1. See also Gentner et al. (1997) for an alternative cognitive account of Kepler's innovations, and Chi (1992) for a general account of conceptual change along somewhat different lines.
2. Gorman and Carlson's work on the creative inventions of Thomas Edison and Alexander Graham Bell had a similar character (Gorman, 1992; Carlson & Gorman, 1992; Gorman & Carlson, 1989). In the study of Bell, the focus was to show how the emergence of particular ideas about mechanical and electrical arrangements can be traced through the series of working drawings made by Bell during the invention of the telephone. They thus revealed a sequence of nonverbal "reasonings" used by Bell to concretize, test, and refine his ideas.
3. Recently, in the context of a graduate course in the history of psychology and two follow-up seminars, I supervised a small number of teams of students in the replication of historically important studies in psychology. These were later published as a special issue of *Science & Education* (Tweney, 2008), partly as contributions to the history of psychology and partly as manifestations of the pedagogical value of replications.
4. In fact, when I began to report my work on Faraday in the early 1980s, a frequent comment went along the lines of, "Well, this is all fine and good, but you won't know anything until you do a similar analysis on a large group of scientists," suggesting that only aggregative approaches had merit. The present review is an argument against such views.

REFERENCES

Andersen, H. (1996). Categorization, anomalies and the discovery of nuclear fission. *Studies in the history and philosophy of science, 27,* 463–492.

Andersen, H., Barker, P., & Chen, X. (2006). *The cognitive structure of scientific revolutions.* Cambridge, MA: Cambridge University Press.

Barker, P. (1999). Copernicus and the critics of Ptolemy. *Journal for the History of Astronomy, 30,* 343–358.

Barsalou, L. W. (2003). Abstraction in perceptual symbol systems. *Philosophical Transactions of the Royal Society of London. B (Biological Sciences), 358,* 1177–1187.

Barsalou, L. W. (2010). Grounded cognition: Past, present, future. *Topics in Cognitive Science, 2,* 716–724.

Bechtel, W. (1988). *Philosophy of science: An overview for cognitive science.* Hillsdale, NJ: Lawrence Erlbaum.

Bradshaw, G. (1992). The airplane and the logic of invention. In R. N. Giere (Ed.), *Cognitive models of science* (Minnesota Studies in the Philosophy of Science, Vol. XV; pp. 239–250). Minneapolis, MN: University of Minnesota Press.

Bruner, J. S. & Postman, L. (1949). On the perception of incongruity: A paradigm. *Journal of Personality, 18,* 206–223.

Buchwald, J. Z. (Ed.) (1995). *Scientific practice: Theories and stories of doing physics.* Chicago, IL: University of Chicago Press.

Caneva, K. L. (2005). "Discovery" as a site for the collective construction of scientific knowledge. *Historical Studies in the Physical and Biological Sciences, 35,* 175–292.

Cantor, G. (1991). *Michael Faraday: Sandemanian and scientist. A study of science and religion in the nineteenth century.* New York, NY: St. Martin's Press.

Carlson, W. B. & Gorman, M. E. (1992). A cognitive framework to understand technological creativity: Bell, Edison, and the telephone. In R. J. Weber & D. N. Perkins (Eds.), *Inventive minds: Creativity in technology* (pp. 48–79). New York, NY: Oxford University Press.

Cavicchi, E. (1997). Experimenting with magnetism: Ways of learning of Joann and Faraday. *American Journal of Physics, 65,* 867–882.

Cavicchi, E. (2003). Experiences with the magnetism of conducting loops: Historical instruments, experimental replications, and productive confusions. *American Journal of Physics, 71 (2),* 156–167.

Cavicchi, E. (2006a). Nineteenth-century developments in coiled instruments and experiences with electromagnetic induction. *Annals of Science, 63(3),* 319–361.

Cavicchi, E. (2006b). Faraday and Piaget: Experimenting in relation with the world. *Perspectives on Science, 14(1),* 66–96.

Chen, X. (2000). *Instrumental traditions and theories of light: The uses of instruments in the optical revolution.* Dordrecht, Holland: D. Reidel Publishing Co.

Chen, X. (2003). Why did John Herschel fail to understand polarization? The differences between object and event concepts. *Studies in the History and Philosophy of Science, 34,* 491–513.

Chi, M. T. H. (1992). Conceptual change within and across ontological categories: Examples from learning and discovery in science. In R. N. Giere (Ed.), *Cognitive models of science* (Minnesota Studies in the Philosophy of Science, Vol. XV, pp. 129–186), Minneapolis, MN: University of Minnesota Press.

Clement, J. (2008). *Creative model construction in scientists and students: Imagery, analogy, and mental simulation.* Dordrecht, Holland: Springer.

De Mey, M. (1982). *The cognitive paradigm.* Dordrecht, Holland: D. Reidel Publishing Co. (Reprinted, 1992, Chicago: University of Chicago Press).

Ebbinghaus, H. (1908). *Psychology: An elementary textbook.* (trans. and ed. by M. Meyer). Boston, MA: D.C. Heath.

Ericsson, K. A. & Simon, H. A. (1993/1984). *Protocol analysis: Verbal reports as data.* Cambridge, MA: MIT Press (Revised Edition).

Ericsson, K. A., Charness, N., Feltovich, P. J., & Hoffman, R. R. (Eds.). (2006). *The Cambridge handbook of expertise and expert performance.* New York, NY: Cambridge University Press.

Forbus, K. (1983). Reasoning about space and motion. In D. Gentner & A. Stevens (Eds.), *Mental models* (pp. 53–74). Hillsdale, NJ: Lawrence Erlbaum.

Gentner, D. & Gentner, D. R. (1983). Flowing waters or teeming crowds: Mental models of electricity. In D. Gentner & A. L. Stevens (Eds.), *Mental models* (pp. 99–130). Hillsdale, NJ: Lawrence Erlbaum Associates.

Gentner, D. & Bowdle, B. (2008). Metaphor as structure-mapping. In R. W. Gibbs, Jr. (Ed.), *The Cambridge handbook of metaphor and thought* (pp. 109–128). Cambridge, MA: Cambridge University Press.

Gentner, D., Brem, S., Ferguson, R. W., Markman, A. B., Levidow, B. B., Wolff, P., & Forbus, K. D. (1997). Analogical reasoning and conceptual change: A case study of Johannes Kepler. *Journal of the Learning Sciences, 6,* 3–40.

Giere, R. (2006). *Scientific perspectivism.* Chicago, IL: University of Chicago Press.

Giere, R. (2008). Cognitive studies of science and technology. In E. J. Hackett, O. Amsterdamska, M. Lynch, & J. Wajcman (Eds.) *Handbook of science and technology studies* (pp. 259–278). Cambridge, MA: MIT Press.

Glasgow, J., Narayanan, N. H., & Chandrasekaran, B. (Eds.). (1995). *Diagrammatic reasoning: Cognitive and computational perspectives.* Menlo Park, CA: AAAI Press & MIT Press.

Gooding, D. (1990). *Experiment and the making of meaning: Human agency in scientific observation and experiment.* Dordrecht, Holland: Kluwer Academic Publishers.

Gooding, D. (1992). Mathematics and method in Faraday's experiments. *Physis, 29* (New Series), 121–147.

Gooding, D., Pinch, T., & Schaffer, S. (Eds.) (1989). *The uses of experiment: Studies in the natural sciences.* Cambridge, MA: Cambridge University Press.

Gorman, M. E. (1992). *Simulating science: Heuristics, mental models, and technoscientific thinking.* Bloomington, IN: Indiana University Press.

Gorman, M. E., & Carlson, W. B. (1989). Interpreting invention as a cognitive process: The case of Alexander Graham Bell, Thomas Edison and the telephone. *Science, Technology, and Human Values, 15,* 131–164.

Gruber, H. E. (1974). *Darwin on man: A psychological study of scientific creativity.* New York, NY: Dutton.

Gruber, H. E. (1989). The evolving systems approach to creative work. In Wallace, D. B. & Gruber, H. E. (Eds.) *Creative people at work: Twelve cognitive case studies* (pp. 3–24). New York, NY: Oxford University Press.

Hanson, N. R. (1958). *Patterns of discovery: An inquiry into the conceptual foundations of science.* Cambridge, MA: Cambridge University Press.

Harman, P. M. (1998). *The natural philosophy of James Clerk Maxwell.* Cambridge, MA: Cambridge University Press.

Heering, P. (1994). The replication of the torsion balance experiment: The inverse square law and its refutation by early 19th-century German physicists. In C. Blondel & M. Dörries (Eds.) *Restaging Coulomb: Usages, controverses et réplications autour de la balance de torsion* (pp. 47–66). Firenze, Italy: Leo S. Olschki.

Holmes, F.L. (1980). Hans Krebs and the discovery of the ornithine cycle. *Federation Proceedings, 39,* 216–225.

Holmes, F. L. (1987). Scientific writing and scientific discovery. *Isis, 78,* 220–235.

Holmes, F. L. (1996). Research trails and the creative spirit: Can historical case studies integrate the long and short timescales of creative activity? *Creativity Research Journal, 9*(2,3), 239–250.

Ippolito, M. F. & Tweney, R. D. (1995). The inception of insight. In R. J. Sternberg & J. E. Davidson (Eds.), *The nature of insight* (pp. 433–462). Cambridge, MA: The MIT Press.

James, F.A.J.L. (2010). *Michael Faraday: A very short introduction.* Oxford, UK: Oxford University press.

Johnson-Laird, P. N. (1980). Mental models in cognitive science. *Cognitive Science, 4,* 71–115.

Klahr, D. (2000). *Exploring science: The cognition and development of discovery processes.* Cambridge, MA: MIT Press.

Klahr, D. & Dunbar, K. (1988). Dual space search during scientific reasoning. *Cognitive Science, 12,* 1–48.

Klahr, D. & Simon, H. A. (1999). Studies of scientific discovery: Complementary approaches and convergent findings. *Psychological Bulletin, 125,* 524–543.

Kuhn, T. S. (1962/1970). *The structure of scientific revolutions* (2nd ed., enlarged). Chicago, IL: University of Chicago Press.

Kulkarni, D., & Simon, H. A. (1990). Experimentation in machine discovery. In J. Shrager & P. Langley (Eds.) *Computational models of scientific discovery and theory formation* (pp. 255–274). San Mateo, CA: Morgan Kaufmann.

Kulkarni, D., & Simon, H. A. (1988). The processes of scientific discovery: The strategy of experimentation. *Cognitive Science, 12,* 139–176.

Kurz-Milcke, E. (1998). Representation, agency, and disciplinarity: Calculus experts at work. In M. A. Gernsbacher & S. J. Derry (Eds.), *Proceedings of the twentieth annual conference of the Cognitive Science Society* (pp. 585–590). Mahwah, NJ: Lawrence Erlbaum Associates.

Kurz-Milcke, E. (2004). The authority of representations. In E. Kurz-Milcke & G. Gigerenzer (Eds.), *Experts in science and society* (pp. 281–302). New York, NY: Kluwer Academic/Plenum.

Kurz-Milcke, E., & Tweney, R. D. (1998). The practice of mathematics and science: From calculus to the clothesline problem. In M. Oaksford & N. Chater (Eds.) *Rational models of cognition* (pp. 415–438). Oxford, UK: Oxford University Press.

Lakoff, G., & Núñez, R. E. (2000). *Where mathematics comes from: How the embodied mind brings mathematics into being.* New York, NY: Basic Books.

Lamiell, J. T. (2003). *Beyond individual and group differences: Human individuality, scientific psychology, and William Stern's personalism.* Thousand Oaks, CA: Sage Publications.

Langley, P. W., Simon, H. A., Bradshaw, G. L., & Zytkow, J. M. (1987). *Scientific discovery: Computational explorations of the discovery process.* Cambridge, MA: MIT Press.

Martin, T. (Ed.) (1932–1936). *Faraday's diary: Being the various philosophical notes of experimental investigation made by Michael Faraday during the years 1820–1862.* (8 volumes). London, UK: G. Bell.

Maxwell, J. C. (1873/1891). *A treatise on electricity and magnetism* (2 volumes). Oxford, UK: Clarendon Press.

Miller, A. I. (1984). *Imagery in scientific thought: Creating 20th-century physics* Boston, MA: Birkhauser.

Miller, A. I. (1989). Imagery and intuition in creative scientific thinking: Albert Einstein's invention of the special theory of relativity. In D. B. Wallace & H. E. Gruber (Eds.), *Creative people at work: Twelve cognitive case studies* (pp. 171–188). New York, NY: Oxford University Press.

Mynatt, C. R., Doherty, M. E., & Tweney, R. D. (1978). Consequences of confirmation and disconfirmation in a simulated research environment. *Quarterly Journal of Experimental Psychology, 30,* 395–406.

Nersessian, N. (1984). *Faraday to Einstein: Constructing meaning in scientific theories.* Dordrecht, Holland: Nijhoff.

Nersessian, N. J. (1986). A cognitive-historical approach to meaning in scientific theories. In N. J. Nersessian (Ed.), *The process of science: Contemporary philosophical approaches to understanding scientific practices* (pp. 161–178). Dordrecht, Holland: Martinus Nijhoff.

Nersessian, N. (1992). How do scientists think? Capturing the dynamics of conceptual change in science. In R. N. Giere (Ed.) *Cognitive models of science* (Minnesota Studies in the Philosophy of Science, Vol. XV, pp. 3–44). Minneapolis, MN: University of Minnesota Press.

Nersessian, N. J. (2005). Interpreting scientific and engineering practices: Integrating the cognitive, social, and cultural dimensions. In M. E. Gorman, R. D. Tweney, D C. Gooding, & A. P. Kincannon (Eds.), *Scientific and technological thinking* (pp. 17–56). Mahwah, NJ: Lawrence Erlbaum.

Nersessian, N. J. (2008). *Creating scientific concepts.* Cambridge, MA: MIT Press.

Newell, A. & Simon, H. A. (1972). *Human problem solving.* Englewood Cliffs, NJ: Prentice-Hall.

Norman, D. A. (1988). *The psychology of everyday things.* New York, NY: Basic Books.

Oaksford, M., & Chater, N. (Eds.) (2010). *Cognition and conditionals: Probability and logic in human thinking.* Oxford, UK: Oxford University Press.

Osbeck, L. M., Nersessian, N. J., Malone, K. R., & Newstetter, W. C. (2011). *Science as psychology: Sense-making and identity in science practice.* Cambridge, MA: Cambridge University Press.

Robbins, P., & Aydede, M. (Eds.) (2009). *The Cambridge handbook of situated cognition.* Cambridge, MA: Cambridge University Press.

Shapin, S. (1994). *A social history of truth: Civility and science in seventeenth-century England.* Chicago, IL: University of Chicago Press.

Siegel, D. M. (1991). *Innovation in Maxwell's electromagnetic theory: Molecular vortices, displacement current, and light.* Cambridge, MA: Cambridge University Press.

Simon, H. A. (1967/1996). *The sciences of the artificial* (3rd ed.). Cambridge, MA: MIT Press.

Smith, C. & Wise, M. N. (1989). *Energy & empire: A biographical study of Lord Kelvin.* Cambridge, MA: Cambridge University Press.

Trickett, S. B., Trafton, J. G., & Schunn, C. D. (2009). How do scientists respond to anomalies? Different strategies used in basic and applied science. *Topics in Cognitive Science, 1*, 711–729.

Tweney, R. D. (1985). Faraday's discovery of induction: A cognitive approach. In D. Gooding & F. A. J. L. James (Eds.), *Faraday rediscovered: Essays on the life and work of Michael Faraday, 1791–1867* (pp. 189–210). New York, NY: Stockton Press/London: Macmillan. (Reprinted, 1990, American Institute of Physics.)

Tweney, R. D. (1989). A framework for the cognitive psychology of science. In B. Gholson, A. Houts, R. M. Neimeyer, & W. Shadish (Eds.), *Psychology of science and metascience* (pp. 342–366). Cambridge, MA: Cambridge University Press.

Tweney, R. D. (1991). Faraday's notebooks: The active organization of creative science. *Physics Education, 26*, 301–306.

Tweney, R. D. (1992). Stopping Time: Faraday and the scientific creation of perceptual order. *Physis: Revista Internazionale di Storia Della Scienza, 29*, 149–164.

Tweney, R. D. (2006). Discovering discovery: How Faraday found the first metallic colloid. *Perspectives on Science, 14(1)*, 97–121.

Tweney, R. D. (Ed.) (2008). Special issue: Studies in historical replication in psychology. *Science & Education, 17*(5), 467–558.

Tweney, R. D. (2009). Mathematical representations in science: A cognitive-historical case study. *Topics in Cognitive Science, 1*, 758–776.

Tweney, R. D. (2011). Representing the electromagnetic field: How Maxwell's mathematics empowered Faraday's field theory. *Science & Education, 20*(7–8), 687–700.

Tweney, R. D. (2012). On the unreasonable reasonableness of mathematical physics: a cognitive view. In R. W. Proctor & E. J. Capaldi (Eds.), *Psychology of science: Implicit and explicit reasoning* (pp. 406–435). Oxford, UK: Oxford University Press.

Tweney, R. D., Doherty, M. E., & Mynatt, C. R. (Eds.). (1981). *On scientific thinking.* New York, NY: Columbia University Press.

Tweney, R. D., & Hoffner, C. E. (1987). Understanding the microstructure of science: An example. In *Program of the ninth annual conference of the cognitive science society* (pp. 677–681). Hillsdale, NJ: Lawrence Erlbaum.

Tweney, R. D., Mears, R. P., & Spitzmüller, C. (2005). Replicating the practices of discovery: Michael Faraday and the interaction of gold and light. In M. Gorman, R. D. Tweney, D. Gooding, & A. Kincannon (Eds.), *Scientific and technological thinking* (pp. 137–158). Mahwah, NJ: Lawrence Erlbaum Associates.

Warwick, A. (2003). *Masters of theory: Cambridge and the rise of mathematical physics.* Chicago, IL: University of Chicago Press.

Wason, P. C. & Johnson-Laird, P. N. (1972). *Psychology of reasoning: Structure and content.* Cambridge, MA: Harvard University Press.

CHAPTER 5

The Scientific Personality

Gregory J. Feist

One of the many building blocks of scientific thought and behavior is personality (Feist, 2006b). Career interests in general and scientific career interest and talent, in particular, stem from personality and individual differences in thought and behavior. The question, therefore, for this chapter is: "Is there such thing as a scientific personality?" That is, are some people more prone to become scientists because of their consistent and unique styles of thought and behavior? As you probably can guess, my answer to this question is "yes, personality shapes scientific interest, ability, talent, and achievement." The purpose of this chapter, however, is to lay bear the evidence for such a claim. I first define personality and then present a biological-functional model of personality before diving into the evidence connecting personality to scientific interest, creativity, and talent. I close the chapter with a review of the literature on the influence of personality on theoretical orientation in science.

PERSONALITY DEFINED

Humans are not alone in their uniqueness and variability between individual members of the species. Individuals within every living species exhibit differences or variability. Indeed, animals such as octopi, birds, pigs, horses, cats, and dogs have consistent individual differences in behavior *within* their species, otherwise known as personality (Dingemanse, Both, Drent, Van Oers, & Van Noordwijk, 2002; Gosling & John, 1999; Morris, Gale, & Duffy, 2002). But the degree to which individual humans vary from one another, both physically and psychologically, is quite astonishing and somewhat unique among species. Some of us are quiet and introverted, others crave social contact and stimulation; some of us are calm and even-keeled, whereas others are high-strung and persistently anxious. In this chapter, I argue that some of these personality differences between individuals make interest in and talent for science more likely.

There is something quite specific in what psychologists mean by the word "personality" (Feist & Feist, 2009; Roberts & Mroczek, 2008). Uniqueness and individuality are one core component; if everyone acted and thought alike, there would be no such thing as personality. This is what is meant by "individual differences." The second major component of personality is "behavioral consistency," which is of two kinds: situational and temporal. Situational consistency is the notion that people behave consistently in different situations, and they carry who they are into most every situation. Temporal consistency, in contrast, is the extent to which people behave consistently over time. To illustrate both forms of consistency as well as individual differences in the context of personality, let us take the trait of friendliness: We would label a person as "friendly" only if we observe her behaving in a friendly manner over time, in many different situations, and in situations where other people were not friendly. In short, personality is what makes us unique and it is what is most stable about who we are.

What are the universal dimensions of personality and how many are there? Although, as with every academic debate, there is some disagreement concerning the answers to these questions, during the last 15 years a surprisingly clear answer has begun to emerge. It has been labeled the "Big-Five" or "Five-Factor" Model (FFM). The FFM is based on factor-analytic studies of personality structure that consistently extract five major factors of personality (Costa & McCrae, 1995; Digman, 1990; Goldberg & Rosolack, 1994; John, 1990; McCrae & John, 1992). The five factors have various labels, depending on the specific researcher, but one of the more common labeling systems, and the one adapted here, is the following: Extraversion (E), Agreeableness (A), Conscientiousness (C), Neuroticism (N), and Openness (O) (Costa & McCrae, 1995).

Moreover, each dimension is both bipolar and normally distributed in the population. That is, for instance, anxiety is one pole of the dimension, with emotional stability being its opposite. Everyone falls somewhere on the continuum from extremely anxious to extremely emotionally stable, with most people falling in between. In addition, just like intelligence, if we were to plot everyone's score on a frequency distribution, we would get a very nice bell-shaped (normal) distribution of scores, with about two-thirds of the scores falling within one standard deviation of the mean. The same is true with extroversion–introversion, openness–closedness, agreeableness–hostility, and conscientiousness–unreliability.

The FFM will be quite helpful when summarizing the literature on personality and scientific interest and creativity, but it is not the only model of personality. Recently, I have developed my own model of personality, one that is both biologically and functionally grounded (Feist, 1999, 2010) and one that I believe sheds light on the origins of the scientific personality.

A BIOLOGICAL-FUNCTIONAL MODEL OF PERSONALITY

To cite Gordon Allport's famous phrase: "Personality *is* something and *does* something" (Allport, 1937, p. 48). What it is and does directly affect behavior.

In response to the infamous "person-situation debate" (Block, 1977; Epstein, 1979; Kenrick & Funder, 1988; Mischel, 1968; Nisbett & Ross, 1980), which contrasted personality and situational forces as more responsible for behavior, many personality psychologists have more recently developed a functional theory of traits (Feist, 1999; Eysenck, 1990; Funder, 1991; Mischel & Schoda, 1999; Rosenberg, 1998). The functional perspective maintains that traits function to lower behavioral thresholds, that is, make particular behaviors more likely in given situations; in short, they raise conditional probabilities (Mischel & Schoda, 1999). The primary function of traits, therefore, is to lower thresholds for trait-congruent behavior (Brody & Ehrlichman, 1998; Ekman, 1984; Eysenck, Mogg, May, Richards, & Mathews, 1991; Rosenberg, 1998). For instance, if a person has the traits of "warm and friendly," this means that in any given situation she is more likely to act in a warm and friendly manner than someone who does not possess that trait. Her threshold for behaving in a friendly manner is lower than if she did not have that trait. Moreover, there are particular situations, such as meeting a new person or being with a group of people, where behaving this way is most likely. That is, in certain situations traited behaviors are most likely for particular people. The function of traits is they raise (or lower) the conditional probability of a given behavior in a given situation.

As I first began arguing in the late 1990s, personality influences thought and action by lowering behavioral thresholds (Feist, 1998, 1999). The idea was and still is that a particular constellation of personality traits function to lower the thresholds of interest and creative behavior, making it more rather than less likely (cf. Allport, 1937; Brody & Ehrlichman, 1998; Ekman, 1984; Feist, 1998; Rosenberg, 1998). The part of the model that has been most intensively investigated over the last decade since the model was first proposed is biological foundations component, especially genetic, epigenetic, and neuropsychological. However, one dimension of the model is new, reflecting even greater growth in research, namely, the clinical personality traits of psychoticism, schizotypal personality, latent inhibition, and negative priming.

The biological-functional model ties biology and variability into personality and argues for the causal primacy of biological factors in personality in general and the scientific personality in particular, much as other personality theorists have done (Eysenck, 1990; Krueger & Johnson, 2008; McCrae & Costa, 2008). By the late 1990s, with the growth in neuroscience and evolutionary perspectives, a clear shift occurred not only in personality research, but in psychological research as a whole. Most models of personality now include some form of neuropsychological or biological component and combined nature and nurture models are more the norm than the exception. The biological-functional model of scientific personality includes three classes of biological precursors of personality, namely genetic influences, epigenetic influences, and brain influences. These, in turn, directly affect personality thresholds for thought and behavior, including scientific thought and behavior.

By combining the biological and the function of traits arguments, Figure 5.1 presents a model for the paths from specific biological processes and mechanisms to psychological dispositions to scientific thought and behavior. The basic idea is that causal influence flows from left to right, with genetic

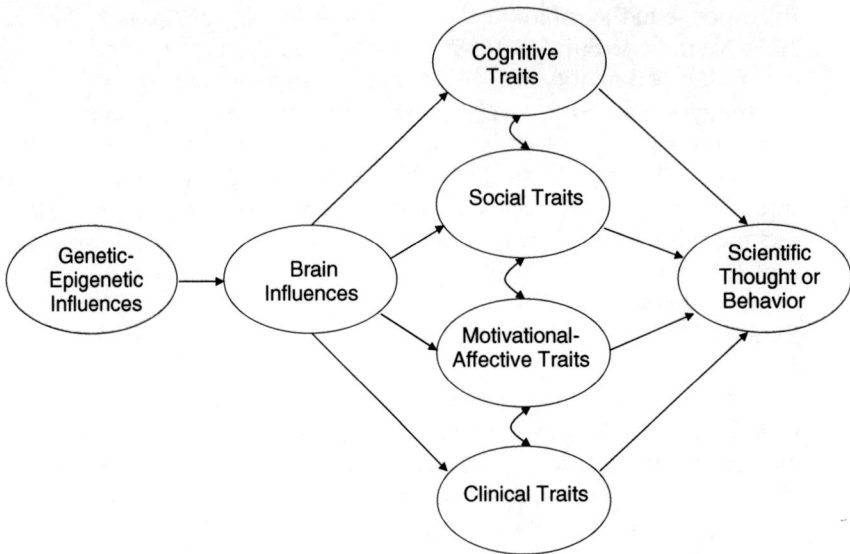

Figure 5.1   *Functional model of the scientific personality.*

From Feist (2010). Reprinted with permission from Cambridge University Press.

and epigenetic influences having a causal effect on brain structures and processes. These neuropsychological structures and processes, in turn, causally affect the four categories of personality influence: cognitive, social, motivational, and clinical. More specifically, as layed out in the rest of the chapter, interest, creativity, and achievement in, as well as talent for math, science, and engineering are a result of different constellations of personality traits (cf. Hany, 1994).

Genetic Influences on Personality

For many people, genetic explanations of personality are thought to be deterministic. Genes were immutable; therefore, if there were a genetic component to thought, behavior, or personality, it is deterministic and unchanging. Both lay people and many scientists eschew the inferred lack of freedom that genetic explanations appeared to have.

We now know this view is outdated and misleading (Krueger & Johnson, 2008; Pinker, 2002). First of all, there is no simple path from genetics to behavior and this is even more true for the path from genetics to personality. Genetic influence on personality is polygenic, meaning dozens if not hundreds of genes are often involved in shaping each trait (Rutter, 2006). There is no such thing as an "extraversion" gene, but there are many genes that are involved in the production of hormones and neurotransmitters that affect extraversion.

Second, the evidence for genetic effects on personality comes primarily from twin-adoption studies. Researchers take advantage of genetically similar and different people by studying twins, siblings, and unrelated individuals reared together or apart. The logic of the twin-adoption approach is simple yet powerful. Identical twins are 100% alike genetically, whereas fraternal twins, like all siblings, share only 50% of their genes. Adopted children and their adoptive parents and siblings share no genes. If genes play a strong role in a trait, then the greater the genetic similarity, the greater the similarity on the trait should be. That is, similarity should be strongest in identical twins reared together and next in identical twins reared apart. It should be modest in siblings reared together and biological parent–offspring. Similarity should be weakest in adopted siblings and adoptive parent–offspring.

Third, a general point to make about the obsolescence of the nature versus nurture debate is modern genetics has uncovered a few discoveries that make clear the old view of genetic and environmental influence being separate and even opposing influences is simply wrong. One such discovery is gene-by-environment research that has found that many behaviors are likely only when particular environmental circumstances occur in conjunction with genetic dispositions (Moffitt, Caspi, & Rutter, 2005; Krueger & Johnson, 2008). One or the other is not enough, but both combined increase the odds of the phenotypic behavior being expressed. For example, depression is most likely to happen not only if someone has short forms of alleles on a gene involved in serotonin production, but only when such a pattern is combined with many major stressful life experiences (Caspi et al., 2003; Wilhelm et al., 2006).

Research on the heritability of talent tends to confirm that genetic factors are necessary for the attainment of talent in various domains, including music, athletics, math, chess, language, and memory (Coon & Carey, 1989; Haworth et al., 2009; Howe, Davidson, & Sloboda, 1999; Lubinski et al., 2006; Ruthsatz et al., 2008; Vinkhuyzen, van der Sluis, Posthuma, & Boomsma, 2009). There are few, if any, direct investigations on the heritability of scientific talent, but there is a literature on two underlying components of scientific talent, namely intelligence in general (mathematical intelligence in particular) and personality.

Scientists, as a group, have one of the highest mean IQs of all occupations. A precondition for being a scientist is being above average in intelligence. The average IQ for scientists and mathematicians approaches 130 (Gibson & Light, 1992; Helson & Crutchfield, 1970; MacKinnon & Hall, 1972). Heritability estimates of g (general intelligence) range from about 0.40 to 0.70, with a typical coefficient being in the 0.50 range (Grigorenko, 2000; Haworth et al., 2009; Lynn, 2006; Plomin & Petrill, 1997). A recent study, for instance, concluded that for those in the 85th percentile and above on IQ, genetics accounted for 50% of the variation in intelligence (95% confidence interval was 0.41–0.60) and shared environmental influences were more moderate, accounting for 28% of the variance in intelligence (Haworth et al., 2009).

The logic, therefore, is straightforward: about 50% of the variation in IQ is attributable to genetics; scientists and mathematicians are roughly two standard deviations higher than the norm on IQ; interest in and talent for science, therefore, is partially attributable to genetic influence. At the present moment, however, this conclusion is more of a logical one than empirical, due to the fact that almost no research has been conducted on the heritability of scientific talent.

There are two main exceptions to the lack of direct evidence of genetic effects on scientific and mathematical talent: work by Benbow, Lubinski, and collagues on mathematically precious youth and a recent paper by Dean Simonton on scientific talent. For example, Chorney et al. (1998) reported differences in DNA markers on the gene for insulin-like growth factor-2 (IGF2R) receptors between children who were extremely gifted in mathematics compared to children extremely gifted in language. These markers have much larger effect sizes in distinguishing children with very high IQ from those with average IQ. Moreover, Simonton (2008) developed a heritability model of scientific talent. Like intelligence, personality tends to have heritability coefficients ranging from 0.40 to 0.60, with a typical coefficient being around 0.45 (Loehlin & Nichols, 1976; Plomin & Caspi, 1999). In short, about 50% of variability in both personality and intelligence can be explained by genetic influences. Simonton asked how much of the variance in scientific talent can be explained by these genetic factors associated with personality and intelligence. Using published effects sizes between personality and creativity (Feist, 1998) and heritability coefficients from twin-studies (Carey, Goldsmith, Tellegen, & Gottesman, 1978), Simonton was able to calculate criterion heritability estimates for the criterion of scientific achievement (talent). For the California Psychological Inventory traits, for instance, the criterion heritability coefficient suggest that 30% of the variance in scientific achievement is attributable to genetic influence on personality. Overall, using personality and intelligence predictors of scientific achievement, Simonton reported that between approximately 10% to 20% of the variance in scientific achievement could be attributed to genetic influence. To put this in context by converting to effect size metrics, Simonton suggests these are medium to large effect sizes in the social sciences. There is evidence, therefore, that scientific talent and achievement are partially attributable to genetic influence. More direct twin-studies and behavioral genetic research, however, needs to be conducted to confirm this conclusion.

Epigenetic Influences

A relatively new field of study has changed our understanding of genetic influence on thought, behavior, and personality: epigenetics. The unique and incomparable genotype with which each of us is born is not the endpoint but the starting point of gene expression. Genes get turned off by many different things, and our experiences and environmental exposure, starting in the womb, are among the off-switches. This means that experience (nurture) shapes our nature. More specifically, methyl-group tags attach to the side of the double-helix and different patterns of tags turn off a gene or leave it on (see Figure 5.2). The incredible fact is that these tags are regulated by environmental events such as diet, drinking, and even exercise (Barres et al., 2012; Carere, Drent, Koolhaas, & Groothuis, 2005; Watters, 2006; Weaver, Cervoni, & Champagne, 2004).

As is the case with genetics, evidence for epigenetic influences on personality is mostly indirect. More and more scholars are arguing that epigenetics is an important mechanism to explain how gene and environment

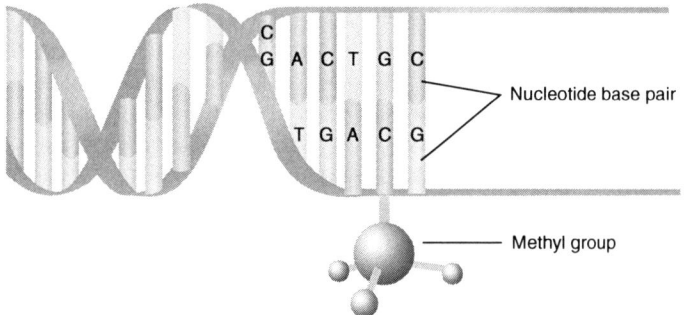

Figure 5.2   *Epigenetic tag (methyl group) on nucleotide base pairs of DNA.*
From Feist and Rosenberg (2012). Reprinted with permission.

interactions work to create much of the developmental and personality traits we observe in individuals (Bagot & Meaney, 2010; Depue, 2009; Kaminsky et al., 2008; Petronis, 2010; Svrakic & Cloninger, 2010; Kaminsky et al., 2008). Indeed, phenotypic differences between identical twins—whether it is in height, weight, eating habits, personality, or disease—is a prime example of epigenetics at work. DNA structures are identical in identical twins, and even though identical twins are always more alike than any two other people, with age and development even identical twins grow further apart. Epigenetics can and does explain these differences. With age twins have ever more diverging sets of life experiences, and hence different events tag different genes and silence them (Martin, 2005). Depending on the genes that get silenced, different phenotypic traits are manifested in one twin compared to the other.

As we have already seen, intelligence is also a necessary but not sufficient condition for scientific ability—the average IQ for scientists is in the 120 to 130 range or 1½ to 2 standard deviations above the mean. Recent empirical evidence demonstrates epigenetic influences on intelligence. For example, toxins ingested by the mother, either intentionally or unintentionally, may epigenetically influence the child's intelligence. Alcohol, drugs, and viral infections in a pregnant woman can seriously lower her child's overall intelligence (Jacobson & Jacobson, 2000; Ruff, 1999; Steinhausen & Spohr, 1998). For example, heavy alcohol consumption during pregnancy can lead to mental retardation in the child (Streissguth, Barr, Sampson, Darby, & Martin, 1989). Prenatal exposure to high levels of lead, mercury, or manganese may lead to serious impairments in a child's intelligence (Dietrich, Succop, Berger, & Hammond, 1991; Jacobson & Jacobson, 2000). The mechanism by which these chemicals alter brain development may well be epigenetic—tags attaching to DNA and muting gene expression.

Brain Influences

Just as epigenetics has revolutionized our view of the interplay between nature and nurture at a genetic level, brain plasticity has done the same at a neuroscientific level. Our brains are very much shaped by and are products of our environment, in particular during fetal development and the first few years of postnatal development (Baltes, Reuter-Lorenz, & Rösler, 2006; Perry, 2002). Personality traits, too, stem from individual differences in central nervous system (CNS) activity, both at a structural and chemical level (Bouchard & Loehlin, 2001; Eysenck, 1990, 1995; Kruger & Johnson, 2008). That is, personality dispositions partly result from differences in brain activity and in neurochemistry.

The history of neurology is replete with examples of personality change as a result of disease or injury to the brain—Phineas Gage being one of the best-known examples. Indeed, Gage's injury to his orbital-frontal brain area was the first clear indication of this region's involvement in personality and impulse control.

Later research expanded our understanding of the frontal area in the higher reaches of human nature, namely consciousness, creativity, personality, and morality (Dunbar, 1993; Fuster, 2002; Krasnegor, Lyon, & Goldman-Rakic, 1997; Miller & Cummings, 1999; Mithen, 1996; Stone, Baron-Cohen, & Knight, 1998; Stuss, Picton, & Alexander, 2001).

Eysenck (1982, 1990) was among the first personality theorists to argue for a biological basis of individual differences, with differences in both CNS anatomy and neurochemistry accounting for the major personality traits of neuroticism and extraversion. One of his fundamental arguments, for example, was that introversion–extraversion stems from differences in CNS sensitivity and responsivity. Compared to extraverts, introverts have higher baseline levels of arousal and are more reactive to sensory stimulation than extraverts and, hence, they regulate their optimal levels of arousal by avoiding and withdrawing from overly stimulating situations. These differences are partially controlled by the reticular activating system in the midbrain, hindbrain region. Eysenck's well-known model of personality proposed psychoticism to be a third super-factor of personality. People high in psychoticism are cold, aloof, eccentric, hostile, impulsive, and egocentric (Eysenck, 1982, 1990).

With the recent ubiquity of brain-imaging technology, neuroscientists have recently turned their attention to the neuroanatomical underpinnings of normal personality functioning. The Big-Five Model is again the personality taxonomy of choice. The dimensions of N, E, and A have the most support. For example, Canli (2004, 2008, 2009) and colleagues have been most active in uncovering the neurological systems most involved in Big-Five personality differences, especially in extraversion and neuroticism. Canli, makes clear, however, that personality traits per se operate at too broad a level to be strongly associated with specific molecular and cellular processes. This rift in level of analysis will necessarily attenuate the association between specific genotype and global phenotype. The solution, therefore, is to examine the more specific cognitive, affective, and motivational underpinnings of personality and how they are connected to specific molecular and cellular variations between individuals. These more specific cognitive and affective processes are dubbed "endophenotypes" (Gottesman & Gould, 2003). When this approach is taken, aspects of neuroticsim (anxiety, depression, emotionality) are associated with the amygdala and anterior cingulate cortex (ACC)

(Canli, 2008). In addition, aspects of extraversion influence reaction times, attention, and preferences for emotionally positive rather than neutral or negative stimuli (Canli, 2008; Derryberry & Reed, 1994).

In addition to neuroanatomical bases of personality, neurochemical variability also accounts for individual differences in people's personality—in particular, the serotonergic and dopaminergic systems. One of the more established neurochemical differences is serotonin and its relative 5-HTT levels and differences on openness to experience (Ebstein et al., 1996; Kalbitzer et al., 2009). Openness to experience is marked by seeking and preferring variety of experience compared to wanting closure. Open people tend to be imaginative, curious, and creative (McCrae, 1987). Different serotonergic markers are associated with anxiety and depression (neuroticism) (Lesch et al., 1996).

Similarly, individual differences in dopamine levels are associated with differences in thrill-seeking and exciting behaviors. People with this trait may seek highly exciting activities like bungee jumping, mountain climbing, or scuba diving. Thrill-seeking activities create a "rush" of excitement—a positive feeling that may be related to the release of dopamine, a neurotransmitter associated with physiological arousal. Given the possible connection between dopamine and thrill-seeking, one theory suggests that people who are deficient in dopamine will tend to seek out exciting situations as a way of increasing their dopamine release and making up for deficient levels of dopamine.

In the mid-1990s, researchers presented the first genetic evidence to support this theory. The gene DRD4 is involved in dopamine production in the limbic system, and the longer the gene sequence, the less efficient dopamine production is. In other words, long versions of the DRD4 gene are associated with less efficient dopamine production. If the theory is correct, people who seek out thrills should have the longer form of this gene, and that is exactly what the research has shown (Ebstein et al., 1996; Hamer & Copeland, 1998). An exciting aspect of this finding is that it was the first to demonstrate a specific genetic influence on a normal (nonpathological) personality trait. I should also point out that neurotransmitters are ultimately products of gene expression and therefore are under genetic control (e.g., Gottschalk & Ellis, 2010; Kim, Lee, Rowan, Brahim, & Dionne, 2006).

With advances in neuroscience, evolutionary psychology, genetics, and now epigenetics, the extent to which the old-fashioned nature versus nurture argument is inherently misguided becomes all the more clear. As Petronis (2010) puts it: "In the domain of epigenetics, the line between 'inherited' and 'acquired' is fuzzy. Stable epigenetic 'nature' merges fluidly with plastic epigenetic 'nurture'" (p. 725).

The biological bases of personality—genetics, epigenetics, brain structure, and neurochemistry—shape individual differences in behavior, that is, personality. These are the biological bases of personality differences, which in turn lay the foundation for differences in people's interest in and talent for science, math, and technology. The biological-functional model of personality is quite useful in interpreting and explaining the personality findings on scientific behavior, interest, and talent. My basic argument is that just as biological mechanisms shape personality differences, these personality differences in turn shape various aspects of scientific thought and behavior. More specifically, it shapes interest in science, achievement, and creativity in science and theoretical orientation. In the following sections, I review the empirical for each of these claims.

PERSONALITY INFLUENCES ON SCIENTIFIC INTERESTS

The first step toward becoming a scientist is simply having a scientific interest in how the world works—be it the physical world, the biological world, or the social world. There are numerous factors that influence interest in science, including, but not limited to: need for cognition, gender, age first developed career interest, and being a recent immigrant (e.g., Ceci & Williams, 2010; Feist, 2006a, 2012; Lippa, 1998; O'Brien, Martinez-Pons, Kopala, 1999; Schoon, 2001). As it turns out, personality dispositions have something to do with whether or not one becomes interested in science as a career choice. Those most interested in science are characterized by higher than normal levels of three personality traits: conscientiousness, openness to experience, and introversion (Bachtold, 1976; Charyton & Snelbecker, 2007; Eiduson, 1962; Feist, 1998, 2012; Wilson & Jackson, 1994). For example, in a sample of more than 600 college students, Feist (2012) found that those most interested in science either as a hobby or a profession were more conscientious, open, and introverted than students who were not interested in science (cf. Albert & Runco, 1987; Bachtold, 1976; Barton, Modgil, & Cattell, 1973, Kline & Lapham, 1992; Perrine & Brodersen, 2005; Wilson & Jackson, 1994).

In addition, in Feist's (1998) meta-analysis of personality and creativity literature, he found the two strongest effect sizes (medium in magnitude) to be for the positive and negative poles of conscientiousness (C). Highly conscientious people are careful, cautious, fastidious, reliable and self-controlled. Low conscientiousness only had five samples compared and consisted of only two personality scales, namely "direct expression of needs" and the MMPI's "psychopathic deviant" scale. Both of these are measures of impulse control or the lack thereof. In addition, scientists tend to be slightly more conventional, rigid, and socialized relative to nonscientists (low in openness). The median effect size was 0.30. There was, however, also a small but robust effect size between science and high openness to experience ($d = 0.11$), which consists of traits such as creative, curious, intelligent, and having wide interests. Similarly, scientists on average were slightly more introverted (deferent, reserved, introverted, and dependent) than nonscientists, with a median effect size of 0.26. Finally, examining the effect sizes of the two subcomponents of extraversion separately (confidence and sociability), the confidence component had a small positive effect, and the sociability component a near zero negative effect.

Recent evidence from more than 2,000 physicists and 78,000 nonscientists confirms the robust role that conscientiousness and introversion (opposite pole of neuroticism) play in shaping interest in science and careers in science (Lounsbury et al., 2012). Physicists were less conscientious and more introverted than nonscientists. But there were some inconsistent results with the meta-analytic findings as well: Physicists were more open and less emotionally stable (i.e., higher in neuroticism) than nonscientists. In short, the FFM dimensions of openness, confidence/dominance (E), introversion, and conscientiousness and discipline appear to be the personality factors that make scientific interest most likely.

One problem with the research on personality and scientific interest is that it is not specific to any specific domain of science but rather covers scientists in general. Very little, if any, research has compared the personality dispositions of physical, biological, and social scientists to examine whether the social scientists have more sociable and extraverted personalities compared to their physical scientist peers. Indeed, I contend that one's preference and orientation toward people or things plays a crucial role in the kind of science in which one becomes interested, especially physical or social science (see Zimmerman and Croker, Chapter 3 in this volume, for a discussion of the development of the social and physical domains of knowledge in children, adolescents, and adults).[1] As discussed, the foundation for the People–Thing orientation comes from the vocational interest in literature. Dale Prediger was the first to modify John Holland's hexagonal model of vocational interests onto two basic dimensions: People–Things and Data–Ideas. The "People" end of the dimension is mapped onto Holland's "Social" career types, whereas the "Thing" end of the dimension is mapped onto "Realistic" career types. According to Holland the social career type prefers occupations that involve informing, training, and enlightening other people. The realistic career type, however, prefers careers that involve manipulating things, machines, objects, tools, and animals (Holland, 1992; Lippa, 1998; Prediger, 1982).

Supporting this People-Thing view of scientific interest, Simon Baron-Cohen and his colleagues have found that engineers, mathematicians, and physical scientists score much higher on measures of high-functioning autism (Asperger's syndrome) than nonscientists, and that physical scientists, mathematicians, and engineers are higher on nonclinical measure of autism than social scientists. Moreover, autistic children are more than twice as likely as nonautistic children to have a father or grandfather who was an engineer (Baron-Cohen et al., 1997, 1998, 1999, 2001). In other words, physical scientists often have temperaments that orient them away from the social and toward the inanimate—their interest and ability in science is then just one expression of this orientation.

Feist, Batey, and Kempel (in preparation) recently replicated these findings and reported higher scores on the Austism Spectrum Scale (AQ) in STEM scientists (science, technology, engineering, and mathematics) than in social scientists (psychologists, sociologists, and anthropologists) in both the United States and the United Kingdom. In particular, STEM scientists were most elevated on behaviors associated on the AQ "Communication" and "Social Skills" subscales. Items on the former included behaviors such as "not enjoying chit-chat," and "not knowing how to keep phone conversations going," whereas items on the latter include behaviors such as "not finding social situations easy," and "preferring to go to the library over a party." Interestingly, cognitive scientists in the United Kingdom were higher than even their STEM peers on these two subscales and on the overall AQ. The U.S. sample confirmed the elevated score for cognitive scientists relative to their peers on the overall AQ scale, but showed only nonsignificant trends ($P. < .10$) on the Communication and Social Skills subscales (Feist et al., in preparation).

These findings confirm the idea that STEM scientists have more difficulty in social situations and social behaviors than social scientists. One point to make, however, is these are subclinical forms of the syndrome and for the most part STEM scientists would not be diagnosed formally with Aspergers' syndrome.

Moreover, becoming even more specific, there is research within rather than between domains of science that demonstrate the importance of personality on distinguishing those who gravitate toward the "harder" and "softer" sides of the profession, namely the research and the applied sides. Clinical psychology, for example, is one of the few disciplines, along with medicine, that has a clear "science-application" split. Indeed, a major concern for PhD programs in clinical psychology is the high rate of students who are not interested in science and research. Clinical psychology students on the whole tend to be more people-oriented (applied) than investigative and research-oriented (Mallinckrodt, Gelso, & Royalty, 1990; Zachar & Leong, 2000). And yet, Clinical Psychology PhD programs are mostly built around the "Boulder Model" which requires a balance in training between the applied-clinical side and the science-research side.

The question, therefore, becomes, "What predicts interest in science-research in these clinical psychology graduate students and can this interest be increased by particular kinds of training environments?" The general conclusion from the studies on this question is that one of the strongest predictors in interest in research (or lack thereof) is personality-vocational interest and that training environment plays but a modest role in increasing interest in research (Kahn & Scott, 1997; Mallinckrodt, et al., 1990; Royalty & Magoon, 1985; Zachar & Leong 1992, 1997). For example, a study by Mallinckrodt and colleagues examined the impact of training environment, personality-vocational interest, and the interaction between the two on increasing research interest. They found that personality-vocational interest was a stronger predictor than research environment in increasing interest in research over the course of graduate training.

Gaps in the People-Thing and domain of science literature do exist. Of most interest would be developmental research that examined whether a preference for things is evident early in life for future physical scientists, and, likewise, whether a preference for people is evident early in life for future social scientists. Similarly, cross-cultural work showing the same association between thing-orientation and physical science and social-orientation and social science the world over would be quite valuable. Therefore, the next line of research for the personality psychology of science is to explore differences in personality between physical, biological, and social scientists. Based on the evidence just cited, my prediction is that the physical scientists as a group will be more introverted and thing-oriented (i.e., have more developed implicit physical intelligence) than the biological scientists, who in turn will be less sociable and extraverted than social scientists (i.e., have more developed implicit social intelligence).

PERSONALITY INFLUENCES ON SCIENTIFIC CREATIVITY AND TALENT

Not only do certain traits lower thresholds for scientific interest, but a somewhat different pattern of traits also lowers thresholds for scientific creativity and talent. The meta-analysis conducted by Feist (1998) also addressed the question of which traits make creativity and eminence in science more likely and what were their magnitudes of effect. The traits can be arranged into four psychologically meaningful categories: cognitive, motivational, social, and clinical.

Cognitive Traits

Creative and eminent scientists tend to be more open to experience and more flexible in thought than less creative and eminent scientists. Many of these findings stem from data on the flexibility (Fe) and tolerance (To) scales of the California Psychological Inventory (Feist & Barron, 2003; Garwood, 1964; Gough, 1961; Helson, 1971; Helson & Crutchfield, 1970; Parloff & Datta, 1965). The Fe scale, for instance, taps into flexibility and adaptability of thought and behavior as well as the preference for change and novelty (Gough, 1987). The few studies that have reported either no effect or a negative effect of flexibility in scientific creativity have been with student samples (Davids, 1968; Smithers & Batcock, 1970).

Feist and Barron (2003), for example, examined personality, intellect, potential, and creative achievement in a 44-year longitudinal study and predicted that personality would explain unique variance in creativity over and above that already explained by intellect and potential. They found that observer-rated potential and intellect at age 27 predicted lifetime creativity at age 72, and yet personality variables (such as tolerance and psychological mindedness) explained up to 20% of the variance over and above potential and intellect. Specifically, two measures of personality—California Psychological Inventory Scales of Tolerance (To) and Psychological Mindedness (Py)—resulted in the 20% increase in variance explained (20%) over and above potential and intellect. The more tolerant and psychologically minded the student was, the more likely he was to make creative achievements over his lifetime. Together, the four predictors (potential, intellect, tolerance, and psychological mindedness) explained a little more than a third of the variance in lifetime creative achievement.

Motivational Traits

The most eminent and creative scientists also tend to be more driven, ambitious, and achievement oriented than their less eminent peers (Adelson, 2003;

Busse & Mansfield, 1984, Helmreich, Spence, Beane, Lucker, & Matthews, 1980). Busse and Mansfield (1984), for example, studied the personality characteristics of 196 biologists, 201 chemists, and 171 physicists, and commitment to work (i.e., "need to concentrate intensively over long periods of time on one's work") was the strongest predictor of productivity (i.e., publication quantity) even when holding age and professional age constant. Helmreich et al. (1980) studied a group of 196 academic psychologists and found that different components of achievement and drive had different relationships with objective measures of attainment (i.e., publications and citations). With a self-report measure, they assessed three different aspects of achievement: "mastery" preferring challenging and difficult tasks; "work" enjoying working hard; and "competitiveness" liking interpersonal competition and bettering others. According to Amabile's (1996) well-known typology, the first two measures could be classified as "intrinsic motives" and the last measure could be an "extrinsic motive." Helmreich and his colleagues found that mastery and work were positively related to both publication and citation totals, whereas competitiveness was positively related to publications but negatively related to citations. Being intrinsically motivated (mastery and work) appears to increase one's productivity and positive evaluation by peers (citations), whereas wanting to be superior to peers leads to an increased productivity, and yet a lower positive evaluation by peers. The inference here is that being driven by the need for superiority may backfire in terms of having an impact on the field. Indeed, in a further analysis of the male psychologists in their 1980 data set, Helmreich and colleagues (Helmreich, Spence, & Pred, 1988) factor analyzed the Jenkins Activity Survey and extracted an achievement striving factor and an impatience/irritability factor. Achievement striving was positively related to both citation and publication counts, whereas impatience/irritability was related to neither publications nor citations. Finally, Heller (2007) recently reported a longitudinal study in which motivation and drive were the strongest 17-year predicts of creativity and success in science ($d = 0.71$).

Social Traits

In the highly competitive world of science, especially big science, where the most productive and influential continue to be rewarded with more and more of the resources, success is more likely for those who thrive in competitive environments, that is for the dominant, arrogant, hostile, and self-confident. For example, Van Zelst and Kerr (1954) collected personality self-descriptions on 514 technical and scientific personnel from a research foundation and a university. Holding age constant, they reported significant partial correlations between productivity and describing oneself as "argumentative," "assertive," and "self-confident." In one of the few studies to examine female scientists, Bachtold and Werner (1972) administered Cattell's 16 Personality Factor to 146 women scientists and found that they were significantly different from women in general on 9 of the 16 scales, including dominance (Factor E) and self-confidence (Factor O). Similarly, Feist (1993) reported a structural equation model of scientific eminence in which the path between observer-rated hostility and eminence was direct and the path between arrogant working style and eminence was indirect but significant.

The scientific elite also tend to be more aloof, asocial, and introverted than their less creative peers. In a classic study concerning the creative person in science, Roe (1952, 1953) found that creative scientists were more achievement oriented and less affiliative than less creative scientists. In another seminal study of the scientific personality, Eiduson (1962) found that scientists were independent, curious, sensitive, intelligent, emotionally invested in intellectual work, and relatively happy. Similarly, Chambers (1964) reported that creative psychologists and chemists were markedly more dominant, ambitious, self-sufficient, and had more initiative compared to less creative peers. Helson (1971) compared creative female mathematicians with less creative female mathematicians, matched on IQ. Observers blindly rated the former as having more "unconventional thought processes," as being more "rebellious and nonconforming," and as being less likely to judge "self and others in conventional terms." Rushton, Murray, and Paunonen (1987) also conducted factor analyses of the personality traits most strongly loading on the "research" factor (in contrast to a "teaching" factor) in two separate samples of academic psychologists. Among other results, they found that "independence" tended to load on the research factor, whereas "extraversion" tended to load on the teaching factor.

Clinical Traits

One of the biggest changes in the field of personality and creativity over the last 10 years—besides the steady rise in neuroscientific studies—is the tremendous growth in research on personality disorders, mental health, and creative thought and behavior. The influences of mental health on creative thought and behavior are so robust now that I must add a new dimension to the three major trait groupings from my previous model. In addition to cognitive, social, and motivational-affective traits, I now include a clinical traits group that includes the normal personality dimension of psychoticism and its related concept of schizotypy. I should make a qualification, however. The evidence for the connection between clinical personality traits and creativity is stronger in the arts than in the sciences (Jamison, 1993; Ludwig, 1995; Rawlings & Larconini, 2008). The question is whether there are any elevated mental health problems in scientists compared to nonscientists. Indeed, historically science is littered with scientists who suffered some kind of psychological disorder, from Newton and Darwin to Tesla and Faraday.

For example, Sir Isaac Newton had what might today be diagnosed as a "schizoid" personality, that is, had a strong disinterest in social relationships and solitary lifestyle, and was emotionally cold and apathetic (Keynes, 1995). What is more certain is that Newton had two nervous breakdowns, one in 1678 and the second, more serious one, in 1693 (Westfall, 1980). "His psychological problems culminated in what would now be called a nervous breakdown in mid-1693, when, after five nights of sleeping 'not a wink', he temporarily lost all grip on reality and became convinced that his friends Locke and Pepys were conspiring against him" (Iliffe & Mandelbrot, 2012).

Charles Darwin apparently suffered a life-long battle with anxiety, agoraphobia, and depression. His anxiety took the form of panic attacks and he struggled with regular bouts of depression (Desmond & Moore, 1991).

In the field of technology, the inventor Nikola Tesla was the quintessence of the so-called "mad scientist." He was oversensitive in pretty much every sensory system—hearing, vision, taste, smell, and even touch (Pickover, 1998). Being a synesthete—someone who experiences secondary senses to primary sensations, such as sensing a smell when seeing a number—may have had something to do with this. He most certainly suffered from obsessive-compulsive disorder, one form of which took toward his compulsive collecting and caring for pigeons. Additionally, he could not eat, for instance, unless he first calculated the volume of the bowl or dish holding his food. It is equally clear, however, that his ideas were both brilliant and quacky. His brilliant ideas included inventing alternating current, microwaves, the loudspeaker, and the radio among other things. His quacky ideas included wanting to get rid of light (and its opposite, loving dark, i.e., scotophilia), extra-sensory perception (ESP), believing he could crack the entire earth like an apple, and that electricity was the best of all medical treatments (he even regularly "treated" himself with electricity and x-rays) (Pickover, 1998).

Anecdotes, however, do not tell us too much by themselves. There will always be individual cases of particular personalities in any and all professions. To answer the question of mental illness and science in a more general fashion we must examine the proportion of scientists who possess various disorders and compare that to the base-rate in the population. Let us, therefore, establish the base-rate of lifetime rate of mental health problem in the general population. That is, what percentage of people will suffer from at least one mental illness during the course of their lifetime?

The most recent survey reported that 46% of the American population in general suffer from at least one diagnosable mental health episode over the course of their lifetime (Kessler et al., 2005). By that standard, natural scientists, however, are less likely than other creative groups to suffer from mental illness (28% versus 59%; Ludwig, 1995; see Figure 5.3). Even social scientists are somewhat less likely to suffer one bout of psychological disorder over the course of their lives than other creative groups (51% versus 59%; Ludwig, 1995). Similarly, Karlson (1970) reported that only about one-fifth to one-quarter of eminent mathematicians came from families whose members exhibit significant levels of psychopathology (Karlson, 1970; Nettle, 2006). Recent self-report research using nonclinical assessment of schizotpy (i.e., eccentricity) also suggests that scientists are lower on unusual experiences and cognitive disorganization than artists and musicians (Rawlings & Locarnini, 2008). A person with schizotypal personality disorder is isolated and asocial, but in addition has very odd or magical thoughts and beliefs (APA, 2000). For instance, people with schizotypal personality disorder may believe that stories on television or in the newspaper were written directly about them or that people they don't know are saying things about them behind their backs. Additionally, Post (1994) studied the biographies of 291 men, world famous for creative

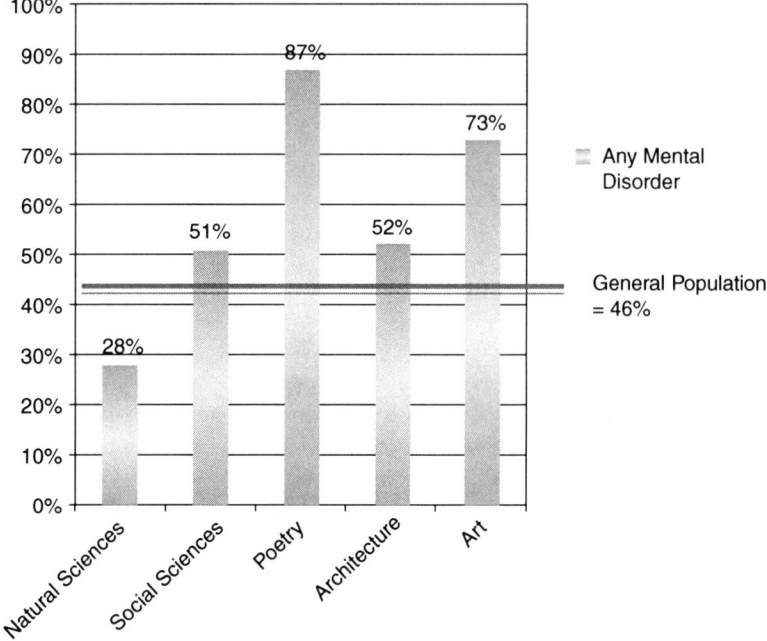

Figure 5.3   *Lifetime rate for any mental disorder for general population and creative professions.*
Adapted from Ludwig (1995).

exploits or political leadership. Among the 291, no group had fewer psychological abnormalities than the scientists. And yet at the same time, Post reported that only one-third of the world famous researchers did not suffer from any psychological disturbances or had only trivial problems. In Post's judgment, 45% of the scientists in his sample suffered from "marked" or "severe" psychopathology—which would put their rate of mental illness to be similar to the general population. Lastly, Ko and Kim (2008) recently reported an interesting Kuhnian twist to the relationship between psychopathology and science. They found that normal scientists who are motivated to preserve a scientific paradigm achieved higher eminence if they exhibited less psychopathology, but revolutionary scientists who rejected a paradigm attained higher eminence if they displayed more psychopathology.

From these results one is tempted to conclude that scientists overall are less prone to serious mental health difficulties than other creative groups (Simonton & Song, 2009), and that may well be the case. Before we draw the conclusion that creative scientists are less likely to suffer mental illness episodes than other creative professions, however, we must point out a caveat. It may be that science necessarily weeds out those with mental health problems in a way that art, music, and poetry do not. In other words, aspiring scientists with mental health problems will be less likely to get hired or finish their degrees than aspiring artists. Science requires regular focus and attention to problems over long periods of time; it also requires one to go into a lab or work on a problem consistently. If a person cannot work methodically and consistently, she or he is not likely to become a productive scientist.

To summarize the distinguishing traits of creative scientists: they are generally more open and flexible, driven and ambitious, and although they tend to be relatively asocial, when they do interact with others, they tend to be somewhat prone to arrogance, self-confidence, and hostility.

PERSONALITY AND THEORETICAL PREDILECTION

Personality qualities consistently lower thresholds for being interested in science in general as well as in particular domains of science. It also lowers thresholds for creative problem solving in science as well as milder forms of mental illness. The last question is whether patterns of personality make theoretical orientations more or less likely. The work on personality can shed light on theory acceptance and even theory creation. Or stated as a question: Do certain personality styles predispose a scientist to create, accept and/or reject certain kinds of theories? The first work on this question was in the mid-1970s by Atwood and Tomkins (1976), who showed through case studies how the personality of the theorist influenced his or her theory of personality. More systematic empirical investigations have expanded this work and have demonstrated that personality influences not only theories of personality, but also how quantitatively or qualitatively oriented and productive psychologists are (Arthur, 2001; Atwood & Tomkins, 1976; Conway, 1988; Costa, McCrae, & Holland, 1984; Hart, 1982; Johnson, Germer, Efran, & Overton, 1988; Simonton, 2000). One general finding from these studies is that psychologists who have more objective and mechanistic theoretical orientations are more rational and extraverted than those who have more subjective and humanistic orientations. For instance, Johnson and colleagues collected personality data on four groups of psychologists (evolutionary-sociobiologists, behaviorists, personality psychologists, and developmental psychologists) and found that distinct personality profiles were evident in the different theoretical groups. That is, scientists who were more holistic, purposive, and constructivist in orientation were higher on the Empathy, Dominance, Intellectual Efficiency, and Flexibility scales of the California Psychological Inventory and the Intuition scale of the Myers-Briggs Type Indicator (MBTI). However, most of these studies have been with psychologists, so answering the question of whether these results generalize to the biological and natural sciences remains a task for future psychologists of science.

William Runyan (Chapter 14, this volume) does a wonderful job in detailing the psychobiographical patterns in the lives of philosopher-mathematicians (Betrand Russell), psychoanalysts (Freud, Horney, and Murray), a behaviorist (Skinner), apsychometrician (Meehl), and statisticians (Fisher and Neyman). The life histories and biographical richness of these scientists and theoreticians makes it clear that personality differences are one way in which scientific interest and achievement are channeled down different paths of science.

SUMMARY AND CONCLUSIONS

In this chapter, I have summarized and integrated key findings from the personality psychology of science.

- Personality functions to lower behavioral thresholds and make trait-consistent behaviors more likely
- Genetic, epigenetic, and neuropsychological mechanisms lay the foundation for individual differences in personality
- Conscientiousness, self-confidence, dominance, and openness are the traits most likely to lower the thresholds for scientific interest
- Being thing-oriented and introverted lowers one's threshold for interest in and talent for physical science, especially cognitive science, whereas being people-oriented, psychologically minded, and extraverted do the same for the social science interest and talent
- Cognitive traits (e.g., tolerance and flexibility), social traits (e.g., dominance, arrogance, hostility), and motivational traits (e.g., driven, intrinsically motivated) each lower thresholds for scientific creativity and achievement
- Finally, creative scientists tend to be more emotionally stable than other creative groups (e.g., artists), but scientists in general are more prone to milder forms on the autism spectrum than nonscientists

A concluding question will have to remain just that, a question: Do these patterns of personality cause scientific interest and creativity or are they effects of them? My biological-functional model implies that personality acts in a causal manner on scientific interest and achievement, but it is also possible that these traits get exacerbated and reinforced by becoming a scientist, especially an eminent scientist. The classic chicken-egg problem awaits solid longitudinal research that measures personality and interest in and talent for science along the life span. Only then will causal directionality have a chance of being determined.

As the results of this chapter should make clear, however, the opening question of whether scientific personality exists can be answered strongly in the affirmative. Scientific traits of personality, as they do in all of us, lower behavior threshold and make behaviors consistent with our most likely traits. Personality is a necessary piece of the larger "psychology of science puzzle" (Feist, 2006b; Feist & Gorman, 1998; Gholson, Shadish, Neimeyer, & Houts, 1999; Roe, 1965; Simonton, 1998).

NOTES

1. The Nobel Prize–winning psychologist Kahneman (2002) offers a nice anecdote for the development of his interest in people:

> I will never know if my vocation as a psychologist was a result of my early exposure to interesting gossip, or whether my interest in

gossip was an indication of a budding vocation. Like many other Jews, I suppose, I grew up in a world that consisted exclusively of people and words, and most of the words were about people. Nature barely existed, and I never learned to identify flowers or to appreciate animals. But the people my mother liked to talk about with her friends and with my father were fascinating in their complexity. Some people were better than others, but the best were far from perfect and no one was simply bad. Most of her stories were touched by irony, and they all had two sides or more. (http://nobelprize.org/nobel_prizes/economics/laureates/2002/kahneman-autobio.html)

REFERENCES

Adelson, B. (2003). Issues in scientific creativity: Insight, perseverance and personal technique profiles of the 2002 Franklin institute laureates. *Journal of the Franklin Institute, 340*(3), 163–189.

Albert, R., & Runco, M. (1987). The possible different personality dispositions of scientists and non-scientists. In D. N. Jackson & J. P. Rushton (Eds.), *Scientific excellence* (pp. 67–97). Beverly Hills, CA: Sage Publication.

Allport, G. (1937). *Personality: A psychological interpretation.* New York, NY: Holt, Rinehart, & Winston.

Amabile, T. (1996). *Creativity in context.* New York, NY: Westview.

Arthur, A. R. (2001). Personality, epistemology and psychotherapists' choice of theoretical model: A review and analysis. *European Journal of Psychotherapy, Counseling and Health, 4*, 45–64.

Atwood, G. E., & Tomkins, S. S. (1976). On subjectivity of personality theory. *Journal of the History of the Behavioral Sciences, 12*, 166–177.

Bachtold, L. M. (1976). Personality characteristics of women of distinction. *Psychology of Women Quarterly, 1*, 70–78.

Bachtold, L. M., & Werner, E. E. (1972). Personality characteristics of women scientists. *Psychological Reports, 31*, 391–396.

Bagot, R. C., & Meaney, M. J. (2010). Epigenetics and the biological basis of gene x environment interactions. *Journal of the American Academy of Child & Adolescent Psychiatry, 49*, 752–771.

Baltes, P. B., Reuter-Lorenz, P. A., & Rösler, F. (Eds.). (2006). *Lifespan development and the brain: The perspective of biocultural co-constructivism.* New York, NY: Cambridge University Press.

Baron-Cohen, S., Wheelwright, S., Stott, C., Bolton, P., & Goodyer, I. (1997). Is there a link between engineering and autism? *Autism, 1*, 101–109.

Baron-Cohen, S. Bolton, P., Wheelwright, S., Short, L., Mead, G., Smith, A., & Scahill, V. (1998). Autism occurs more often in families of physicists, engineers, and mathematicians. *Autism, 2*, 296–301.

Baron-Cohen, S., Wheelwright, S., Stone, V., & Rutherford, M. (1999). A mathematician, a physicist, and a computer scientist with Asperger syndrome: Performance on folk psychology and folk physics tests. *Neurocase, 5*, 475–483.

Baron-Cohen, S., Wheelwright, S., Skinner, R., Martin, J., & Clubley, E. (2001). The Autism-Spectrum Quotient (AQ): Evidence from Asperger syndrome/high-functioning autism, males and females, scientists and mathematicians. *Journal of Autism & Developmental Disorders, 31*, 5–17.

Barres, R., Yan, J., Egan, B., Treebak, J. T., Rasmussen, M., Fritz, T.,…Zierath, J. R. (2012). Acute exercise remodels promoter methylation in human skeletal muscle. *Cell Metabolism, 15*, 405–411.

Barton, K., Modgil, S., & Cattell, R. B. (1973). Personality variables as predictors of attitudes toward science and religion. *Psychological Reports, 32*, 223–228.

Batey, M., & Furnham, A. (2006). Creativity, intelligence, and personality: A critical review of the scattered literature. *Genetic, Social and General Psychology Monographs, 132*, 355–429.

Benbow, C. P., Lubinski, D., Shea, D. L., & Eftekhari-Sanjani, H. E. (2000). Sex differences in mathematical reasoning ability at age 13: Their status 20 years later. *Psychological Science, 11*, 474–480.

Block, J. (1977). Recognizing the coherence in personality. In D. Magnusson & N. D. Endler (Eds.). *Personality at the crossroads: Current issues in interactional psychology.* Hillsdale, NJ: Erlbaum & Associates.

Brody, N., & Ehrlichman, H. (1998). *Personality psychology: The science of individuality.* Upper Saddle River, NJ: Prentice Hall.

Bouchard, T. J., Jr., & Loehlin. (2001). Genes, evolution, and personality. *BehavioralGenetics, 31*, 243–273.

Brody, N., & Ehrlichman, H. (1998). *Personality psychology: The science of individuality.* Upper Saddle River, NJ: Prentice Hall.

Busse, T. V. & Mansfield, R. S. (1984). Selected personality traits and achievement in male scientists. *The Journal of Psychology, 116*, 117–131.

Canli, T. (2004). Functional brain mapping of extraversion and neuroticism: Learning from individual differences in emotion processing. *Journal of Personality, 72*, 1105–1132.

Canli, T. (2008). Toward a "molecular psychology" of personality. In O. P. John, R. W. Robins, & L. A. Pervin (Eds.,). *Personality handbook: Theory and research* (pp. 311–327). New York, NY: Guilford Press.

Canli, T. (2009). Neuroimaging of personality. In P. J. Corr, G. Matthews, P. J. Corr, & G. Matthews (Eds.), *The Cambridge handbook of personality psychology* (pp. 305–322). New York, NY: Cambridge University Press.

Carey, G., Goldsmith, H. H., Tellegen, A., & Gottesman, I. I. (1978). Genetics and personality inventories: The limits of replication with twin data. *Behavior Genetics, 8*, 299–313.

Carere, C., Drent, P. J., Koolhaas, J. M., & Groothuis, T. G. G. (2005). Epigenetic effects on personality traits: Early food provisioning and sibling competition. *Behaviour, 142*, 1329–1355.

Caspi, A., Sugden, K., Moffitt, T. E., Taylor, A., Craig, I. W., Harrington, H., McClay, J.,…Poulton, R. (2003). Influence of life stress on depression: Moderation by a polymorphism in the 5-HTT gene. *Science, 301*, 386–389.

Ceci, S. J., & Williams, W. M. (2010). *The mathematics of sex: How biology and society conspire to limit talented women and girls.* New York, NY: Oxford University Press.

Chambers, J. A. (1964). Relating personality and biographical factors to scientific creativity. *Psychological Monographs: General and Applied, 78*, 1–20.

Charyton, C., & Snelbecker, G. E. (2007). Engineers' and musicians' choices of self-descriptive adjectives as potential indicators of creativity by gender and domain. *Psychology of Aesthetics, Creativity, and the Arts, 1*, 91–99.

Chorney, M. J., Chorney, K., Seese, N., Owen, M. J., Daniels, J., McGuffin, P., Thompson, L. A.,…Plomin, R. (1998). A quantitative trait locus associated with cognitive ability in children. *Psychological Science, 9*, 159–166.

Conway, J. B. (1988). Differences among clinical psychologists: Scientists, practitioners, and science-practitioners. *Professional Psychology: Research and Practice, 19*, 642–655.

Coon, H., & Carey, G. (1989) Genetic and environmental determinants of musical ability in twins. *Behavioral Genetics, 19*, 183–193.

Costa, P., & McCrae, R. R. (1995). Solid ground in the wetlands of personality: A reply to Block. *Psychological Bulletin, 117*, 216–220.

Costa, P. T., McCrae, R. R., & Holland, J. L. (1984). Personality and vocational interests in an adult sample. *Journal of Applied Psychology, 69*, 390–400.

Crowley, K., Callanan, M. A., Tenenbaum, H. R., & Allen, E. (2001). Parents explain more often to boys than to girls during shared scientific thinking. *Psychological Science, 12,* 258–261.

Davids, A. (1968). Psychological characteristics of high school male and female potential scientists in comparison with academic underachievers. *Psychology in the Schools, 3,* 79–87.

Depue, R. A. (2009). Genetic, environmental, and epigenetic factors in the development of personality disturbance. *Development and Psychopathology, 21,* 1031–1063.

Derryberry, D., & Reed, M. A. (1994). Temperament and attention: Orienting towards and away from positive and negative signals. *Journal of Personality and Social Psychology, 66,* 1128–1139.

Desmond, A., & Moore, J. (1991). *Darwin: The life of a tormented evolutionist.* New York, NY: Warner Books.

Dietrich, K., Succop, P., Berger, O., & Hammond, P. (1991). Lead exposure and the cognitive development of urban preschool children: The Cincinnati Lead Study cohort at age 4 years. *Neurotoxicology and Teratology, 13*(2), 203–211.

Digman, J. M. (1990). Personality structure: Emergence of the five-factor model. *Annual Review of Psychology, 41,* 417–440.

Dingemanse, N. J., Both, C., Drent, P. J., Van Oers, K., & Van Noordwijk, A. J. (2002). Repeatability and heritability of exploratory behaviour in great tits from the wild. *AnimalBehaviour, 64,* 929–938.

Dunbar, R. I. M. (1993). Coevolution of neocortical size, group size and language in humans. *Behavioral and Brain Sciences, 16,* 681–735.

Ebstein, R. P., Novick, O., Umansky, R., Priel, B. Osher, Y., Blaine, D., ... Belmaker, R. H. (1996). Dopamine D4 receptor (D4DR) exon III polymorphism associated with the human personality trait of novelty seeking. *Nature Genetics, 1,* 78–80.

Eiduson, B. T. (1962). *Scientists: Their psychological world.* New York, NY: Basic Books.

Ekman, P. (1984). Expression and the nature of emotion. In K. R. Scherer & P. Ekman (Eds.). *Approaches to emotion,* (pp. 319–344). Hillsdale, NJ: Erlbaum.

Epstein, S. (1979). The stability of behavior: I. On predicting most of the people much of the time. *Journal of Personality and Social Psychology, 37,* 1097–1126.

Eysenck, H. J. (1982). *Personality, genetics and behavior: Selected papers.* New York, NY: Praeger.

Eysenck, H. J. (1990). Biological dimensions of personality. In L.A. Pervin (Ed.), *Handbook of personality theory and research* (pp. 244–276). New York, NY: Guilford.

Eysenck, H. J. (1995). *Genius: The natural history of creativity.* New York, NY: Cambridge University Press.

Eysenck, M. W., Mogg, K., May, J., Richards, A., & Mathews, A. (1991). Bias in interpretation of ambiguous sentences related to threat in anxiety. *Journal of Abnormal Psychology, 100,* 144–150.

Feist, G. J. (1993). A structural model of scientific eminence. *Psychological Science, 4,* 366–371.

Feist, G. J. (1998). A meta-analysis of the impact of personality on scientific and artistic creativity. *Personality and Social Psychological Review, 2,* 290–309.

Feist, G. J. (1999). Personality in scientific and artistic creativity. In R. J. Sternberg (Ed.). *Handbook of human creativity* (pp 273–296). Cambridge, England: Cambridge University Press.

Feist, G. J. (2006a). The development of scientific talent in Westinghouse finalists and members of the National Academy of Sciences. *Journal of Adult Development, 13,* 23–35.

Feist, G. J. (2006b). *The psychology of science and the origins of the scientific mind.* New Haven, CT: Yale University Press.

Feist, G. J., & Barron, F. X. (2003). Predicting creativity from early to late adulthood: Intellect, potential and personality. *Journal of Research in Personality, 37,* 62–88.

Feist, G. J. (2010). The function of personality in creativity: The nature and nurture of the creative personality. In J. C. Kaufman & R. J. Sternberg (Eds.), *Cambridge handbook of creativity* (pp. 113–130). New York, NY: Cambridge University Press.

Feist, G. J. (2012). Predicting interest in and attitudes toward science from personality and need for cognition. *Personality and Individual Differences, 52,* 771–775.

Feist, J., & Feist, G. J. (2009). *Theories of personality.* New York, NY: McGraw-Hill.

Feist, G. J., & Gorman, M. E. (1998). Psychology of science: Review and integration of a nascent discipline. *Review of General Psychology, 2,* 3–47.

Feist, G. J., & Rosenberg, E. L. (2012). *Psychology: Perspectives and connections* (2nd ed.). New York, NY: McGraw-Hill

Feist, G. J., Batey, M., & Kempel, C. (in preparation). Psychological profiles of STEM and social scientists in the United States and in the United Kingdom. Department of Psychology, San Jose State University.

Funder, D. C. (1991). Global traits: A neo-Allportian approach to personality. *Psychological Science, 2,* 31–39.

Fuster, J. M. (2002). Frontal lobe and cognitive development. *Journal of Neurocytology, 31,* 373–385.

Garwood, D. S. (1964). Personality factors related to creativity in young scientists. *Journal of Abnormal and Social Psychology, 68,* 413–419.

Gholson, B., Shadish, W. R., Neimeyer, R. A., & Houts, A. C. (Eds.). (1999). *Psychology of science: Contributions to metascience.* Cambridge, England: Cambridge University Press.

Gibson, J., & Light, P. (1992). Intelligence among university scientists. In R. S. Albert (Ed.), *Genius and eminence* (2nd ed., pp. 109–111). Elmsford, NY: Pergamon Press.

Goldberg, L. R., & Rosolack, T. K. (1994). The Big Five factor structure as an integrative framework: An empirical comparison with Eysenck's P-E-N model. In C. F. Halverson, Jr., G. A. Kohnstamm, & R. P. Martion (Eds.). *The developing structure of temperament and personality from infancy to adulthood* (pp. 7–35). Hillsdale, NJ: Erlbaum.

Gosling, S. D., & John, O. P. (1999). Personality dimensions in non-human animals: A cross-species review. *Current Directions in Psychological Science, 8,* 69–75.

Gottesman, I. I., & Gould, T. D. (2003). The endophenotype concept in psychiatry: Etymology and strategic intentions. *American Journal of Psychiatry, 160,* 636–645.

Gottschalk, M., & Ellis, L. (2010). Evolutionary and genetic explanations of violent crime. In C. J. Ferguson (Ed.). *Violent crime: Clinical and social implications* (pp. 57–74). Thousand Oaks, CA: Sage Publications.

Gough, H. G. (1961, February). *A personality sketch of the creative research scientist.* Paper presented at 5th Annual Conference on Personnel and Industrial Relations Research, UCLA, Los Angeles, CA.

Gough, H. G. (1987). *California Psychological Inventory: Administrators guide.* Palo Alto, CA: Consulting Psychologists Press.

Grigorenko, E. (2000). Heritability and intelligence. In R. J. Sternberg (Ed.), *Handbook of intelligence* (pp. 53–91). New York, NY: Cambridge University Press.

Hamer, D., & Copeland, P. (1998). *Living with our genes.* New York, NY: Anchor Books.

Hany, E. A., (1994). The development of basic cognitive components of technical creativity: A longitudinal comparison of children and youth with high and average intelligence. In R. F. Subotnik & K. D. Arnold (Eds.) *Beyond Terman: Contemporary longitudinal studies of giftedness and talent* (pp. 115–154). Norwood, NJ: Ablex.

Hart, J. J. (1982). Psychology of the scientists: XLVI: Correlation between theoretical orientation in psychology and personality type. *Psychological Reports, 50,* 795–801.

Haworth, C. M. A., Wright, M. J., Martin, N. W., Martin, N. G., Boomsma, D. I., Bartels, M., Posthuma, D.,…Plomin, R. (2009). A twin study of the genetics of high cognitive ability selected from 11,000 twin pairs in six studies from four countries. *Behavioral Genetics, 39,* 359–370.

Helmreich, R. L., Spence, J. T., & Pred, R. S. (1988). Making it without losing it: Type A, achievement motivation and scientific attainment revisited. *Personality and Social Psychology Bulletin, 14,* 495–504.

Helmreich, R. L., Spence, J. T., Beane, W. E., Lucker, G. W., & Matthews, K. A. (1980). Making it in academic psychology: Demographic and personality correlates of attainment. *Journal of Personality and Social Psychology, 39,* 896–908.

Helson, R. (1971). Women mathematicians and the creative personality. *Journal of Consulting and Clinical Psychology, 36,* 210–220.

Helson, R., & Crutchfield, R. (1970). Mathematicians: The creative researcher and the average PhD. *Journal of Consulting and Clinical Psychology, 34,* 250–257.

Holland, J. L. (1992). *Making vocational choices* (2nd ed.). Odessa, FL: Psychological Assessment Resources.

Howe, M. J., Davidson, J. W., & Sloboda, J. A. (1999). Innate talents: reality or myth? *Behavorial Brain Science, 21,* 399–407.

Heller, K. A. (2007). Scientific ability and creativity. *High Ability Studies, 18*(2), 209–234.

Iliffe R., & Mandelbrot, S. (2012). *The Newton Project: Isaac Newton's personal life.* Retrieved on March 2, 2012 from http://www.newtonproject.sussex.ac.uk/prism.php?id=40

Jacobson, S., & Jacobson, J. (2000). *Teratogenic insult and neurobehavioral function in infancy and childhood.* Mahwah, NJ: Lawrence Erlbaum Associates Publishers.

Jamison, K. R. (1993). *Touched with fire: Manic-depressive illness and the artistic temperament.* New York, NY: The Free Press.

John, O. P. (1990). The "Big-Five" factor taxonomy: Dimensions of personality in the natural language and in questionnaires. In L. A. Pervin (Ed.), *Handbook of personality theory and research* (pp. 66–100). New York, NY: Guilford.

Johnson, J. A., Germer, C. K., Efran, J. S., & Overton, W. F. (1988). Personality as the basis for theoretical predilections. *Journal of Personality and Social Psychology, 55,* 824–835.

Kahn, J. H., & Scott, N. A. (1997). Predictors of research productivity and science-related career goals among counseling psychology doctoral students. *Counseling Psychologist, 25,* 38–67.

Kalbitzer, J., Frokjaer, V. G., Erritzoe, D., Svarer, C., Cumming, P., Nielsen, F. Å., & ... Knudsen, G. M. (2009). The personality trait openness is related to cerebral 5-HTT levels. *Neuroimage, 45*(2), 280–285.

Kahneman, D. (2002). *Autobiography.* Retrieved online June 2, 2011 at http://www. nobelprize.org/nobel_prizes/economics/laureates/2002/kahneman-autobio.html

Kaminsky, Z., Petronis, A., Wang, S., Levine, B., Ghaffar, O., Floden, D., & Feinstein, A. (2008).Epigenetics of personality traits: An illustrative study of identical twins discordant for risk-taking behavior. *Twin Research and Human Genetics, 11,* 1–11.

Karlson, J. I. (1970). Genetic association of giftedness and creativity with schizophrenia. *Hereditas, 66,* 177–182.

Kenrick, D. T., & Funder, D. C. (1988). Profiting from controversy: Lessons from the person-situation debate. *American Psychologist, 43,* 23–34.

Kessler, R. C., Berglund, P., Demler, O., Jin, R., Merikangas, K. R., & Walters, E. E. (2005). Lifetime prevalence and age-of-onsetdistributions of *DSM-IV* disorders in the National Comorbidity Survey replication. *Archives of General Psychiatry, 62,* 593–602.

Keynes, M. (1995). The personality of Isaac Newton. *Notes and Records of the Royal Society of London, 49,* 1–56.

Kim, H., Lee, H., Rowan, J., Brahim, J., Dionne, R. A. (2006). Genetic polymorphisms in monoamine neurotransmitter systems show only weak association with acute postsurgical pain in humans. *Molecular Pain, 2,* 24.

Kline, P., & Lapham, S. L. (1992). Personality and faculty in British universities. *Personality and Individual Differences, 13,* 855–857.

Ko, Y., & Kim, J. (2008). Scientific geniuses' psychopathology as a moderator in the relation between creative contribution types and eminence. *Creativity Research Journal, 20,* 251–261.

Krasnegor, N. A., Lyon, G. R., & Goldman-Rakic, S. (Eds.). (1997). *Development of the prefrontal cortex: Evolution, neurobiology, and behavior.* Baltimore, MD: Paul H. Brookes Publishing.

Krueger, R. F., & Johnson, W. (2008). Behavioral genetics and personality: A new look at the integration of nature and nurture. In O. P. John, R. W. Robins, & L. A. Pervin (Eds.), *Personality handbook: Theory and research* (pp. 287–310). New York, NY: Guilford Press.

Lesch, K. P., Bengel, D., Heils, A., Sabol, S. Z., Greenburg, B. D., Petri, S.,…Murphy, D. L. (1996). Association of anxiety-related traits with a polymorphism in the serotonin transporter gene regulatory region. *Science, 274,* 1527–1531.

Lippa, R. (1998). Gender-related individual differences and the structure of vocational interests: The importance of the people-things dimension. *Journal of Personality and Social Psychology, 74,* 996–1009.

Loehlin, J. C., & Nichols R. C. (1976). *Heredity, environment, and personality: A study of 850 sets of twins.* Austin, TX: University of Texas Press.

Lounsbury, J. W., Foster, N., Patel, H., Carmody, P., Gibson, L. W., & Stairs, D. R. (2012). An investigation of the personality traits of scientists versus nonscientists and their relationship with career satisfaction. *R&D Management, 42,* 47–59.

Lubinski, D., Benbow, C. P., Webb, R. M., & Bleske-Rechek, A. (2006). Tracking exceptional human capital over two decades. *Psychological Science, 17,* 194–199.

Ludwig, A. M. (1995). *The price of greatness.* New York, NY: Guilford Press.

Lynn, R. (2006). *Race differences in intelligence: An evolutionary analysis.* Augusta, GA: National Summit.

MacKinnon, D. W., & Hall, W. B. (1972). Intelligence and creativity. *Proceedings of the XVIIth International Congress of Applied Psychology, Liege Belgium* (Vol. 2, 1883–1888). Brussels, EDITEST.

Mallinckrodt, B., Gelso, C. J., & Royalty, G. M. (1990). Impact of the research training environment and counseling psychology students' Holland personality type on interest in Research Professional Psychology. *Research and Practice, 21,* 26–32.

Martin, G. M. (2005). Epigenetic drift in aging identical twins. *Proceedings of the National Academy of Sciences of the United States of America, 102,* 10413–10414.

McCrae, R. R. (1987). Creativity, divergent thinking, and openness to experience. *Journal of Personality and Social Psychology, 52,* 1258–1265.

McCrae R. R. (1991). The Five-Factor Model and its assessment in clinical settings. *Journal of Personality Assessment, 57,* 399–414.

McCrae, R. R., & Costa, P. T., Jr. (2008). The five-factor theory of personality. In O. P. John, R. W. Robins, & L. A. Pervin (Eds.), *Personality handbook: Theory and research* (pp. 159–181). New York, NY: Guilford Press.

McCrae, R. R., & John, O. P. (1992). An introduction to the Five-Factor Model and its applications. *Journal of Personality, 60,* 175–215.

Miller, B. L., & Cummings, J. L. (Eds.). (1999). *The human frontal lobes: Functions and disorders.* New York, NY: The Guilford Press.

Mischel, W. (1968). *Personality and assessment.* New York, NY: Wiley.

Mischel, W., & Shoda, Y. (1999). Integrating dispositions and processing dynamics within a unified theory of personality: The Cognitive-Affective Personality System. In L. A. Pervin & O. P. John (Eds.). *Handbook of personality theory and research* (pp. 197–218). New York, NY: Guilford Press.

Mithen, S. (1996). *The prehistory of the mind: The cognitive origins of art and science.* London, England: Thames and Hudson.

Moffitt, T., Caspi. A., & Rutter, M. (2005). Strategy for investigating interactions between measured genes and measured environments. *Archives of General Psychiatry, 62,* 473–481.

Morris, P. H., Gale, A., & Duffy, K. (2002). Can judges agree on the personality of horses? *Personality and Individual Differences, 33,* 67–81.

Nettle, D. (2006). Schizotypy and mental health amongst poets, visual artists, and mathematicians. *Journal of Research in Personality, 40,* 876–890.

Nisbett, R. E., & Ross, L. (1980). *Human inference: Strategies and shortcomings of social adjustment.* New York, NY: Prentice-Hall.

O'Brien, V., Martinez-Pons, M., & Kopala, M. (1999). Mathematics self-efficacy, ethnic identity, gender, and career interests related to mathematics and science. *Journal of Educational Research, 92,* 231–235.

Parloff, M. B., & Datta, L. (1965). Personality characteristics of the potentially creative scientist. *Science and Psychoanalysis, 8,* 91–105.

Perrine, N., & Brodersen, R. M. (2005). Artistic and scientific creative behavior: Openness and the mediating role of interests. *Journal of Creative Behavior, 39,* 217–236.

Perry, B. D. (2002). Childhood experience and the expression of genetic potential: What childhood neglect tells us about nature and nurture. *Brain and Mind, 3,* 79–100.

Petronis, A. (2010). Epigenetics as a unifying principle in the aetiology of complex traits and diseases. *Nature, 465*(7299), 721–727.

Pickover, C. A. (1998). *Strange brains and genius: The lives of eccentric scientists and madmen.* New York, NY: Plenum Trade.

Pinker, S. (2002).*The blank slate: The modern denial of human nature.* New York, NY: Viking.

Plomin, R., & Caspi, A. (1999). Behavioral genetics and personality. In L. A. Pervin & O. P. John (Eds.), *Handbook of personality theory and research* (pp. 251–276). New York, NY: Guilford Press.

Plomin, R., & Petrill, S. A. (1997). Genetics and intelligence: What's new? *Intelligence, 24,* 53–77.

Post, F. (1994). Creativity and psychopathology: A study of 291 world-famous men. *British Journal of Psychiatry, 165,* 22–34.

Prediger, D. J. (1982). Dimensions underlying Holland's hexagon: Missing link between interest and occupations? *Journal of Vocational Behavior, 21,* 259–287.

Rawlings, D., & Locarnini, A. (2008). Dimensional schizotypy, autism, and unusual word associations in artists and scientists. *Journal of Research in Personality, 42,* 465–471.

Roberts, B. W., & Mroczek, D. (2008). Personality trait change in adulthood. *Current Directions in Psychological Science, 17,* 31–35.

Roe, A. (1952). *The making of a scientist.* New York, NY: Dodd, Mead.

Roe, A. (1953).A psychological study of eminent psychologists and anthropologists, and a comparison with biological and physical scientists. *Psychological Monographs: General and Applied, 67,* 1–55.

Roe, A. (1965). Changes in scientific activities with age. *Science, 150,* 313–318.

Rosenberg, E. L. (1998). Levels of analysis and the organization of affect. *Review of General Psychology, 2,* 247–270.

Royalty, G. M., & Magoon, T. M. (1985). Correlates of scholarly productivity among counseling psychologists. *Journal of Counseling Psychology, 32,* 458–461.

Ruff, H. (1999). Population-based data and the development of individual children: The case of low to moderate lead levels and intelligence. *Journal of Developmental & Behavioral Pediatrics, 20*(1), 42–49.

Rushton, J. P., Murray, H. G., & Paunonen, S. V. (1987). Personality characteristics associated with high research productivity. In D. Jackson & J.P. Rushton (Eds.), *Scientific excellence* (pp. 129–148). Beverly Hills, CA: Sage.

Ruthsatz, J., Detterman, D., Griscom, W., & Cirullo, B. (2008). Becoming an expert in the musical domain: It takes more than just practice. *Intelligence, 36*(4), 330–338.

Rutter, M. (2006). *Genes and behavior: Nature-nurture interplay explained.* Malden, MA: Blackwell Publishing.

Schoon, I. (2001). Teenage job aspirations and career attainment in adulthood: A 17-year follow-up study of teenagers who aspired to become scientists, health professionals, or engineers. *International Journal of Behavioral Development, 25,* 124–132.

Simonton, D. K. (1988). *Scientific genius: A psychology of science.* Cambridge, England: Cambridge University Press.

Simonton, D. K. (1988). *Scientific genius: A psychology of science.* Cambridge, UK: Cambridge University Press.

Simonton, D. K. (2000). Methodological and theoretical orientation and the long-term disciplinary impact of 54 eminent psychologists. *Review of General Psychology, 4,* 13–21.

Simonton, D. (2008). Scientific talent, training, and performance: Intellect, personality, and genetic endowment. *Review of General Psychology, 12*(1), 28–46.

Simonton, D. K., & Song, A. V. (2009). Eminence, IQ, physical and mental health, and achievement domain: Cox's 282 geniuses revisited. *Psychological Science, 20,* 429–434.

Smithers, A. G., & Batcock, A. (1970).Success and failure among social scientists and health scientists at a technological university. *British Journal of Educational Psychology, 40,* 144–153.

Steinhausen, H., & Spohr, H. (1998). Long-term outcome of children with fetal alcohol syndrome: Psychopathology, behavior, and intelligence. *Alcoholism: Clinical and Experimental Research, 22*(2), 334–338.

Stone, V. E., Baron-Cohen, S., & Knight, R. T. (1998). Frontal lobe contributions to theory of mind. *Journal of Cognitive Neuroscience, 10,* 640–656.

Streissguth, A., Barr, H., Sampson, P., Darby, B., & Martin, D. (1989). IQ at age 4 in relation to maternal alcohol use and smoking during pregnancy. *Developmental Psychology, 25*(1), 3–11.

Stuss, D. T., Picton, T. W., & Alexander, M. P. (2001). Consciousness and self-awareness, and the frontal lobes. In S. P. Salloway, P. F., Malloy, & J. D. Duffy (Eds.). *The frontal lobes and neuropsychiatric illness* (pp. 101–109).Washington, DC: American Psychiatric Publishing.

Svrakic, D. M., & Cloninger, R. C. (2010). Epigenetic perspective on behavior development, personality, and personality disorders. *PsychiatriaDanubina, 22,* 153–166.

Van Zelst, R. H., & Kerr, W. A. (1954). Personality self-assessment of scientific and technical personnel. *Journal of Applied Psychology, 38,* 145–147.

Vinkhuyzen, A., van der Sluis, S., Posthuma, D., & Boomsma, D., (2009). The heritability of aptitude and exceptional talent across different domains in adolescents and young adults. *Behavior Genetics, 39,* 380–392.

Watters, E. (2006, November 22). DNA is not destiny. *Discover,* Retrieved from http://discovermagazine.com/2006/nov/cover

Weaver, I. C. G., Cervoni, N., Champagne, F. A. (2004). Epigenetic programming by maternal behavior. *Nature Neuroscience, 7,* 847–854.

Westfall, R. (1980). *Never at rest.* Cambridge, England: Cambridge University Press.

Wilhelm, K., Mitchell, P. B., Niven, H., Finch, A., Wedgewood, L., Scimone, A., Blair, I.P.,…Schofield, P. R. (2006). Life events, first depression onset and the serotonin transporter gene. *British Journal of Psychiatry, 188,* 210–215.

Wilson, G. D., & Jackson, C. (1994). The personality of physicists. *Personality and Individual Differences, 16,* 187–189.

Zachar, P., & Leong, F. T. L. (1992). A problem of personality: Scientist and practitioner differences in psychology. *Journal of Personality, 60,* 667–677.

Zachar, P., & Leong, F. T. L. (1997). General versus specific predictors of specialty choice in psychology: Holland codes and theoretical orientations. *Journal of Career Assessment, 5,* 333–341.

Zachar, P., & Leong, F. T. L. (2000). A 10-year longitudinal study of scientists and practitioner interests in psychology: Assessing the Boulder model. *Professional Psychology: Research and Practice, 31,* 575–580.

CHAPTER 6

Scientific Communication in the Process to Coauthorship

Sofia Liberman and Kurt Bernardo Wolf

Scientific communication has been the subject of several fields of research and has been studied from diverse points of view. Historians have examined the formation and disintegration of scientific societies in the Greek, Roman, and Arab civilizations, and the impact of lone geniuses in Renaissance and early modern Europe, who realized the importance of communicating through visit and correspondence with their peers. Books were written by Copernicus, Galileo, Newton, Descartes, and many others, who nurtured the coalescence of learned societies in Italy, England, and France, under royal or civil patronage. This process accelerated in the 19th century with the formalization of academic bodies in universities and research groups hired by the chemical, optical, and military establishments. A memorable boost to scientific activity occurred at the dawn of the 20th century, when the fundamental theories of Nature were discussed at the first pan-European meetings in Solvay, Heidelberg, and Göttingen. The growth of science and the establishment of its present patterns of communication can be considered to have begun after World War II, when the number of scientists reached the thousands, with specialized journals and standard refereeing processes. By then, periodic face-to-face meetings became part and parcel of the activity of scientists, who spent much of their time engaged in conversations about their work, exchanging opinions and results, striving for increased productivity, personal gain, and prominence in their fields. Presently, the trend for intensive communication is accelerating, with large groups partaking in multinational experimental installations and active research societies, as well as in interested companies promoting topical workshops and massive meetings.

Nevertheless, the production of scientific knowledge is a personal affair (Polanyi, 1958) that takes place in the individual mind of each scientist, in the surroundings of a small society of colleagues who will understand, validate, or refute the various conjectures, assertions, and conclusions of his work. Scientists spend much time engaged in conversations about their work, exchanging data and opinions. Considering that science is objective and should exclude all personal influence, it is widely accepted that scientists need to communicate with other scientists for the advancement of their field, their research group, and their own interest in prominence.

The purpose of this chapter is to present a model for interpersonal scientific communication in the framework of psychology of science. Some of the postulates for this conceptual model are drawn from previous approaches, and others from our own research in social psychology. Here we describe a model for the informal-interpersonal communication among scientists that leads to coauthorship in publications and propose a measure for the patterns of interpersonal communication in science. In this model we consider the process of communication as part of the *inner cycle* of scientific activity, and coauthorship as the public manifestation of its *outer cycle*. We consider here a third cycle of communication: that of popularization, that is, the divulgation of the results and benefits of scientific knowledge to the society at large, which in turn funds scientific endeavor. Although we do not develop the third cycle in this chapter, we hold that it is a major concern for scientists to have effect on the society at large. We hold that this model can provide a connection between studies of scientific communication processes in small groups, describe their resulting structure, and contribute to determine their place within the hierarchy of larger communication networks of science. We see scientific communication as a multidisciplinary concept with a very wide scope.

RESEARCH ON INTERPERSONAL COMMUNICATION

In this section we give a succinct description of the current trends of research in interpersonal communication, from both the psychological and the sociological points of view, with emphasis on the relevance of informal communication for the advancement of science. We assume for this that social interaction plays a major role in the creation of scientific knowledge, as Price (1963) and Crane (1969, 1972) did when grounding the concept of "invisible colleges," thus demonstrating that scientists talk to each other informally in any possible occasion.

Positioning scientific conversations and gathering practices are considered necessary for communicating and furthering research. Krauss and Fussell (1996) described a variety of disciplines from natural sciences, social sciences, and engineering, which have used the concept of communication with little agreement on its precise meaning. Scientific communication can be studied within two realms, the informal and the formal. Because these are not hypothetical constructs but rather behavioral acts, we focus on the personal and interpersonal aspects of this communication, laying the idea that communication is personal. In a metaphorical sense we can say that *informal communication* deals with the interpersonal expectations, choices, and acts between individual scientists, while *formal and published communications* are mediated, public, and belong to humanity as open information. We have named these two realms of activity the *inner* and the *outer* cycles of communication, respectively (Liberman & Wolf, 1997).

The inner cycle resides in personal acts of communication with other scientists, sometimes face to face, and sometimes through some medium.

Interpersonal communication is preponderant because it fulfills psychological needs such as self-confirmation, inclusion, and affection (Schutz, 1966). It facilitates social interaction to understand the behavior of others and our own, it yields fresh information to reduce uncertainty, and promotes opportunity for influence. Studies on this topic range from conversations in a dyad up to larger group interaction. Often these studies distinguish between structure and process, and some recognize that the process modifies the structure in a reciprocal way. Structure refers to fixed characteristics of the group, as how many members are present, who originated the idea, or who is the appointed leader, whereas process description implies the dynamic changing aspects of the relationships involved, known as relational communication in groups, bearing the transfer of symbolism and meaning in communication (Keyton, 1999). In some situations of knowledge creation and/ or acquisition, the dynamic processes may differ, but the structural characteristics remain. For example, a professor meets a student in the corridor and reminds him of some reading, but if the student has already read this article and found a new one that is more precise, he will tell the professor, and the professor will have to find it and know what is in it, so the conversation can continue. The professor has learned something from the student, but he stays the professor. The structure of communication is illustrated by the roles held in the conversation and the acceptance that this conversation is based on some issue that concerns both of them. We can estimate that the relational character of this particular communication episode has changed temporarily, but the task-structure and roles have been kept by the structure of the relationship. Some of the characteristics of interpersonal relationships can be directly observed, while others can be accounted for by the outcome of the relationship. Relational communication helps members of a group to find out where they stand in the context of their relationship, to share meaning and recognize intentions.

Substantial communications enriched by visual cues from nonverbal messages are antecedent to interpersonal behavior that can bring engagement (Barker et al., 2000). Conditions where scientists can meet in a direct manner, either in a casual encounter at the university, in a class, or in a meeting, give an opportunity for bridging cultural diversity, inspiration, motivation, recognition, and relationship building that are considered valuable. For scientists, the benefit of face-to-face communication lies in that it is the most efficient way to advance personal objectives, create social bonding, and allow for information exchange and symbolic expressions of commitment through verbal and nonverbal messages in a visual context. Bavelas and Chovil (2000, p. 167) stated that "in face-to-face dialogue, many visible behaviors act with words to convey meaning"; in addition, they maintained the necessity to study the use of language with accompanying nonverbal cues (i.e., the visible acts of meaning), instead of studying language as an abstract system or as a set of separate categories, since influence tactics and style are permeated by verbal (e.g., phonetic) and nonverbal cues. In short, influence is more effective when there are visual cues.

Because of the search for originality in the findings and specialization of scientific tasks and competition, scientists will attempt to relate and influence colleagues and others in their field, attract those that are doing similar studies or hold similar goals. It has been confirmed that relational communication is perceived as informal and multidimensional, where relational messages play an important role in task groups (Keyton & Beck, 2009). There are various distinct ways in which informal-relational communication influences other people. Kelman (1958) distinguished three different processes of social influence: compliance, identification, and internalization. These three processes produce different influence patterns of social involvement: either people settle, resolve, or reconcile a solution in a decision, or negotiate the solution of a dispute. As Kelman (2006) has put forth, reactions to influence may be motivated by instrumental concerns or by concerns of self-maintenance; in the first case the context is very important, while in the second the inner psychological process of the individual is affected. By reacting in terms of interests, relationships, or intellectual identity, scientists can advance arguments in favor or against a theory, acknowledge what others are doing, and if they reach a commitment, become coauthors.

The unfolding of interpersonal communication in science is also reflected in the development of scientific theories and controversies, giving way to different interpretations, the validation of new or rejection of old theories embodying the continuing process of communication. Integrative models such as those described by Krauss and Fussell (1996) have emphasized the need to account for both, process and content in the symbolic and expressive behaviors of intrapersonal and interpersonal communication, and recognized that the description of context has been neglected to a certain point. Studying groups in context has had a significant effect on theoretical interpretations in group communication studies (Frey, 2003). The debate between theoretical and applied research in communication is context dependent, and suggests an integral perspective through the use of applied findings for theory development (Keyton et al., 2009).

The inner cycle of scientific communication can therefore be defined as the interaction process through which scientists affect and are affected by other scientists, with a purpose and symbolic expressions of attachment, involving intention, interpretation, and feedback, in a context of a social structure; it is tacitly assumed that shared meanings will allow them to communicate and breed agreement and, if possible, further the relationship into a stable structure such as coauthorship. This definition must acknowledge that the interacting scientists will have instrumental, self-presentation, and relational goals to be also included in the episode.

INFORMAL COMMUNICATION IN SCIENCE

Informal communication in science is distinguished from formal communication because it is performed in face-to-face or directly mediated interpersonal contacts, such as e-mail, telephone, or a group of mutual friends who act as messengers; the latter is mediated by scientific literature and institutional relationships. Informal communication is random and presumably very difficult to scrutinize; however, it is justified for the most part by an instrumental goal. Everyday work in scientific research implies an understanding of the premises under which scientists are responsible for the advancement, transference, and diffusion of scientific knowledge. Since knowledge is produced in the midst of overlapping social structures, if scientists are to succeed in their personal careers, they must perform a series of specific communication activities, displaying social and professional capabilities, and conforming to certain norms of participation and reciprocity manifested in the process of communication within the research group. The emergence of new fields and specialties, novel paradigms, or simply the improvement and validation of existing theories, generates this social activity whose essential mechanism is the process of communicating, validating, and establishing the priorities or property of ideas, projects, and results. Most scholars agree that communication lies at the heart of science, allowing for the scrutiny and acceptance of research results (Meadows, 1998).

As an example of how communication and organization advance science, James McKeen Cattell (known for his theory on individual differences) acquired the title of the journal *Science*, after it ceased to be published at the end of the 19th century. He had to seek the cooperation of outstanding scientists of that time to re-start the publication of *Science* in January 4, 1895. As in this example, we know that starting a society or the publication of a journal is usually done in the midst of a series of agreements and consensus between scientists that takes place through a sequence of episodes of interpersonal informal communication (Liberman, 2011). Mullins has described this process as the development of scientific specialties by describing the way an informal group of scientists follow a leader in their normal activities (Mullins, 1983).

There is no question that the genesis of the collaboration process takes place in the midst of interpersonal informal communication. Often collaboration starts because of geographical proximity (Katz, 1994); when scientists look for a suitable partner in a personal direct/meditated encounter, they establish contact with relevant others to exchange information, share knowledge, and procure confirmation. It has been proven that the transference of tacit knowledge requires face-to-face contact (Aydogan & Lyon, 2004). Moreover, when scientists need help, feedback, or recognition, they personally communicate gaining fresh and actual information they can use in the development of their research and for advancing their career. Numerous studies have focused on communication taking place at different levels within scientific organizations (Pelz & Andrews, 1976) and in close collaboration, both at the relationship and task levels (Kraut, Galegher, & Egido, 1987).

In the process of communicating, scientists search for information either using the public archives of science, the information transmitted from person to person, or throughout memos and mail (Pelz & Andrews, 1976). Scientists collaborate for many reasons; still, although it is not obligatory, it is clearly desirable that scientists share information in their communication practices; subsequently, they get involved in professional societies and attend meetings. Scientists will always have the desire and need to meet their colleagues; hence we assert that social interaction between scientists is part of the very nature of science. This implies that the psychological aspect of scientific communication can be construed as the development of common understanding of the subject, which is not separate from its rational or epistemic development (Mitroff, 1974). Even if we were seduced by the fact that written and oral communication among scientists is undergoing a transformation with the advent of new technologies, interpersonal communication will continue to take place, either face to face or through written and electronic media, the latter being also personal mediated contact.

Interpersonal communication between scientists includes not only the flow of professional information in the narrow field of their expertise, but the process of discussion itself sharpens the problem at issue and validates or refutes the result (Liberman, Krötzsch, & Wolf, 2003). Interpersonal communication also teaches through practice the tacit skills that every successful scientist is expected to display (Thagard, 2005). The flow of knowledge that takes place between scientists clarifies what is assumed to be tacit knowledge in the field (Collins, 2001). For example, because they are so well known, Pythagoras' theorem or Newton's laws need no longer be referred to explicitly, but recently produced knowledge by a living scientist does require reference to its discoverer. Communication among scientists thus also brings its rewards, sometimes in pecuniary form when a legal patent is involved, but mostly in the recognition of his scientific prominence by his peers, which opens the road to invited lectures at prestigious international conferences.

Communication operates at various levels according to the context in which it is established. Different contexts create different patterns of communication and different outputs (Homans, 1950; Bavelas, 1950; Kraut et al., 1987). Face-to-face interaction in workshops and small conferences will lead to social proximity (Kraut, Egido, & Galegher 1990). Social proximity occurs even if the two scientists do not exactly share the same type of expertise (Gorman, 2002) as, for example, when a physicist and a mathematician discuss a problem in optics, which results in two quite different papers announcing the solution in different contexts. However, if the communication takes place in written form, as when two or more scientists plan to write a joint article to be submitted to a credited journal, the communication process will be precise, but need not entail social contact, or even proximity.

Garvey and Griffith (1968) described how scientists are willing to do anything to get fresh information, even before it is published. These authors describe the advantages of informal communication and how scientists sometimes "*informalize*" formal channels by publishing lists of papers under review, preprints, previous publications, and oral presentations by the author and registering requests prior to publication. In their search for information relevant to current work, scientists today have much easier access to information than before electronic databases and e-mail became available, and they have learned to search and find new ways to communicate with those who are working in similar or connected problems, as well as to continue the practice of going to meetings to expose their work to their peers.

Other psychological processes are involved in scientific communication: Andrews (1979) investigated the inner commitment to do research and science, paying attention to the degree of enthusiasm and motivation as individual characteristics having an effect on tasks but showing a systematic variation in different research units. There is also some evidence that enthusiasm and trust are related in temporary teams (Jarvenpaa & Leidner, 1999). Motivation is based on personal interests, as shown by Fortes and Lomnitz (1991): scientists also value their personal interests and needs, which originate from observing the mores of their mentors from the time they were students. In this regard, the search for recognition and propriety become part of the communication system having an effect on productivity and career. Other psychological variables may be present in the informal cycle of scientific communication, such as the prominence of the communicator, reflecting on the existence of hierarchy in science or elite scientists in terms of credibility, perceived higher status or intellectual authority, or the promise of positioning oneself in a new paradigm that may overthrow the previous schema. For instance, Mahoney (1976) describes the passions and emotions of scientists as having a repercussion on their career in relation to credit and recognition based on labor division and trust pursued in many occasions through means like procuring specific informal encounters with other scientists.

As we have pointed out earlier, scientific communication is more than the exchange of data and pieces of information. Scientists do have particular development histories and personality traits (Feist, 2006), and are involved in a social process where there are assorted roles for charisma (Gustin, 1973), emotional intelligence (Goleman, 1995), moods, emotional contagion, and affective manipulation (Kelly & Barsade, 2001). These features combine and interact to influence the leadership in science. Describing influence, Westrum (1989) developed a model for scientific dialogues in small group interactions, introducing a taxonomy of types of conversational episodes in science, which translate into the roles that scientists play when interacting, and the changes in their ways of thinking. The influential actions of individuals are guided by their ability to define the situation and establish a positional identity or a common ground. The perception of interdependence has a direct effect on building trust and goal achievement. The acceptance of interdependence between collaborators is related to the assumption of trust that is obtained in a face-to-face situation by touch as a "social lubricant" needed for sharing knowledge (Gallie & Guichard, 2005).

A THREE-STAGE MODEL OF SCIENTIFIC COMMUNICATION

In the model that we propose, there are three cycles in scientific communication, according to their purpose and timing. Initially, scientists communicate directly in informal situations, either *face-to-face* or through informally written messages (mail or instant messaging), and this information is collected individually and kept for possible future use in research or publications. These acts of communication are part of the private mental archive of scientists or of their immediate workgroup. We consider these to be the *inner cycle* of scientific communication (Liberman & Wolf, 1997), assuming that there is social or geographical proximity and/or intellectual identification (Mathieu et al., 2000).

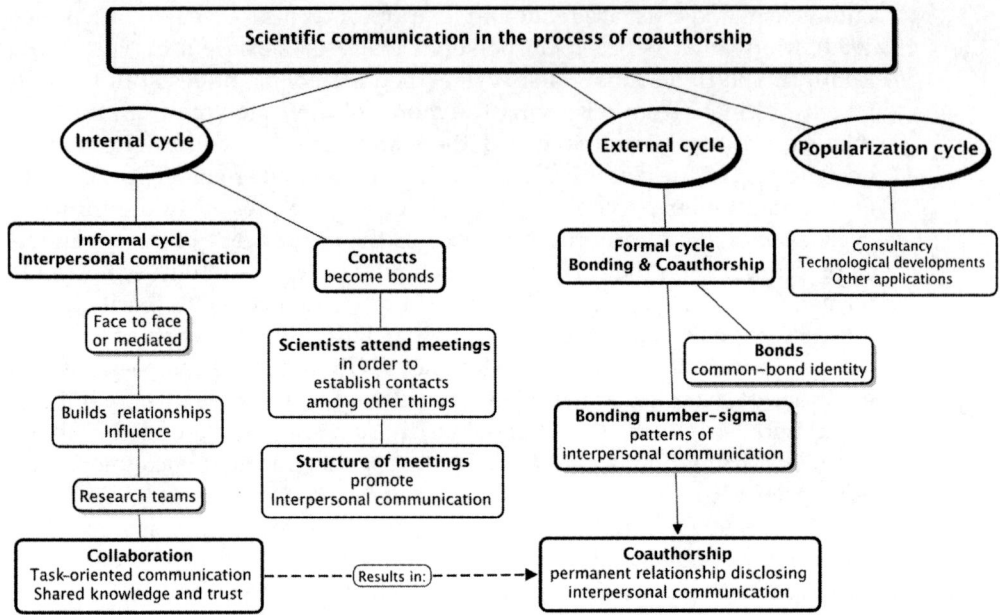

Figure 6.1   *Three-cycle model of scientific communication in the process of coauthorship.*

Created with Cmaps Tools. Copyright © 1998 to 2010. Institute for Human & Machine Cognition (IHMC).
Downloaded from: http://cmap.ihmc.us

Then comes the second, or *outer cycle* of scientific communication, when the outcome of a scientist's research enters into the formal process of written communications, published as books, articles in scientific journals and/or electronic databases. These hold a validated record of the scientist's work, and of those who have collaborated in the project as coauthors, and who are supposed to be and are universally accessible.

One may take a broader point of view and incorporate society at large as a recipient of scientific communication, introducing a third cycle of popularization (Lievrouw, 1990), where scientific knowledge comes to serve humanity in the form of common knowledge of the natural world, improved organization, more efficient work practices, and invention of everyday labor-saving gadgets. We will not describe this third stage in this chapter, but we claim that popularization of science is based in the actual effort of bridging science and society (see Figure 6.1).

SCIENTIFIC CONTACTS

Contacts are the initial act in developing relationships for future collaboration. We have defined *contact* as "the exchange of a useful piece of information, as communication bounded in time, retained in the researcher's memory or briefcase, which generates some later action, ranging from immediately influencing his/her research work to expanding his/her general scientific culture" (Liberman & Wolf, 1997, p. 278). In time, contacts further relationships into the formation of research groups or invisible colleges, and eventually turn into long-lasting relationships, which may lead scientists to participate in joint publications in the written scientific literature.

An invisible college is a set of informal communication relations among scholars or researchers who share a specific common interest or goal (Lievrouw, 1990, p. 66). Invisible colleges are built through personal contact. In the midst of a meeting on informal communication in science, Price (1971, p. 5) claimed that "invisible college research is rather like hunting the unicorn," referring to the intangible communicative relations between scientists in a small group. Storer (1966), however, described invisible colleges as reference groups comprising specialized relationships. Zuccala defined invisible colleges as "a set of interacting scholars or scientists who share similar research interests concerning a subject specialty, who often produce publications relevant to this subject and who communicate both formally and informally with one another to work towards important goals in the subject, even though they may belong to geographically distant research affiliates" (Zuccala, 2006, p. 155). We may conceive any single episode of spontaneous interpersonal communication between two scientists as an evanescent chance event, because that helps to explain why invisible colleges are invisible. The network of acquaintances that weaves an invisible college is a tapestry of personal contacts that form communication patterns within the group of participating scientists.

The contacts that create and sustain invisible colleges are activated in various ways, through personal encounters or through electronic means, and they can be spontaneous or planned. Invisible colleges now also exist through electronic media; thus, Kwon (1998) has proven that communication through computers can be useful in group problem solving, and Anderson (1998) has shown that communication through computer networks increases the ability to collaborate, increases productivity, and furthers the general level of exchanges among scientists, promoting their affiliation to "invisible" colleagues. What Granovetter (1983) called *weak ties* are created through contacts, which he presumes to be beneficial for the individual career and mobility of scientists, as well as for the advancement of science. The social network approach to scientific communication has advanced the description of the structure in scientific groups (Borgatti et al., 2009).

Scientific contacts can change scientific practice. Coleman, Katz and Menzel (1957) demonstrated that medical doctors were influenced by those with whom they established social contact, and confirmed this hypothesis by emphasizing the importance of socialization in the early use of a new drug. Social influence in adopting drugs exemplifies how intense integration among local colleagues is strongly related to the adoption of an innovation as it occurs in social influence networks (Kalkhoff et al., 2010). Technologically oriented groups will rely more on personal communication because the published literature cannot include many timely aspects of practical work and tacit knowledge, since much of the written knowledge in their field has to be decoded or explained by some specialist. This is one of the reasons why applied scientists make a special effort to be aware of what is happening in their field through personal contact with other professionals. Pelz and Andrews (1976) studied contact via memos, meetings, and direct conversation with colleagues, and established that scientists interacting with one another are more productive. When scientists work together knowing that their results may be published and thus become part of the universal database of scientific knowledge, they must go through the agreement process of accepting the content of every assertion included in their publication.

Janet Bavelas et al. (1997) substantiated that face-to-face communication can be a standard for evaluating other communication systems. Unrestricted and meaningful verbal acts and voluntary collaboration in such dialogues contrast with other communication systems, where they may encounter limitations in language or other formats of communication. Although there may be social restrictions in the freedom of a face-to-face dialogue, these can be ameliorated by nonverbal communication or by the development of a relationship between speaker and listener. There is also the possibility of meta-communication and reinterpretation of what is said, which can occur at the same time communication takes place (Renz & Greg, 2000).

Kyvik and Larsen (1994) established a strong correlation between frequency of contact and research performance. To some extent there is a need to explain why scientists become socially connected and conscious of their need to interact with colleagues whom they identify as related to their own work; for this kind of communication they will undertake travel and travail. Scientists eagerly search for opportunities to meet kindred souls to exchange ideas, to receive feedback or criticism, to improve their methodology, and even to make friends. They will also try to find collaborators, students, and postdoctoral associates who can develop, evaluate, and validate what they are doing. Interpersonal communication in science is always purposeful and in the context of the scientist's professional interests, the interaction between social and cognitive conditions involved in establishing contacts is a result of the curiosity and expectation of response from other scientists (Snyder & Stukas, 1999).

If scientists are aware of the extent of their contacts, they may succeed in becoming more visible in the community and become more widely read in the circulating archives of scientific literature; contacts are related to influence (Pool & Kochen, 2006). Pelz and Andrews (1976) concluded that the contacts that were purposefully originated had the highest payoff for the scientists; contacts were positively related to highest performance even when the scientist preferred working alone. The anatomy of a contact can be characterized as the initiation of a social relationship in which there is an intentional exchange of meaning in the form of concepts, gestures, needs, experience, and information. Contacts are fragments in a relationship; often they can be registered at the beginning, with the expectation of becoming a stable relationship that will affect intrinsically those involved. In science contacts are mostly related to work, and in this sense it is feasible to account for their existence and reciprocation. Gamble and Gamble (1998) have set forth the idea that interpersonal communication is a dynamic process in which "every encounter is a point of departure for a future encounter" (1998, p. 17).

On a similar thread, Granovetter (1973, p.1361) established that "the strength of a tie is a (probably linear) combination of the amount of time, the emotional intensity, the intimacy (mutual confiding), and the reciprocal services which characterize the tie. Each of these is somewhat independent of the other, though the set is obviously highly intra-correlated." Extremely productive people seem to have a considerably larger amount of contacts (Griffith et al., 1971). Grannoveter (1973) established that more contacts or weak ties are correlated with a higher probability of mobility between institutions.

Attending scientific meetings is the ideal form for establishing contacts. Following White (1994), who proposed that scientific communication can be divided into four modes (informal oral, informal written, formal oral, and formal written), we suggest that in meetings there is another form of scientific communication that is informal–formal. This occurs when the presentation of a paper is in a formal setting, one person talks to an audience, but where the content is acknowledged not to be final, and may be subject to changes based on new information elicited by informal communications during the meeting.

SCIENTIFIC MEETINGS: THE ESTABLISHMENT OF CONTACTS

The attendance of scientists in professional meetings has grown from hundreds to tens of thousands in disciplines such as biology and chemistry (Söderqvist & Silverstein, 1994), and this increase has had a definitive impact on the growth of fields and forms of organization in science. When scientists go to meetings, they search through the program to select those presentations that are related to their work. Sometimes they establish contact with the speaker, ask for a copy of his presentation, and start a relationship to collaborate and share his knowledge. In other cases, if they cannot attend a meeting, they scan the proceedings of the meeting and may establish contact with the presenters through the Internet. Scientists have a strong motivation to talk about their work, and if they want to find somebody to whom their ideas have meaning, they will find the channels for competent and effective interpersonal communication. Their exchange of knowledge may later give rise to a process where the scientist identifies with a group of his peers who work in the same field. These groups develop a strong identity, or collective self-conception, which allows for self-identification, group comparison, and out-group evaluation (Abrams et al., 2005), in which the attendance to the same meetings reinforces their identity.

Scientific meetings are emergent communication processes whose results are not easily predictable; Goffman (1967) called the week-long conferences; "interactional mastodonts, that push to the limit what can be called a social occasion" (p. 1). We have found that scientists participate on average in three or four meetings per year, depending on their scope and location, and they claim to establish at those meetings an average of three to six *contacts* (Liberman & Wolf, 1997). There are of course large differences between disciplines: mathematicians claim to establish some 20% more contacts than physicists, for example. Scientists in Mexico report that meetings abroad provide some 30% more contacts than national meetings (Liberman & Wolf, 1997). To communicate, a scientist must pay attention to the mechanisms of the informal machinery of science; in many instances, a collaboration with a scientific leader will start from an established contact through an adequate interpersonal context, even between strangers and independently of their specialty (Genuth, Chompalov, & Shrum, 2000), or through an invitation when somebody is considered an expert in his field (Hagstrom, 1965). Both acknowledgment and collaboration in science are based on preferential relationships (Caplow & McGee, 1958).

Meetings are organized situations where scientists can meet in a legitimate context to expose their recent and contemplated research and obtain direct and frank feedback; they provide a specific environment to foster subsequent contacts in context. The social life among scientists is not a casual event since they do not attend parties or single-bars, but have a professional interest in gaining acquaintance with their peers and leaders. Scientific communication is a group process that is privy to the group or community that takes part in it (Liberman & López, in press). The process of communication in science is thus characteristic of the scientific activity by field, community, culture, and nationality. Language differences may sometimes weigh on this process, but are not an insurmountable obstacle for communication, because most scientists speak English. Presentations have to be prepared in advance, with the same structure and language as other papers in that meeting, and it is expected that their content will be new knowledge, compatible with the prevailing paradigm and in dialogue with the published work of peers.

Lomnitz (1983) described scientific meetings as rituals similar to those of the Eskimo community form of organization; meetings provide the chance to acquire a feeling of belonging, sometimes solving conflicts, getting prizes, and symbolic awards: "The use of symbolic titles and honorific duties emphasizes the social importance of participation in these collective rituals and consolidates the communal values internalized by each individual thus strengthening the sense of identity and of belonging" (Lomnitz, 1983, p. 5). Participation in a scientific meeting is full of expectations on what will take place based on previous experiences; it is counted on being the ideal marketplace for the exchange of specialized and timely knowledge, confronting mental models and emotional states, ways to understand the world, and prompting plans for future research (Liberman et al., 2003).

Bales (1955) described the ways people interact in conferences and meetings by analyzing acts and communicative behaviors and attitudes, such as asking and voicing opinions and useful information; his method was later developed into an instrument to measure forms of participation in group dynamics (Bales et al., 1979). A vast number of studies in group communication have relied on this method. For example, Hirokawa and Poole (1996) created a list of functions that group communication performs in relation to decision making. Hirokawa (1988) suggested that the method for interaction analysis provides information about the distributional structure of categorized acts and their sequence, offering a possibility to interpret the dynamics of interpersonal relations.

Meetings provide a good framework for interacting with peers and lead to the formation of organized scientific groups, for example, the International Society for the Psychology of Science and Technology (ISPST), formed after a meeting in Zacatecas, Mexico, in September 2006. The most important psychological function of meetings is the stimulation of face-to-face contacts and exchange of information, as well as other aspects in the psychological world of scientists, such as the chance to earn inclusion and receive recognition and affection (Schutz, 1966). Meetings also foster consensus around a research field and create criteria for the acceptance or rejection of topics that may be judged to belong to another domain of knowledge; they delimit cognitive territories and give rise to shared opinions on what constitutes interesting research and what does not. Söderqvist and Silverstein (1994) studied the dynamics within a specific discipline to establish leadership, comparing these results with citation counts and attendance overlap in meetings; they found a strong correlation between frequent participation, citation, and reputation. The only exceptions were for some highly regarded scientists who do not attend but nevertheless are highly cited, and also of scientists who attend many meetings but are not frequently cited.

At a certain stage of their careers, some scientists may become organizers of meetings. For this activity they need to exhibit leadership, coordination skills, and informed knowledge of their specialty. Scientists also organize meetings for several other reasons: on occasion for the acquisition of academic prestige, power over a group of peers, or plainly for pecuniary gain. This attainment of social and political prominence is not separate from scientific practice, and meetings provide ideal situations to make them explicit. Being an invited speaker to a meeting places a clear tag on the prestige and reputation of a scientist, which make financial and peer support more accessible. Organizing a meeting also requires some tacit abilities to create an atmosphere for effective communication. The structure of a meeting plays an important role for actual communication to take place; good scientific meetings allow the participants to attend all the talks in their sphere of interest and promote personal exchange with the main speakers. Different kinds of meeting structures occupy specific niches in the activation and extension of communication links in the larger community, ranging from local and national to international.

Among the typical meeting configurations of scientific communication, we highlight the following recognizable interaction scenarios:

a. **Congresses and conventions**, where the number of participants is unlimited and often ranges from 10,000 to 30,000; simultaneous sessions must necessarily take place during the typical 1-week duration of the event.
b. **Conferences**, where 100 to 350 participants establish various contacts with specific colleagues, which can be easily pursued during the successive days of the meeting.
c. **Symposia and colloquia** assemble usually no more than some 40 participants, who share a single auditorium where all attendees can listen to all speakers.
d. **Workshops** on a specific research topic, where the number of participants is below 25, including graduate students.
e. **Research group gatherings**, where 3 to 15 participants of a research project or specialized group meet during 1 or 2 weeks, attendance being by invitation only.

The size and structure of a meeting has a strong effect on the dynamics of the communication processes taking place. Meetings where all attendants are present in every presentation promote personal acquaintance, leaving room for argumentation, group formation, and consensus on topics and issues.

Giant meetings with many simultaneous sessions are usually deaf to younger researchers whose presentations can be easily ignored when they coincide with an important lecture in some other session. Different groups can prefer different kinds of meetings. In an ongoing study, for example, we have found that most scientists in quantum mechanics prefer type-*(c)* meetings (Liberman, 2009). The motivation to participate in a given meeting is based on the expected feedback, establishment of contacts, visibility and involvement in the scientific world that allocates leadership and status, recognition and salience in their field, and also the capacity for promoting students and raising funds. To achieve prominence, scientists have to acquire tacit skills both for presenting relevant research and for organizing successful meetings.

In peripheral countries, international contacts are regarded as an important condition for producing visible and relevant science. Scientists from these countries attend meetings trusting the universality of science, the norm being that they are recognized by the excellence of their research, and not necessarily for belonging to any nationality or ethnic group. Mobility is thus a relevant issue. Scientists from Latin America or South Asia will often invest effort and money to travel to a meeting where they may get recognition, attend the invitation to a famous institute, or find a more profitable job. The interorganizational mobility of scientists constitutes one of the most important channels for the flow of knowledge in science (Laudel, 2003), for there is a strong correlation between contact frequency and international publishing activity in all fields of learning (Kyvik & Larsen, 1994). For example, for scientists in México, collaboration means the establishment of international contacts (Liberman & Galán, 2005), and in practice it has led to the foundation of institutions such as the Centro Internacional de Ciencias (1986) in Cuernavaca, whose structure was based on the International Centre for Theoretical Physics (1964) in Trieste, conducted for many years by Nobel Laureate Abdus Salam. The Centro in Cuernavaca initially organized conferences and symposia with success, and after a critical examination of the cost versus benefit for the organizing research groups, the board concluded that small research group gatherings with international participation provide the best scheme for significant scientific interaction and knowledge transfer (Liberman, Wolf, & Seligman, 1991).

Meetings are often organized and associated with academic societies, national, or international, which are also responsible for keeping a formal memory of its resulting communications, usually through the publication of proceedings that become part of the public archive of science, and therefore constitute the formal communication result of the series of informal presentations. The informal communications and the transference of tacit knowledge and know-how, generally mediated by the physical proximity of the participants, remain within the community and in the long run contribute to the diffusion of science among the nations.

COAUTHORSHIP: BONDING BY PUBLICATIONS

Although scientific contacts may become personal bonds between individuals, the bonds evinced by coauthorship in refereed published articles become public. Moreover, while personal contacts may be ephemeral and their existence and estimated number must be retrieved through personal interviews or questionnaires, the coauthorship of scientific publications is documented in journal indices and in the yearly reports of research institutes, and can thus be subject to statistical analysis.

The term "coauthorship" has been used interchangeably with collaboration (Beaver, 2001), and has also been equated with scientific communication or as a measure of collaboration (Melin & Persson, 1996), or at least as a partial indicator of collaboration (Katz & Martin, 1997). Laudel (2002) defined collaboration as "a system of research activities by several actors related in a functional way and coordinated to attain a research goal corresponding with these actors' research goals or interests" (p. 5). Laudel's definition, therefore, restricts collaboration to research that includes personal interactions, where these actions have to be coordinated. In short, communication scholars cannot agree completely on the differences between collaboration and coauthorship. Our model is an attempt to discriminate between the relational process of a group of individuals where person-to-person communications (which could be mediated by e-mail) take place as required from the resulting coauthorship agreement, the placing of signatures and the name ordering to appear in the publication.

Many scientists will argue that anyone directly related to a research product, either generating the basic idea, choosing the method of inquiry, using an instrument, performing the measurements, interpreting the results, or writing the final draft of the paper, should be considered a collaborator (Fine & Kurdek, 1993). In theory, there is an agreement on the ethical norms implicit in collectively signing a written communication, with the names of all participating scientists; but these norms vary widely between the various fields of science, showing the different interpersonal processes in the context of the tasks involved. Publications in mathematics, theoretical or experimental physics, biotechnology, anthropology, engineering, biomedicine, and further afield, will include and list coauthors and acknowledge collaborators in very different ways, which readers in their field will understand (Liberman, 2008). Explanations about the order of names, which may place the scientific leader in the first or in the last position, or maintain a strict alphabetical order, provide little homogeneity for scientometers to study the underlying creation process. Moreover, publications in languages different from English, including those of social scientists and psychologists who tend to publish in regional journals that are not registered in the most accessible international databases (Russell & Liberman, 2002), will not reveal who is the leader and who are the subordinates in a research team. The existing theoretical knowledge of the process of interpersonal communication in psychology can help to explain why and how scientists gather around a prominent researcher working on a given topic, with whom they identify, collaborate, work, and publish together. For example, it is very common that directors of institutes are elected because they hold the generalized approval of colleagues in the same institute or field of specialization, but at the same time this approval is based on that person's visibility in science. We may presume that colleagues working with these scientists identify themselves as an in-group. This does not mean that all projects were originally his ideas, but his coworkers will comply to share recognition. In this process, there are differences between disciplines: physicists claim to be democratic in their decisions on research, taking turns depending on whose ideas generated the project, and claiming at the same time that the order of authors in a publication be alphabetical, whereas biotechnologists tend to identify the leader by his position as the last author in the list (Liberman, 2008).

Scientific research generally involves collaboration of an interacting group, where duties and responsibilities are broadly structured, and where an agreement exists on sharing the resulting credits. The process itself is dynamic; emerging results and roadblocks require frequent adaptation of the working plan that the group leader must tackle and communicate to his associates. Moreover, in this process of collaboration there are internal and external aspects that require distinct tacit skills on the part of the leader. Internal aspects include keeping the group functioning with optimum productivity and fairly apportioning rewards and credit within the team, whereas external tasks include negotiating the continuation of contracts and funds, publication channels, and institutional support while following the rules set by the wider community, which may not be aware at all of the unique inner group process of communication.

There are some clear differences between collaboration, coauthorship, and coordination in science. On the one hand, "collaboration is an intense form of interaction that allows for effective communication as well as sharing competence and other resources" (Melin & Persson, 1996, p. 363; Melin, 2000). On the other hand, coauthorship is the result of an interpersonal, bonded relationship between two or more scientists who have agreed to sign together the written outcome of the results of their communal work. Moreover, coordination is the activity of directing individual efforts toward achieving a common explicitly recognized goal (Blau & Scott, 1962).

Collaboration, coauthorship, and coordination have been studied empirically. In a recent study, Liberman and López (in press) describe how scientists understand coauthorship using a technique to represent a network of semantic meanings of a concept, based on the premise that semantic meaning is the basis for inferring intentions to perform action. The scientists confirmed that coauthorship means collaboration and teamwork, but also means common interest, discussion, the sharing of ideas, active participation, friendship, and commitment; they associate learning, influence, adaptation, and resilience to the concept of coauthorship. Because collaboration and teamwork are the central meanings of coauthorship, we may infer that interpersonal relations develop since scientists first meet and agree to work together, up to when they can distribute to their satisfaction the rewards and credits of their coauthored paper. Coordination and interdependence are basic for the development of scientific projects, even if not all members belong permanently to the same group or have the same long-range goals, but as long as they all agree that the particular research result will be coauthored jointly.

THE BONDING *SIGMA* NUMBER

Related to coauthorship and collaboration we have set forth a model of contacts and social bonds. As stated previously within the informal communication framework, we have studied contacts taking place at scientific gatherings and have established the difference between informal contacts and formal coauthorship bonds in publications (Liberman & Wolf, 1997, 1998). Contacts belong to the *inner cycle* of communication in science, whereas bonds are part of the *outer cycle*. We analyzed bonds based on multiple coauthorships from local publication databases for mathematics, physics, biotechnology, and anthropology. Our main interest was to generate a measure that would help to determine pattern differences of interpersonal communication among these disciplines. Our results showed that in theoretical sciences such as mathematics, scientists tend to work in greater isolation than in physics or biotechnology, where often the results will depend on experimental laboratory teamwork.

A *bond* is a form of attachment to a person or a group. Common-identity theory can make predictions on how individuals go through this process, "while common-bond theory makes predictions about the causes and consequences of people's attachment to individual group members" (Ren et al., 2007, p. 377), and explain how the two concepts illustrate different forms of attachment in groups. These two dimensions show that there are different attachment strengths in a group (Prentice et al., 1994; France et al., 2010). Commitment to a person is not the same as commitment to a group; an individual may see a group as attractive because he shares the same worldview or goals, or because he likes the atmosphere and/or social support offered. Common-bond groups demand personal engagement with other individuals in the group, while common-identity groups are based on the identity being related to the group as a whole. In a similar way, bonds in science are established person to person, with an intention, an explicit exchange, and future involvement, because each one hopes to advance his career by engaging with a well-known leader or collaborator.

In every case, contacts take place during the informal-inner cycle of communication process, whereas coauthorship results from this process and develops into the formal-outer communication cycle of science. We state that the difference between contacts and bonds is that the former are private while the latter are public, part and parcel of the corpus of scientific literature (Liberman & Wolf, 1998). The preceding contacts are part of the personal recipient's memory as enlargement and sharpening of his own working knowledge, and are generally intended for future use, while the bonds revealed by coauthorship are permanent, and can be accessed from the yearly reports and digital databases. This communication model was inspired by the early work of Bavelas (1950), who established that the centrality and frequency of communication acts performed by individuals in a group would reveal the structure of that group and the process of communication.

We defined a bond as the link between a pair of coauthors (Liberman & Wolf, 1998). The number of bonds contained in an article published by n authors is thus the number of pairs of authors in that collection of token names: the combinations of n objects in sets of two. Each of the n authors has $n-1$ coauthors, so there will be $n(n-1)$ directional links; but since coauthorship is a reciprocal property, the number of bonds in that article is half of this, and thus given by the following *binomial coefficient:*

$$\frac{1}{2}n(n-1) = \binom{n}{2}$$

This number is the same as the number of glass clicks heard in any n-person banquet where each participant toasts with each one of his friends. We use this coefficient to quantify with a single number the interaction contained in any one article. Single-authored articles ($n = 1$) contain zero bonds; an article with two authors ($n = 2$) evinces one bond; three coauthors ($n = 3$) imply three bonds; four coauthors engage in six bonds, and so on.

From the outset, it was clear that different disciplines have rather distinct characteristic numbers of bonds per average article. This characteristic number is the *mean number of coauthors* in the published corpus of refereed articles produced by that community of scientists in some given time interval; we called this the *bonding number*, indicating it by *sigma* (σ), to whose definition we now turn, together with other quantities that are based on the count of bonds.

Let A_n be the number of n-author articles published by the researchers in a given community and year, for $n = 1, 2, 3, \ldots$ Then, the number of token authors is the sum (indicated by S_n) over all values of n, of n times A_n, namely $N_{authors} = S_n\, n\, A_n$. Purportedly, each coauthor has established a *bond* of interaction with each other coauthor of the same article, and the total number of such bonds is $N_{bonds} = \frac{1}{2} S_n\, n\, (n{-}1)\, A_n$. Hence, the mean number of bonds in that community is their sigma number, $s = N_{bonds} / N_{authors}$.

The four research communities we studied were: mathematicians, physicists, biotechnologists, and anthropologists, from four institutes of the Universidad Nacional Autónoma de México during the years 1982 to 1994. In this study we found an average of 2.91 bonds in physics, 0.9 in mathematics 3.9 in biotechnology, and 0.69 in anthropology (Liberman & Wolf, 1998). In a following study we found that the bonding number for biotechnologists was 4 in 1982 and 4.4 in 2004, with an average of 3.9 in the period; for physicists 2.89 in 1982 and 3.07 in 2004, with an average of 3.05; and for mathematicians in 1982 it was 1 while in 2004 the bonding number was 2, with an average of 1.18 for this period. There is a significant difference between the *sigma* (*bonding*) values between disciplines but these had not changed appreciably over those years. Inclusion of one more decade of data in physics revealed a certain discrete increase not only in productivity per researcher, but also in their bonding number (Liberman & Wolf, 2008). The latter study had to include a few articles with 100 plus authors in the last years, which put the spotlight again on the question of validity of the bonding number for such articles, which increases asymptotically as $\approx n^2/2$. There are obvious reasons to doubt the appropriateness of assigning one bond to each pair of coauthors when their number is large. Such multiauthored articles usually include researchers from different institutes in various countries, many of whom have never met in face-to-face scientific conversation. The interaction present in such works takes place mostly within the subsets of theoreticians, engineers, software developers, and among colleagues of the same institution. A simile would be a banquet with very many guests: while in a small party each toast indeed produces $n(n-1)/2$ glass clicks, in a large gathering each guest will click glasses only with his n_s immediate neighbors, where n_s is the *saturation number* of the mean participant, beyond which no more bonds can be established.

There are various estimates on the optimum size of a group, all of whose members will interact with all others. James (1951) concludes that groups larger than five members tend to be unstable, and that freely forming groups are stable when they average three members. Hare (1952) adds that in larger groups there is less chance for speaking and factionalism is evident, while Kretschmer (1985) finds that the maximum level of information processing in a research group occurs at sizes between 6 and 12 members. However, in the practice of group psychology, Slater (1958) demonstrated that the number of meaningful pairwise interactions in a gathering is understood to be saturated with around five participants, and Qurashi (1993) established that it appears to be close to multiples of 8.5, indicating the possibility that a subgroup of eight to nine persons could be forming the basic unit of interaction in these particular research groups. Eliminating from the data set the A_n articles with $n > n_s$ authors are not the best solution, because the bonds are there and need to be counted, albeit differently. Our proposition here is to model the interaction among the authors of a corpus of literature by bonds between pairs of coauthors *up to a saturation number n_s*, is to count them by the original quadratic growth $n(n-1)/2$ in the range $1 < n \le n_s$, and with *linear* growth $n_s(n-1)/2$ beyond that saturation number, for $n_s \le n$, up to the maximum number of authors $n = z$. Thus corrected, the number of bonds in the collection of articles is then counted by

$$N'_{bocdx} := \frac{1}{2} \sum_{n=2}^{n_s} n(n-1)A_n + \frac{1}{2} \sum_{n=n_s+1}^{z} n_s(n-1)A_n$$

where we used $n_s = 8$.

In that work we were interested in proposing indicators of scientific communication that are independent of the size of the researcher population, total number of articles it produces, and indeed their academic excellence. For those qualities other indicators exist, such as the number of citations, the impact number of a journal, or the h-index of a researcher. Said in the language of thermodynamics, we wanted to extract the *intensive* variables (such as temperature or density) from the *extensive* ones (such as total heat or mass). The sigma number, $s = N_{bonds} / N_{authors}$, being the mean number of bonds of coauthors in each scientific community, satisfies the criterion of being a proper intensive indicator, which can be claimed to be characteristic of a research field, an institute, a journal, or indeed a single individual, over a given period of time.

Cross-checks on the consistency of the bonding number with other commonly used concepts such as cohesiveness were undertaken (Lima, Liberman, & Russell, 2005), showing that they can characterize the disciplines of mathematics, physics, and biotechnology, where distinct dynamic configurations exist within the typical research groups. The concept of bonding number can also be applied to an individual, as we noted above, representing the history of a scientist's bonding tactics along his or her professional development.

SUMMARY AND CONCLUSIONS

In this chapter we have described how informal-interpersonal communication is present at all levels of the social organization of science. Interpersonal communication leads to the establishment of contacts in informal situations between scientists in any accessible occasion and during scientific meetings. Through informal-interpersonal communication scientists build relationships that later become collaboration. The result of the development of relationships out of contacts can be extracted and measured from the publications in the archives of science. We develop a three-cycle model from individual to group and to societal levels of scientific communication.

The first cycle describes the psychological aspects of interpersonal communication in informal settings, such as personal encounters and meetings. The second cycle describes how common-bond groups are formed based on the task in the process of collaboration. The third cycle, which is not described in this chapter, brings attention to the popularization, applications, and relations between science to society at large.

Considering that scientific communication takes place in a context, the organization of successful meetings should include the variables mentioned above, namely face-to-face interaction and all instances of informal-interpersonal communication. We distinguished between collaboration and coauthorship as different social-psychological behaviors; collaboration is an interpersonal process characterized by group processes, while coauthorship is the signature and credit shown in publications that may be considered an indicator of the resulting permanent relationships of group collaboration.

We have described these processes to contextualize our model of scientific communication within the larger body of studies on scientific communication. This model can be useful to further the quality of interaction between scientists and to underline the importance of the size of scientific groups and meetings. New information technologies will probably accelerate the cycles of scientific communication but they will not modify its actors.

We propose that a closer systematic look into interpersonal communication in science from the psychological perspective can be useful to develop a theory and to establish, first, the deeper and not so visible internal aspects of scientific communication; and second, the psychological variables involved in the formation of scientific groups and collaboration. The bonding number we described is a measure of patterns of interaction for group formation in science, where common-bond attachment to groups in science can be revealed.

Studies in computer-mediated communication are concerned with the development of rich media technologies to facilitate informal communications, either in local groups or in "collaboratories," where scientists from various countries work on one project and where the technology enhances connectedness (Nardi, 2005).

Bringing back our initial statement that scientific knowledge is personal (Polanyi, 1958), the exchange of information and practical knowledge takes place mostly at the interpersonal zone. Socializing knowledge is one of the most important activities that scientists perform; it can be considered a tacit skill. Once we have described the psychological features of informal and formal communication in science, we have a basis to differentiate the diverse strata of theoretical approaches and perspectives to promote interdisciplinary studies of scientific communication.

ACKNOWLEDGMENT

Sponsored by DGAPA-UNAM Project IN303310, and the Centro Internacional de Ciencias AC.

REFERENCES

Abrams, D., Hogg, M. A., Hinkle, S., & Otten, S. (2005). The social identity of small groups. In M. S. Poole & A. B. Hollingshead (Eds.), *Theories of small groups: Interdisciplinary perspectives* (pp. 99–137). Thousand Oaks, CA: Sage.

Andrews, F. M. (1979). Motivation, diversity, and performance of research units. In F. M. Andrews & G. Aichholzer (Eds.), *Scientific productivity, the effectiveness of research groups in six countries* (pp. 253–289). Cambridge, UK: Cambridge University Press.

Anderson, C. M. (1998). The use of computer technology by academics to communicate internationally: Computer mediated communication and the invisible college (Doctoral dissertation, University of Massachusetts, Amherst). *Dissertation Abstracts International, 59*(7-A), 2375.

Aydogan, N., & Lyon, T. P. (2004). Spatial proximity and complementarities in the trading of tacit knowledge. *International Journal of Industrial Organization, 22*(8), 1115.

Bales, R. F. (1955). *How people interact in conferences*. San Francisco, CA: Freeman.

Bales, R. F., Cohen, S. P., & Williamson, S. A. (1979). *SYMLOG: A system for the multiple level observation of groups*. New York, NY: Free Press.

Barker, V. E., Abrams, J. R., Tiyaamornwong, V., Seibold, D. R., Duggan, A., Park, H. S., & Sebastian, M. (2000). New contexts for relational communication in groups. *Small Group Research, 31*(4), 470–503.

Bavelas, A. (1950). Communication patterns in task oriented groups. *Journal of the Acoustical Society of America, 22*, 271–282.

Bavelas, J., Hutchinson, S., Kenwood, C., & Matheson, D. (1997). Using face-to-face dialogue as a standard for other communication systems. *Canadian Journal of Communication, 22*, 5–24.

Bavelas, J. B., & Chovil, N. (2000). Visible acts of meaning: An integrated message model of language in face-to-face dialogue. *Journal of Language and Social Psychology. 19*(2), 163–194.

Beaver, D. D. (2001). Reflections on scientific collaboration (and its study): Past, present, and future. *Scientometrics, 52*(3), 365–377.

Blau, P. M., & Scott, W. R. (1962). *Formal organizations*. San Francisco, CA: Scott, Foreman.

Borgatti, S. P., Mehra, A., Brass, D. J., & Labianca, G. (2009). Network analysis in the social sciences. *Science, 323*(5916), 892–895.

Caplow, T., & McGee, R. J. (1958). *The academic marketplace*. New York, NY: Basic Books.

Coleman, J. S., Katz, E., & Menzel, H. (1957). *The diffusion of an innovation among physicians.* Bobbs-Merrill reprint series in the social sciences., S-48. Indianapolis, IN: Bobbs-Merrill.

Collins, H. M. (2001). Tacit knowledge, trust and the Q of sapphire. *Social Studies of Science.* 31(1), 71–85.

Crane, D. (1969). Social structure in a group of scientists: A test of the "Invisible College" hypothesis. *American Sociological Review.* 34(3), 335–352.

Crane, D. (1972). *Invisible colleges. Diffusion of knowledge in scientific communities.* Chicago, IL: Chicago University Press.

Feist, G. J. (2006). How development and personality influence scientific thought, interest, and achievement. *Review of General Psychology: Journal of Division 1, of the American Psychological Association.* 10(2), 163–182.

Fine, M. A., & Kurdek, L. A. (1993). Reflections on determining authorship credit and authorship order on faculty-student collaborations. *American Psychologist, 48*(11), 1141–1147.

Fortes, J., & Lomnitz, L. (1991). *La formación del científico en México: Adquiriendo una nueva identidad.* Educación. México, D.F.: Siglo Veintiuno Ed.

France, M., Finney, S., & Swerdzewski, P. (2010). Students' group and member attachment to their university: A construct validity study of the university attachment scale. *Educational and Psychological Measurement, 70*(3), 440–458.

Frey, L. R. (2003). *Group communication in context: Studies in bona fide groups.* Mahwah, NJ: L. Erlbaum.

Gallie, E. P., & Guichard, R. (2005). Do collaboratories mean the end of face-to-face Interactions? An evidence from the ISEE project. *Economics of Innovation and New Technology, 14*(6), 517–532.

Gamble, T. K., & Gamble, M. (1998). *Contacts: Communicating interpersonally.* Boston, MA: Allyn and Bacon.

Garvey, W. D., & Griffith, B. C. (1968). Informal channels of communication in the behavioral sciences: Their relevance in the structuring of formal or bibliographic communication. In E. B. Montgomery (Eds.), *The foundations of access to knowledge:* A symposium (pp. 129–146). Frontiers of librarianship, no. 8. Syracuse, NY: Division of Summer Sessions, Syracuse University.

Genuth, J., Chompalov, I., & Shrum, W. (2000). How experiments begin: The formation of scientific collaborations. *Minerva, 38*(3), 311–348.

Goffman, E. (1967). *Interaction ritual; essays on face-to-face behavior.* Garden City, NY. Doubleday.

Goleman, D. (1995). *Emotional intelligence.* New York, NY: Bantam Books.

Gorman, M. E. (2002). Levels of expertise and trading zones: A framework for multidisciplinary collaboration. *Social Studies of Science, 32*(6), 933–938.

Granovetter, M. S. (1973). The strength of weak ties. *The American Journal of Sociology, 78*(6), 1360–1380.

Granovetter, M. (1983). The strength of weak ties: A network theory revisited. *Sociological Theory.* 1, 201–233.

Griffith, B. C., Jahn, M., & Miller, A. J. (1971). Informal contacts in science: A probabilistic model for communication processes. *Science, 173*(3992), 164–165.

Gustin, B. H. (1973). Charisma, recognition, and the motivation of scientists. *American Journal of Sociology, 78*(5), 1119–1134.

Hagstrom, W. O. (1965). *The scientific community.* New York, NY: Basic Books.

Hare, A. P. (1952). A study of interaction and consensus in different sized groups. *American Sociological Review, 17*(3), 261–267.

Hirokawa, R. Y. (1988). Group communication research: Considerations for the use of interaction analysis. In C. H. Tardy (Eds.), *A handbook for the study of human communication: Methods and instruments for observing, measuring, and assessing communication processes. Communication and information science* (pp. 229–245). Norwood, NJ: Ablex.

Hirokawa, R. Y., & Poole, M. S. (1996). *Communication and group decision making.* London, UK: Sage.

Homans, G. C. (1950). *The human group*. New York, NY: Harcourt, Brace.

James, J. (1951). A preliminary study of the size determinant in small group interaction. *American Sociological Review, 16*(4), 474–477.

Jarvenpaa, S. L., & Leidner, D. E. (1999). Communication and trust in global virtual teams. *Organization Science, 10*(6), 791–815.

Kalkhoff, W., Friedkin, N. E., Johnsen, E. C. (2010). Status, networks, and opinions: A modular integration of two theories. In S. R. Thye & E. J. Lawler (Eds.), *Advances in Group Processes* (pp. 1–38). Bradford, UK: Emerald Group Publishing Limited.

Katz, J. S. (1994). Geographical proximity and scientific collaboration. *Scientometrics, 31*(1), 31.

Katz, J. S., & Martin, B. R. (1997). What is research collaboration? *Research Policy, 26*(1), 1–18.

Kelly, J., & Barsade, S. (2001). Mood and emotions in small groups and work teams. *Organizational Behavior and Human Decision Processes, 86*(1), 99–130.

Kelman, H. C. (1958). Compliance, identification, and internalization: Three processes of attitude change. *Journal of Conflict Resolution, 2*(1), 51–60.

Kelman, H. C. (2006). Interests, relationships, identities: Three central issues for individuals and groups in negotiating their social environment. *Annual Review of Psychology, 57*, 1.

Keyton, J. (1999). Relational communication in groups. In L. R. Frey, D. S. Gouran, & M. S. Poole (Eds.), *The handbook of group communication theory and research* (pp. 192–222). Thousand Oaks, CA: Sage Publications.

Keyton, J., & Beck, S. J. (2009). The influential role of relational messages in group interaction. *Group dynamics: Theory, research, and practice, 13*(1), 14–30.

Keyton, J., Bisel, R. S., & Ozley, R. (2009). Recasting the link between applied and theory research: Using applied findings to advance communication theory development. *Communication Theory, 19*(2), 146–160.

Krauss, R. M., & Fussell S. R. (1996). Social psychological models of interpersonal communication. In E. T. Higgins & A. W. Kruglanski (Eds.), *Social psychology: Handbook of basic principles* (pp. 655–701). New York, NY: Guilford Press.

Kraut, R. E., Egido, C., & Galegher, J. R. (1990). Patterns of contact and communication in scientific collaborations. In J. R. Galegher, R. E. Kraut, & C. Egido (Eds.), *Intellectual teamwork: social and technological foundations of cooperative work* (pp. 149–171). Hillsdale, NJ: Erlbaum.

Kraut, R. E., Galegher, J., & Egido, C. (1987). Relationships and tasks in scientific research collaboration. *Human-Computer Interaction, 3*(1), 31–58.

Kretschmer, H. (1985). Cooperation structure, group size and productivity in research groups. *Scientometrics, 7*(1–2), 39–53.

Kwon, H. I. (1998, December). The effects of computer-mediated communication on the small-group problem-solving process. *Dissertation Abstracts International A: Humanities and Social Sciences, 59*(6-A).

Kyvik, S., & Larsen, M. I. (1994). International contact and research performance. *Scientometrics, 29*(1), 161.

Laudel, G. (2002). What do we measure by coauthorships? *Research Evaluation, 11*, 3–16.

Laudel, G. (2003). Studying the brain drain: Can bibliometric methods help? *Scientometrics, 57*(2), 215–237.

Liberman, S. (2008, July). *Norms of integration, roles and tasks in scientific groups in relation to name ordering in publications*. Paper presented at 2nd Bi-Annual Conference for the International Psychology of Science & Technology (ISPST), Berlin, Germany.

Liberman, S. (2009, October 30). *The relation between the individual motivation of scientists to attend meetings and the structure of scientific gatherings*. Paper presented at the meeting of the Society for Social Studies of Science, Washington, DC.

Liberman, S. (2011). Psychology of science: Notes on starting a new discipline in psychology. In N. Milgram, A. M. O'Roark, and R. Roth (Eds.), N. Milgram, A. M. O'Roark, and R. Roth (Eds), *Scientific Psychology: New Developments Internationally: Proceedings of* the 67th Annual Convention, International Council of Psychologists, July 3–8, 2009, Mexico City, México. Aachen, Germany: Shaker-Verlag.

Liberman, S. S., & Galán D. C. R. (2005). Shared semantic meaning of the concept of international collaboration among scientists. *Journal of Information Management and Scientometrics*, 2(2), 27–34.

Liberman, S., & López, O. R. (in press). Semantic meaning of the concept of coauthorship among scientists. *Journal of Psychology of Science and Technology*, ISPST.

Liberman, S., & Wolf, K. B. (1997). The flow of knowledge: Scientific contacts in formal meetings. *Social Networks*, 19(3), 271.

Liberman, S., & Wolf, K. B. (1998). Bonding number in scientific disciplines. *Social Networks*, 20(3), 239.

Liberman, S., & Wolf, K. B. (2008). Patrones de interacción en 50 años de la Revista Mexicana de Física. *Boletín de la Sociedad Mexicana de Física*, 22(2), 97–100.

Liberman, S., Krötzch G., & Wolf, K. B. (2003). Porqué discuten los científicos?. *CIENCIA: Revista de la Academia Mexicana de Ciencias*, Abril-junio, 54. N° 2. ISSN 1405–6550.

Liberman, S. S., Wolf, K. B., & Seligman P. (1991). Costos de la transferencia Internacional del conocimiento científico. *Ciencia y Desarrollo*. CONACYT. Nov-dic. XVII, num. 101, 56–66.

Lievrouw, L. A. (1990). Communication and the social representation of acientific knowledge. *Critical Studies in Mass Communication*, 7, 1–10.

Lima, M., Liberman S., & Russell M. J. (2005). Scientific group cohesiveness at the National University of México. *Scientometrics*, 64, 55–66.

Lomnitz, L. (1983). The scientific meeting: An anthropologist's point of view. *4S Review*, 1(2), 2–7.

Mahoney, M. J. (1976). *Scientist as subject: The psychological imperative*. Cambridge, MA: Ballinger.

Mathieu, J. E, Heffner, T. S., Goodwin, G. F., Salas, E, & Cannon-Bowers, J. A. (2000). The influence of shared mental models on team process and performance. *The Journal of Applied Psychology*, 85(2), 273–283.

Meadows, A. J. (1998). *Communicating research*. San Diego, CA: Academic Press.

Melin, G., & Persson, O. (1996). Studying research collaboration using coauthorships. *Scientometrics*, 36(3), 363–378.

Melin, G. (2000). Pragmatism and self-organization—Research collaboration on the individual level. *Research Policy*, 29(1), 31–40.

Mitroff, I. I. (1974). Integrating the philosophy and the social psychology of science or a plague on two houses divided. *PSA: Proceedings of the Biennial Meeting of the Philosophy of Science Association*, 1974, 529–548.

Nardi, B. (2005). Beyond bandwidth: dimensions of connection in interpersonal communication. *Computer Supported Cooperative Work*, 14(2), 91–130.

Pelz, D. C., & Andrews, F. M. (1976). *Scientists in organizations: Productive climates for research and development*. Ann Arbor, MI: Inst. for Social Research.

Polanyi, M. (1958). *Personal knowledge; towards a post-critical philosophy*. Chicago, IL: University of Chicago Press.

Pool, I. D. S., & Kochen, M. (2006) Contacts and influence. In M. E. J. Newman, A. L. Barabási, & D. J. Watts (Eds.), *The structure and dynamics of networks* (pp. 83–129). Princeton, NJ: Princeton University Press.

Prentice, D. A., Miller, D. T., & Lightdale, J. R. (1994). Asymmetries in attachments to groups and to their members: Distinguishing between common-identity and common-bond groups. *Personality and Social Psychology Bulletin*, 20(5), 484.

Price, D. J. S. (1963). *Little science, big science*. New York, NY: Columbia University Press.

Price, D. (1971). Invisible college research: State of the art. In S. Y. Crawford, F. L. Strodtbeck, & D. J. D. S. Price (Eds.), *Informal communication among scientists: Proceedings of a conference on current research* (pp. 3–14). Chicago, IL: American Medical Association. ERIC #:ED056697.

Qurashi, M. M. (1993). Dependence of publication-rate on size of some university groups and departments in U. K. and Greece in comparison with N. C. I., USA. *Scientometrics*, 27(1), 19–38.

Ren, Y., Kraut, R., & Kiesler, S. (2007). Applying common identity and bond theory to design of online communities. *Organization Studies, 28*(3), 377–408.

Renz, M. A., & Greg, J. B. (2000). *Effective small group communication in theory and practice.* Boston, MA: Allyn and Bacon.

Russell, J. M., & Liberman, S. (2002). Desarrollo de las bases de un modelo de comunicación de la producción científica de la Universidad Nacional Autónoma de México. *Revista Española de Documentación Científica, 25*(4), 361–370.

Schutz, W. (1966). *The interpersonal underworld.* Palo Alto, CA: Science & Behavior Books.

Slater, P. H. (1958). Contrasting correlates of group size. *Sociometry, 21*(2), 129–139.

Snyder, M., & Stukas, A. A. (1999). Interpersonal processes: The interplay of cognitive, motivational, and behavioral activities in social interaction. *Annual Review of Psychology, 50,* 273–304.

Söderqvist, T., & Silverstein, A. M. (1994). Participation in scientific meetings: A new prosopographical approach to the disciplinary history of science—The case of immunology, 1951–1972. *Social Studies of Science, 24*(3), 513–548.

Storer, N. W. (1966). *The social system of science.* New York, NY: Holt, Rinehart and Winston.

Thagard, P. (2005). How to be a successful scientist. In M. E. Gorman, R. D. Tweney, D. C. Gooding, & A. P. Kincannon. *Scientific and technological thinking* (pp. 159–171). Mahwah, NJ: L. Erlbaum.

Westrum, R. (1989). The psychology of scientific dialogues. In B. Gholson, W. R. Shadish, R. A. Neimeyer, & A. C. Houts (Eds.), *Psychology of science: Contributions to meta-science* (pp. 370–382). Cambridge, UK: Cambridge University Press

White, H. D. (1994). Scientific communication and literature retrieval. In Cooper, H. M., Hedges, L. V., & Valentine, J. C. *The handbook of research synthesis and meta-analysis.* (pp. 41–55). New York: Russell Sage Foundation.

Zuccala, A. (2006). Modeling the invisible college. *Journal of the American Society for Information Science and Technology: Jasist, 57,* 2, 152.

Development and Theory Change

CHAPTER 7

Scientific Reasoning: Explanation, Confirmation Bias, and Scientific Practice

Barbara Koslowski

One of the reasons there are so many approaches to studies of scientific reasoning is not necessarily because researchers disagree about what constitutes sound scientific reasoning (although disagreements certainly exist), but because there are many aspects to sound scientific reasoning, and different researchers focus on different aspects. In the present chapter, the focus will be on issues in psychological and, to some extent, sociological approaches to scientific reasoning. Readers interested in the large and exciting body of research on issues related to education would be well served by beginning with the recent report by the National Research Council (2007). Similarly, readers might also find interesting studies of the reasoning and discovery processes of individual scientists (e.g., Tweney, 1985; Gruber, 1981), and some of this work, along with the work of others (e.g., Feist, 2006; Simonton, 2010) will also be relevant to those interested in the creative process.

The present chapter will focus on explanation, broadly construed. The reason is straightforward: explanation is central to, and the ultimate goal of, scientific inquiry. In the psychological literature, this is reflected in work on causal reasoning and on the role of explanation (Zimmerman, 2000, 2007). The two are inextricably intertwined; one way of explaining an event is to identify its the cause. For example, one way of explaining why someone died is to say that her death was caused by tuberculosis (TB). In sociological/ anthropological literature, the focus on explanation is reflected in research on the broader social context in scientific inquiry. The present chapter will focus on questions in which there is relevant overlap among the disciplinary emphases.

The present chapter will also include research on children as well as college students and practicing scientists. The reason for studying the first two populations is that many researchers are interested in what the cognitive precursors are of scientific reasoning (e.g., Carruthers, Stich, & Siegal, 2002). What are children and college students able—and not able—to do that affects their later ability to engage in scientific inquiry?

Finally, for reasons discussed in the section "Different Views of Science and of Explanation," because this chapter focuses on the ways in which explanation is approached in actual scientific *practice*, it includes very little coverage of formal models of thinking. Moreover, a central argument in this chapter is that, although one can *frame* principles of scientific inquiry as though they were formal, the principles can be *applied successfully* only when background information (including information about mechanism or explanation) is taken into account. Thus, in addition to research on the role of explanation in scientific reasoning, this chapter will also focus on two other types of background information: information about alternative explanations, and information about anomalies.

DIFFERENT VIEWS OF SCIENCE AND OF EXPLANATION

One's view of causal explanation and of how causal reasoning ought to be studied depends, in part, on one's view of science. In psychological literature, typically (though not necessarily), different views of science map onto different approaches to research.

Explanations Can Identify Causal Events or Causal Mechanisms

One way of distinguishing psychological research strategies (and the views of science that they represent) is in terms of the type of explanation that psychologists study. The type most prominently studied consists of identifying the causal event in a situation, asking, for example, whether it is insufficient folic acid that is the cause of neural tube defects, or whether flipping one switch rather than another causes a light to turn on. For the most part, researchers who study how people identify causal events rely on the philosophical contributions of David Hume and ask about the extent to which people rely on Humean indices (of priority, contiguity, and covariation) to do so. For example, if X causes Y, then X is temporally prior to Y, is temporally and spatially contiguous with Y, and covaries with Y. (Some of this research, including research with infants, is summarized in Haith & Benson, 1998, and in Koslowski & Masnick, 2002; Koslowski & Thompson, 2002). In addition, researchers who study people's reliance on Humean indices often (though not always) at least tacitly assume that relying on Humean indices is, and ought to be, sufficient for identifying causal events (Gopnik & Schulz, 2007; Kuhn, Amsel, & O'Loughlin, 1988; Schulz & Gopnik, 2004).

Of course, at least since Kant, there have been serious philosophical criticisms of Hume's approach. Although some current philosophers of science describe themselves as advancing a Humean conception of causation, the idea that relying on Humean indices, independently of theoretical considerations, is an important tool in scientific investigations is no longer credible (for discussion from a range of perspectives see, e.g., Boyd, 1989; Fine, 1984; Harman, 1965; Kitcher, 1981; Kuhn, 1970; Lipton, 1993; Putnam, 1972, 1975; Salmon, 1984, 1990; van Fraassen, 1980). Furthermore, although identifying causal events is often a first step in explaining an effect, scientific (as well as lay reasoning) is often concerned with identifying the mechanism or process by which an effect is brought about; people want to know *how* or *why* insufficient folic acid causes neural tube defects.

Formal Rules Are Applied Successfully Only When Background Information Is Considered

Researchers who focus on the role of mechanism in causal reasoning also point out that even when rules can be *framed* as though they are formal (as the Humean rules often are), they cannot be *applied successfully* unless background information (including information about mechanism or explanation) is taken into account (see immediately preceding references). Philosophers of science of all persuasions (Humeans as well as nonHumeans) agree that in actual scientific *practice*, scientists choose explanations and theories from among a handful of relevant alternatives, rather than all possible alternatives, and that the relevant alternatives are suggested, not just by Humean indices, but also by scientists' theoretical conceptions of the relevant phenomena, including mechanism information (Boyd, 1989; Fine,1984; Harman, 1965; Kuhn, 1970; Lipton, 1993; Van Fraassen, 1980). For an especially thorough discussion of the role of mechanism considerations, see Darden (2006).

By "formal rules," I mean rules that can be applied regardless of content or background information. A clear example of a formal rule is that 2 + 2 = 4, regardless of whether one is adding elephants or airplanes. In psychology, rules that have been *treated* as formal include parsimony (Bonawitz & Lombrozo, 2012), internal consistency (Samarapungavan, 1992), probability (Lombrozo, 2007), and (as already noted) Humean indices such as covariation, priority, and contiguity (Bullock, Gelman, & Baillargeon, 1982). When rules are treated as formal, the idea is that, for example, if the form of bread (rolls vs. slices) covaries with susceptibility to colds, then form of bread should be treated as causal, regardless of whether one's background information, broadly understood, provides many reasons why this conclusion is very unlikely.

In addition, even in the early stages, when one is trying only to identify the causal event in a situation and is not yet concerned about the mechanism by which it might operate, people (including scientists) do and should rely on mechanism information to decide which potentially causal events are also likely to be actually causal. Thus, if insufficient folic acid and a preference for the Beatles rather than Bach both covary with neural tube defects, one would likely attribute the defects to the former rather than the latter. The reason is that there is a plausible mechanism by which folic acid can bring about certain defects, but no (currently known) plausible mechanism by which preference for the Beatles can do so. Put differently, even though we do and should rely on Humean principles to some extent, we nevertheless rely on mechanism information to decide when to take the results seriously. (This is the basis for the standard warning that correlation does not guarantee causation.) Humean indices might reflect an underlying causal relation, but they cannot by themselves be sufficient to identify a causal relation.

The disagreement about the role that should be played by Humean indices is often reflected not only in how research is designed, but also in how the results are interpreted. In psychological research, treating the Humean indices as the sole indicators of causation means that, sometimes, experiments are designed so that certain events covary with effects (even though they do not cause the effects in the actual world), and "correct" performance is stipulated to consist of treating the covarying events as causing the effect. For example, one experiment was designed so that, for children three to eleven years of age, dropping a marble into one side of a box was followed by the ringing of a bell from inside the box. An inhibitory "cause" consisted of a toy car placed on top of the box, which "prevented" the bell from ringing (Shultz & Mendelson, 1975). In another study, for example, eating bread in the form of rolls rather than slices was presented as affecting susceptibility to colds (Kuhn, Amsel, & O'Loughlin, 1988). In such experiments, the participants are often required to do two things to be counted as having performed "correctly." One is that they need to take account of the Humean indices with which they are presented; the other is that they often have to ignore or override background information acquired outside of the laboratory, namely, that proximity of toy cars typically has no effect on how a mechanical apparatus will work, and that the form in which bread is eaten does not change the composition of bread and is thus unlikely to affect illness. That is, they need to treat Humean information as the only information that matters. However, as already noted, in actual scientific practice, applying rules framed to appear formal, such as the Humean indices, is done in conjunction with, rather than by ignoring, background information, including information about mechanism or explanation.

This point can also be seen with respect to the rule that one should consider alternative hypotheses. In practice, one never considers all *possible* alternative hypotheses (such as, cloud cover while the experiment was running, or the experimenter's hair color). Rather, one tries to control for *plausible* alternative hypotheses, and one relies on background information (often about explanation or theory) to decide which are plausible. If one can specify the process or mechanism by which cloud cover might affect a chemical reaction (by noting, e.g., that a biochemical process is affected by sunlight), then cloud cover might be worth controlling for—especially if the results are not consistent with what else we know about such reactions. However, in many cases (such as whether cloud cover or experimenter's hair color affects memory for nonsense syllables) such possible alternative hypotheses neither are, nor ought to be, taken seriously.

One could, of course, consider controlling for all *possible* alternative hypotheses, but actually to do so would be *pragmatically* impossible, because the number of possible (as opposed to plausible) alternatives is extremely large and the amount of time that would be required to digest the output is indeed long. Thus, one must rely on background information (including information about theory or mechanism) to decide which of the possible alternative hypotheses should also be treated as plausible and are thus worth considering. (In some cases, it might appear that alternative hypotheses need not be considered. For example, to determine whether drug X cures illness Y, one might administer the drug to half of the victims and withhold it from the other half. However, when this procedure is effective, it is typically because it is combined with random assignment and a large enough sample that possible alternatives—including those that have not yet been discovered—will be ruled out.

Note that researchers have documented that sheer familiarity (or a sense of "reality") can also make it easier to apply formal rules; for example, it is easier for people to complete Wason's (1968) 4-cards task (which is a formal rule that is not causal), when the task is conceptualized as, "When a letter is sealed, then it has a 50-lire stamp on it" (Johnson-Laird, Legrenzi, & Legrenzi, 1972). In the present chapter, the concern is not with arbitrary rules but with causal explanations. In addition, the concern is not with familiarity, but with three particular types of background information: information about explanation or mechanism, about alternative explanations, and about anomalous information.

Idealized or Formal Views of Science Differ From Actual Scientific Practice

Because applying various formal rules cannot, by itself, guarantee good science, philosophers of science have been interested in describing actual scientific practice—in which background information plays a crucial role. The argument is that, at least in the long run, scientific practice is effective: bridges get built, astronauts land on the moon, and illnesses are cured. Since scientific practice relies on background information, including information about explanation or mechanism, the present chapter preferentially includes research on the role of explanation in scientific reasoning for two reasons. One is that it is one of the areas in which psychological research on scientific inquiry overlaps with research in sociology and anthropology. The other is that the role of explanation has come to be seen as increasingly important in a variety of cognitive activities (Ahn, Novicki, & Kim, 2003; Brem & Rips, 2000; Gelman & Coley, 1990; Giere, 1992; Keil, 2006; Koehler, 1991; Lombrozo, 2007; Lombrozo & Carey, 2006; Murphy & Medin, 1985; Proctor & Capaldi, 2006).

The most comprehensive description of actual scientific practice is termed "inference to the best explanation" (IBE), sometimes called "abduction." IBE argues that an explanation is chosen over its competitors because it provides a better causal account of the data (Harman, 1965; Lipton, 1991; Magnani, 2001; Proctor & Capaldi, 2006; Thagard, 1989; Thagard & Verbeurgt, 1998). Roughly, IBE argues that an explanation is preferred to its plausible competitors because it provides a more plausible account of, or is more plausibly causally consistent with, what else we know about the world.

The classic example of IBE is evolutionary theory. The argument that different species evolved has been judged, not in isolation, but with respect to the alternative that the individual species arose by divine design. In addition, IBE also notes that whether a piece of information is treated as evidentially relevant may well depend on whether an explanation or theory makes it so. Evolutionary theory (in contrast to the alternative) makes the existence of intermediate fossil forms and niche-specific adaptations evidentially relevant, because it incorporates them into a broader causal framework (in a way that the spontaneous rise of species does not) and illustrates the interdependence of theory and evidence. To be sure, the observations can be distorted to be consistent with each of the explanations. One could, for example, attribute all of the niche-specific adaptations to coincidence. However, the evidence is more *plausibly* (and causally) consistent with evolution. That is, IBE suggests that we evaluate competing explanations by assessing the *relative weight* of the evidence in favor of each and judging, which explanation the evidence is more plausibly consistent. Furthermore, evolution (but not the alternative) is causally consistent with the web of related background information (Quine & Ullian, 1970), that is, with well-established background beliefs about, for example, plate tectonics, population genetics, and animal breeding. In addition, IBE draws attention not only to the fact that, in actual scientific practice, theory and data are interdependent, but also to the fact that related background information can involve theory (or explanation) as well as data. Finally, the fact that evidence for an explanation is cumulative over time acknowledges that initial hypotheses are often working hypotheses that do not specify all the relevant variables (often, because they have not yet been discovered) and that they are modified (or rejected) as additional data become available. For example, initial hypotheses about change in species over time did not specify punctuated equilibrium, because the data for it only recently became salient (e.g., Eldredge & Gould, 1972). Clearly, idealized or formal models of scientific inquiry would fail to address many of the issues reflected in the controversy about how different species arose.

Finally, although many models of disconfirmation that are formal do not deal with explanation and are, thus, beyond the scope of this chapter, an analogous point applies. In Wason's (1960) triples task, for example, the gold standard of performance consists of disconfirming the hypothesized rule that is, testing a triple expected *not* to conform to it. However, the reason this strategy is successful is because the rule that people typically hypothesize (numbers increasing by two) is a specific instance of the more general rule (increasing numbers) designated by the experimenter as correct (Klayman & Ha, 1987; Popper, 1959). However, testing cases that *are* expected to conform to a rule, although it can lead to systematic errors, can in many realistic conditions, be a good heuristic for discovering evidence that is disconfirming (Klayman & Ha, 1987). In addition, when there is less pressure for a quick evaluation, people are less likely to restrict themselves to testing cases expected to conform to the rule (Gorman & Gorman, 1984). Analogously, testing positive cases can also be useful when there is a possibility for error in the data (Gorman, 1992). Indeed, when college students and adolescents were asked to give justifications for why they had chosen seemingly confirming tests, they sometimes noted that they wanted to see whether an event could be replicated, whether it happened all or only some of the time (see Motivation Cannot Always Be Inferred From Behavior, below).

Scientific Practice Treats Principles of Scientific Inquiry as Heuristics (Which Make a Good Outcome More Likely) Rather Than Algorithms (Which Guarantee a Good Outcome)

Another consequence, related to changes with historical time and to the importance of background information in applying seemingly formal rules, is that strategies for doing sound science do not guarantee success; there is no set of rules for doing sound science such that applying them, without considering background information, will guarantee the right answer. Sometimes the background information in the relevant field is not accurate enough or (probably more often) not complete enough that the correct solution would occur even to the best informed researchers. In addition, in any particular science at a particular time, the actually relevant alternative hypotheses might not yet have been discovered. This means that, because the relevant content might not be currently known, it might not occur to researchers to take it into account by considering it as a possible alternative and either examining it or controlling for it. The obvious way of seeing this is to consider that, if there were such rules, then there would already be a cure for cancer, and someone would already have landed on Jupiter. The fact that such strategies are applied successfully only when background information is both considered and is pretty nearly accurate means that, in practice, such rules are heuristics or rules of thumb that are *frequently* useful; they make a good outcome increasingly likely, but they do *not* guarantee it. This point becomes especially important when one considers that technology can often suggest new research (Nersessian, 2005) and technology changes with historic time.

Put differently, one might consider scientific theorizing or explanation to consist of a series of working hypotheses that are successively revised (or sometimes rejected) as additional information and additional alternative hypotheses become available.

The limitation of treating formal rules, mistakenly, as though they guaranteed good science has implications beyond the academy. For example, consider the seemingly formal rule that one should prefer explanations that are internally consistent over those that are not. Until the flurry of research (Block & Dworkin, 1974; Block, 1995; Fraser, 1995; Kamin, 1974; Steele, 1997; Steele & Aronson, 1995) carried out as a response to Herrnstein's (1973) and later Murray and Herrnstein's (1994) claims, it was easy to generate an internally consistent argument that blacks were genetically inferior to whites in terms of intelligence. The argument noted, roughly, that Blacks in contrast to Whites had, on average, less education; earned less, when one controlled for number of years of schooling; scored lower on standardized tests; had lower paying jobs; and so forth. That is, there was an internally consistent argument congruent with the claim of genetic inferiority. However, though internally consistent, this argument failed to take account of several alternative hypotheses, the most obvious being pervasive and institutionalized racism (see the section "The Broader Context"). One might argue that, at the time, scientists should have applied two formal rules, supplementing the rule about internal consistency with one about alternative hypotheses. However, even adding the rule, consider alternative hypotheses, would not have guaranteed success, because awareness of institutionalized racism was, at the time, so much outside the prevailing cultural zeitgeist—even for people who did not agree with Herrnstein's conclusions—that it would likely not have occurred to many people. That is, the psychology of individual scientists intersects with broader social factors. The importance of the alternatives that the culture, broadly understood, makes available at different times in history is vividly demonstrated by recent feminist approaches to sociology and anthropology and will be described below. It is worth noting here that when Samarapungavan (1992) studied internal consistency, she asked children about unfamiliar events and was thus able to control for background information. Similarly, when Lombrozo (2007) studied parsimony, she too controlled for background information, as did Proctor and Ahn (2007), when they examined the effect of causal knowledge on assessments of likelihood of unknown features.

There is an understandable and serious reason for wanting to treat the principles of scientific inquiry as though they were formal rules and, thus, applicable without relying on background information: background information includes theories (or explanations), and some explanations or theories can be at best misleading and, at worst, pernicious. If theory guides which data one treats as relevant and which data one searches for, then an incorrect theory can easily lead to fruitless searches. Even worse, as noted above, theories about, for example, the inferiority of entire groups of people, can lead to conclusions that are insidious. However, given the amount of potentially relevant data in the world, doing science without relying on theory is not pragmatically possible; trying to navigate without a guiding theory or explanation can easily leave one rudderless and overwhelmed. Thus, the only pragmatic option is to question theories, to subject them to evaluation, in particular by relying on (at least) three strategies: consideration of anomalies; consideration of the extent to which the theory is consistent with well-established background information; and consideration of competing alternative theories, which, as work in the anthropology and sociology of science demonstrates (see The Broader Context, below) sometimes means questioning the received framework that guides research in an entire discipline.

THE ROLE OF THEORY OR EXPLANATION

Even Children Have a Strong Tendency to Consider Explanations—Including Alternative Explanations

Concern with explanations begins early in development. Children take account of an explanation if one is presented to them, and in other situations, they either request or provide it spontaneously. For example, when shown a display in which dropping a marble into one side of a box was followed by a bell ringing from inside the box, four- to seven-year-olds were more likely to accept a longer delay between the marble dropping and the bell ringing if there was a mediating tube that connects the two boxes (Mendelson & Shultz, 1976). The tube was seen as *explaining* the temporal delay; the children assumed the marble had to travel through the tube, and the travel took time. In a quite different task, four- and five-year-olds were shown fictitious animals, each of which had various characteristics (sharp claws, horns, spiky tail, wings, large ears, etc.). Even though each animal had only some characteristics in the set, the children's ability to classify them improved when the experimenters explained the mechanism by which the characteristics enabled the animals to be either better fighters or better hiders (Krascum & Andrews, 1998).

In addition to making use of explanatory information when it is presented to them, children also seek, and in some cases provide it, spontaneously. Although the frequency of causal questions certainly increases over the preschool years (Chouinard, 2007), Callanan and Oakes (1992) found that even three-year-olds asked *why* and *how come* questions about social, natural, and mechanical phenomena. Frazier, Gelman, and Wellman (2009) found that children not only asked questions, but also that they repeated their questions or provided their own explanations when their initial questions were met with nonexplanatory answers. That is, their behavior indicated a genuine concern for receiving explanatory information, rather than merely engaging in conversation.

Furthermore, although the concern for explanation does occur spontaneously, it is also likely a by-product of a more general tendency to be curious (Berlyne, 1954). One way curiosity manifests itself is in a concern about explanations for violated expectations. For example, researchers have documented such questions and explanations as, "Why doesn't the ink run out when you hold up a fountain pen?" (Isaacs, 1930, p. 308; Harris, 2000); when told that a friend's dead cat had gone up to heaven, a child responded that the cat "did not go *up* to heaven when it died, because bones are heavy and clouds are just water and bones are heavy and would fall right through the clouds"; "some trees (evergreens) keep their leaves (needles) in winter, because they're put on with glue (sap)" (Koslowski & Pierce, 1981).

Children also generate explanations even when doing so fails to enhance—and actually reduces—correct performance. When presented with weighted and nonweighted blocks to balance, even though children in the earliest phase were able to balance almost all of the blocks correctly by relying on proprioceptive cues, they, nevertheless, in the next phase, generated a rule—albeit the only partially correct rule, *balance blocks at the geometric center* (Karmiloff-Smith & Inhelder, 1974/1975). This rule, of course, caused the children no longer to correctly balance the weighted blocks, but it did enable them to impose an explanatory framework to account for why some blocks did balance. (The point about a temporary reversal in performance with development will be noted again in the section on "Conceptual Change.")

Children as well as adults also take account of alternative explanations. For example, situations in which possible causes are confounded prompt children to engage in more search than situations in which confounding is absent (Schulz & Bonawitz, 2007). Furthermore, even children actually propose collecting controlled data, which rule out alternative possible causes, to answer causal questions. They realize, for example, that to test whether a lantern with a roof and few small holes will prevent a candle inside from blowing out, they need a contrast lantern with few small holes, but no roof (Bullock, 1991; Bullock & Ziegler, 1999). At least by sixth-grade, when trying to identify causal events, people treat confounded variables as more problematic than controlled variables (Koslowski, 1996, Chap. 6).

Finally, when college students were presented with possible mechanistic explanations for various events and asked what information they would like to have to find out whether the explanation was likely to be correct, the overwhelming number of participants' responses dealt with possible alternative causes that the students wanted to know about (Koslowski, P'Ng, Kim, Libby, & Rush, 2012). Similarly, when asked how they would find out whether hospitalized children recovered faster when their parents stayed with them overnight, all but 1 of 48 adolescents and college students asked about possible alternative causes with which parental presence might have been confounded (e.g., quality of doctors, severity of diagnosis, etc.) (Koslowski, 1996, ch. 12). However, being aware of the *conceptual* importance of alternative explanations does not necessarily mean that someone has the technical competence to deal with them. In the study just mentioned, participants were not given data to deal with. In a subsequent study, they were given "data" along with pencil and paper. The data included whether or not parents stayed with the children, age of child, gender, and diagnosis. Again, all of the participants said that one would need to take into account possible alternative causes besides parental presence. However, when asked how they would do that, roughly half of the participants were unable to propose a way. Instead, they relied on such strategies as "eyeballing" the data, looking at extreme values, examining the ranges, and so on. As in the first study, although participants had *conceptual* knowledge of the importance of considering alternatives, they lacked the *technical* ability to arrange the data to do so (Koslowski, Cohen, & Fleury, 2010). That is, the presence of background knowledge does not guarantee the technical wherewithal to be able to apply it correctly.

Even Children Can Distinguish Theory or Explanation From Evidence

The philosophical literature argues that distinctions among theory (or mechanism or explanation), data (or observation or evidence), and background information are difficult to pin down because of the way scientific inquiry works (Boyd, 1983, 1985, 1989; Kuhn, 1970; Psillos, 1999). Evidence is not independent of theory; a theory or explanation can affect whether we treat a particular observation or data point as evidence (see Considerations of Theory or Explanation Also Help Us Realize That Information Can Be Evidence, below). Furthermore, background information can include theory as well as data or observations. Therefore, the present chapter uses different terms, not because the terms refer to clearly delineated constructs, but because in psychological studies, the terms are typically operationalized in different ways.

One way to frame the distinction between theory and evidence is to ask whether children recognize that a contrast is more useful in answering questions than is a demonstration. The answer is that even preschoolers do. They realize, for example, that to find out whether blue cars or yellow cars go down a ramp faster, it is better to have one car of each color go down a ramp than to have two yellow cars go down (Brenneman et al., 2007; Massey, Roth, Brenneman, & Gelman, 2007).

Another way of asking about the theory versus evidence distinction is to ask whether participants differentiate the beliefs that they bring to a situation and the data or observations relevant to assessing the belief. Using this distinction, Kuhn et al. (1988) have argued that children and adolescents cannot distinguish theory and evidence. In contrast, other researchers have argued that they can (Sodian, Zaitchik, & Carey, 1991). The reason for the difference lies in the way that different researchers construe sound reasoning, in particular, sound scientific reasoning. In Kuhn et al., children were asked about variables that they believed to be either causal or noncausal. For each type of belief, one was confirmed; one, disconfirmed. Thus, if the participant held the belief that eating chocolate but not carrot cake made people ill, then in the disconfirm condition, she would be shown that type of cake did not matter; that children eating carrot cake and those eating chocolate cake were equally likely to be ill. To be counted as performing correctly, participants needed to reject their initial hypothesis that type of cake mattered. That is, participants were treated as having confused theory and evidence, because they failed to reject their beliefs in the face of covariation data inconsistent with the beliefs and, instead, often simply modified the beliefs to take account of the disconfirming evidence.

However, when children identified variables as causal or noncausal, they sometimes had mechanistic reasons for doing so (e.g., eating chocolate cake is worse than eating carrot cake, *because chocolate cake has more sugar, and sugar increases blood pressure*). Nevertheless, as noted earlier, the information presented in the experimental task disconfirmed only the covariation component of the belief and left the mechanism component intact. In contrast, when disconfirming information undermined both components of a belief (the covariation along with the mechanism component), then the disconfirming evidence had a greater effect than when only the covariation component was disconfirmed (Koslowski, 1996, Chap. 10; also "Attempts to Disconfirm a Belief Might Fail to Deal With all Aspects of the Belief").

Furthermore, the strategy of modifying rather than rejecting a belief in the face of inconsistent data is one that is often followed in legitimate scientific practice (see Anomalous Information and Alternative Explanations, below). For example, if one believed that penicillin killed germs and then learned that it was pragmatically effective only against gram-positive bacteria, one would not reject the hypothesis that penicillin kills germs. Rather, one would modify it by restricting it to refer only to gram-positive bacteria.

Other tasks, not rooted in assessments of covariation data, also suggest that children can distinguish theory and evidence. Sodian, Zaitchik, and Carey (1991) asked children to decide whether a mouse living in a house was large or small and were shown two boxes, one with a large and one with a small opening. When asked which box they should put some food in to determine the mouse's size, the children chose the box with the small opening, since the large opening could accommodate mice of both sizes, while the small opening could accommodate only the small mouse. Although seeming to be straightforward to an adult, note that the children's decision relies on at least two pieces of background information. One is that mice do not change their girth (as do some snakes when they detach their jaws); the other is that the door openings do not change in size (as they might if the walls were made of rubber). That is, the children were able to choose the correct bit of evidence precisely because, like adults, they (tacitly) relied on an important though admittedly small web of related information. Their tacit explanation of why the small opening would be the better choice was consilient with their background information about both mice and about rigid openings.

Explanation Can Function as Evidence

Although even children often distinguish theory and evidence, theory or explanation can also sometimes be evidential. One of the standard canons of scientific inquiry is that correlation does not guarantee causation. The reason is that, for some correlations between X and Y, there is no plausible mechanism that is currently known and that can account for the process by which X might have brought Y about. Put differently, the presence of a plausible possible mechanism or explanation that mediates X and Y is some evidence of a causal relation between X and Y. This principle can also be seen in the reasoning of individual nonscientists.

For example, when college students (as well as practicing clinicians) provided an explanation, or a mechanism, by which a behavior might have come about, they came to view the behavior as increasingly normal (Ahn, 2003). Furthermore, plausible explanations had a stronger effect than those that were implausible. In the same vein, when reasoning about causal attributions, people do not restrict themselves to asking about covariation; they also ask about possible mechanisms (Ahn, 1995).

What might seem to indicate an inability to distinguish theory and evidence is that, in some cases, people offer explanations to justify an opinion. However, this is more likely to happen when other, nonexplanatory evidence is lacking or limited (Brem & Rips, 2000). Furthermore, being able to generate a mechanism by which an event might have been brought does not necessarily constitute evidence that the mechanism is correct. However, if the mechanism is at least moderately plausible and if competing mechanisms are absent, this could indicate that the generated mechanism might be worth a second look.

Finally, when a potential causal event is initially *im*plausible (even though it does covary with the effect), the event is nevertheless increasingly likely to be treated as causal when it is paired with mechanism information that can account for the process by which it might have brought about the effect. Thus, car color as a possible—though implausible—cause of differences in gas mileage was increasingly seen as affecting gas mileage if participants were told that red cars lead to alertness, which in turn leads to the sort of driving that conserves gas (Koslowski, 1996, Chap. 8). In such situations, people might assume that an explanation (or mechanism) exists that has yet to be discovered.

Not Every Explanation Is Understood in Depth

Concern with explanation does not mean that explicit consideration or knowledge of mechanism underwrites every causal judgment. For many situations that require identifying a causal event, people (including children) have at least a limited "catalogue of likely causes" that they invoke to account for effects, noting, for example, that hospitalized children might recover faster if they have good doctors, are in good hospitals, have mild illnesses, and so on. (Koslowski, 1996, Chap. 12). Furthermore, the catalogue of likely causes is often based on distributed cognition, or a division of cognitive labor; we believe that penicillin cures infections, not necessarily because of personal experience or because we understand the mechanism involved, but because experts have told us that it does (Giere, 2002; Koenig & Harris, 2005; Lutz & Keil, 2002). In addition, people (including children) may well not even think about (at least early in learning about an apparatus) the mechanism by which some causes operate (Koslowski, Spilton, & Snipper, 1981); we turn on the TV without necessarily wondering why clicking the remote control makes a picture appear. Furthermore, sometimes even adults have only a very rudimentary or shallow understanding of particular mechanisms (Mills & Keil, 2004; Rozenblit & Keil, 2002).

Nevertheless, it is worth noting that Keil and his colleagues typically asked their participants about fairly complicated (though admittedly familiar) entities, such as helicopters, quartz watches, and so on. Thus, although people may be unable to explain the *particular* mechanism by which complicated things such as helicopters work, they doubtless believe that there is some sort of mechanism that makes them work and that some expert knows what the mechanism is (Koslowski, 1996, Chap. 5). The point is that, in addition to a catalogue of likely causes—whether or not based on other people's expertise—background information about mechanisms plays a role in causal judgments, and sometimes the background information involves relying on a division of cognitive labor to infer that mechanisms power various devices, even when the details of the mechanism are not known.

Considerations of Theory or Explanation Help Us Realize That Information Can Be Evidence

It is easy to think of evidence as a fairly straightforward notion: either information is evidence or it is not. Furthermore, it is easy to think of theory and evidence as two distinct entities. After all, we certainly talk about collecting evidence to evaluate theories, as though the two constructs were separable. What also fosters this notion is that, often, we forget that the reason why certain information is in the catalogue of likely causes is grounded in theory (or at least in the statements of experts, as noted above). For example, alcohol consumption is in the catalogue of likely causes of auto accidents because of our background knowledge of how it is that alcohol consumption affects attention and, thus, driving.

However, in some cases, the fact that information is evidence is *not* obvious. In such cases, information is often not recognized as evidentially relevant unless there is a theory or explanation that can incorporate the information into a broader causal framework. For example, college students were asked to decide whether two pieces of information was relevant to explaining why mountain dwellers are shorter than people who live nearby at sea level. One piece of information was that the oral history of the mountain dwellers told of them migrating to the mountains to escape invaders from across the sea;

the other was that the languages of the two peoples have different grammatical structures and expressions. All participants were told of a target possible explanation (different gene pools) and a control (access to different foods). When the possible target explanation mentioned different gene pools, both pieces of information were rated as more relevant to explaining the height differences than when told of the control alternative that turned on access to different foods. Possible genetic differences made causal sense of flight in the face of invaders and with different language characteristics, while access to different foods did not (Koslowski, Marasia, Chelenza, & Dublin, 2008; Koslowski, Hildreth, Fried, Waldron, 2012).

Now consider an example from life outside the laboratory. Japan has very little in the way of natural resources, including oil. This might be seen as an interesting aspect of geography not very relevant to why the Japanese government attacked Pearl Harbor. If, however, one learns that the Japanese wanted to end the pre–World War II U.S. blockade that prevented shipments of oil and other natural resources from reaching Japan, then the fact that Japan has very little oil becomes very relevant, indeed (Worth, 1995). This is not an argument that either the attack on Pearl Harbor or the blockade of Japan were justified. Rather, it is to make the point that, once one considers an explanation that consists of access to raw materials, then information about the lack of natural resources in Japan becomes relevant in a way that it might otherwise not be. In addition, whether one has access to this information often depends on culture, in particular, whether one has read about this particular aspect of global history—or whether one was educated in Japan.

Theory/Framework Can Also Affect Which Information People Observe and Search

Klahr has demonstrated that, in generating a hypothesis (in Klahr's terms, in searching the hypothesis space) the initial state "consists of some knowledge about a domain, and the goal state is a hypothesis that can account for some or all of that knowledge." This is followed by a search in experiment space for an experiment that can be informative, that can, for example, in the case of two competing experiments, help decide between the two. Finally, the evidence is evaluated by comparing the predictions derived from the hypotheses with the results obtained by the experiments. Put differently, the search for hypotheses is constrained by background knowledge, and the search for experiments is constrained by the hypotheses (Klahr, 2000). (However, there are individual differences in whether people search preferentially in hypothesis space or experiment space.) There are also developmental changes: children's limited background information can restrict the hypotheses that they find plausible, which in turn can affect the information they are likely to gather (Klahr, 2000).

Theory or background information can affect the search for other sorts of information as well. College students and college-bound adolescents were told that a hospital administrator wanted to find out whether having parents stay overnight with their hospitalized children would make the children recover faster. The participants were asked, "What sorts of information or evidence would you consider?" Ninety-four percent of the participants explicitly verbalized mechanism information as the motivation for the information they wanted. "You could have some parents stay overnight and some not, *because parents might help their kids by telling the doctors if there were problems*," or "I'd look for age, *because young kids might need their parents more than older kids*" (Koslowski, 1996, Chap. 12). Note that the participants were not asked to consider mechanism; they did so because they had been asked what sort of evidence they would consider, and considerations of mechanism—or theory, broadly construed—motivated their decisions about whether some information would be evidentially relevant.

Outside of the laboratory, the rise of feminist social theory in archaeological and anthropological research (Gero & Conkey, 1991) led people to look for information that had been largely ignored. One of the most salient had to do with relative contributions of men and women to food intake among hunter-gatherers. It was (and probably still is) easy to think of groups of males providing the bulk of the food for hunter-gatherer societies, relying on primitive tools and traps to bring down mammals large enough to feed an entire village—a picture that is central to many museum dioramas of early hunter-gatherers. It is also a picture that is internally consistent with other beliefs; it "made sense" to think that women, because they would be pregnant and nursing, would be tied to a home settlement and thus dependent on males, who were not only mobile but also stronger, for the bulk of their food. However, Slocum's (1975) work demonstrated that, in extant hunter-gatherer societies, women are at least as mobile as men and, more to the point, provide the bulk of the group's dietary intake. That is, doing anthropology from a feminist perspective (an explanation in the broad sense of the term) led to a search for data that might otherwise have remained in the shadows.

Theory Can Also Affect How Information Is Framed in the Broader Culture

A classic example of this from the sociology of science is Martin's (1996) survey of the metaphors used in biology and medical textbooks. In the sections on reproduction, the sperms are presented as active agents that "penetrate" the egg, while the egg is presented as a passive recipient—in line with standard stereotypes of competitive males and reserved, retiring females. Martin also notes that this metaphor continued to be used even after researchers at Johns Hopkins found evidence that "the forward thrust of sperm is extremely weak" and that, in fact, "sperm and egg stick together because of adhesive molecules on the surfaces of each."

In the same vein, Longino notes that the "man-the-hunter" view suggests that males drive evolution, because an upright stance enhanced the ability to manipulate tools, which in turn "is understood to be a consequence of the development of hunting by males" (Longino, 1990, p. 107). However, Longino points out that equally compatible with the evidence is the idea that women drive evolution. Female gatherers' ability to collect food from grasslands and to defend themselves against small, aggressive animals would also have been facilitated by an upright stance that permitted more effective tool use with tools such as sticks and reeds). In short, in the absence of observations of the prehistoric hunter-gatherers, and given that some tools such as sticks and reeds would not have survived, the question of which gender drives evolution cannot be definitively answered and is consistent with both a male- and a female-centered view. That is, the view that men drive evolution reflects a cultural assumption, rather than any actual data.

Furthermore, one can argue that misleading frameworks are not necessarily simple infelicities of speech that have negative effects; Nersessian (2005) has demonstrated that what is available in the laboratory—including various models and presumably metaphors—can actually have positive effects by helping scientists generate, as well as answer, questions. Thus, misleading metaphors can actually hinder the search for new questions, while approximately accurate metaphors can help.

Theory or Explanation Can Also Lead People Astray

Note that the fact that science is theory-laden also makes salient that relying on theory can have bad as well as good outcomes. Earlier sections have already made the point that theories or explanations can be based on incomplete background information. Thus, the theories themselves can be at best incomplete and at worst downright wrong. Sometimes, for example, what is required to make explanations more accurate is the discovery of new information, that is, information that is not currently available. Furthermore, in some cases, mistaken theories can not only make it difficult to consider alternative explanations (e.g., alternatives to the "man-the-hunter" view, or the notion that sperm alone determine whether an egg is fertilized), but can also have pernicious effects (such as "explanations" based on putative racial differences in intelligence). In short, theories or explanations can lead people astray.

For example, children make incorrect predictions when they treat such constructs as current and heat as being in the ontological category of substance rather than process or event (Chi, 2008). Children who have an incomplete understanding of what makes something alive can produce animistic overextensions, arguing, for example, that a table is alive because it has legs (Carey, 1985).

The Role of Theory in Scientific Reasoning Is Also Relevant to How the Term "Bias" Is Used

If one treats, as biased, any research that is informed by theory or background information, then all science is biased; it is impossible to do good science without relying on background information, including theory. However, when "biased" is used as a pejorative, it typically means that there is some reason to question the accuracy or the comprehensiveness of the theory or background information, or that there is some reason to believe that the explanation has failed to take account of plausible alternative explanations. If one has an algorithmic conception of science, the two uses of "bias" can become conflated. It is very important to distinguish between them, for pragmatic as well as conceptual reasons. If one acknowledges that relying on theory and background information is crucial for doing science, then counteracting assumed bias will not consist of trying to eliminate theory; rather, it will consist of trying to make the theories as likely as possible (given currently available information) to be correct and of trying to consider plausible relevant alternatives. Recall, also, that the theory-laden aspect of science also makes salient that relying on theory can have bad as well as good outcomes, depending on the extent to which the background information is complete, and on the extent to which the theory approximates a correct explanation that is also more plausible than the available, plausible alternatives. That is, to come full circle, science does not offer guarantees; the principles of scientific inquiry are heuristics, rather than algorithms.

Theories, and often especially, alternative theories, also affect the way in which anomalous data are detected and evaluated, and this will be discussed in the following section. In addition, the section, "Confirmation Bias," examines the role of anomalous data and of alternative explanations as they refer to confirmation bias.

ANOMALOUS INFORMATION AND ALTERNATIVE EXPLANATIONS

Given the extent to which, for better or worse, science relies on theory or explanation, it is crucial that an explanation be evaluated—especially in light of data that are anomalous to it.

Kuhn (1970) drew attention to the importance of anomalies in prompting the scientific community to move from one paradigm to another (e.g., from a geocentric to a heliocentric model of the universe). He argued that, when anomalies were few in number, the scientific community first tried to modify the existing paradigm to account for the anomalies. As anomalies mounted up, and the modifications became increasingly unwieldy, the situation became ripe for a paradigm shift. Before and after a paradigm shift, "normal" science took place. Another aspect of Kuhn's argument is that paradigms were incommensurable, because they relied on different concepts, different notions of what constituted data, and that scientists in the two paradigms could not argue about a common ground—a notion that echoes concerns in some of the developmental literature, especially in physics—although others have criticized the notion of incommensurability. In using Kuhn's work as a model of theory change in other situations, it is important to keep in mind that genuine paradigm shifts are rare and occur in a scientific community, rather than in the thinking of an individual person. Nevertheless, there are several parallels.

Researchers have documented the role of anomalies in the thinking and discoveries of individual scientists (e.g., Holmes, 1998; Thagard, 2005; Tweney, 1985; etc.), in laboratory groups (Dunbar, 1995), and in experimental studies. A common theme is that, in normal science, anomalous (or at least unexpected) data are not rare (Dunbar, 1995; Dunbar & Fugelsang, 2005; Thagard, 2005), even though they do not always rise to the level of prompting a Kuhnian paradigm shift. Indeed, they accounted for over half the findings reported in 12 laboratory meetings that Dunbar and Fugelsang (2005) analyzed. When unexpected findings occurred, one of the initial approaches was to question the methodology or to attribute the anomaly to chance. However, when the anomalies occurred in a series, even though the methodology had been modified, Dunbar and Fugelsang found that a major shift in reasoning took place. The scientists also often searched for common features of the anomalies. In addition, the scientists drew analogies to other research and generated new hypotheses (which often allowed them to capture other unexpected findings), and individual scientists in the group proposed individual aspects of the new model. That is, the reasoning became what Dunbar and Fugelsang termed "group reasoning." Note that this is relevant to the earlier question (see Even Children Can Distinguish Theory or Explanation From Evidence, above) of whether sound scientific reasoning consists of always rejecting, rather than modifying, an explanation in light of anomalous information. It does not, if sound scientific reasoning consists of what scientists actually do when they practice.

In laboratory studies, college students (as well as 11-year-olds) also take account whether anomalies share common features. When the anomalies do share common features, they are treated as less problematic for an explanation (Koslowski, 1996, ch. 9). The common features suggest that one should probably search for another factor that might be constraining the situations in which the target factor operates. For example, if drug X produces an effect except in certain populations, and if those populations have respiratory problems, one might modify the working hypothesis to conclude that X produces an effect, but only if respiratory problems are not present. One would likely also search for what it is about respiratory problems that limits the effectiveness of X.

In addition to whether anomalies share common features, anomalies can also be distinguished in terms of whether they are weak or strong. When presented with anomalies previously identified as either weak or strong, and asked to generate resolutions for them, college students generated comparable *numbers* of resolutions for each anomaly type. However, they rated resolutions for strong anomalies as *less plausible* than those for weak anomalies (Koslowski, Libby, O'Connor, Rush, & Golub, 2012). In short, it is as easy to generate resolutions for strong as for weak anomalies; the difference is that the resolutions for strong anomalies are not very compelling, even to the people who generated them.

Reactions to anomalies also span a range. Abelson (1959) identified four strategies that people might, in principle, use to resolve what he termed "belief dilemmas." The strategies included denial (e.g., arguing that the anomaly resulted from sloppy data collection); bolstering (finding additional evidence to support the explanation); differentiation (e.g., arguing that the theory is still correct, but only under certain circumstances); and transcendence (building up elements into a larger unit by, e.g., dealing with the conflict between science and religion by arguing that the rational and the spiritual man must both be cultivated.)

In more recent research, Chinn and Brewer (1992, 1993a, 1993b, 1998) asked college students specifically about responses to anomalies, and did so with respect to mechanistic explanations (about, e.g., the mass extinction of dinosaurs and explanations for whether dinosaurs were cold- or warm-blooded). Strategies included: ignoring the anomalous data; rejecting the data because of, for example, measurement error; arguing that the data are irrelevant to the theory; holding anomalous data in abeyance (e.g., by withholding judgment about it; reinterpreting the data to make it consistent with the theory; making peripheral theory change in response to the anomalies; and changing to a new theory). College students were found to deploy all of these strategies. (One can see parallels between some of these strategies and those identified by Abelson.)

In addition, when assessing explanations in light of anomalous data, people also take into account the extent to which the explanation is causally consistent with (and the anomalous data are causally *in*consistent with) related background information. For example, Chinn and Brewer (1988, 1998) asked college students to assess the theory that dinosaurs were cold-blooded in light of the anomalous data that the bone density of dinosaurs was comparable to the bone density of extant warm-blooded animals. In reasoning about the anomalies, the students relied on their related knowledge of, for example, the effects of dietary calcium on bones and the possibility that fossilization might have altered the structure of bone tissue. (Of course, as already noted, although one can suggest ways of resolving anomalies to an explanation, the resolutions might not necessarily have been seen as plausible.) The point about networks of related information is also relevant to questions about confirmation bias (see "Attempts to Disconfirm a Belief Might Fail to Deal With All Aspects of the Belief").

Among children, some researchers have also found an additional wrinkle in responses to anomalous data, namely, whether anomalies can be detected at all. For example, Chinn and Malhotra (2002) found that elementary school children, who often believe that a heavy object falls faster than a lighter one, might actually fail to see that a heavy and a light object both land at the same time. This is analogous to children in Karmiloff-Smith and Inhelder's (1974/1975) study who, having generated an only approximately correct rule for balancing blocks, dismissed blocks that could not be balanced by relying on the rule as being "impossible to balance." However, difficulty detecting anomalous data does not always occur. When Penner and Klahr (1996) interviewed children who believed that heavier objects sink faster than lighter objects, they found that the children actually did notice the unexpected finding and tried to reconcile it with their initial belief. Furthermore, the children tried to account for the anomaly by doing what practicing scientists do: they generated possible mechanisms to account for the results, suggested the anomalous observation might not be true of all instances, or identified a possible factor such as shape that might be bringing about the anomalous result in a particular instance.

In addition, children can sometimes actually benefit from anomalies, using them to generate additional hypotheses and experiments to further their knowledge. However, their ability to do this is also a function of how much information they bring to the situation in the first place (Chinn & Malhotra, 2002; Echevarria, 2003). This is echoed in research with adults in which the adults were also able to recognize evidence inconsistent with their beliefs (Vallee-Tourangeau, Beynon, & James, 2000).

Finally, in the research described above, the anomalies were presented to the participants. However, there is also evidence that even children often spontaneously notice disconnects between two sets of information—rendering one set anomalous to the other and prompting the children to ask questions (Harris, 2000; Isaacs, 1930). Consider a preschool child, recorded by Isaacs, who asks, "Why doesn't the ink run out when you hold up a fountain pen?" (Isaacs, 1930, p. 308). That child was trying to reconcile the lack of ink falling out with the usual effects of gravity. Similarly, consider a preschool child who had been told that a friend's dead cat had gone up to heaven. The child responded that the cat "did not go *up* to heaven when it died, because bones are heavy and clouds are just water and bones are heavy and would fall right through the clouds" (Koslowski & Pierce, 1981). This child argued that the lack of consistency between the friend's claim and a network of related background information (about the composition of clouds, the relative weight of bones and of clouds, gravity—though likely not under that description—the belief that heaven is up, and the permeability of clouds to bones, etc.) made the friend's claim not very credible.

CONFIRMATION BIAS

There is much research on the myriad ways in which people are irrational (e.g., Gilovich, 1991; Tversky & Kahneman, 1977) However, people do nevertheless manage occasionally to do things—including science—in ways that are successful. Thus, a complete picture of human reasoning in general and scientific reasoning in particular requires documenting rational, as well as irrational, behavior. Because a standard way in which people are said to be irrational is that they engage in confirmation bias, it is important to ask when behaviors thought to reflect confirmation bias are rational and when they are not.

Before addressing confirmation bias in causal explanations, consider the large literature on testing hypotheses about noncausal rules. Confirmation bias is sometimes operationalized in terms of how people identify the set of which certain instances are members. For example, such tasks might present participants with three numbers (2, 4, 6) and ask them to discover the set (increasing numbers, even numbers, etc.) that the numbers represent (Wason, 1960). Similarly, such tasks might present participants with three cities and ask them to discover the set (coastal cities, etc.) to which the cities belong. A frequently cited argument in both psychology and (previously) in philosophy is that the gold standard for evaluating an explanation is to try to disconfirm it. However, the standard conclusion in the literature on confirmation bias suggests that attempts to disconfirm rather than to confirm a belief are the exception rather than the rule; that is, people are said to suffer from a widespread confirmation bias. However, there are reasons for thinking that the standard conclusion ought to be framed in a more nuanced way (Capaldi & Proctor, 2008; Klayman & Ha, 1987; Koslowski et al., 2008). A fairly basic point is that, as already noted, tests likely to confirm a hypothesis can, nevertheless, often yield data that disconfirm it (Klahr, 2000; Klayman & Ha, 1987). Furthermore, a positive test strategy can be especially useful in the early stages of an investigation, because it can help people identify those findings that merit further examination (Klahr, 2000).

Now consider causal explanations. Confirmation bias is said to be widespread, and conclusions from this literature are often treated as being applicable to thinking in general (such as judicial reasoning, rationalization of public policy, etc.) and to scientific reasoning in particular (Kuhn et al., 1988; Nickerson, 1998). Confirmation bias is said to be a "generic concept that subsumes several more specific ideas that connote the inappropriate bolstering of hypotheses or beliefs whose truth is in question" (Nickerson, 1998, p. 175). In the context of scientific reasoning, an assumed confirmation bias is said to blind us to alternative explanations, to blind us to evidence that disconfirms or is inconsistent with an explanation, and to make us reluctant to reject the explanation in light of evidence that it calls into question.

However, there are conceptual as well as empirical arguments suggesting that a more qualified view of confirmation bias might be more accurate (Koslowski, 2012). In addition, even if we assume for the sake of argument that confirmation bias is as broad a phenomenon as is often suggested, that would still leave the question of what underwrites it. Viewed from another perspective, the question is a pragmatic one: if one wanted to attenuate confirmation bias, what strategy might one follow?

Two of the reasons for thinking that some of the conclusions about confirmation bias ought to be qualified were discussed earlier in this section. There are additional reasons, as well, and the following are the most relevant to the present chapter.

There Are Conceptual Reasons for Questioning Confirmation Bias

People Do Consider Alternatives

Recall that even when people have been asked to generate their own explanations for events, they have a strong tendency to consider alternative explanations, which is not what a confirmation bias would suggest (see earlier sections on the role of theory or explanation).

What Appears to Be Confirmation Bias Might Actually Be a Bookkeeping Strategy for Dealing With Information Overload

For example, in Chapman and Chapman's (1967a, 1967b) classic studies of illusory correlations, participants were shown "patient cards" on which putative diagnoses were paired—at random—with sets of symptoms. Despite the randomness, clinicians and college students reported that there was a correlation, for example, between drawing atypical eyes and a diagnosis of paranoia. However, the incidence of reported correlations decreased by almost half when participants were allowed to organize the stimulus cards and use a pencil and paper to keep track of the records (Chapman & Chapman, 1967b). Given how difficult it would be to detect a pattern in data that were randomly arranged, it is not surprising that people relied on their default knowledge to try to keep track of, and impose some organization—however flawed—on, the data. It is also not surprising that this tendency was attenuated when people were given other means of keeping track of the randomly arranged data. That is, it is likely that, in the absence of pencil, paper, and the opportunity to rearrange the patient cards, the beliefs brought into the experimental situation served as a mechanism (albeit a flawed one) for bookkeeping. More to the point, in the absence of paper and pencil, it is not at all clear how one could have kept track, and made sense, of random data.

Modifying a Hypothesis in Light of Inconsistent Data Is Often Treated as Less Desirable Than Rejecting It, and This May Not Be Appropriate

Recall also that even practicing scientists do not automatically cling to an explanation in the face of unexpected or anomalous data. Rather, they take account of the actual pattern formed by the anomalous data to decide how problematic the anomalies are (see Anomalous Information and Alternative Explanations, above).

"Confirmation bias" sometimes refers to situations in which people refine rather than reject a hypothesis in light of data that do not fit. The argument is that rejection is what would be warranted by the data (i.e., the particular data presented in the confines of the experimental situation), while modification is motivated by a desire to cling to the initial hypothesis. The view that rejection in light of anomalous data is preferable to modification often goes along with the tendency, mentioned earlier, to design experiments so that what is treated as sound scientific reasoning consists of taking into account only the information presented in the experiment and of actually ignoring background information that the participant brings to the experimental situation. For example, as already mentioned, participants who believed that, for example, type of bread (rolls vs. slices) played no role in whether children would become ill were presented, during the experiment, with "data" in which children who ate bread in one form did become ill while children who ate bread in another form remained healthy (Kuhn et al., 1988). In this experiment, sound reasoning was stipulated to consist of changing one's mind in the face of the disconfirming data and concluding that type of bread was, indeed, causal. However, participants often failed to reject their background information when they entered the laboratory. Thus, participants continued to rely on it to maintain their belief that form of bread did not matter, and their decision was treated as flawed reasoning.

I would argue that the participants' strategy was, in fact, consistent with sound scientific reasoning in two ways. One is that, when doing research, practicing scientists continue to rely on the background knowledge they have accumulated. For example, an epidemiologist, beginning a research project to track the global incidence of cancer, would likely also keep in mind what she had learned about earlier, namely, the role played by smoking. The other reason is that rejecting an explanation at the first sign of anomalous data fails to describe actual scientific practice (recall the Dunbar et al. research). That is, refinement rather than rejection is often sound science—especially since much of scientific practice consists of generating working hypotheses, based on incomplete data, and then modifying the working hypotheses as additional data become available. Put differently, because science is cumulative, background knowledge does and should (indeed, must) play a role and (working) hypotheses are often modified rather than rejected in light of data that do not quite fit.

There Are Also Empirical Reasons for Questioning Confirmation Bias

Confirmation Bias Is Often Only a Transitory State

There is a difference between treating confirmation bias as an enduring tendency and treating it, instead, as a transitory state that reflects what may well be healthy initial skepticism about an anomaly. Anomalous data sometimes do result from measurement or other methodological errors. When practicing scientists encounter such data, not only do they look for patterns in the anomalies, but they also question whether measurement or other methodological errors might have been at fault (Dunbar & Fugelsang, 2005). One would, and should, think twice before rejecting a finding from a single study that eating fatty foods contributes to heart disease. However, if evidence were to mount up that the real culprits in heart disease are nitrates in processed meats, then sound scientific reasoning would require at least modifying and possibly rejecting the fatty foods claim in favor of a claim based on nitrates. In fact, this is what actually happens when nonscientists (adults as well as children) reason about issues.

For example, in the Karmiloff-Smith and Inhelder task mentioned earlier, the children in phase II (who held onto their claim despite disconfirming evidence) did eventually move to phase III. Similarly, Schauble (1990) found that children's causal and noncausal beliefs about whether, for example, engine size and presence of a muffler would affect the speed of a car were initially resistant to disconfirmation. However, Schauble notes that, "although the children displayed the typical belief bias, it was not altogether resistant to the cumulative weight of the evidence" (Schauble, 1990, p. 54).

The presence of alternative hypotheses also plays a role in how anomalies are treated. Klahr (2000) found that the tendency to maintain hypotheses in the face of disconfirming data is attenuated when participants are forced to generate some alternative hypotheses at the outset of an experiment. Presumably, generating an alternative helps people realize that there is another option for responding to the anomalies—an option that does not involve modifying the working hypothesis to account for the anomalies. Vallee-Tourangeau, Beynon, and James (2000) asked college students to reason about contingencies (whether various teams were effective in detecting comets). The presence of an alternative did not affect participants' likelihood of recognizing negative evidence, but did make it less likely that they would persist in basing their predictions on discredited contingencies. Finally, although practicing scientists were often initially reluctant to consider disconfirming data, often attributing them to a methodological problem, the scientists did not simply ignore the data but typically did follow up on them, often using the same methodology as the one that resulted in the problematic data (Fugelsang, Stein, Green, & Dunbar, 2004). Although people can certainly use the possibility of error to make their hypotheses resistant to disconfirmation (Gorman, 1986), sometimes the initial attribution of disconfirming instances to error makes scientific sense; human error can be involved in collecting scientific data, and even formal rules (such as those in the "triples" tasks) might, from the participants' perspective, be probabilistic—which can only be ascertained by repeating a test in which the outcome is expected to conform to a rule.

A Failure to Generate Alternatives or Diagnostic Evidence May Reflect Not a Confirmation Bias, but Rather a Lack of Background Information

As noted earlier, people have a strong tendency, when evaluating explanations, to consider alternatives—even when they are not explicitly asked to do so. However, in the experiments that demonstrated this, the events being explained were, if not familiar to the participants, at least analogous to events that were familiar. Thus, participants presumably were able to rely on their background information to generate possible alternatives.

Consider the cases when people might lack the necessary background information to be able to generate an alternative, or for that matter, to generate information that distinguishes between two explanations. An easy way to see this (at least for some of us) is to imagine being transported to a research team that is designing experiments to test relativity theory. I suspect that few of us would have the background information that would enable us to generate (plausible) alternatives.

Recall also an earlier example. Many college students, when questioned, are not able to generate (or remember) a reason why the Japanese government bombed Pearl Harbor other than that they wanted to initiate a war. When pressed, they were not able to answer the question of *why* the Japanese government would have wanted to start a war in the first place. Lacking information about Japan's scarcity of resources and about the early blockade of Japan, the students are often at a loss for generating any explanation, let alone both a target and an alternative explanation.

Now consider an experimental study. An experimental group of college students was presented with background information about, for example, oriental carpets (their history, how they came to be exported, etc.). For the experimental group, some of the information was replaced by information that the type of wool produced by sheep depends on the sheep's diet and on how much lanolin they produce. In a seemingly unrelated study, both groups were then told of reports that, for example, carpets seem to last longer when people remove their shoes before walking on them. All participants generated alternative explanations. However, people in the experimental condition were more likely than in the control condition to generate an alternative that dealt with the quality of the wool from which the carpets had been made. When people have background information, they use it to generate alternative explanations, but they cannot always generate an alternative (or at least a plausible one) if they lack the relevant background information (Koslowski, Rosenblum, Masnick, Barnett, 2012).

Motivation Cannot Always Be Inferred From Behavior

It is not always possible to tell whether a participant's motivation for conducting a particular test is to see whether the result of the test would confirm the hypothesis or would disconfirm it (McKenzie, 2006; Wetherick, 1962). This is related to the more general point that it is results, not tests, that confirm or disconfirm (Baron, 1985). For example, in the classic study in which it seemed to be triangles (but was really figures of a certain brightness) that were repelling a moving particle, people typically tested the triangle hypothesis by testing additional instances of triangles, and the tests were treated as instances of confirmation bias (Mynatt, Doherty, & Tweney, 1977). However, in a separate study, participants were asked to explain their choices (Koslowski & Maqueda, 1993). With one exception, none of the participants reported that they wanted to confirm their hypothesis. Rather, they reported that, for example, they wanted to see whether the relation could be replicated; whether it was probabilistic (whether triangles repelled all or only some of the time); whether all or only some triangles repelled; whether triangles repelled only when in certain quadrants of the computer screen; and so on. That is, their motivation was not to confirm the initial triangle hypothesis. Instead, it was to find out how general the triangle hypothesis was; they were treating their initial hypothesis as a working hypothesis and searching for data that would enable them to replicate and refine it.

To avoid inferring motivation from behavior, a separate study manipulated motivation experimentally; participants were provided with either one or two possible explanations for several events and were explicitly told to confirm (support), disconfirm (undermine), or evaluate an explanation for the target explanation. They were then asked to decide which of several pieces of information was relevant to each motivational condition. Some pieces of information were neutral, while other pieces were more plausibly consistent with one explanation rather than the other. That is, such information was diagnostic, in that it distinguished between the two explanations.

Unlike what confirmation bias would have predicted, participants told to disconfirm were more likely than those told to confirm to identify as relevant information that was in fact causally neutral. In addition, being told to confirm an explanation did not confer an advantage; it did not make people more likely to discern information that was supportive. Finally, as was the case with the task involving contingencies (Vallee-Tourangeau, Beynon, & James, 2000), the presence of an alternative explanation reduced the likelihood of choosing information inappropriate to the motivational condition. In terms of an earlier point, the suggestion was that an alternative made it clear that, although information could be distorted to be consistent with one explanation, it was more *plausibly* consistent with the alternative. That is, the presence of an alternative provided a framework for assessing the *relative* weight of the evidence in favor of each of the two explanations. In short, when motivation is explicitly manipulated, a motivation to disconfirm is not necessarily the gold standard of reasoning (Koslowski, Marasia, & Liang, 2008).

People Often Do Treat Diagnostic Information as More Useful Than Nondiagnostic Information

One criterion of sound scientific reasoning is the ability to realize that diagnostic information, which distinguishes between two competing explanations, is more useful than information that is compatible with both. However, the literature on confirmation bias typically concludes that people often have difficulty generating diagnostic information. For example, when primed to ask questions to decide whether someone was an introvert (or an extrovert), people often asked questions whose answers would fail to distinguish between the two possibilities (Snyder & Swann, 1978; Snyder, 1981). (Of course, in terms of an earlier suggestion, one possibility is that people were using the primed personality type as a bookkeeping aid, that is, as a framework for charting which questions to ask.)

To remove the demands of bookkeeping, college students in a separate study were presented with (rather than asked to generate) three types of information: diagnostic (which distinguished between two explanations by being more plausibly consistent with one than the other); nondiagnostic (which was causally consistent with both explanations); and, causally neutral. The higher people rated the target explanation, the more likely they were to find diagnostic information more useful than nondiagnostic information (Koslowski et al., 2008). Confirmation bias would have suggested that, the more one agreed initially with an explanation, the more one would be tempted to find relevant information that was consistent with it. And, since nondiagnostic information was compatible with both explanations, it was compatible with the target.

Attempts to Disconfirm a Belief Might Fail to Deal With All Aspects of the Belief

Often, when we have beliefs about what is causal, the belief consists of two components: a belief that two events covary and a belief that there is a mechanism makes the covariation causal (e.g., that type of cake is causally related to illness, because chocolate cake has more sugar than carrot cake and sugar is not healthy) (see "Even Children Have a Strong Tendency to Consider Explanations—Including Alternative Explanations"). Thus, to disconfirm a belief, one might need to disconfirm both components. To test this hypothesis, interviews were done with college students and adolescents who held two widespread beliefs about covariation: that eating high-sugar rather than low-sugar candy before bedtime makes it difficult for children to fall asleep, and that drinking low-fat rather than whole milk has no effect on sleeplessness. The mechanism that participants gave for believing that sugar caused sleeplessness was some variant of an increase in energy ("Candy gives you a second wind" or, "It revs you up.") For both the causal and the noncausal belief, we confirmed and disconfirmed it in two conditions: when covariation information was presented alone and when it was combined with mechanism information. Covariation information was more effective when combined with mechanism information than when presented alone. Furthermore (and surprisingly, in light of the assumed prevalence of confirmation bias) was the fact that it did not matter whether the evidence confirmed the initial belief or disconfirmed it; people were receptive to mechanism information when one of their beliefs was disconfirmed as well as when it was confirmed.

Two other studies are also relevant to the question of why people often fail to reject theories in light of anomalous data. The reasoning behind both studies was that, if any single belief is embedded in, and judged with respect to, a network of related beliefs, then to disconfirm the general belief, one might need to chip away at the network of related beliefs that support it. To some extent, this was examined in the previous study, by taking account of the mechanism information that supported a belief. In the following two studies, the beliefs were embedded in much wider networks and were of substantially more importance than beliefs about sugar and fat.

Masnick (1999) and Swiderek (1999) interviewed college students about their opinions on affirmative action and capital punishment, respectively. In Swiderek's study, participants were also interviewed about several subsidiary beliefs that underlie the general belief (e.g., whether capital punishment in fact reduces the incidence of capital crimes, is cost-effective, applied in a nonbiased way, etc.). Regardless of whether participants agreed or disagreed with each of the sub-beliefs, they were presented with disconfirming evidence. For example, participants who believed that capital punishment did not reduce the incidence of capital crimes were shown data that the rate of capital crimes was reduced after a state introduced capital punishment. Those who believed it did reduce capital crimes were given the same information but told that the decrease only held for the first three months after capital punishment was introduced. In her study, Masnick noted the inconsistencies in each participant's responses. In the experimental group, each inconsistency was pointed out (as something that "some people" have identified as inconsistent) and the participant was asked "to give some thought to that." In the control condition, inconsistencies were not pointed out. Instead, participants were asked whether they might have "anything further to say after they spent a bit more time thinking about some of the initial questions." However, this question was not framed in the context of mulling over inconsistencies.

A week after the first interview, participants were brought back for a second interview and were asked again about their general belief as well as about the sub-beliefs. The sum of the change on the individual sub-beliefs predicted 79% of the variance in change in the general belief about affirmative action, and 85% of the variance of the change on the general belief in capital punishment. The results echo the point that any explanation or belief is embedded in a network of related information and, thus, to change any particular belief, one must also change the related beliefs in which it is embedded.

It would appear that changing someone's mind is, first of all, not the sort of thing that one can do with a single piece of information (however compelling) and, second, that it requires trying to establish what all the information is that supports a particular belief, so that each piece can be undermined.

Why Reasoning, Including Scientific Reasoning, Can Be Flawed

People have a tendency to generate and to consider alternative explanations; to realize that confounded data are more problematic than controlled data; to take account of both background information and seemingly formal rules for doing science; and, to base decisions on whether to refine or reject working hypotheses on the structure of anomalous data. Why then is scientific reasoning so often flawed?

Some of the answers have been alluded to earlier. Any particular person (or member of a culture) might have insufficient background information to be able to generate the relevant alternative hypothesis. This could reflect a personal shortcoming, or could reflect what the culture makes (or fails to make) available. Analogously, someone might lack the background information to be able to generate (or recognize) evidence that would distinguish between two hypotheses. In addition, behavior that is treated in the psychological literature as flawed reasoning (such as a tendency to modify rather than reject beliefs in light of anomalous data, or to ask for tests that appear to be confirmatory but that can yield disconfirming data) is often, in fact, sound scientific practice.

Furthermore, the suggestion that confirmation bias should be treated in a more nuanced way does not mean that confirmation bias does not exist. In psychological studies of individuals, confirmation bias can be the motivational hope that a test will yield confirmatory rather than disconfirmatory results, or it can be the filter (based on background beliefs, including theory) that makes one ignore or distort anomalous results. It can also consist of modifying an initial hypothesis to take account of anomalous data when the data (initial and anomalous) are more plausibly consistent with an available alternative. Finally, it can be the information made available by a culture that restricts the alternatives deemed to be plausible and the background information that would be necessary to evaluate the explanations. To deal with confirmation bias, one needs to understand what underlies it.

As already noted, the approach to scientific reasoning in the present chapter is informed by IBE, a description of actual scientific practice, rather than idealized versions of how science might be done. The research described thus far highlights causal mechanistic explanations, the role of alternative explanations, the importance of plausibility considerations, the extent to which explanation can render information causally relevant, and the importance of the network of related beliefs in which an explanation is embedded. The following sections deal with related aspects of scientific reasoning.

CONCEPTUAL CHANGE

Despite confirmation bias—however broadly one understands the term—people do change their beliefs and explanations and, of course, sometimes the change comes as a result of anomalous data. Research on how such change occurs addresses two overlapping questions: whether beliefs are coherent or fragmented, and whether they are internally consistent or contradictory. Furthermore, these questions have been examined with respect to the development that comes with increasing age as well as the development that takes place as novices become experts.

Chi (2008) notes that existing knowledge can be missing, incomplete, or in conflict with the knowledge to be acquired. Beliefs that are either missing or incomplete can fairly easily be refuted—or, more accurately, enhanced—simply by providing the child with additional information (e.g., that whales are mammals, not fish). The most difficult change to effect occurs when existing knowledge conflicts with the knowledge to be acquired, and it is when this happens that conceptual change is said to occur (Carey, 1985; Chi, 2008). Furthermore, conceptual change can occur not only as children develop, but also as novices become experts.

Consider first the question of whether beliefs present early in development are organized into fairly cohesive theories (Carey, 1985; Gopnik & Meltzoff, 1997; Vosniadou & Brewer, 1992, 1994), or whether, instead, they consist of clusters of beliefs that are fragmented (diSessa, Gillespie, & Esterly, 2004; Siegal, Butterworth, & Newcombe, 2004). For example, some researchers have argued that children hold a succession of fairly coherent "mental models" of the earth and of its place in the universe. Furthermore, such researchers claim that it is coherent but mistaken mental models that might make it difficult for children to make sense of scientific information about the earth when it is presented to them (Vosniadou & Brewer 1992, 1994). However, more recently, researchers (Nobes & Panagiotaki, 2007; Panagiotaki, Nobes, & Potton, 2009; Siegal et al., 2004) have noted that the seeming coherence results, in part, from ambiguous questions and misleading instructions. They point out that when the same questions are given to adults, many give responses as naïve as those of the children and, when asked about them, attribute the naïve responses to the ambiguous and confusing nature of the questions. The researchers note further that, when cultures do expose children to scientific information about a spherical earth, children subsequently hold a mixture of scientific and nonscientific beliefs that are fragmented rather than organized into a coherent whole.

As another example of the coherent versus fragmented debate, Carey's (1985) research on children's understanding of what it means to be alive suggests that there is a central concept that renders children's knowledge coherent, namely, humanity. Thus, cats are treated as alive because of their resemblance to humans, but worms and plants are not. An additional organizing principle involves explaining living processes in terms of psychology (people have children because they want to) rather than biology. In contrast, diSessa et al. (2004), who studied children's understanding of force, argue that children's beliefs consist of "knowledge in pieces." For example, studying children's concepts of force, they found that children's beliefs depend heavily on context or task and that, for children, "force" has a fairly broad meaning. His data suggest that, although students rely on certain "explanatory primitives," which indicates the beliefs are not random, the primitives are activated in specific contexts and are not strongly systematic.

The extent to which notions of force are nonrandom but context dependent is also reminiscent of Gelman and Weinberg's (1972) work on children's concepts of compensation and conservation. The beliefs are fragmented in the sense that whether children can be said to have such beliefs depends dramatically on the measures used to assess them.

Whether it makes sense to characterize children as having fairly cohesive theories or knowledge in pieces, it is clear that it makes sense not only to focus on what children lack, but also on what sorts of alternative beliefs or conceptual frameworks they do harbor. As Carey (2000a, 2000b) notes, this is especially important to keep in mind when trying to design curricula. In that regard, Chi (2008) has examined how naïve conceptions of physics conflict with, and are eventually overturned in favor of, more sophisticated conceptions. Chi (1992, 2008) distinguishes three types of ontological categories: matter (or substance), events or processes, and abstractions, with the different categories having different attributes or properties. For example, matter such as furniture can have mass and color, while events such as war cannot. Conceptual change within an ontological category can be a matter of addition or deletion, such as learning that whales do give birth to live young and do not have scales—information that moves whales from the category fish to the category mammal, but still keeps them in the ontological category of matter. (In Chi's taxonomy, children's understanding of the concept of alive, studied by Carey, would be classified as change that takes place within an ontological category.) In contrast, Chi argues that there are no physical operations that can transform a member of one ontological category into a member of another.

More recently, Chi (2008) has added two subcategories to the process or event category: direct versus emergent. An emergent event occurs when, for example, a crowd forms a bottleneck, not because any single agent tells them to do so, but because everyone rushes to an opening at roughly the same speed, pushing whichever other people happen to be in the way. If the process were direct, it would be guided by an outside agent. Chi argues that, to acquire a correct understanding of physics, one must also understand that the process of, for example, heat transfer is emergent rather than direct. It is "not caused by hot molecules moving from one location to another" as they would in a direct way, but by "the collisions of faster jostling 'hotter' molecules into slower-moving molecules," as in an emergent way (Chi, 2008, p. 76). It is the crossing of ontological boundaries from direct to emergent processes, Chi argues, that may account for the difficulty students have in learning physics.

Chi also notes that, often, an ontological boundary cannot be crossed until one learns about the properties of the *new* ontological category—either by explicit instruction or by induction (typically, during instruction). Refutation of a single belief (treated in an isolated way) leaves many of the old beliefs intact. Thus, the strategy of refuting a single belief is ineffective for conceptual change, when the new concept consists of complex representational structures in which individual concepts are connected. Furthermore, refutation of individual beliefs is also often ineffective, partly because the knowledge that contradicts the mistaken belief cannot always be assimilated into the new set of beliefs, because the new beliefs have not yet been acquired. Therefore, Chi argues, an effective way (perhaps the only successful way) to help students cross ontological boundaries is first to teach the students the properties of the new set of beliefs in the to-be-learned category. New knowledge can then be assimilated into the new, rather than the old, ontological category.

Changes from novice to expert understandings of physics are also reflected in several other ways. For example, increasing expertise in physics comes along with the ability to store knowledge in large, interconnected equations for solving a problem, rather than in a fragmented set of individual equations (Larkin et al., 1980), and to classify problems on the basis of the principles of physics required to solve the problem (e.g., whether the solution rests on invoking Newton's second law) rather than on the basis of surface similarities (e.g., grouping all the pulley problems in one category) (Chi, Feltovich & Glaser, 1981).

However, on the way to acquiring expertise in physics, people often hold beliefs that are often contradictory, even after instruction. Often, this involves a conflict between "everyday knowledge" and physics (e.g., McDermott, Reif, & Allen, 1992). Put differently, there are often two systems of organization, and while beliefs within a system may be internally consistent, beliefs in one system may well contradict beliefs in the other system (McDermott, 1984; Rozin & Nemeroff, 1990).

Organizing beliefs into systems that are held simultaneously and that contradict each other is not restricted to physics. Legare and Gelman (2008) found evidence for the coexistence of natural and supernatural explanatory frameworks for illness and disease transmission in two South African communities where Western and traditional frameworks are both available. Children and adults in these communities invoke both biological and witchcraft explanations for the same phenomena, and the bewitchment explanations are surprisingly highest among adults. Evans found that even nonfundamentalist parents hold mixed beliefs, espousing evolution for nonhuman species and creationism for humans. Luhrmann (1989) documented people who practiced Wicca during evenings and weekends, but who held middle-class professional jobs during the work week. Thus, during the work week, they functioned in accord with norms of standard physical rather than metaphysical beliefs, while during their nonwork hours, they, for example, performed the "water ritual."

Dunbar, Fugelsang, and Stein (2007) used brain imaging to study students' change from impetus to Newtonian views of physics. Half of the students who had taken no physics courses in high school or college nevertheless (correctly) judged it to be normal when two balls of different mass fell at the same rate. However, activation of the anterior cingulate cortex (indicating detection of information inconsistent with a belief) suggested that the students viewed the event as inconsistent with their theoretical perspective. That is, even when Newtonian principles seemed to have replaced impetus theories, the brain-imaging data suggested that the students still had access to their naïve, impetus theories, which had simply been inhibited, rather than replaced by, the correct theory.

Finally, conceptual development—including conceptual development that is science-related—is not always linear. For example, when asked to balance both weighted and unweighted blocks, young children do extremely well because they balance one block at a time, without relying on a general rule. When they eventually do generate a rule (balance at the geometric center), their performance deteriorates, since the rule can be applied only to unweighted blocks (Karmiloff-Smith & Inhelder, 1974/1975). (In terms of an earlier issue, one might think of the geometric-center rule as an example of confirmation bias, and that may well be. However, doing so might make every situation in which a child fails to rely on an alternative that has not yet been acquired an instance of confirmation bias—which would render the term not very useful.) Siegler and Jenkins (1989) also found discontinuity in the development of children's addition strategies. Children often revert to an earlier strategy even after they have discovered a more advanced strategy. Analogously, when designing experiments, children might progress to the point of using a valid (controlled) strategy that changes one thing at a time, but then revert to using a previous, confounded strategy (Kuhn & Phelps, 1982; Schauble, 1990, 1996; Siegler & Crowley, 1991). With regard to the shift from novice to expert, Lesgold et al. (1988) found an analogous result, namely, that third- and fourth-year medical residents often do worse, not only than experts, but also than first- and second-year medical residents. If beliefs were organized coherently, one would not expect this sort of regression.

THE BROADER CONTEXT

Science is not carried out in a vacuum. Individual scientists often work in groups and across disciplines, and groups of scientists work in a broader social and cultural context. As already noted, Dunbar and Fugelsang (2005) documented the "group reasoning" that takes place after an unexpected result has been noticed in a lab, with individual members contributing individual aspects of a new model. Nersessian (2005), in trying to coordinate cognitive science and sociological approaches, reminds us that thinking occurs not only in the brain, but also in the environment, and suggests that one way of studying this is to examine what she terms, "evolving distributed cognitive systems." For example, she notes that the laboratory is more than a physical space; it is also a problem space that is reconfigured as the research program develops and changes paths. Thus, the technology, people, knowledge sources, mental models, and so forth change to meet changes in research agendas. In addition, representations and models both enhance and constrain cognitive models; they allow the reasoner to go beyond "natural" capabilities and to stimulate problem solving by instantiating thought. Indeed, one of the changes that takes place as novices become experts is realizing that technology does not consist only of material objects, but also of objects that can help answer questions and, more importantly, perhaps, generate new ones. Recall, also, the work of Slocum (1975) and of Martin (1996) cited earlier, on the effects of feminist perspectives in science. In addition, earlier sections offered examples in which the particular alternative explanations that an individual considers depend, in part, on the possible explanations that a culture makes available.

Other researchers have focused on the social interactions that take place when scientists work in groups. For example, Collins (1974, 1985) found that even when a source laboratory provides a second lab with detailed instructions on how to perform a particular procedure, the relevant knowledge is transferred to the second lab only when the second lab has contact with the source lab, either by "personal visits and telephone calls or by transfer of personnel." Collins argues that there is a tacit knowledge that is only learned by actually "doing" something in an interactive situation, rather than simply hearing or reading verbal descriptions of what to do. Collins and Evans (2002) argue that knowledge in science proceeds from social interaction. In their framework, when scientists from different disciplines collaborate, they initially communicate in only the most superficial ways. Next, they become "interactional experts"; they learn to "talk the talk" enough to be able to engage in thoughtful conversation with members of the different discipline. They can understand (and even make) jokes, for example. Eventually, they become "contributory experts," having learned enough about the other discipline to be able to make genuine contributions. (What would likely be a source of controversy is whether becoming an interactional expert involves only learning the language of the other discipline, or whether learning the language is able to occur, because the interactional expert also acquires some notion of to what the language refers.)

Galison (1997) used the metaphor of trading zones to describe situations in which scientists and engineers collaborating on developing radar and particle detectors can become trading partners—which enables them to achieve "local coordination" despite vast "global differences," by, in part, developing a new language—analogous to a pidgin or creole language. Gorman and Mehalik (2002) integrated the notion of trading zones with three network states that characterize the ways that people and technology are linked. The first state is one in which there is no trading; the network is controlled by an elite; the nonelite either adopt the dictates of the elite or their views are dismissed. In the second state, systems serve as "boundary objects" in the sense that they are flexible enough to serve the needs of the various parties using them. They are likely to exist where trading partners are relatively equal. In the case of boundary systems, mental models are incomplete, because each trading partner views it through her own lens. In contrast, in the third state, the network has to develop a shared mental model to tackle a problem that involves multiple disciplines. In the third state, the goal of the trading partners is a common one that often results in the development of a new language, and hierarchy plays almost no role.

More recently, Collins, Evans, and Gorman (2007) have elaborated Galison's (1997) concept of trading zones and Gorman's (2002) classification of trading zones based on the notion of interactional expertise. They argue that there are other strategies besides devising a new language that can resolve communication problems between people from different groups. They argue that the strategies depend on two axes: collaboration/coercion ("the extent to which power is used to enforce trade") and homogeneity/heterogeneity ("the extent to which trade leads to a homogenous new culture"). Biochemistry is a prototypic example of mutual agreement to trade in which the merged culture tends toward homogeneity. At the opposite extreme, slavery involves coerced trading without the goal of a new homogenous culture. It is not conducive to communication across or between disciplines.

SUMMARY AND CONCLUSIONS

Using actual scientific practice (rather than idealized formal models) as a framework for what ought to count as sound scientific reasoning, one can argue that the reasoning of both children and (nonscientist) adults has much in common with sound scientific practice. However, one can also argue that there are also certainly differences.

Most generally, in terms of similarities, both children and adults do not restrict themselves to identifying possible causal agents, but concern themselves as well with explanation—including alternative explanations. Furthermore, like scientists, when people deploy scientific principles, they treat them as heuristics rather than algorithms. That is, they rely on background information (including information about explanation) when making judgments about, for example, covariation and confounding. Analogously, they rely on theory or explanation to decide whether information is evidence.

Furthermore, like scientists, people do have the conceptual skills to evaluate explanations by relying on: consistency of the explanation with the network of related information, information that is anomalous to the explanation, and the presence of plausible alternative explanations. In addition, like scientists, people do not jettison an explanation immediately when anomalies come to light; rather, they raise questions about methodology, they seek patterns in the anomalies, and they treat patterns as possibly indicating a mechanism that has yet to be discovered.

However, despite the many similarities between the reasoning of scientists and nonscientists, there are also salient differences. Unlike scientists, people do not always have the technical skills to be able to evaluate explanations. In addition, reliance on background information, including explanation, can also lead people astray: the relevant background information can be either faulty or lacking. When this happens, people can be prevented from assimilating the information necessary for conceptual change to take place. Furthermore, explanations (especially when they involve conceptual change) can be fragmented and new and old explanations can co-exist, even when they are incompatible, and that can also hinder conceptual change.

For scientists and nonscientists, the broader social context in which scientific inquiry takes place can restrict the alternative explanations or frameworks that are available for consideration. In addition, science often takes place in the context of social interaction. And, when the social interaction takes place across disciplines, collaboration can be hindered until a new language can be devised.

Finally, while some behaviors that interfere with sound reasoning (such as faulty or incomplete information) can, in principle, be remedied by acquiring additional data or information, other problems cannot be, namely, the tendency to hold nonrational or irrational beliefs that either co-exist with or actually trump rational considerations.

ACKNOWLEDGMENTS

Tremendous thanks to Richard N. Boyd for discussion, comments, and suggestions.

REFERENCES

Abelson, R. P. (1959). Modes of resolution of belief dilemmas. *Journal of Conflict Resolution, 3*(4), 343–352.

Ahn, W., Kalish, C. W., Medin, D. L., & Gelman, S. A. (1995). The role of covariation versus mechanism information in causal attribution. *Cognition, 54,* 299–352.

Ahn, W., Novicki, L., & Kim, N. S. (2003). Understanding behavior makes it more normal. *Psychonomics Bulletin & Review, 10*(3), 746–752.

Baron, J. (1985). *Rationality and intelligence.* Cambridge, MA: Cambridge University Press.

Berlyne, D. E. (1954). A theory of human curiosity. *British Journal of Psychology, 45,* 180–191.

Block, N. J. (1995). How heritability misleads about race. *Cognition, 56*(2), 99–128.

Block, N. J., & Dworkin, G. (1974). IQ: Heritability and inequality. *Philosophy and Public Affairs, 3*(4), 331–409.

Bonawitz, E. B., & Lombrozo, T. (2012). Occam's rattle: Children's use of simplicity and probability to constrain inference. *Developmental Psychology, 48*(4), 1156–1164.

Boyd, R. N. (1983). On the current status of the issue of scientific realism. *Erkenntnis, 19,* 45–90.

Boyd, R. N. (1985). Observations, explanatory power, and simplicity. In P. Achinstein & O. Hannaway (Eds.), *Observation, experiment, and hypothesis in modern physical science* (pp. 47–94). Cambridge, MA: MIT Press.

Boyd, R. N. (1989). What realism implies and what it does not. *Dialectica, 43*(1–2), 5–29.

Brem, S., & Rips, L. (2000). Explanation and evidence in informal argument. *Cognitive Science, 24*(4), 573–604.

Brenneman, K., Gelman, R., Massey, C., Roth, Z., Nayfield, I., & Downs, L. (2007, October). *Methods for assessing scientific reasoning in preschool children.* Poster session presented at the annual meeting of the Cognitive Development Society, Santa Fe, NM.

Bullock, M. (1991). *Scientific reasoning in elementary school: Developmental and individual differences.* Paper presented at the biennial meeting of the Society for Research in Child Development, Seattle, WA.

Bullock, M., Gelman, R., & Baillargeon, R. (1982). Development of causal reasoning. In J. W. Friedman (Ed.), *The developmental psychology of time* (pp. 209–254). New York, NY: Academic Press.

Bullock, M., & Ziegler, A. (1999). Scientific reasoning: Developmental and individual differences. In. F. E. Weinert & W. Schneider (Eds.), *Individual development from 3 to 12. Findings from the Munich Longitudinal Study* (pp. 38–44). Cambridge, MA: Cambridge University Press.

Callanan, M. A., & Oakes, L. M. (1992). Preschoolers' questions and parents' explanations: Causal thinking in everyday activity. *Cognitive Development, 7*(2), 213–233.

Capaldi, E. J., & Proctor, R. W. (2008). Are theories to be evaluated in isolation or relative to alternatives? An abductive view. *The American Journal of Psychology, 121*(4), 617–641.

Carey, S. (1985). *Conceptual change in childhood.* Cambridge, MA: MIT Press.

Carey, S. (2000a). Science education as conceptual change. *Journal of Applied Developmental Psychology, 21,* 13–19.

Carey, S. (2000b). The origin of concepts. *Journal of Cognition and Development, 1,* 37–41.

Carruthers, P., Stich, S., & Siegel, M. (Eds.). (2002). *The cognitive bases of science: Multidisciplinary approaches.* Cambridge, MA: Cambridge University Press.

Chapman, L. J. (1967). Illusory correlation in observational report. *Journal of Verbal Learning and Verbal Behavior, 6*(1), 151–155.

Chapman, L. J., & Chapman, J. P. (1967). Genesis of popular but erroneous psychodiagnostic observations. *Journal of Abnormal Psychology, 3,* 193–204.

Chi, M. T. H. (2008). Three types of conceptual change: Belief revision, mental model transformation, and categorical shift. In S. Vosniadou (Ed.), *Handbook of research on conceptual change* (pp. 61–82). Hillsdale, NJ: Erlbaum.

Chi, M. T. H., Feltovich, P. J., & Glaser, R. (1981). Categorization and representation of physics problems by experts and novices. *Cognitive Science, 5,* 121–152.

Chinn, C. A., & Brewer, W. F. (1988). An empirical test of a taxonomy of responses to anomalous data in science. *Journal of Research in Science Teaching, 35*(6), 623–654.

Chinn, C. A., & Brewer, W. F. (1992). Psychological responses to anomalous data. *Proceedings of the Fourteenth Annual Conference of the Cognitive Science Society,* 165–170.

Chinn, C. A., & Brewer, W. F. (1993a). Factors that influence how people respond to anomalous data. *Proceedings of the Fifteenth Annual Conference of the Cognitive Science Society,* 318–323.

Chinn, C. A., & Brewer, W. F. (1993b). The role of anomalous data in knowledge acquisition: A theoretical frame-work and implications for science instruction. *Review of Educational Research, 63,* 1–49.

Chinn, C. A., & Malhotra, B. A. (2002). Children's responses to anomalous scientific data: How is conceptual change impeded. *Journal of Educational Psychology, 94,* 327–343.

Chouinard, M. (2007). Children's questions: A mechanism for cognitive development. *Monographs of the Society for Research in Child Development, 72,* 1–57.

Collins, H. M. (1974). The TEA set: Tacit knowledge and scientific networks. *Social Studies of Science, 4*(2), 165–185.

Collins, H. M. (1985). *Changing order: Replication and induction in scientific practice.* Chicago, IL: University of Chicago Press.

Collins, H. M., & Evans, R. (2002). The third wave of science studies: Studies of expertise and experience. *Social Studies of Science, 32*(2), 235–296.

Collins, H. M., Evans, R., & Gorman, M. (2007). Trading zones and interactional expertise. *Studies in History and Philosophy of Science, 38*(4), 657–666.

Darden, L. (2006). *Reasoning in biological discoveries: Essays on mechanisms, interfield relations and anomaly resolution.* Cambridge, MA: Cambridge University Press.

DiSessa, A., Gillespie, N. M., & Esterly, J. B. (2004). Coherence versus fragmentation in the development of the concept of force. *Cognitive Science, 28,* 843–900.

Dunbar, K. (1995). How scientists really reason: Scientific reasoning in real-world laboratories. In R. J. Sternberg & J. E. Davidson (Eds.), *The Nature of insight* (pp. 365–395). Cambridge, MA: MIT Press.

Dunbar, K., & Fugelsang, J. (2005). Causal thinking in science: How scientists and students interpret the unexpected. In M. E. Gorman, R. D. Tweney, D. Gooding, & A. Kincannon (Eds.), *Scientific and technical thinking* (pp. 57–79). Mahwah, NJ: Lawrence Erlbaum Associates.

Dunbar, K. N., Fugelsang, J. A., & Stein, C. (2007). Do naïve theories ever go away? Using brain and behavior to understand changes in concepts. In M. Lovett &

P. Shah (Eds.), *Thinking with Data: 33rd Carnegie Symposiums on Cognition*. Mahwah, NJ: Erlbaum.

Duschl, R. A., Schweingruber, H. A., Shouse, A. W., National Research Council (U.S.), Committee on Science Learning, Kindergarten Through Eighth Grade. (2007). *Taking science to school: Learning and teaching science in grades K-8*. Washington, DC: National Academies Press.

Echevarria, M. (2003). Anomalies as a catalyst for middle school students' knowledge construction and scientific reasoning during science inquiry. *Journal of Educational Psychology, 95,* 357–374.

Eldredge, N., & Gould, S. J. (1972). Punctuated equilibria: An alternative to phyletic gradualism. In T. J. M. Schopf (Ed.), *Models in paleobiology* (pp. 82–115). San Francisco, CA: Freeman Cooper. Reprinted (1985) in N. Eldredge (Ed.), *TimeFrames*, (pp. 193–223). Princeton: Princeton Univ. Press.

Feist, G. (2006). *The psychology of science and the origins of the scientific mind*. New Haven, CT: Yale University Press.

Fraser, S. (1995). *The bell curve wars: Race, intelligence and the future of America*. New York, NY: Basic Books.

Frazier, B. N., Gelman, S. A., & Wellman, H. M. (2009). Preschoolers' search for explanatory information within adult-child conversation. *Child Development, 80,* 1592–1611.

Fine, A. (1984). The natural ontological attitude. In J. Leplin (Ed.), *Scientific realism*. Berkeley, CA: University of California Press.

Fugelsang, J., Stein, C., Green, A., & Dunbar, K. (2004). Theory and data interactions of the scientific mind: Evidence from the molecular and the cognitivive laboratory. *Canadian Journal of Experimental Psychology, 58,* 132–141.

Galison, P. L. (1997). *Image and logic: A material culture of microphysics*. Chicago, IL: University of Chicago Press.

Gelman, S. A., & Coley, J. D. (1990). The importance of knowing a dodo is a bird: Categories and inferences in 2-year-old children. *Developmental Psychology, 26,* 796–805.

Gelman, R., & Weinberg, D. H. (1972). The relationship between liquid conservation and compensation. *Child Development, 43,* 371–383.

Gero, J. M. & Conkey, M. W. (Eds.). (1991). *Engendering archaeology*. Oxford, UK: Basil Blackwell.

Giere, R. (Ed.). (1992). *Minnesota studies in the philosophy of science, Volume XV*, Minneapolis, MN: University of Minnesota Press.

Giere, R. (2002). Scientific cognition as distributed cognition. In P. Carruthers, S. Stich, & M. Siegel (Eds.), *The cognitive basis of science: Multidisciplinary approaches*. Cambridge, MA: Cambridge University Press.

Gilovich, T. (1991). *How we know what isn't so: The fallibility of human reason in everyday life*. New York, NY: Free Press.

Gopnik, A., & Meltzoff, A. (1997). *Words, thoughts and theories*. Cambridge, MA: MIT Press.

Gopnik, A., & Schulz, L. (2007). *Causal learning*. Oxford, UK: Oxford University Press.

Gorman, M. E. (1984). A comparison of disconfirmatory, confirmatory and control strategies on Wason's 2–4-6 task. *The Quarterly Journal of Experimental Psychology, 36A,* 629–648.

Gorman, M. E. (1986). How the possibility of error affects falsification on a task that models scientific problem solving. *British Journal of Psychology, 77,* 85–96.

Gorman, M. E. (1992). *Simulating science*. Bloomington, IN: Indiana University Press.

Gorman, M. E., & Mehalik, M. M. (2002). Turning good into gold: A comparative study of two environmental invention networks. *Science, Technology & Human Values, 27,* 499–529.

Gruber, H. E. (1981). *Darwin on man: A psychological study of scientific creativity*. Chicago, IL: University of Chicago Press.

Haith, M. M., & Benson, J. B. (1998). Infant cognition. In D. Kuhn, & R. Siegler (Eds.), *Handbook of child psychology* (5th ed., Vol. 2). Hoboken, NJ: John Wiley & Sons.

Harman, G. (1965). The inference to the best explanation. *Philosophical Review, 74,* 88–95.

Harris, P. L. (2000). Children's metaphysical questions. In K. S. Rosengren, C. N. Johnson, & P. L. Harris (Eds.), *Imagining the impossible: Magical, scientific, and religious thinking in children* (pp. 130–156). Cambridge, MA: Cambridge University Press.

Herrnstein, R. J. (1973). *I.Q. in the meritocracy.* Atlantic Monthly Press. Boston, MA: Little, Brown.

Herrnstein, R. J., & Murray, C. (1994). *The bell curve: intelligence and class structure in American life.* New York, NY: Free Press.

Holmes, F. L. (1998). *Antoine Lavoisier, the next crucial year, or the sources of his quantitative method in chemistry.* Princeton, NJ: Princeton University Press.

Isaacs, S., & Isaacs, N. (1970). *Intellectual growth in young children.* London, UK: Routledge & Kegan Paul.

Johnson-Laird, P. N., Legrenzi, P., & Legrenzi, M. S. (1972). Reasoning and a sense of reality. *British Journal of Psychology, 63*(3), 395–400.

Kamin, L. J. (1974). *The science and politics of I.Q.* Potomac, MA: Lawrence Erlbaum Associates.

Kaplan, A. (1964). *The conduct of inquiry: Methodology for behavioral science.* Scranton, PA: Chandler.

Karmiloff-Smith, A., & Inhelder, B. (1974–1975). If you want to get ahead, get a theory. *Cognition, 3*(3), 195.

Keil, F. C. (2006). Explanation and understanding. *Annual Review of Psychology, 57,* 227–254.

Kitcher, P. (1981, December). Explanatory unification. *Philosophy of Science, 48*(4), 507–531.

Klahr, D. (2000). *Exploring science: The cognition and development of discovery processes.* Cambridge, MA: MIT Press.

Klayman, J., & Ha, Y. W. (1987). Confirmation, disconfirmation, and information in hypothesis testing. *Psychological Review, 94*(2), 211–228.

Koehler, D. J. (1991). Explanation, imagination, and confidence in judgment. *Psychological Bulletin, 110*(3), 499–519.

Koenig, M., & Harris, P. L. (2005). The role of social cognition in early trust. *Trends in Cognitive Science, 9*(10), 457–459.

Koslowski, B. (1996). *Theory and evidence: The development of scientific reasoning.* Cambridge, MA: MIT Press.

Koslowski, B. (in press). Inference to the best explanation (IBE) and the causal and scientific reasoning of non-scientists. In R. W. Proctor & E. J. Capaldi (Eds.), *The psychology of science: Implicit and explicit reasoning.* Oxford, UK: Oxford University Press.

Koslowski, B., Beckmann, L., Bowers, E., DeVito, J., Wonderly, B., & Vermeylan, F. (2008, July). *The cognitive basis for confirmation bias is more nuanced than one might expect.* Paper presented at the meetings of the International Society of the Psychology of Science and Technology, Berlin, Germany.

Koslowski, B., Cohen, L., & Fleury, J. (2010). *The conceptual understanding of ruling out alternatives does not guarantee the technical ability to do so.* Paper presented at the biennial meetings of the International Society for the Psychology of Science and Technology, Berkeley, CA.

Koslowski, B., Hildreth, A., Fried, L., & Waldron, C. (2012). *Whether information is seen as evidence depends on what the competing alternative explanation is.* Manuscript in preparation.

Koslowski, B., Libby, L. A., O'Connor, K., Rush, K., & Golub, N. (2012). *What makes some anomalies more problematic than others.* Manuscript in preparation.

Koslowski, B., & Maqueda, M. (January 01, 1993). What is confirmation bias and when do people actually have it? *Merrill-Palmer Quarterly, 39*(1), 104–130.

Koslowski, B., Marasia, J., Chelenza, M., & Dublin, R. (2008). Information becomes evidence when an explanation can incorporate it into a causal framework. *Cognitive Development, 23*(4), 472–487.

Koslowski, B., Marasia, J., Liang, V., & Vermeylen, F. M. (2008, July). *Confirming a belief is not necessarily problematic; disconfirming a belief is not necessarily good.* Paper presented at the meetings of the International Congress of Psychology, Berlin, Germany.

Koslowski, B., & Masnick, A. M. (2002). Causal reasoning. In U. C. Goswami, (Ed.), *Blackwell handbook of child cognitive development*. Oxford, UK: Blackwell Publishers.

Koslowski, B., & Maqueda, M. (1993). What is confirmation bias and when do people actually have it? *Merrill-Palmer Quarterly, 39*(1), 104–130.

Koslowski, B., & Pierce, A. (1981). *Children's spontaneous explanations and requests for explanations.* Paper presented at the Society for Research in Child Development, Boston, MA.

Koslowski, B., P'Ng, J., Kim, J., Libby, L. A., & Rush, K. (2012). *When told to evaluate an explanation, people request information about alternatives.* Manuscript in preparation.

Koslowski, B., Spilton, D., & Snipper, A. (1981). Children's beliefs about instances of mechanical and electrical causation. *Journal of Applied Developmental Psychology, 2,* 189–210.

Koslowski, B., Susman, A., & Serling, J. (1991, April). *Conceptual vs. technical understanding of evidence in scientific reasoning.* Paper presented at the biennial meeting of the Society for Research in Child Development, Seattle, WA.

Koslowski, B., & Thompson, S. L. (2002). Theorizing is important, and collateral information constrains how well it is done. In P. Carruthers, S. Stitch, & M. Siegal (Eds.), *The cognitive bases of science: Multidisciplinary approaches*. Cambridge, MA: Cambridge University Press.

Koslowski, B., & Winsor, A. P. (1981, April). *Preschool children's spontaneous explanations and requests for explanations: A non-human application of the child-as-scientist metaphor.* Unpublished manuscript, Cornell University, Ithaca, NY.

Krascum, R. M., & Andrews, S. (1998). The effects of theories on children's acquisition of family resemblance categories. *Child Development, 69*(2), 333–346.

Kuhn, D., Amsel, E., & O'Loughlin, M. (1988). *The development of scientific thinking skills.* Orlando, FL: Academic Press.

Kuhn, D., & Phelps, E. (1982). The development of problem-solving strategies. In H. Reese (Ed.), *Advances in child development and behavior* (pp. 1–44). New York, NY: Academic Press.

Kuhn, T. (1970). *The structure of scientific revolutions* (2nd ed.). Chicago, IL: University of Chicago Press.

Larkin, J. H., McDermott, J., Simon, D. P., & Simon, H. A. (1980). Expert and novice performance in solving physics problems. *Science, 208,* 1335–1342.

Legare, C. H., & Gelman, S. A. (2008). Bewitchment, biology, or both: The co-existence of natural and supernatural explanatory frameworks across development. *Cognitive Science: A Multidisciplinary Journal, 32,* 607–642.

Lesgold, A., Rubinson, H., Feltovich, P., Glaser, R., Klopfer, D., & Wang, Y. (1988). Expertise in a complex skill: Diagnosing x-ray pictures. In T. H. Michelene, R. Chi, K. Glasser, & M. J. Farr (Eds.), *The nature of expertise* (pp. 311–342). Hillsdale, NJ: Lawrence Erlbaum Associates.

Lipton, P. (1991). *Inference to the best explanation.* London, UK: Routledge.

Lipton, P. (January 01, 1993). Is the best good enough? *Proceedings of the Aristotelian Society, 93,* 89–104.

Lombrozo, T. (2007). Simplicity and probability in causal explanation. *Cognitive Psychology, 55,* 232–257.

Longino, H. E. (1990). *Science as social knowledge: Values and objectivity in scientific inquiry.* Princeton, NJ: Princeton University Press.

Luhrmann, T. M. (1989). *Persuasions of the witch's craft: Ritual magic in contemporary England.* Cambridge, MA: Harvard University Press.

Lutz, D., & Keil, F. C. (2002). Early understanding of the division of cognitive labor. *Child Development, 73,* 1073–1084.

Magnani, L. (2001). *Abduction, reason, and science: Processes of discovery and explanation.* Dordrecht, the Netherlands: Kluwer Academic Publishers.

Martin, E. (1996). The egg and the sperm: how science has constructed a romance based on stereotypical male-female roles. In B. Laslett, S. G. Kohlsted, H. Longino, & E. Hammonds (Eds.), *Gender and scientific authority* (pp. 323–339). Chicago, IL: University of Chicago Press.

Masnick, A. M. (1999). *Belief patterns and the intersection of cognitive and social factors.* Unpublished doctoral thesis, Cornell University, Ithaca, NY.

Massey, C., Roth, Z., Brenneman, K., & Gelman, R. (2007). *By way of comparison: Scientific reasoning in preschool and early elementary school children.* Poster presented at the biennial meeting of the Society for Research in Child Development, Boston, MA.

McDermott, L. C. (January 01, 1984). Research on conceptual understanding in mechanics. *Physics Today, 37,* 7.

McKenzie, C. R. M. (2006). Increased sensitivity to differentially diagnostic answers using familiar materials: Implications for confirmation bias. *Memory & Cognition, 34*(3), 577–588.

Mendelson, R., & Shultz, R. R. (1976). Covariation and temporal contiguity as principles of causal inference in young children. *Journal of Experimental Child Psychology, 22,* 408–412.

Michotte, A. E. (1963). *The perception of causality.* Oxford, UK: Basic Books.

Mills, C. M., & Keil, F. C. (2004). Knowing the limits of one's understanding: The development of an awareness of an illusion of explanatory depth. *Journal of Experimental Child Psychology, 87,* 1–32.

Murphy, G., & Medin, D. (1985). The role of theories in conceptual coherence. *Psychological Review, 92*(3), 289–316.

Mynatt, C. R., Doherty, M. E., & Tweney, R. D. (February 01, 1977). Confirmation bias in a simulated research environment: An experimental study of scientific inference. *Quarterly Journal of Experimental Psychology, 29,* 85–95. Excerpts reprinted in P.N. Johnson-Laird and P.C. Wason (Eds.), *Thinking: Readings in cognitive Science.* Cambridge, MA: Cambridge University Press.

Nersessian, N. J. (2005). Interpreting scientific and engineering practices: Integrating the cognitive, social, and cultural dimensions. In M. E. Gorman, R. D. Tweney, D. C. Gooding, & A. P. Kincannon (Eds.), *Scientific and technological thinking* (pp. 17–56). Mahwah, NJ: Lawrence Erlbaum Associates.

Nickerson, R. S. (1998). Confirmation bias: A ubiquitous phenomenon in many guises. *Review of General Psychology, 2,* 175–220.

Nobes, G., & Panagiotaki, G. (2007). Adults' representations of the Earth: Implications for children's acquisition of scientific concepts. *British Journal of Psychology, 98,* 645–665.

Panagiotaki, G., Nobes, G., & Potton, A. (2009). Mental models and other misconceptions in children's understanding of the earth. *Journal of Experimental Child Psychology, 104,* 52–67.

Penner, D. E., & Klahr, D. (1996). When to trust the data: Further investigations of system error in a scientific reasoning task. *Memory & Cognition, 24*(5), 655–668.

Popper, K. R. (1959). *The logic of scientific discovery.* London, UK: Hutchinson.

Proctor, C., & Ahn, W. (2007). The effect of causal knowledge on judgments of the likelihood of unknown features. *Psychonomic Bulletin and Review, 14*(4), 635–639.

Proctor, R. W., & Capaldi, E. J. (2006). *Why science matters.* Malden, MA: Blackwell.

Psillos, S. (1999). *Scientific realism: How science tracks truth.* London, UK: Routledge.

Putnam, H. (1972). Explanation and reference. In G. Pearce & P. Maynard (Eds.), *Conceptual change* (pp. 199–221). Dordrecht, Netherlands: D. Reidel.

Putnam, H. (1975). *Mind, language and reality.* New York, NY: Cambridge University Press.

Quine, W., & Ullian, J. (1970). *The web of belief.* New York, NY: Random House.

Reif, F., & Allen, S. (1992). Cognition for interpreting scientific concepts: A study of acceleration. *Cognition and Instruction, 9,* 1–44.

Rozenblit, L., & Keil, F. (2002). The misunderstood limits of folk science: An illusion of explanatory depth. *Cognitive Science, 92,* 1–42.

Rozin, P., & Nemeroff, C. J. (1990). The laws of sympathetic magic: A psychological analysis of similarity and contagion. In J. Stigler, G. Herdt, & R. A. Shweder (Eds.), *Cultural psychology: Essays on comparative human development* (pp. 205–232). Cambridge, England: Cambridge.

Salmon, W. (1984). *Scientific explanation and the causal structure of the world.* Princeton, NJ: Princeton University Press.

Salmon, W. (1990). Rationality and objectivity in science or Tom Kuhn meets Tom Bayes. In C. W. Savage (Ed.), *Scientific theories: Minnesota studies in the philosophy of science* (Vol. 14). Minneapolis, MN: University of Minnesota Press.

Samarapungavan, A. (1992). Children's judgments in theory choice tasks: Scientific rationality in childhood. *Cognition, 45*, 1–32.

Schauble, L. (1990). Belief revision in children: The role of prior knowledge and strategies for generating evidence. *Journal of Experimental Child Psychology, 49*, 31–57.

Schauble, L. (1996). The development of scientific reasoning in knowledge-rich contexts. *Developmental Psychology, 32*(1), 102–119.

Schulz, L., & Bonawitz, E. B. (July, 2007). Serious fun: Preschoolers play more when evidence is confounded. *Developmental Psychology, 43*(4), 1045–1050.

Schulz, L. E., & Gopnik, A. (2004). Causal learning across domains. *Developmental Psychology, 40*, 162–176.

Shultz, T. R., & Mendelson, R. (1975). The use of covariation as a principle of causal analysis. *Child Development, 46*, 394–399.

Siegal, M., Butterworth, G., & Newcombe, P. A. (2004). Culture and children's cosmology. *Developmental Science, 7*(3), 308–324.

Siegler, R. S., & Crowley, K. (1991). The microgenetic method: A direct means for studying cognitive development. *American Psychologist, 46*, 606–620.

Siegler, R. S., & Jenkins, E. A. (1989). *How children discover new strategies.* Hillsdale, NJ: Erlbaum.

Simonton, D. K. (2010). Creativity in highly eminent individuals. In J. C. Kaufman & R. S. Sternberg (Eds.), *The Cambridge handbook of creativity.* Cambridge, MA: Cambridge University Press.

Slocum, S. (1975). Woman the gatherer: Male bias in anthropology. In R. Reiter (Ed.), *Toward an anthropology of women* (pp. 36–50). New York, NY: Monthly Review Press.

Sloman, S. A., Love, B. C., & Ahn, W. (1998). Feature centrality and conceptual coherence. *Cognitive Science, 22*, 189–228.

Snyder, M. (1981). Seek and ye shall find: Testing hypotheses about other people. In E. T. Higgins, C. P. Heiman, & M. P. Zanna (Eds.), *Social cognition: The Ontario symposium on personality and social psychology* (pp. 277–303). Hillsdale, NJ: Erlbaum.

Snyder, M., & Swann, W. B. (1978). Hypothesis testing processes in social interaction. *Journal of Personality and Social Psychology, 36*, 1202–1212.

Sodian, B., Zaitchik, D., & Carey, S. (1991). Young children's differentiation of hypothetical beliefs from evidence. *Child Development, 62*, 753–766.

Steele, C. M. (1997). A threat in the air: How stereotypes shape the intellectual identities and performance of women and African-Americans. *American Psychologist, 52*, 613–629.

Steele, C. M., & Aronson, J. (1995). Stereotype threat and the intellectual test performance of African-Americans. *Journal of Personality and Social Psychology, 69*, 797–811.

Swiderek, M. R. (1999). *Beliefs can change in response to disconfirming evidence and can do so in complicated ways, buy only if collateral beliefs are disconfirmed.* Unpublished doctoral dissertation, Cornell University, Ithaca, NY.

Thagard, P. (1989). Explanatory coherence. *Behavioral and Brain Sciences, 12*, 435–502.

Thagard, P. (2005). How to be a successful scientist. In M. E. Gorman, R.D. Tweney, D.C. Gooding, & A.P. Kincannon (Eds.), *Scientific and technological thinking.* Mahwah, NJ: Erlbaum.

Thagard, P., & Verbeurgt, K. (1998). Coherence as constraint satisfaction. *Cognitive Science, 22*, 1–24.

Tversky, A., & Kahneman, D. (1977). Judgement under uncertainty: Heuristics and biases. In P. N. Johnson-Laird & P. C. Wason (Eds.), *Thinking, readings in cognitive science* (pp. 326–337). Cambridge, MA: Cambridge University Press.

Tweney, R. D. (1985). Faraday's discovery of induction: A cognitive approach. In D. Gooding & F. James (Eds.), *Faraday re-discovered: Essays on the life and work of Michael Faraday* (pp. 2792–1867). New York, NY: Stockton Press.

Vallee-Tourangeau, F., Beynon, D. M., & James, S. A. (2000). The role of alternative hypotheses in the integration of evidence that disconfirms an acquired belief. *European Journal of Cognitive Psychology, 12*(1), 107–129.

Van Fraassen, B. (1980). *The scientific image.* Oxford, UK: Oxford University Press.

Vosniadou, S., & Brewer, W. F. (1992). Mental models of the earth: A study of conceptual change in childhood. *Cognitive Psychology, 24,* 535–585.

Wason, P. C. (1960). On the failure to eliminate hypotheses in a conceptual task. *Quarterly Journal of Experimental Psychology, 12,* 129–140.

Wason, P. C. (1968). Reasoning about a rule. *Quarterly Journal of Experimental Psychology, 20,* 271–281.

Wetherick, N. E. (1962). Eliminative and enumerative behaviour in a conceptual task. *Quarterly Journal of Experimental Psychology, 14,* 246–249.

Worth, Jr., & Roland, H. (1995). *No choice but war: The United States embargo against Japan and the eruption of war in the pacific.* Jefferson, NC: McFarland.

Zimmerman, C. (2000). The development of scientific reasoning skills. *Developmental Review, 20,* 99–149.

Zimmerman, C. (2007). The development of scientific thinking skills in elementary and middle school. *Developmental Review, 27,* 172–223.

CHAPTER 8

The Nature of Scientific Reasoning and Its Neurological Substrate

Anton E. Lawson

Scientific reasoning is used during the creation of scientific knowledge, more specifically, during the generation and test of proposed explanations of nature. This chapter raises two central questions: What inferences and patterns of reasoning do scientists use to generate and test their proposed explanations? And, what is currently known about the neurological substrate that drives those inferences and reasoning patterns?

To begin to answer the first question, the chapter's initial section will identify the reasoning that was used to explain a puzzling observation made while my wife and I were taking an evening walk. Considering successful reasoning in a familiar and relatively simple context should aid in its explication. Next the reasoning that was presumably used by Galileo Galilei during his discovery of Jupiter's moons in 1610 will be considered. The reasoning in both contexts will be discussed in terms of the inferences of abduction, deduction, and induction as identified and defined by Charles Sanders Peirce (Peirce, 1903, 1905). Peirce is generally credited as the originator of pragmatism, a philosophical view opposing logical positivism and favored by contemporaries William James and John Dewey. Pragmatism, with its roots in Darwinian natural selection theory, argues against the existence of absolute or transcendental truth and in favor of a more ecological account of knowledge generation grounded in inquiry and in the testing and retention of ideas that work.

The chapter will then turn to the second question, that is, what is currently known about the neurological substrate that drives those inferences and reasoning patterns? Before doing so, I must first introduce some basics of brain development and then discuss the neurological basis of learning, including neurological models of idea generation and testing. In this way the chapter intends to not only outline a psychology of science but also to present a cognitive neuroscience of science. To close, I discuss common reasoning errors and draw some educational implications.

AN "EVERYDAY" EXAMPLE OF SUCCESSFUL REASONING

The Events

Recently my wife and I took an evening walk around the block. When we were part-way around, we approached an unlit street light. As we approached, the street light suddenly went on. Then as we walked a bit further, just as suddenly, it went off. At which time I exclaimed: "It's a motion detector light"—meaning the light went on when it sensed we were near and went off when we moved away. Upon hearing this, my wife replied: "It may just be a bum light in need of repair. It may simply be a coincidence that the light went on and off as we walked past." So I asked: "Suppose we turn around and walk back past the light. If I am right what should happen?" And she answered: "If you are right it should again go on and off as we pass by." But she added that if she were right, the light may go on and off, but not necessarily as we pass by. So I said: "Let's turn around and see what happens." But as we were now quite far from the light and had already been walking quite awhile, she said no. She wanted to head home. So we headed home.

The next night we went out to dinner with some friends. While driving home my wife said: "Let's drive around the block and see if that street light goes on and off again." So we did. However, while nearing the light, but still presumably well out of its motion detection range, we could see that it was already on. We both commented that this meant that my wife's bum-light idea was probably right. But we kept driving toward the light, and just after we had driven past, the light suddenly went off. We both laughed at this because, in addition to evidence that my motion detector idea was wrong, we now also had evidence that it was right!

So which was it? To find out we quickly turned around and drove back past the still dark light. Unfortunately for my motion detector idea, as we approached and as we drove past, the light remained off. Just to be sure we turned around and drove past again, with the same result. At this point we were both quite sure that we were dealing with a bum light.

Explicating the Underlying Reasoning

In the order of events, here is an explication of the underlying reasoning in terms of its basic inferences:

Initial Exploration

We take a "random" walk around the block. A street light goes on and then off while we walk past. Thus a *puzzling observation* is made and a *causal question* is raised, namely: Why did the light go on and then off?

Generating a Possible Explanation—A Causal Hypothesis

Next I generated a possible explanation (a *causal hypothesis*) that the street light has a motion detector that senses when people are near, that is, the street light may be *like* our neighbor's motion detector porch light, which goes on only when people are on the porch. Peirce referred to this spontaneous and creative act of hypothesis generation as *abduction* (or sometimes as *retroduction*) because the puzzling observation is seen as similar to, or analogous to, previously explained observations that have been stored in long-term memory (LTM), thus get abducted/stolen/transferred from that store to tentatively explain the puzzling observation. Peirce (1903) described abduction as an inference as follows:

> A puzzling observation C is made. However, if...A were true, then...C would be a matter of course.
> Therefore...there is reason to believe that A is true. (Peirce, 1903, p. 168)

And by philosopher Norwood Hanson (1958):

> Before Peirce treated retroduction as an inference logicians had recognized that the reasonable proposal of an explanatory hypothesis was subject to certain conditions. The hypothesis cannot be admitted, even as a tentative conjecture, unless it would account for the phenomena posing the difficulty—or at least some of them. (p. 86)

In the present example we get this:

> A street light goes on and then off as we pass by (C). However, if..., like our neighbor's porch light, it contains a motion detector (A), then...seeing the street light go on and off as we pass by (C) would be a matter of course. Therefore...there is reason to believe that the street light contains a motion detector (A).

Thus abduction (retroduction) results in the generation of hypotheses. Interestingly, the inference represents the logical fallacy called affirming the consequent (i.e., $p \supset q$, q $\therefore$ p where p is the hypothesized cause and q is the puzzling observation). Thus the best that abduction can do is to suggest that something might be. During a public lecture, the famous Nobel Prize–winning physicist Richard Feynman explained this limitation in his typical humor as follows:

> You know the most amazing thing happened to me tonight. I was coming here, on the way to the lecture, and I came in through the parking lot. And you won't believe what happened. I saw a car with the license plate ARW 357. Can you imagine? Of all the millions of license plates in the state, what was the chance that I would see that particular one tonight? Amazing! (1995, p. xxi)

Generating an Alternative Hypothesis

Subsequently, my wife drew on a different analogy to abductively generate an alternative bum light hypothesis. That is, the street light is *like* the old worn out lamp in our bedroom, which goes on and off at random or when its cord is jiggled. Thus it was just a coincidence that it went on and off as we walked past, that is:

> A street light goes on and then off as we pass by (C). However, if...the street light is worn out (A), then...seeing it go on and off seemingly at random as we pass by (C) would be a matter of course. Therefore...there is reason to believe that the street light is worn out (A).

Or in the form of affirming the consequent:

> If...the street light is worn out (p) then it should go on and off seemingly at random (q; $p \supset q$). The street light does go on and off seemingly at random (q). Therefore...perhaps the street light is worn out (p).

Using Deduction as a Further Test of the Alternatives

I then asked my wife how we might find out which hypothesis is correct. In response to this question, she proposed a test with *future* expectations/ predictions, that is:

> *If*...the motion detector hypothesis is correct, *and*...we turn around and walk past it again (proposed test), *then*...it should go on when we get near it and it should go off after we pass by (future expectations). Alternatively, *if*...the bum light hypothesis is correct, *then*...it should not necessarily go on when we get near and should not necessarily go off after we pass by (future expectation).

Peirce (1905) referred to this inferential process as *deduction*, which he described as follows:

> Abduction having suggested a theory, we employ deduction to deduce from that ideal theory a promiscuous variety of consequences to the effect that if we perform certain acts, we shall find ourselves confronted with certain experiences. We then proceed to try these experiments, and if the predictions of the theory are verified, we have a proportionate confidence that the experiments that remain to be tried will confirm the theory. (p. 209)

In other words, deduction involves assuming that the hypothesis in question is correct, imagining how it could be tested, and then inferring how the test *should* turn out when conducted. Deduction may appear to operate "automatically" and "logically." This, however, is an illusion, as there is no universal deductive logic that works in all situations. Instead, links to content-specific knowledge are needed to know what *predictions* (i.e., expectations) should follow. In short, deduction uses *if/and/then* reasoning to derive predictions from hypotheses and planned tests in a *hypothetico–deductive* manner.

Conducting the Test and Making the Subsequent Observations

After deriving the above predictions via deduction, the next night we conducted our test by driving to and then past the street light. The light was initially on and then it went off after we passed by. We then turned around and drove past the light again, and then again. Each time the light remained off. These are our *observed results*.

Using Induction to Draw a Conclusion

Consequently, we then completed the previous hypothetico–deductive argument by including our *observed results* to draw the following *conclusion*:

> *If…* the motion detector hypothesis is correct, *and…* we drive past the street light, *then…* it should go on when we get near it and it should go off after we pass by. Alternatively, *if…* the bum light hypothesis is correct, *then…* it should not necessarily go on when we get near and should not necessarily go off after we drive past. *And…* that is what happened (observed results). *Therefore…* the bum light hypothesis is supported and the motion detector hypothesis is contradicted because the observed results match bum light predictions but not the motion detector predictions (conclusion).

According to Peirce (1903), a third inference, which he called *induction,* was used to draw the above conclusion. More generally:

> *If …* the predicted and observed results match, *then…* the hypothesis is supported. Alternatively, *if…* the predicted and observed results do not match, *then…* the hypothesis is contradicted.

Although Peirce referred to this inference as induction, it is not the form of induction that some have claimed generates general conclusions from limited cases (e.g., this crow is black, so is this one, and so on—therefore *all* crows are black)—a form of "enumerative" induction that people probably do not use (Lawson, 2005; Popper, 1965). Rather, enumerative induction can at best suggest descriptive claims in need of deductive test (e.g., all of the crows I have seen thus far are black; thus *perhaps* all crows are black. *If…* all crows are black, *and…* this new bird is a crow, *then…* I deduce/predict that it will also be black).

Peirce's form of induction can be characterized as an inference that leads to increased confidence in one's conclusions with each additional supporting or contradicting result. In Peirce's (1903) words:

> If that supposition be correct, a certain sensible result is to be expected under certain circumstances which can be created, or at any rate are to be met with. The question is, Will this be the result? If Nature replies "No!" the experimenter has gained an important piece of knowledge. If Nature says "Yes," the experimenter's ideas remain just as they were—only somewhat more deeply engrained. If Nature says "Yes" to the first twenty questions, although they

were so devised as to render that answer as surprising as possible, the experimenter will be confident that he is on the right track, since 2 to the 20th power exceeds a million. (p. 168)

In summary, reasoning is provoked by puzzling observations that are subsequently explained by the cyclic and repeated use of abduction, deduction, observation, and induction in an *If/then/Therefore* manner. Again in Peirce's (1905) words:

Abduction furnishes all our ideas concerning real things, beyond what are given in perception, but is mere conjecture, without probative force. Deduction is certain but relates only to ideal objects. Induction gives us the only approach to certainty concerning the real that we can have. In forty years diligent study of arguments, I have never found one which did not consist of these elements. (p. 209)

We next turn to an example of scientific discovery to see if the same inferences and *If/then/Therefore* reasoning pattern can be identified. The example is of Galileo Galilei's (1564–1642) discovery of Jupiter's moons in 1610 (Galilei, 1610/1954, as initially interpreted by Lawson, 2002).

GALILEO'S DISCOVERY OF JUPITER'S MOONS

Initial Puzzling Observation and Abduction

In January 1610, Galileo had recently invented a new and improved telescope and had began using it to explore the "heavens." During his initial exploration, Galileo was puzzled by his observation of three previously unseen points of light near the planet Jupiter. On the basis of their similar appearance to fixed stars (i.e., stars that lie in the celestial sphere beyond Jupiter), he initially generated the hypothesis that that they were in fact fixed stars. Putting Galileo's hypothesis generation in the form of Peirce's abductive inference we get:

Galileo observes three puzzling points of light in the night sky near Jupiter (C). However, if... the three points of light are fixed stars (A), then... seeing the three new points of light (C) would be a matter of course. Therefore... there is reason to believe they are fixed stars (A).

Or in the form of affirming the consequent we get:

If... fixed stars embedded in the celestial sphere (p) cause points of light in the night sky (q; p ⊃ q), *and*... three new points of light are seen in the night sky (q), *then*... perhaps they are caused by fixed stars (p).

Similarly:

If... an artificial selection process (p) causes gradual changes in captive plants and animals (q; p ⊃ q), *and*... an analogous selective process occurs in nature (q), *then* ... perhaps that analogous "natural" selection process causes gradual changes in wild organisms (p).

If ... sensing the Earth's magnetic field (p) enables homing pigeons to navigate home (q; p ⊃ q), *and* ... salmon can also sense that magnetic field (q), *then* ... perhaps they also use it to return to their home streams (p).

An Initial Test of the Fixed-Star Hypothesis Using Deduction and Induction

Interestingly, Galileo's further reasoning led him to doubt his fixed-star hypothesis. As he put it:

> ... although I believed them to belong to the number of the fixed stars, yet they made me somewhat wonder, because they seemed to be arranged exactly in a straight line, parallel to the ecliptic, and to be brighter than the rest of the stars, equal to them in magnitude. (p. 59)

Galileo's expressed doubt can be cast more explicitly as follows:

> *If* ... the three points of light near Jupiter are fixed stars, *and* ... their position and brightness are compared to each other and to nearby fixed stars, *then* ... variations in position and brightness should be random, as is the case for other fixed stars. *But* ... "they seem to be arranged exactly in a straight line, parallel to the ecliptic, and to be brighter than the rest of the stars." *Therefore* ... the fixed-star hypothesis is contradicted. Or as Galileo put it, "yet they made me wonder somewhat."

This argument goes beyond abduction. Here, Galileo's presumed reasoning has tested his previously abducted fixed-star hypothesis and led him to doubt its veracity based on the fact that the variations in position and brightness of the points of light are *not* random as expected. In other words, his observed results did not match his predicted results. Therefore, via deduction and induction, the fixed-star hypothesis was not supported.

Also note that the above argument includes observations that were made *prior to* Galileo's generation of the fixed-star hypothesis. Accordingly, two types of arguments can be identified: (a) those that test hypotheses by using previously gathered data—an argumentative pattern that is sometimes called *retrodiction*, and (b) those that test hypotheses by conducting tests that produce new data. Both types of arguments involve abduction, deduction, and induction. And in philosopher Carl Hempel's (1966) words:

> ... from a logical point of view, the strength of the support that a hypothesis receives from a given body of data should depend only on what the hypothesis asserts and what the data are: the question of whether the hypothesis or the data were presented

first, being a purely historical matter, should not count as affecting the confirmation of the hypothesis. (p. 38)

Generating an Alternative Hypothesis and Deducing Predictions

Consequently, Galileo went back to his LTM to abductively generate another hypothesis. Perhaps, thought Galileo, the points of light are moons orbiting Jupiter—*like* the moon that orbits Earth, or *like* the planets that orbit the Sun. Presumably after using abduction to generate an orbiting-moons hypothesis, he sought a deductive way to test it. More specifically, Galileo generated an argument that deductively led from his new hypothesis and planned test to two *future* predictions, that is:

> *If* ... the three points of light are moons orbiting Jupiter (orbiting-moons hypothesis), *and* ... I observe them over the next several nights (planned test), *then* ... some nights they should appear to the east of Jupiter and some nights they should appear to the west. And they should always appear along a straight line on either side of Jupiter (predictions).

Making the Necessary Observations

After presumably deriving the above two predictions via deduction, Galileo remarked, "I therefore waited for the next night with the most intense longing, but I was disappointed of my hope, for the sky was covered with clouds in every direction" (p. 60). So due to cloud cover, Galileo was unable to make the necessary observations. Fortunately, the next night and several subsequent nights were clear, and sure enough, the points of light appeared just as Galileo's orbiting-moons hypothesis and planned test led him to expect.

Using Induction to Draw a Conclusion

Consequently, we can complete the argument with Galileo's observed results and conclusion in this manner:

> *And* ... some nights they appeared to the east of Jupiter and some nights they appeared to the west. Further, they always appeared along a straight line on either side of Jupiter (observed results). *Therefore* ... the orbiting-moons hypothesis is supported (conclusion).

In Galileo's (1610) words:

> I, therefore, concluded, and decided unhesitatingly, that there are three stars in the heavens moving about Jupiter, as Venus and Mercury round the sun; ... These observations also established

that there are not only three, but four, erratic sidereal bodies performing their revolutions round Jupiter … These are my observations upon the four Medicean planets, recently discovered for the first time by me. (pp. 60, 61)

Summary of Galileo's Reasoning and Discovery

To summarize, three inferences can be identified in Galileo's reasoning and discovery (see Table 8.1):

1. First, thanks to his new and improved telescope (the role of technology) Galileo undertook a new exploration that led to a *puzzling observation* (the three unexplained points of light near Jupiter)
2. Then, thanks to his prior store of knowledge, Galileo used *abduction* to generate a causal hypothesis (a tentative explanation) for the points of light (i.e., perhaps they are fixed stars)
3. Next, Galileo used *deduction* and *induction* to subconsciously test his fixed-star hypothesis, which due to a mismatch between predictions and prior observations led to some doubt and then to rejection (via *retrodiction*)
4. Then he once again used *abduction* to generate another hypothesis (an orbiting-moons hypothesis)

Table 8.1 *Inferences Involved in Scientific Hypothesis Generation and Test*

Inference	Question	Example
Abduction	What caused the puzzling observation (e.g., the three new points of light near Jupiter)?	*If…*fixed stars embedded in the celestial V cause points of light in the night sky, *and*…three new points of light are seen in the night sky, *then…* perhaps they are caused by fixed stars.
Deduction	What does the proposed cause lead us to predict about future and/or prior observations?	*If…* the three points of light are moons orbiting Jupiter, *and…* I observe them over the next several nights, *then…* some nights they should appear to the east of Jupiter and some nights they should appear to the west. Further, they should appear along a straight line on either side of Jupiter.
Induction	How do the predictions and observations compare?	*If…* the observations match the predictions based on the orbiting-moons hypothesis, as they do in this case (e.g., some nights the lights appeared to the east of Jupiter and some nights they appeared to the west), *then…* the hypothesis is supported.

5. He then planned a test and used *deduction* to generate future predictions
6. Subsequently, after the cloud cover dissipated he made the necessary *observations*, which matched his predictions
7. Lastly, based on this match, he used *induction* to draw the *conclusion* that his orbiting-moons hypothesis had been supported. Therefore, he was able to proudly proclaim to the world that he was the first to discover "four, erratic sidereal bodies performing their revolutions round Jupiter."

Viewed in this way, scientific reasoning and discovery consist of undertaking novel explorations that lead to puzzling observations that are subsequently explained by the cyclic and repeated use of abduction, deduction, observation, and induction. Sometimes the observations used to test proposed explanations have already been made. And sometimes they have yet to be made. Although experimental data are highly valued, circumstantial and correlational data can also be used to test proposed explanations.

MODELING THE ELEMENTS AND INFERENCES OF SCIENTIFIC REASONING

Figure 8.1 models the basic elements and inferences involved in hypothesis generation and test. The figure distinguishes between declarative and

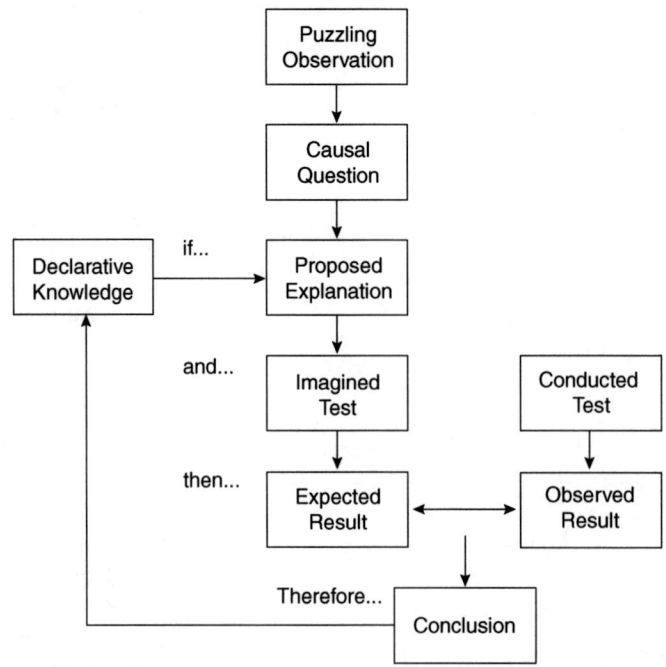

Figure 8.1   *A model of the elements, inferences, and if/then/therefore pattern of reasoning and argumentation used during the generation and subsequent test of proposed explanations.*

procedural knowledge. The distinction is respectively between "knowing that" (e.g., I declaratively know that animals inhale oxygen and expel carbon dioxide, and that Phoenix is the capital of Arizona) and "knowing how" (e.g., I procedurally know how to ride a bicycle, how to count, and how to conduct a controlled experiment). Thus, declarative knowledge, sometimes called explicit knowledge, comprises the facts/concepts that we know, whereas procedural knowledge, sometimes called tacit knowledge, comprises the guidelines/plans/strategies that we know how to carry out (e.g., Anderson, 1980; Collins, 2001; Nonaka & von Krogh, 2009; Piaget, 1970; Tsoukas, 2003). Procedural knowledge, which is expressed through performance, is often implicit or tacit in the sense that we may not be conscious that we have it or precisely when or how it was acquired. The word "development" is often used in conjunction with the acquisition of procedural knowledge. However, declarative knowledge is explicit—that is, we often know that we have it and when and how it was acquired. The word "learning" is often used in conjunction with the acquisition of declarative knowledge.

Neurological research indicates that the procedural knowledge, once acquired, resides in neural networks that are hierarchical in nature. The hierarchical networks culminate in single neurons located in the brain's prefrontal cortex (Wallis, Anderson, & Miller, 2001). Alternatively, declarative knowledge resides in associative memory, which is located primarily in the hippocampus, the limbic thalamus, and the basal forebrain (Kosslyn & Koenig, 1995; Squire, 2004). Further, it appears that the conscious recollection of procedural knowledge is independent of the medial temporal lobe, and thus depends on other brain systems such as the neo-striatum, whereas the storage and recollection of declarative knowledge depends on the functional integrity of the medial temporal lobe (Squire & Zola-Morgan, 1991).

Both types of knowledge operate together during cognition. For example, one needs to procedurally "know how" to count to declaratively "know that" there are 10 marbles on a table. Alternatively, declarative knowledge is required not only as a source of abductively derived hypotheses, but also as a source of deductively derived predictions. Consequently, skill in identifying causes is dependent in part on procedural knowledge (i.e., guidelines/plans/ strategies) for testing hypotheses and for evaluating results, and in part on declarative knowledge (i.e., a store of hypotheses to be tested and means of testing them).

Once the test(s) have been conducted and results obtained, induction is used to draw a conclusion about the relative truth or falsity of the hypothesis under test, based on the extent to which predicted and observed results match. Consistent with Peirce's underlying pragmatism, a good match supports but does not "prove" the hypothesis because one or more unstated and perhaps unimagined alternative hypotheses may give rise to the same prediction under similar test conditions—that is, false positives can occur (labeled a type II error in statistics). Similarly, a poor match does not "disprove" the hypothesis, because a poor match can arise from a faulty deduction or a faulty test instead of a faulty hypothesis. Thus, false negatives are also possible (labeled a type I error in statistics).

Also, as pointed out by authors such as Brannigan (1981) and Collins (1985), scientific explanations are evaluated within social and cultural contexts that play a role in their acceptance or rejection. Recall the words of Charles Darwin written in the concluding chapter of *The Origin of Species* (initially published in 1859):

> Although I am fully convinced of the truth of the views given in this volume under the form of an abstract, I, by no means expect to convince experienced naturalists whose minds are stocked with a multitude of facts all viewed during a long course of years, from a point of view directly opposite to mine....but I look with confidence to the future, to young and rising naturalists, who will be able to view both sides of the question with impartiality. (1898 ed., pp. 294,295)

Similarly, in his autobiography, the physicist Max Plank (Planck, 1949, pp. 33,34) wrote: "A scientific truth does not triumph by convincing its opponents and making them see the light, but rather because its opponents eventually die, and a new generation grows up that is familiar with it."

And finally, as mentioned in the previous section, not all data that may bear on the validity of hypotheses must be gathered after the hypotheses have been generated. In fact, an advantage of using previously gathered data is that such data are more likely to be free of experimenter bias. A classic case is Erwin Chargaff's rules about the relative amounts of nucleotide bases in DNA known in the 1940s (i.e., adenine equals thymine, guanine equals cytosine). An explanation for Chargaff's rules was not advanced until the spring of 1953 when a young James Watson searching for the structure of DNA pushed cardboard models of its bases together in various combinations until: "Suddenly I became aware that an adenine–thymine pair held together by two hydrogen bonds was identical in shape to a guanine–cytosine pair held together by at least two hydrogen bonds" (Watson, 1968, p. 123). This was a key puzzle piece that enabled Watson to construct a new and better double-helical model of DNA (i.e., a puzzle piece that gave him a new and better hypothesis). Only then did Watson realize that his new model implied/predicted that the amount of adenine in DNA should equal the amount of thymine, and that the amount of guanine should equal the amount of cytosine. As Watson put it, "Chargaff's rules then suddenly stood out as a consequence of a double-helical structure for DNA" (Watson, 1968, p. 125). More explicitly:

> *If...* DNA is a double helix with adenine always paired with thymine and with guanine always paired with cytosine, *and...* the relative amounts of the bases in DNA are determined, *then...* the amount of adenine should equal the amount of thymine and the amount of guanine should equal the amount of cytosine (prediction derived by Watson in 1953). *And...* the amounts are equal (as determined in the 1940s by Chargaff). *Therefore...* the double-helix hypothesis is supported.

The following section will consider the roles played by different brain regions during scientific reasoning and discovery. We will begin with some basics of brain development.

SOME BASICS OF BRAIN DEVELOPMENT

The neocortex, which is the most recently evolved part of the brain (Jerison, 2001; Jerison & Jerison, 1988), has a full compliment of brain cells (neurons) at birth—some 100 billion. Yet the most rapid growth of the neocortex takes place during the first 10 years of life. This growth is primarily due to proliferation of the branching projections (dendrites) that connect with and receive input, via synapses, from nearby neurons. Importantly, the number of dendrites varies depending on use or disuse (e.g., Baltes, Reuter-Lorenz, & Rosler, 2006; Greenough, Volkmar, & Juraska, 1973; Perry, 2002). For example, the neurons in the brain area that deal with word understanding (Wernicke's area) have more dendrites in college-educated people than in people with only a high school education (Diamond, 1996).

A classic study of the effect of disuse of neurons was conducted during the 1970s by Wiesel and Hubel. They covered one eye of newborn kittens at birth. When the covered eyes were uncovered 2 weeks later, the eyes were unable to see. Presumably the lack of environmental input prevented the deprived neurons from developing dendrite connections (for a review of this and related work, see Hubel & Wiesel, 2005). As Diamond (1996) put it, the phrase "use it or lose it" definitely applies in the case of dendrites, and that passive observation is not enough for these new connections. One has to actively interact with the environment.

How does environmental interaction lead to an increase in the number of functional dendrites? Based on the pioneering research of Eric Kandel in the early 1960s on the giant marine snail *Aplysia* (see Kandel, 2006), dendrites become functional when synaptic strengths increase occurs due to the pairing of electrical signals along two different neural pathways. The proper pairing increases neuro-transmitter release rate at synaptic knobs and this makes signal transmission from one neuron to the next easier. Hence learning is understood as an increase in the number of "operative" synaptic connections among neurons. That is, learning occurs when transmitter release rate at synaptic knobs increases so that the signals can be easily transmitted across synapses that were previously there, but inoperative.

Experience Strengthens Connections

Grossberg (1982, 2005) has proposed and tested equations describing the basic interaction of the key neural variables involved in learning. Of particular significance is his learning equation, which describes changes in transmitter release rate (i.e., Z_{ij}). The learning equation identifies factors that modify the synaptic strengths of knobs N_{ij}. Z_{ij} represents the initial synaptic strength. B_{ij} is a constant of decay. Thus $B_{ij} Z_{ij}$ is a forgetting or decay term. $S'_{ij} [X_j] +$ is the learning term as it drives increases in Z_{ij}. S'_{ij} is the signal that has passed from node V_i to knob N_{ij}. The prime reflects the fact that the initial signal, S_{ij}, may be slightly altered as it passes down e_{ij}. $[X_j]+$ represents the activity level of postsynaptic nodes V_j that exceeds the firing threshold. Only activity above threshold can cause changes in Z_{ij}. In short, the learning

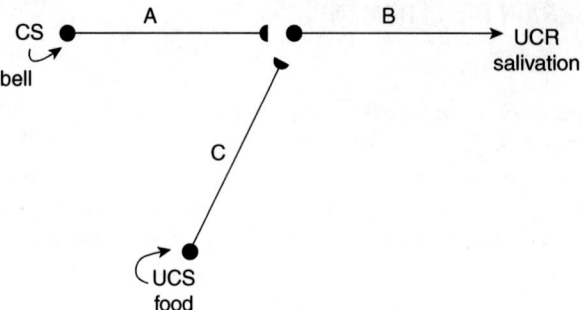

Figure 8.2   *Classical conditioning in a simple neural network. Cells A, B, and C represent layers of neurons.*

term indicates that for information to be stored in LTM, two events must occur simultaneously. First, signals must be received at N_{ij}. Second, nodes V_j must receive inputs from other sources that cause the nodes to fire. When these two events drive activity at N_{ij} above a specified constant of decay, the Z_{ij}'s increase; and the network learns. For a network with n nodes, the learning equation is as follows:

$$\dot{Z}_{ij} = -B_{ij}Z_{ij} + S'_{ij}[X_j]^+$$

where the over dot represents a time derivative and i, j,= 1, 2,...n.

For example, consider Pavlov's classical conditioning experiment in which a dog is stimulated to salivate by the sound of a bell. When Pavlov first rang the bell, the dog, as expected, did not salivate. However, upon repeated simultaneous presentation of food, which did initially cause salivation, and bell ringing, the ringing alone eventually caused salivation. Thus, the food is the unconditioned stimulus (US). Salivation upon presentation of the food is the unconditioned response (UCR). And the bell is the conditioned stimulus (CS). Pavlov's experiment showed that when a CS (e.g., a bell) is repeatedly paired with a US (e.g., food), the CS alone will eventually evoke the UCR (e.g., salivation). How can the US do this?

Figure 8.2 shows a simple neural network capable of explaining classical conditioning. Although the network is depicted as just three cells, A, B, and C, each cell represents many neurons of the type A, B, and C. Initial food presentation causes cell C to fire. This creates a signal down its axon that, because of prior learning (i.e., a relatively large Z_{cb}), causes the signal to be transmitted to cell B. Thus, cell B fires, and the dog salivates. At the outset, bell ringing causes cell A to fire and send signals toward cell B. However, when the signal reaches knob N_{AB}, its synaptic strength Z_{AB} is not large enough to cause B to fire. So the dog does not salivate. However, when the bell and the food are paired, cell A learns to fire cell B according to Grossberg's learning equation. Cell A firing results in a large S'_{AB} and the appearance of food results in a large $E[X_B]+$. Thus the product $S'_{AB}[X_B]+$ is sufficiently large to drive an increase in Z_{AB} to the point at which it alone causes node V_B to fire and evoke salivation. Food is no longer needed. The dog has learned to salivate at the ringing of a bell. The key point is that learning is driven by simultaneous activity of pre- and postsynaptic neurons, in this case, activity of cells A and B.

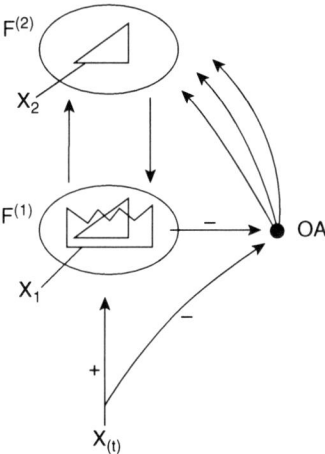

Figure 8.3   *Adaptive resonance occurs when a match of activity patterns occurs on successive slabs of neurons.*

Adapted from Grossberg (1982); Carpenter & Grossberg (2003).

Adaptive Resonance: Matching Input With Expectations

Another key aspect of neural network theory explains how the brain processes a continuous steam of sensory input by matching sensory input with expectations derived from prior experience. Grossberg's mechanism for this, called adaptive resonance, is shown in Figure 8.3.

The process begins when sensory input $X_{(t)}$ is assimilated by a slab of neurons designated as $F^{(1)}$. Due to prior experience, a pattern of activity, X_1, then plays at $F^{(1)}$ and causes a firing of pattern X_2 at another slab of neurons, $F^{(2)}$. X_2 then excites a pattern X on $F^{(1)}$. The pattern X is compared with the input following X_1. Thus, X is the expectation. X will be X_1 in a static visual scene and the pattern to follow X_1 in a temporal sequence. If the two patterns match, then you see what you expect to see. This allows an uninterrupted processing of input and a continued quenching of nonspecific arousal. One is only aware of patterns that enter the matched/resonant state. Unless resonance occurs, coding in LTM is not likely to take place. This is because only in the resonant state is there both pre- and postsynaptic excitation of the cells at $F^{(1)}$ (i.e., see Grossberg's learning equation).

Now suppose the new input to $F^{(1)}$ does not match the expected pattern X from $F^{(2)}$. Mismatch occurs and this causes activity at $F^{(1)}$ to be turned off by lateral inhibition, which in turn shuts off the inhibitory output to the nonspecific arousal source. This turns on nonspecific arousal and initiates an internal search for a new pattern at $F^{(2)}$ that will match X_1.

Such a series of events explains how information is processed across time. The important point is that stimuli are considered familiar if a memory record of them exists at $F^{(2)}$ such that the pattern of excitation sent back to $F^{(1)}$ matches the incoming pattern. If they do not match, the incoming stimuli are unfamiliar and orienting arousal (OA) is turned on to allow an unconscious search for another pattern. If no such match is obtained, then no coding in

LTM will take place unless attention is directed more closely at the object in question. Directing careful attention at the unfamiliar object may boost pre-synaptic activity to a high enough level to compensate for the relatively low postsynaptic activity and eventually allow a recording of the sensory input into a set of previously uncommitted cells.

How Visual Input Is Processed in the Brain

As reviewed by Kosslyn and Koenig (1995), the ability to visually recognize objects requires participation of the six major brain areas. As shown in Figure 8.4, sensory input from the eyes passes from the retina to the back of the brain and produces a pattern of electrical activity in the visual buffer (located in the occipital lobe). This activity produces a spatially organized image within the visual buffer. Next, a smaller region within the visual buffer (called the attention window) performs additional processing. The processed electrical activity is then simultaneously sent along two pathways on each side of the

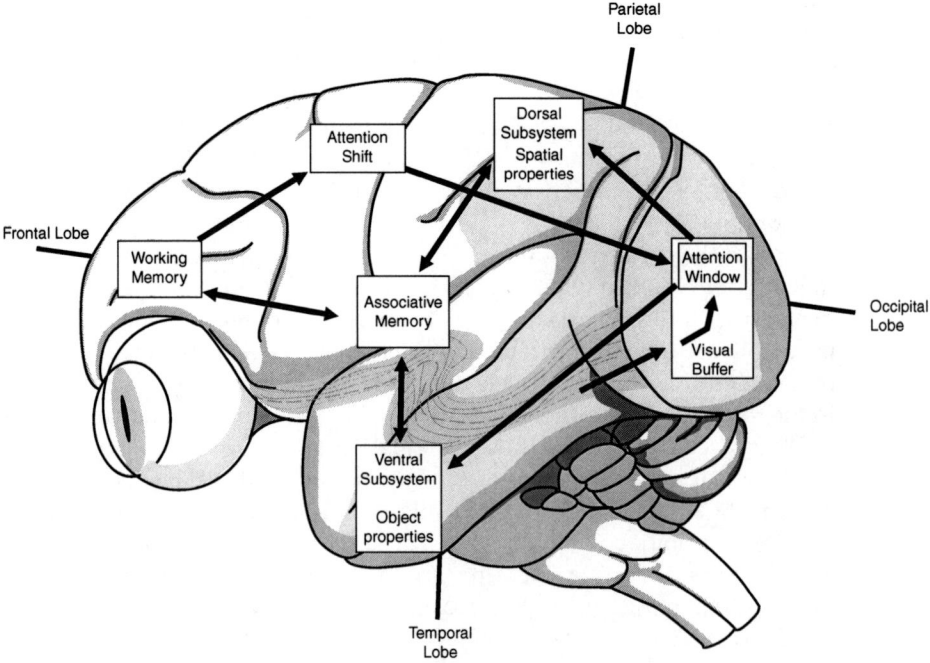

Figure 8.4 *Kosslyn and Koenig's model of the visual system consists of six major subsystems that spontaneously and subconsciously generate and test hypotheses about what is seen.*

brain; two pathways run down (to the ventral subsystem in the lower temporal lobes), and two run up (to the dorsal subsystem in the parietal lobes). The ventral subsystem analyzes object properties, such as shape, color, and texture. The dorsal subsystem analyzes spatial properties, such as size and location. Patterns of electrical activity within the ventral and dorsal subsystems are then sent and matched to visual patterns stored in associative memory, which is located primarily in the hippocampus, the limbic thalamus, and the basal forebrain. If a good match is found (i.e., an adaptive resonance), the object is recognized and the observer knows the object's name, categories to which it belongs, sounds it makes, and so on.

However, if a good match is not obtained, the object remains unrecognized and additional sensory input must be obtained. The search for additional input is not random. Rather, stored patterns are used to make a second hypothesis about what is being observed; this hypothesis leads to new observations and to further encoding. In the words of Kosslyn and Koenig, when additional input is sought, "One actively seeks new information that will bear on the hypothesis... The first step in this process is to look up relevant information in associative memory" (p. 57). Information search involves activity in the prefrontal lobes in an area referred to as working memory. Activating working memory causes an attention shift of the eyes to a location where an informative component *should be* located. Once attention is shifted, the new visual input is processed in turn. The new input is then matched to shape and spatial patterns stored in the ventral and dorsal subsystems and kept active in working memory. Again, in Kosslyn and Koenig's words, "The matching shape and spatial properties may in fact correspond to the hypothesized part. If so, enough information may have accumulated in associative memory to identify the object. If not, this cycle is repeated until enough information has been gathered to identify the object or to reject the first hypothesis, formulate a new one, and test it" (p. 58).

In other words, as one seeks to identify objects, the brain generates and tests stored patterns selected from memory. Kosslyn and Koenig even speak of these stored patterns as hypotheses, where the term hypothesis is used in its broadest sense. One looks at part of an unknown object and the brain spontaneously and immediately generates an idea of what it is—a hypothesis. Thanks to links in associative memory, the hypothesis carries implied consequences (i.e., expectations/predictions). Consequently, to test the hypothesis one can carry out a simple behavior to see if the prediction does in fact follow. If it does, one has support for the hypothesis. If it does not, then the hypothesis is not supported and the cycle repeats. Thus, brain activity during visual processing utilizes the same *If/then/Therefore* pattern that we identified previously when discussing scientific reasoning.

Auditory Input Processing

The visual system is only one of several of the brain's information processing systems. However, information seems to be processed in a similar manner by other brain systems. For example, with respect to learning the meaning of spoken words, Kosslyn and Koenig (1995) state: "Similar computational analyses can be performed for visual object identification and spoken word identification, which will lead us to infer analogous sets of processing subsystems" (p. 213).

Like visual recognition, word recognition involves brain activity in which hypotheses arise immediately, unconsciously, and before any other activity. In other words, the brain does not make several observations before it generates a hypothesis of what it thinks is out there. Instead, from the slimmest piece of input, the brain immediately generates an idea of what it "thinks" is out there. The brain then acts on that initial idea until subsequent behavior is contradicted. In other words, the brain is not an inductivist organ. Rather it is an idea-generating and testing organ that works in a hypothetico–deductive way. There is good reason in terms of human evolution why this would be so. If you were a primitive person and you look into the brush and see stripes, it would certainly be advantageous to get out of there quickly, as the consequences of being attacked by a tiger are dire. Anyone programmed to look, look again, and look still again in an "inductivist" way before generating the tiger hypothesis would most likely not survive long enough to pass on his plodding inductivist genes to the next generation.

The important point is that learning in general does not happen the way you might think. Your brain does not prompt you to look, look again, and look still again until you somehow internalize a successful behavior from the environment. Rather, your brain directs you to look, and as a consequence of that initial look, the brain generates an initial hypothesis that then drives behavior, behavior that carries with it a specific expectation. Hopefully, the behavior is successful in the sense that the prediction is matched by the outcome of the behavior. But sometimes it is not. So the contradicted behavior then prompts the brain to generate another hypothesis and so on until eventually the resulting behavior is not contracted.

NEURAL NETWORKS AND HIGHER LEVELS OF REASONING AND LEARNING

Researchers have shown that the previous neural network principles can be successfully applied to explain more complex learning. For example, Levine and Prueitt (1989) developed and tested a neural network model to explain performance of normal persons and those with frontal lobe damage on the Wisconsin Card Sorting Task. Jani and Levine (2000) developed a neural network model that simulates the learning involved in proportional analogy-making, and more recently Levine (2009) modeled the networks involved in decision making. But what about the networks involved in scientific discovery? Lawson (2002) applied neural network theory to the case of Galileo Galilei during his discovery of Jupiter's moons in 1610. Following Lawson (2002), let's turn to that report and analyze Galileo's reasoning in terms of *If/then/Therefore* reasoning and the previously introduced neural network principles.

Galileo's Reasoning Within Grossberg's Theory

Grossberg's theory can be used to understand what might have been going on in Galileo's mind in terms of neurological events. Figure 8.5 depicts two

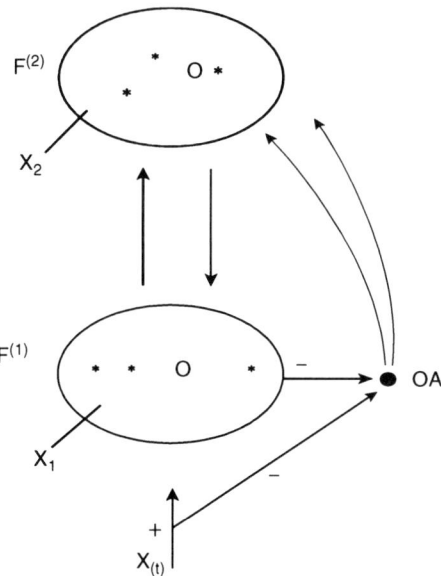

Figure 8.5   *Grossberg's model of the match and mismatch of activity patterns on successive slabs of neurons in the brain. Input $X_{(T)}$ (e.g., three spots of light near Jupiter) excites an activity pattern at slab $F^{(1)}$ and inhibits OA. The pattern at slab $F^{(1)}$ excites a pattern at slab $F^{(2)}$, which feeds back to $F^{(1)}$. A mismatch (i.e., a new observation that does not match an expectation), causes quenching of activity at $F^{(1)}$ and shuts off inhibition of OA. Oa is then free to search for another pattern (i.e., another hypothesis) to match the input.*

successive slabs of neurons in the brain, $F^{(1)}$ and $F^{(2)}$. According to Grossberg, sensory input $X_{(t)}$, (e.g., light coming from the three objects near Jupiter on the night of January 7th) excites an electrical pattern of activity at slab $F^{(1)}$ and sends a signal to inhibit nonspecific OA. The electrical pattern X_1 at $F^{(1)}$ then excites another electrical pattern, X_2, at the next slab of neurons at $F^{(2)}$, which feeds signals back to $F^{(1)}$. In the case of Galileo's initial observations, the pattern at $F^{(2)}$ corresponds to his star category and initially matches the pattern at $F^{(1)}$. Thus, all is well both neurologically and conceptually.

But as reported, Galileo's continued deductive and inductive reasoning led to a partial mismatch (e.g., his star category deductively predicted that stars should *not* be lined up along a straight line and should not be equidistant from each other). This partial mismatch led Galileo to "somewhat wonder." Neurologically speaking, a mismatch (i.e., a new observation that does not match an expectation), causes quenching of activity at $F^{(1)}$ and shuts off inhibition of OA. OA is then free to search for another pattern (i.e., another hypothesis) to match the input. In other words, with Galileo's continued observations and reasoning, the mismatch between the patterns at $F^{(1)}$ and $F^{(2)}$ presumably became so great that activity at $F^{(1)}$ was quenched. Thus, inhibition of OA was shut down. OA was then free to excite $F^{(2)}$ leading to a search for another pattern of activity to hopefully match the input pattern at $F^{(1)}$. On the conceptual level, Galileo's mind was now free to search for alternative hypotheses (e.g., the planet hypothesis, the moon hypothesis) to replace (via abduction) the rejected fixed-star hypothesis. Once an activity pattern at $F^{(2)}$ was found that actually matched the input pattern at $F^{(1)}$, OA was shut down and Galileo's search was complete. He had "discovered" four new moons orbiting Jupiter.

Galileo's Thinking Within Levine and Preuitt's Model

A more detailed way to model Grossberg's basic ideas has been proposed by Levine and Prueitt (1989). Lawson (1993) used Levine and Prueitt's model to account for the processes involved in simple hypothesis testing and in descriptive concept acquisition. Figure 8.6 suggests how Levine and Prueitt's model may also be used in the present context.

As you can see, the model includes feature nodes referred to as F_1. These nodes code input features (e.g., number of spots of light, the sizes of those

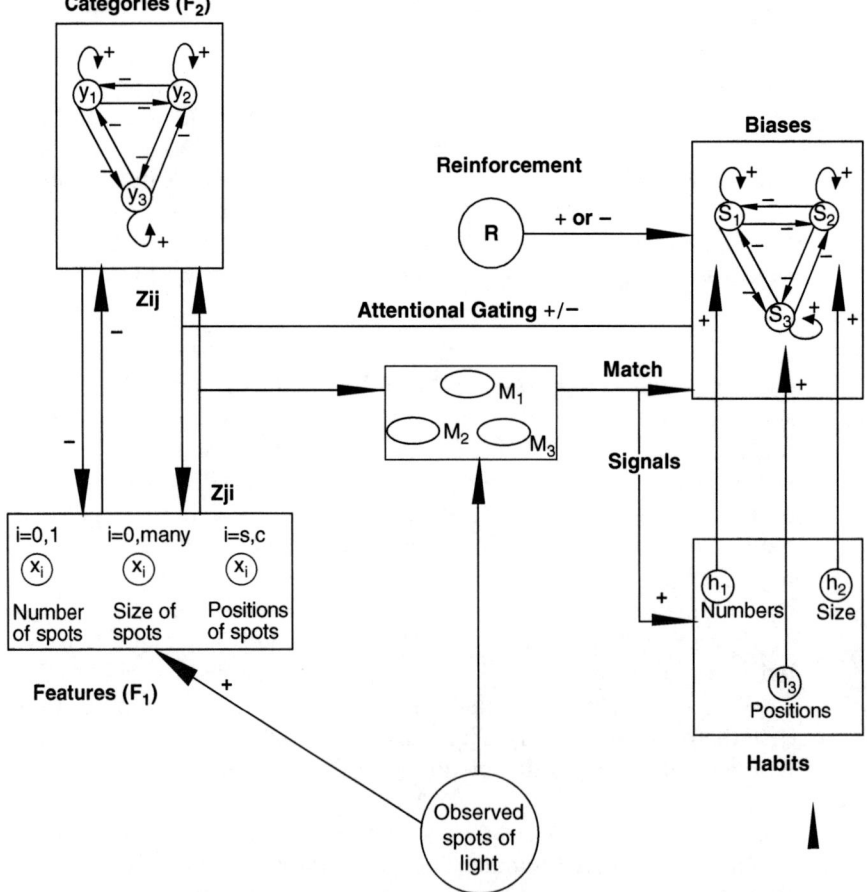

Figure 8.6   *Levine and Prueitt's model of neural activity including feature, category, bias, and habit nodes. The feature nodes code observable input features (e.g., numbers, sizes, positions). The category nodes represent prior knowledge categories into which the input may fit (e.g., star category, planet category, moons category). The habit and bias nodes keep track of past decisions and influence subsequent decisions.*

spots, their positions). Nodes in F_2 code for categories into which the input can be placed (e.g., fixed stars, planets, moons). Once again, these categories serve as alternative hypotheses. The model also includes habit and bias nodes. Habit nodes detect how often prior classifications have correctly and incorrectly been made. The bias nodes are affected by activity in the habit nodes and by reinforcement. Details of network function can be found in Levine and Prueitt (1989) and Lawson (1993). The important point in terms of the present argument is that information processing, whether it involves basic descriptive concept formation, simple hypothesis testing, or the "discovery" of Jupiter's moons, is basically a process of matching new input with prior categories stored in memory. To state differently, scientific reasoning is hypothetico–deductive in nature because the brain spontaneously processes new input in a hypothetico–deductive way.

Galileo's Thinking Within Kosslyn and Koenig's Model

Kosslyn and Koenig's description of brain subsystem functioning is about recognizing objects present in the visual field during a very brief time period—not distant spots of light seen through a telescope. Nevertheless, the hypothetico–deductive nature of this system functioning is clear. All one need do to apply the same principles to Galileo's case is to extend the time frame over which observations are made—observations that either match or mismatch expectations. For example, Figure 8.7 shows how the brain subsystems may have been involved in Galileo's reasoning as he tests his moons hypothesis.

The figure highlights the contents of Galileo's working memory, which is seated in the lateral prefrontal cortex, in terms of one cycle of *If/then/Therefore* reasoning. As shown, to use *If/then/Therefore* reasoning to generate and test his moon hypothesis, Galileo must not only allocate attention to it and its predicted consequences, he must also inhibit his previously generated fixed stars and astronomers-made-a-mistake hypotheses. Thus, working memory can be thought of as a temporary network to sustain information while it is processed. During reasoning, one must pay attention to task-relevant information and inhibit task-irrelevant information. Consequently, working memory involves more than simply allocating attention and temporarily keeping track of it. Rather, during the reasoning process, working memory actively selects information relevant to one's goals and actively inhibits irrelevant information.

LOCATION, FUNCTION, AND DECLINE OF WORKING MEMORY

Working memory is thought of primarily as a temporary network to sustain information while it is processed. However, as we have seen, during reasoning one must pay attention to task-relevant information and inhibit task-irrelevant information. Consequently, working memory involves more than

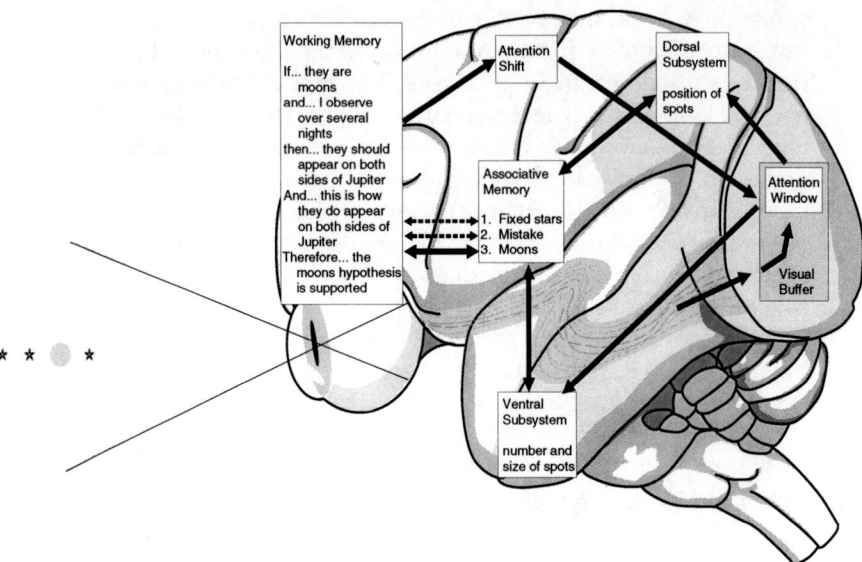

Figure 8.7   *What might have been in Galileo's working memory when he tested the moons hypothesis.*

simply allocating attention and temporarily keeping track of it. Rather, working memory actively selects information relevant to one's goals and actively inhibits irrelevant information. Although working memory is seated in the lateral prefrontal cortex, research indicates that working memory cannot be pinned down to a single prefrontal region. Rather, its location depends in part on the type of information being processed (e.g., Friedman & Goldman-Rakic, 1994; Fugelsang & Dunbar, 2004; Fugelsang, Roser, Corballis, Gazzaniga, & Dunbar, 2004; Funster, 1989; Green, Fugelsang, Kraemer, Shamosh, & Dunbar, 2006; Kwon, Lee, Shin, & Jeong, 2009). With its many projections to other brain areas, working memory plays a crucial role in keeping representations active while it coordinates mental activity.

Following Baddeley (1995, 1996), working memory was conceived of in terms of three main components, (a) a component that activates representations of objects and their properties called the *visuo-spatial scratchpad*, (b) a component that does the same for linguistic representations, called the *phonological loop*, and (c) a catch-all component for processes such as reasoning and problem solving that do not fit neatly into the other two components. This third component is called the *central executive*. Research by Smith and Jonides (1994) and Paulesu, Frith, and Frackowiak (1993) suggest a respective right and left hemisphere specialization for the scratch pad and phonological loop.

Research on the decline of reasoning abilities with advancing age during adulthood suggests that the decline may be due to a loss of working memory efficiency, specifically to a decrease in inhibitory functioning (Hasher & Zacks, 1988; Shimamura, 2000; Viskontas, Morrison, Holyoak, Hummel, & Knowlton, 2004) coupled with a loss of selective attention (Craik & Byrd, 1982). Such declines appear to be linked to what has been called fluid intelligence (e.g., Isingrini & Vazou, 1997), but not to crystallized intelligence and semantic knowledge, which appear to be largely preserved with age (Ackerman & Rolfhus, 1999; Lindenberger & Baltes, 1997). In a functional imaging study, Rypma, Prabhakaran, Desmond, and Gabrieli (2001) found that older adults showed less activation of the dorsolateral prefrontal cortex when performing working memory tasks than did younger people. Additionally, several studies have found a decline in blood flow and metabolism in frontal cortex with advancing age (Gur, Gur, Obrist, Skolnick, & Reivich, 1987; Mielke, Herholz, Grond, Kessler, & Heiss, 1992).

REASONING BIASES AND ERRORS

A discussion of the reasoning driving hypothesis generation and test would not be complete without mention of common reasoning biases and errors. In the context of medical diagnosis, which involves the generation and test of hypotheses to explain patients' symptoms, Croskerry (2003) lists some 32 biasing tendencies that may lead to errors. These include the tendency to unduly favor a hypothesis based on the ease with which past examples are recalled, the tendency to look for and/or overemphasize (cherry pick) supporting evidence and not seek, ignore, or rationalize contradictory evidence, and the tendency to accept a hypothesis before adequate testing. The large majority of these subconscious tendencies, as summarized in Table 8.2, operate to limit the number of alternatives generated or to limit their adequate consideration and test.

Although one typically generates and tests one hypothesis before moving on to another, consideration of the biasing tendencies listed in Table 8.2 suggests a better way. That better way involves an explicit attempt very early on to resist the temptation to go with the first hypothesis that comes to mind and perhaps obtain some initial support. Instead, as geologist Chamberlain argued long ago (Chamberlain, 1965, original publication 1890), one should immediately and consciously brainstorm as many hypotheses as possible, as this not only increases the odds of eventually finding a solution, but also decreases the likelihood of becoming unduly biased by an early hypothesis. This view is supported by a recent study of diagnostic errors that found that by far the most common major error (24%) resulted from a failure to or delay in considering the correct hypothesis, whereas another common error was placing too much weight on a competing or coexisting incorrect hypothesis (12%) (Schiff et al., 2009).

EDUCATIONAL IMPLICATIONS

Neurologically speaking, there are two quite different ways to learn, that is, to transfer new input into LTM. One way is through shear repetition and/or via emotionally charged contexts. Repetition and emotion can "burn" new input into one's synapses essentially by boosting presynaptic activity to a high enough level to create new functional synaptic connections. For

Table 8.2   *Cognitive Tendencies (Biases) That Limit the Number of Hypotheses Generated and Adequately Tested in the Context of Medical Diagnosis*

Tendencies that limit the number of hypotheses generated

Tendency to unduly favor a hypothesis based on the ease with which past examples are recalled (availability bias)

Tendency to focus solely on the obvious hypothesis and not give serious consideration to alternatives (Sutton's slip)

Tendency to allow prior categories of cases to shape thinking (e.g., stereotyping, gender bias)

Tendency to favor the tools of one's discipline, thus ignore alternatives (discipline bias)

Tendency to zero in on typical symptoms of a disease and ignore, or not seek, atypical symptoms that may lead to an alternative diagnosis. The saying, "If it looks like a duck, quacks like a duck, then it is a duck," could be wrong (representative error)

Tendency to underestimate the actual base-rate of a disease, thus fail to consider it as a possibility (base-rate neglect)

Tendency to not consider rare diagnoses. Doctors who hunt "zebras" are often ridiculed by their peers (zebra retreat)

Tendency to "frame" patients: "I'm sending you a case of diabetes and renal failure," or "I have a drug addict here in the emergency room (ER) with fever and a cough from pneumonia." Accepting the frame can keep one from considering alternatives (framing error, diagnosis momentum, triage cueing)

Tendency to recall information presented at the beginning (primacy effect) and at the end (recency effect) of a report. Thus, when information is transferred from patients, nurses, and other physicians, care must be taken to consider all of the information (order effects)

Tendency to blame patients for their illness rather than seek alternatives (attribution error)

Tendencies that limit adequate consideration and testing of the alternative hypotheses

Tendency to accept a hypothesis before adequate testing. Good thinking takes time. Working in haste and cutting corners are the quickest routes to cognitive errors (premature closure)

Tendency to look for and/or overemphasize supporting evidence and not seek, ignore, or rationalize contradictory evidence (cherry picking, confirmation bias, anchoring)

Tendency to see what you want or expect to see (perception bias)

Tendency to call off the search once support for a single hypothesis has been found (search satisfaction)

Tendency to favor hypotheses that lead to more favorable outcomes (e.g., less serious diseases) because they produce less patient chagrin (outcome bias, wishful thinking, affect bias) or because they are more likely than more serious and rare diseases (frequency gambling)

Tendency to hold on to a particular hypothesis when considerable time, energy, money, and ego have been invested in its testing (sunk costs bias)

Tendency to believe that we know more than we do, thus act on incomplete information, intuitions, and hunches instead of carefully gathered evidence (overconfidence bias)

Tendency toward action rather than inaction. This bias is more likely to be present in a person who is overconfident, whose ego is inflated, but can also occur when a person is desperate and gives in to the urge to "do something" (commission bias)

Tendency to accept test results when they should be doubted (i.e., false positives, false negatives). Arriving at the correct hypothesis may not require repeating tests, only doubting them (acceptance bias)

Tendency to give up the search and believe that nothing more can be done when several hypotheses have been thoroughly tested and contradicted, thus a diagnosis has not been made. The patient has been "worked up the Ying-Yang," but to no avail (Ying-Yang out bias, surrender error)

Adapted from Croskerry, 2003.

example, students can memorize their multiplication tables and the positions of letters on a keyboard in this "rote" way. They can also learn to solve proportions problems in a rote way by use of a "cross-multiplication" algorithm (e.g., $4 / 6 = 6 / X$, $(4)(X) = (6)(6)$, $(4)(X) = 36$, $X = 36 / 4$, $X = 9$). Unfortunately, in spite of the fact that such students can cross multiply and "solve" rote problems, they typically have no idea why the algorithm works or how to solve "real" problems involving proportional relationships. For example, most 12-year-olds in the United States can easily tell you that X equals nine in the previous equation, but when given the "Cylinders" problem shown in Figure 8.8, they incorrectly predict that water will rise to the 8th mark "... because it rose two more before, from 4 to 6, so it will rise 2 more again, from 6 to 8."

Fortunately, as we have seen, there is a second way to transfer information into LTM. That is to form new functional synaptic connections by linking new input with prior ideas. When neural activity is simultaneously boosted by new input and by prior ideas, the resulting pre- and postsynaptic activities combine to create new functional connections. This connectionist (or constructivist) way of learning has several advantages, not the least of which is that learning is not rote, in the sense that it is connected to what you already know, and thus becomes much more useful in reasoning and problem solving. In the case of proportions this means that students not only know how to solve for X, they also know when to use a proportions strategy and when not to, that is, they know when other strategies, such as addition and subtraction, should be used instead. The point is that if we want students to become good problem solvers and good scientific thinkers, we cannot teach in ways that lead to rote learning. Instead, we need to become connectionist teachers.

Thus, the key point in terms of instruction is that for meaningful and lasting learning to occur, students must personally and repeatedly engage in the generation and test of their own self-generated ideas. This means that

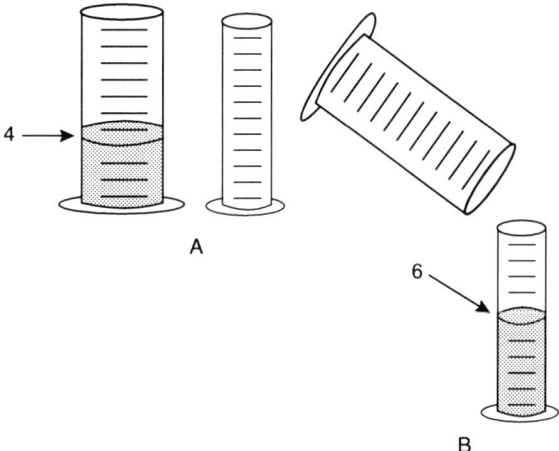

Figure 8.8   *The cylinders problem.*

Note: To the right are drawings of a wide and a narrow cylinder. The cylinders have equally spaced marks on them. Water is poured into the wide cylinder up to the 4th mark (see A). This water rises to the 6th mark when poured into the narrow cylinder (see B). Both cylinders are emptied, and water is poured into the wide cylinder up to the 6th mark. How high will this water rise when poured into the empty narrow cylinder?

personal experiences become the main instructional vehicles. But such experiences cannot be "cookbook" in nature. Instead, they should allow students the freedom to openly inquire and raise puzzling observations. The puzzling observations should then prompt students to generate and test *several* alternative explanations (note the biasing tendencies listed in Table 8.2) with the following sorts of questions becoming the central focus of instruction: What is puzzling about what you observed? What questions are raised? What are some possible answers/explanations? How could these possibilities (alternative hypotheses) be tested? What does each hypothesis and planned test lead you to expect to find? What are your results? How do your results compare with your predictions? And lastly, what conclusions can be drawn?

SUMMARY OF CHAPTER'S MAIN POINTS

1. In response to puzzling observations, successful reasoning used to explain those puzzling observations first involves use of the inference of abduction (retroduction), which is used to generate tentative explanations (hypotheses). The hypotheses are then tested using the inference of deduction to derive predictions, and the inference of induction to derive a conclusion based on the match or mismatch of predictions with subsequent observations.

2. The reasoning that presumably guided Galileo's discovery of Jupiter's moons in 1610 is explicated in terms of the above inferences. The chapter assumes that such a reasoning pattern is general to the process of scientific discovery. Consequently, the chapter presents a model of that general pattern—a pattern that can be verbally characterized in an *If/and/then/ Therefore* (i.e., hypothetico–deductive) manner.

3. The chapter then presents some basics of brain development and brain functioning, noting that the brain is "hardwired" to function in the same hypothetico–deductive manner via a process known as adaptive resonance. Thus the chapter links the psychology of reasoning and learning with the physiology of the brain.

4. The "hypothetico–deductive" neural networks involved in both visual and auditory processing can also be successfully applied to explain higher levels of reasoning and learning, including performance on the Wisconsin Card Sorting Task as well as proportional reasoning, decision making, and scientific discovery, more specifically Galileo's discovery of Jupiter's moons.

5. The role of working memory during successful reasoning and scientific discovery is explicated as is the decline of working memory with advancing age due to a decrease in inhibitory functioning and selective attention—both aspects of what has been referred to as fluid intelligence.

6. Next the chapter briefly introduces several sources of reasoning biases and errors. These include the tendency to unduly favor a particular hypothesis based on the ease with which past examples can be recalled, the cherry picking of supporting evidence, and the outright failure to consider

contradictory evidence. Most of these errors operate subconsciously and operate to limit the number of alternatives generated and successfully tested.

7. Thus the primary educational implication is that instruction must help raise the pattern of hypothetico–deductive reasoning to the conscious level by allowing, indeed requiring, that students personally confront puzzling observations and then repeatedly engage in the generation and test of alternative explanations by comparing their deduced consequences with subsequent observations.

REFERENCES

Ackerman, P. L., & Rolfhus, E. L. (1999). The locus of adult intelligence: Knowledge, abilities and nonability traits. *Psychology and Aging, 14,* 314–330.

Anderson, J. R. (1980). *Cognitive psychology and its implications.* San Francisco, CA: Freeman.

Baddeley, A. (1995). Working memory. In M. S. Gazzaniga (Ed.), *The cognitive neurosciences* (pp. 755–764). Cambridge, MA: MIT Press.

Baddeley, A. (1996). Exploring the central executive. *Quarterly Journal of Experimental Psychology: Human Experimental Psychology, 49,* 5–28.

Baltes, P. B., Reuter-Lorenz, P. A., & Rosler, F. (2006). *Lifespan development and the brain.* Cambridge, MA: Cambridge University Press.

Brannigan, A. (1981). *The social basis of scientific discoveries.* Cambridge, MA: Cambridge University Press.

Carpenter, G. A., & Grossberg, S. (2003). Adaptive resonance theory. In M. A. Arib (Ed.), *The handbook of brain theory and neural networks* (2nd ed., pp. 87–90). Cambridge, MA: MIT Press.

Chamberlain, T. C. (1965). The method of multiple working hypotheses. *Science, 148,* 754–759. Originally published 1897.

Collins, H. M. (1985). *Changing order.* London, UK: Sage Publications.

Collins, H. M. (2001). Tacit knowledge, trust and the Q of sapphire. *Social Studies of Science, 31*(1), 71–85.

Craik, F. I. M., & Byrd, M. (1982). Aging and cognitive deficits: The role of attentional resources. In F. I. M. Craik & S. Trehub (Eds.), *Aging and cognitive processes* (pp. 191–211). New York, NY: Plenum Press.

Croskerry P. (2003). The importance of cognitive errors in diagnosis and strategies to minimize them. *Academic Medicine, 78,* 775–780.

Darwin, C. (1898). *The origin of species* (7th ed.). New York, NY: Appleton and Company.

Diamond, M. C. (1996). The brain…use it or lose it. *Mind Shift Connection,* 1(1). Retrieved from www.newhorizons.org/neuro/diamond

Feynman, R. P. (1995). *Six easy pieces.* Reading, MA: Addison-Wesley.

Friedman, H. R., & Goldman-Rakic, P. S. (1994). Coactivation of prefrontal cortex and inferior parietal cortex in working memory tasks revealed by 2DG functional mapping in the rhesus monkey. *Journal of Neuroscience, 14,* 2775–2788.

Fugelsang, J. A., & Dunbar, K. N. (2004). Brain-based mechanisms underlying complex causal thinking. *Neuropsychologia, 43,* 1204–1213.

Fugelsang, J. A., Roser, M. E., Corballis, P. M., Gazzaniga, M. S., & Dunbar, K. N. (2004). Brain mechanisms underlying perceptual causality. *Cognitive Brain Research, 24,* 41–47.

Funster, J. M. (1989). *The prefrontal cortex: anatomy, physiology, and neuropsychology of the frontal lobe* (2nd ed.). New York, NY: Raven Press.

Galilei, G. (1610/1954). The sidereal messenger. In H. Shapley, S. Rapport, & H. Wright (Eds.), *A treasury of science*. New York, NY: Harper & Brothers.

Green, A. E., Fugelsang, J. A., Kraemer, D. J. M., Shamosh, N. A., & Dunbar, K. N. (2006). Frontopolar cortex mediates abstract integration in analogy. *Brain Research, 1096*, 125–137.

Greenough, W. T., Volkmar, F. R., & Juraska, J. M. (1973). Effects of rearing complexity on dendritic branching in frontolateral and temporal cortex of the rat. *Experimental Neurology, 41*, 371–378.

Grossberg, S. (1982). *Studies of mind and brain*. Dordrecht, Holland: D. Reidel.

Grossberg, S. (2005). Linking attention to learning, expectation, competition, and consciousness. In L. Itti, G. Rees, & J. Tsolsos (Eds.), *Neurobiology of attention* (pp. 652–662). San Diego, CA: Elsevier.

Gur, R. C., Gur, R. E., Obrist, W. D., Skolnick, B. E., & Reivich, M. (1987). Age and regional cerebral blood flow at rest and during cognitive activity. *Archives of General Psychiatry, 44*, 617–621.

Hanson, N. R. (1958). *Patterns of discovery*. London, UK: Cambridge University Press.

Hasher, L., & Zacks, R. T. (1988). Working memory, comprehension, and aging: A review and a new view. In G. H. Bower (Ed.), *The psychology of learning and motivation* (Vol. 22, pp. 193–225). San Diego, CA: Academic Press.

Hempel, C. (1966). *Philosophy of natural science*. Upper Saddle River, NJ: Prentice-Hall.

Hubel, D. H., & Wiesel, T. N. (2005). *Brain and visual perception: the story of a 25-year collaboration*. New York, NY: Oxford University Press.

Isingrini, M., & Vazou, F. (1997). Relation between fluid intelligence and frontal lobe functioning in older adults. *International Journal of Aging and Human Development, 45*, 99–109.

Jani, N. G., & Levine, D. S. (2000). A neural network theory of proportional analogy making. *Neural Networks, 13*, 149–183.

Jerison, H. J. (2001). The study of primate brain evolution: Where do we go from here? In D. Palk & K. R. Gibson (Eds.), *Evolutionary anatomy of the primate cerebral cortex* (pp. 305–333). Cambridge, UK: Cambridge University Press.

Jerison, H. J., & Jerison, H. (Eds.). (1988). *Intelligence and evolutionary biology*. Berlin, Germany: Springer-Verlag.

Kandel, E. R. (2006). *In search of memory*. New York, NY: Norton.

Kosslyn, S. M., & Koenig, O. (1995). *Wet mind: the new cognitive neuroscience*. New York, NY: The Free Press.

Kwon, Y. J., Lee, J. K., Shin, D. H., & Jeong, J. S. (2009). Changes in brain activation induced by the training of hypothesis generation skills: An fMRI study. *Brain and Cognition, 69*, 391–397.

Lawson, A. E. (1993). Deductive reasoning, brain maturation, and science concept acquisition: Are they linked? *Journal of Research in Science Teaching, 30*(9), 1029–1052.

Lawson, A. E. (2002). What does Galileo's discovery of Jupiter's moons tell us about the process of scientific discovery? *Science & Education, 11*(1), 1–24.

Lawson, A. E. (2005). What is the role of induction and deduction in reasoning and scientific inquiry? *Journal of Research in Science Teaching, 42*(6), 716–740.

Levine. D. S. (2009). Brain pathways for cognitive-emotional decision making in the human animal. *Neural Networks, 22*(3), 286–293.

Levine, D. S., & Prueitt, P. S. (1989). Modeling some effects of frontal lobe damage: Novelty and perseveration. *Neural Networks, 2*, 103–116.

Lindenberger, U., & Baltes, P. B. (1997). Intellectual functioning in old and very old age: Cross-sectional results from the Berlin Aging Study. *Psychology and Aging, 12*, 410–432.

Mielke, R., Herholz, K., Grond, M., Kessler, J., & Heiss, W. D. (1992). Differences in regional cerebral glucose metabolism between presenile and senile dementia of Alzheimer type. *Neurobiology of Aging, 13*, 93–98.

Nonaka, I., & von Krogh, G. (2009). Perspective-tacit knowledge and knowledge conversion: Controversy and advancement in organizational knowledge creation theory. *Organizational Science, 20*(3), 635–652.

Paulesu, E., Frith, D. D., & Frackowiak, R. S. J. (1993). The neural correlates of the verbal component of working memory. *Nature, 362,* 342–345.

Peirce, C. S. (1903/1997). *Pragmatism as a principle and method of right thinking. The 1903 Harvard lectures on pragmatism* (P. A. Turrisi, Ed.). Albany, NY: State University of New York Press (see also Bergman, M., & Paavola, S. (Eds.). (2003). *The Commens dictionary of Peirce's terms.* Retrieved from www.helsinki.fi/science/commens/dictionary.html).

Peirce, C. S. (1905/2003). A letter to Calderoni. In M. Bergman & S. Paavola (Eds.), *The Commens dictionary of Peirce's terms.* Retrieved from www.helsinki.fi/science/commens/dictionary.html

Perry, B. D. (2002). Childhood experience and the expression of genetic potential: What it tells us about nature and nurture. *Brain and Mind, 3,* 79–100.

Piaget, J. (1970). *Genetic epistemology.* New York, NY: Norton.

Planck, M. (1949). *Scientific autobiography* (E. Guynor, Trans.). New York, NY: Philosophical Library.

Popper, K. (1965). *Conjectures and refutations. The growth of scientific knowledge.* New York, NY: Basic Books.

Rypma, R., Prabhakaran, V., Desmond, J. E., & Gabrieli, J. D. E. (2001). Age differences in prefrontal cortical activity in working memory. *Psychology and Aging, 16,* 371–384.

Schiff, G. D., Hasan, O., Kim, S., Abrams, R., Cosby, K., Lambert, B. L., … McNutt, R. A. (2009). Diagnostic error in medicine. *Archives of Internal Medicine, 169*(20), 1881–1887.

Shimamura, A. (2000). The role of prefrontal cortex in dynamic filtering. *Psychobiology, 28,* 207–218.

Smith, E. E., & Jonides, J. (1994). Working memory in humans: Neuropsychological evidence. In M. S. Gazzaniga (Ed.), *The cognitive neurosciences* (pp. 1009–1020). Cambridge, MA: MIT Press.

Squire, L. R., & Zola-Morgan, S. (1991). The medial temporal lobe memory system. *Science, 253,* 1380–1386.

Squire, L. R. (2004). Memory systems and the brain: A brief history and current perspective. *Neurobiology of Learning and Memory, 82,* 171–177.

Tsoukas, H. (2003). Do we really understand tacit knowledge? In M. Easterly-Smith & M. A. Lyles (Eds.), *The Blackwell handbook of organizational learning and management* (pp. 411–427). Cambridge, MA: Blackwell Publishing.

Viskontas, I. V., Morrison, R. G., Holyoak, K. K., Hummel, J. E., & Knowlton, B. J. (2004). Relational integration, inhibition, and analogical reasoning in older adults. *Psychology and Aging, 19*(4), 581–591.

Wallis, J. D., Anderson, K. C., & Miller, E. K. (2001). Single neurons in prefrontal cortex encode abstract rules. *Nature, 411,* 953–956.

Watson, J. D. (1968). *The double helix.* New York, NY: Penguin Books.

CHAPTER 9

Children as Scientific Thinkers

David Klahr, Bryan Matlen, and Jamie Jirout

Children are engaged in scientific thinking when they attempt to understand and explain entities and processes in the natural world. One type of scientific thinking involves thinking about the *content* of science. The other type involves the process of *doing* science.

THINKING ABOUT THE CONTENT OF SCIENCE

Scientific content includes a very large set of specific domains, traditionally arranged by professional organizations, university departments, text book publishers, and state curriculum designers into categories such as physics, chemistry, biology, and earth sciences (e.g., the alphabetical listing of the National Academy of Sciences' 31 divisions ranges from "Animal, Nutritional, and Applied Microbial Sciences" and "Anthropology," to "Social and Political Science" and "Systems Neuroscience"). Moreover, each of these broad domains of science has many subdivisions (both in university organizational structures and the table of contents of science text books).

The continually expanding range of such substantive topics in science is daunting. By some estimates, there are thousands of possible science concepts that could be taught (Roth, 2008) and the science standards proposed by each of the 50 states in the United States often run to hundreds of pages, and a typical middle-school textbook in a broad topic like biology or earth science can run to 700 or 800 pages. Consequently, science educators have lamented the fact that the K-12 science curriculum is "a mile wide and an inch deep." One proposed remedy to this problem is to eschew topical breadth and instead focus on depth, a strategy that appears to be effective at improving long-term learning in science (Schwartz, Sadler, Sonnert, & Tai, 2008).

Of course, when children first begin to think about the content of science, they are not influenced, for better or worse, by the kinds of categories listed above. Instead, they simply notice and wonder about clouds, bugs, bubbles, food, people, dreams, and an endless array of other entities that they encounter in their world. For example, our own research on children's scientific curiosity (Jirout, 2011) has recorded questions from 4- and 5-year-old children such as "Can bees be killed by electricity?, "Do clouds make wind?", "How do leaves change their colors?", and "What do worms' eyes look like?" Research on children's thinking about science content ranges from studies of infants' understanding of causality and simple physical regularities (Baillargeon, 2004; Gopnik, Meltzoff, & Kuhl, 1999), to young children reasoning about the sun–moon–earth system (Vosniadou & Brewer, 1992), to college students reasoning about chemical equilibrium (Davenport, Yaron, Klahr, & Koedinger, 2008). These studies reveal a protracted period during which children often develop deeply entrenched misconceptions and preconceptions. In many cases children's developing understanding of the natural world recapitulates the history of scientific discovery, and in other cases, children's misconceptions turn out to be remarkably resistant to instruction (see "Problem-Solving Methods" section).

THINKING ABOUT THE PROCESS OF SCIENCE

The second kind of scientific thinking—*doing* science—includes a set of reasoning processes that can be organized into three broad, but relatively distinct, categories: formulation of hypotheses, design of experiments and observations, and evaluation of evidence. In turn, both the formation of hypotheses and the design and execution of experiments and observations can be viewed as types of problem solving, involving search in problem spaces (Newell & Simon, 1972). Scientific reasoning, under this view, consists of the coordination of this "dual search" in the experiment space and the hypothesis space (Klahr, 2000; Klahr & Simon, 1999; Klahr, Dunbar, & Fay, 1990). However, as in the case of thinking about the content of science, when children are engaged in these different aspects of the process of science, they are unaware of the distinctions between, for example, forming hypotheses and designing experiments, even though these distinctions are useful for psychologists who investigate the properties of these different aspects of scientific thinking.

The characterization of scientific reasoning as a kind of general problem solving raises the question about whether scientific thinking is qualitatively different from other types of reasoning. We believe that the answer is "no." That is, we will argue that the reasoning processes used in scientific thinking are not unique to scientific thinking: they are the very same processes involved in everyday thinking, so that questions about children's ability to think scientifically are inexorably linked to broad and general questions about cognitive development. Before turning to a summary of the relevant literature, it is worth noting that we are in good company in basing our review on the "nothing special" claim about the nature of scientific reasoning.

> The scientific way of forming concepts differs from that which we use in our daily life, not basically, but merely in the more precise

definition of concepts and conclusions; more painstaking and systematic choice of experimental material, and greater logical economy. (Einstein, 1950, p. 98)

Nearly 40 years after Einstein's remarkably insightful statement, Francis Crick offered a similar perspective: that great discoveries in science result from commonplace mental processes, rather than from extraordinary ones. The greatness of the discovery lies in the thing discovered.

I think what needs to be emphasized about the discovery of the double helix is that the path to it was, scientifically speaking, fairly commonplace. What was important was *not the way it was discovered*, but the object discovered—the structure of DNA itself. (Crick, 1988, p. 67; emphasis added)

The assumption in this chapter is that the literature on cognitive development can contribute to our understanding of children's scientific thinking because scientific thinking involves the same general cognitive processes—such as induction, deduction, analogy, problem solving, and causal reasoning—that children apply in nonscientific domains. Thus we get a window into children's *scientific* thinking by a better understanding of their thinking in more general terms. This view is simply a developmental version of Herbert Simon's insight about the psychology of scientific discovery:

It is understandable, if ironic, that 'normal' science fits…the description of expert problem solving, while 'revolutionary' science fits the description of problem solving by novices. It is understandable because scientific activity, particularly at the revolutionary end of the continuum, is concerned with the discovery of new truths, not with the application of truths that are already well-known…it is basically a journey into unmapped terrain. Consequently, it is mainly characterized, as is novice problem solving, by trial-and-error search. The search may be highly selective—but it reaches its goal only after many halts, turnings, and back-trackings. (Simon, Langley, & Bradshaw, 1981, p. 5)

Perhaps this view overstates the case for all revolutionary discoveries (see e.g., Holton, 2003 on how Einstein's development of relativity departed from Simon's account of revolutionary ideas in science). Nevertheless, Simon's view remains an interesting and important characterization of discovery in the absence of a well-developed scientific schema.

SCIENTIFIC THINKING IN CHILDREN

General Overview of Research on Children's Scientific Thinking

A recent expert panel report from the National Research Council (NRC) (Duschl, Schweingruber, & Shouse, 2007), summarizes the research literature on children's scientific thinking from kindergarten to 8th grade. The report notes that young children have substantial knowledge of the natural world, much of which is implicit, and that scientific knowledge at any particular age is the result of a complex interplay among maturation, experience, and instruction. Moreover, general knowledge and experience play a critical role in children's science learning, influencing several key processes associated with science:

a. Knowing, using, and interpreting scientific explanations of the natural world
b. Generating and evaluating scientific evidence and explanations
c. Understanding how scientific knowledge is developed in the scientific community
d. Participating in scientific practices and discourse

Our emphasis in this chapter will be on the cognitive psychology end of the spectrum (items a through c), while acknowledging other aspects of "scientific practice" and "scientific discourse" that are important but not discussed much in this chapter.

Over the past half century, there has been a wealth of research in cognitive development on the processes and content of children's scientific thinking. While early theories of children's thinking characterized children as egocentric, perception-bound, and developing in relatively distinct stages (e.g., Piaget's theory), more recent theories have stressed children's development as a continuous process, with an emphasis on the ways in which children acquire new knowledge rather than on what they can and can't do at certain ages (e.g., Siegler's [1995] Overlapping Waves theory). One of the most pervasive findings in cognitive development research is that children have a large base of knowledge and abilities that are available from as early as the time children enter formal schooling, and this knowledge has a profound impact on what and how children subsequently learn when studying science.

From as early as infancy children begin to learn about the natural world in ways that will influence their later scientific thinking. Researchers have identified a number of distinct domains in which infants appear to develop specific knowledge, including psychology, biology, and physics. Another powerful achievement in the early years of life is the ability to think representationally—that is, being able to think of an object as both an object in and of itself as well as a representation of something else. Starting as early as three years, children are able to use scale models to guide their search for hidden objects in novel locations (DeLoache, 1987). This fundamental ability lies at the heart of important scientific skills such as analogical reasoning, and lays the ground-work for later scientific tasks such as interpreting and reasoning with models, maps, and diagrams.

The way in which this knowledge arises is a matter of fierce contention in cognitive development research. Some researchers have argued that children's knowledge acquisition is primarily "theory-based"—guided by top-down, domain-specific learning mechanisms for evolutionarily privileged domains (e.g., physics, biology, psychology, etc.)—while other researchers have argued that more attention-driven, domain-independent learning mechanisms are sufficient to account for the vast amount of information that children acquire in the early years of life. These contrasting views are fueled in part by findings that, on the one hand, young children are capable of exhibiting adult-like performance in a variety of higher order reasoning tasks (Gelman & Coley, 1990; Gopnik & Sobel, 2000; Goswami & Brown, 1990; Keil, Smith, Simons, & Levin, 1998) while, on the other hand, children's performance on such tasks is often highly dependent on low-level perceptual, memory, and attentional factors (Fisher, Matlen, & Godwin, 2011; Rattermann & Gentner, 1998; Sloutsky, Kloos, & Fisher, 2007; Rakison & Lupyan, 2008; Smith, Jones, & Landau, 1996). Regardless of the cognitive mechanisms at play in the early years of life, however, it is clear that children have substantial knowledge about the physical world by the time they enter formal schooling. A major challenge in science education is to build on students' existing knowledge of the natural world to help them think and reason about the scientific phenomena.

Problem-Solving Methods

Given our view that scientific problem solving is a special case of general problem solving, we will summarize some key ideas about the psychology of problem solving. Newell and Simon (1972) define a problem as comprising of an initial state, a goal state, and a set of operators that allow the problem solver to transform the initial state into the goal state through a series of intermediate states. Operators have constraints that must be satisfied before they can be applied. The set of states, operators, and constraints is called a "problem space," and the problem-solving process can be characterized as a search for a path that links the initial state to the goal state.

In all but the most trivial problems, the problem solver is faced with a very large set of alternative states and operators, so the search process can be demanding. For example, if we represent the problem space as a branching tree of m moves with b branches at each move, then there are bm moves to consider in the full problem space. As soon as m and b get beyond very small values, exhaustive search for alternative states and operators is beyond human capacity, so effective problem solving depends in large part on how well the search is constrained. There are two broad categories—or methods—of search constraint: *strong methods* and *weak methods*. Strong methods are algorithmic procedures, such as those for long division or for computing means and standard deviations. The most important aspect of strong methods is that—by definition—they guarantee a solution to the problem they are designed to solve. However, strong methods have several disadvantages for human problem solvers. First, they may require extensive computational resources. For example, a strong method for minimizing cost (or maximizing protein) of a list of grocery items subject to other dietary and budget constraints is to apply a standard linear-programming algorithm. Of course, doing this in one's head while pushing a shopping cart is hardly feasible. Second, strong methods may be difficult to learn because they may require many detailed steps (e.g., the procedure for computing a correlation coefficient by hand). Finally, strong methods, by their very nature, tend to be domain specific and thus have little generality.

Weak methods are heuristic: they may work or they may not, but they are highly general. The trade-off for the lack of certainty associated with weak methods is that they make substantially lower computational demands, are more easily acquired, and are domain general. Of particular importance for this chapter is the possibility that some of the weak methods are innate, or, at the least, that they develop very early without any explicit instruction or training. Newell and Simon (1972) describe several kinds of weak methods, but here I will briefly describe only three.

Generate and Test

The generate and test method is commonly called "trial and error." Its process is simply applying some operator to the current state and then testing to determine whether the goal state has been reached. If it has, the problem is solved. If it has not, then some other operator is applied. In the most primitive generate and test methods, the evaluation function is binary: either the goal has been reached or it has not, and the next move does not depend on any properties of the discrepancy between the current state and the goal state or the operator that was just unsuccessfully applied. An example of a "dumb" generating process is searching in a box of keys for a key to fit a lock, and sampling with replacement: tossing failed keys back into the box without noting anything about the degree of fit, the type of key that seems to fit partially, and so forth. A slightly "smarter" generator would, at the least, sample from the key box without replacement.

Hill Climbing

The hill climbing method gets its name from the analogy of attempting to reach the top of a hill whose peak cannot be directly perceived (imagine a foggy day with severely limited visibility). One makes a tentative step in each of several directions, then heads off in the direction that has the steepest gradient. Hill climbing utilizes more information about the discrepancy between the current state and the goal state than does generate and test. Instead of a simple all-or-none evaluation, it computes a measure of goodness of fit between the two and uses that information to constrain search in the problem space.

Means–Ends Analysis

Perhaps the best-known weak method is means–ends analysis (Duncker, 1945; Newell & Simon, 1972). Means–ends analysis compares the current state with the goal state and notes the relevant differences. Then it searches for operators that can reduce those differences and selects the one that will reduce the most important differences and attempts to apply it to the current state. However, it may be that the operator cannot be immediately applied because the conditions for doing so are not met. Means–ends analysis then formulates a sub-problem in which the goal is to reduce the difference between the current state and a state in which the desired operator can be applied, and then recursively attempts to solve the sub-problem.

An Empirical Study of Preschoolers' Problem-Solving Ability

With these brief descriptions of different problem-solving methods, we now turn to the question of the extent to which young children can actually employ them. In particular, we focus on the extent to which preschoolers can go beyond simple trial and error and use the much more powerful means–ends analysis method. In a study of preschool children's ability to "think ahead" while solving puzzles requiring means–ends analysis, Klahr and Robinson (1981) found that children showed rapid increases in their ability between the ages of 4 and 6 years. Children were presented with puzzles in which they had to describe a series of multiple moves...solving the problem "in their heads"...while describing the solution path. The main question of interest is how far into the future a child could "see" in describing move sequences. To avoid overestimating this capacity on the basis of a few fortuitous solutions, a very strict criterion was used: A child was scored as able to solve n-move problems only after proposing the minimum path solution for four different problems of length n. For example, to be classified as having the capacity to see five moves into the future, a child would have to produce the minimum path solution for four five-move problems.

The proportion of children in each age group producing correct solutions for all problems of a given length is shown in Figure 9.1. Note that the abscissa in the figure is not overall proportion correct, but rather a much

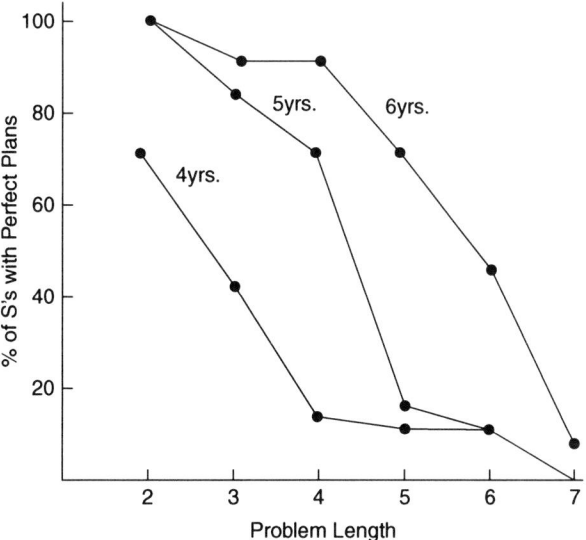

Figure 9.1   *The proportion of children in each age group producing correct solutions for all Tower of Hanoi problems of a given length.*

Adapted from Klahr and Robinson (1981).

more severe measure: the proportion of children with perfect solutions on all problems of a given length. For example, 69% of the 6-year-olds were correct on all four of the five-move problems, while only 16% of the 5-year-olds and 11% of the 4-year-olds produced four flawless five-move solutions.

The absolute level of performance was striking: over two-thirds of the 5-year-olds and nearly all of the 6-year-olds consistently gave perfect four-move solutions, and over half of the 6-year-olds gave perfect six-move solutions. Almost half of the 4-year-olds could do the three-move problems. Note that these solutions required that the child manipulate mental representations of future states.

Children's and Adult's Problem Solving on a Scientific Reasoning Task

Klahr and Dunbar (1988) extended the search in a problem space approach and proposed that scientific thinking can be thought of as a search through two related spaces: an hypothesis space and an experiment space, with search in one of the spaces constraining and informing search in the other. Klahr, Fay, and Dunbar (1993) presented children (third and sixth graders) and adults with a complex scientific reasoning task in which they had to figure out the rules underlying the operation of a programmable toy robot. They discovered several important differences between the way that the children and adults approached the task.

(a) *Children and adults respond differently to plausible and implausible hypotheses.* One of the most robust findings in the literature on scientific reasoning in adults is that they attempt to confirm, rather than disconfirm, their hypotheses (Klayman & Ha, 1987). Similarly, developmental studies show that even when explicitly instructed to generate evidence that could potentially falsify a rule, children at the sixth-grade level or below perform very poorly (Kuhn, 1989; Ward & Overton, 1990). However, Klahr, Fay, and Dunbar (1993) report a more flexible kind of response: When hypotheses were *plausible*, subjects at all levels tended to set an experimental goal of demonstrating key features of a given hypothesis, rather than conducting experiments that could discriminate between rival hypotheses. However, adults' response to *implausibility* was to propose a counter-hypotheses and then to conduct experiments that could discriminate between the two. In contrast, the third graders' general response to implausible hypotheses was to simply ignore them while attempting to demonstrate the correctness of a more plausible one of their own creation. By sixth grade, subjects appeared to understand how to generate informative experiments.

(b) *Focus on only a few features of hypotheses.* Experiments and hypotheses, are both complex entities having many aspects on which one could focus. Adults, but not children, were very conservative in the way they generated different experiments: they tended to design sequences of experiments that differed on only a single feature, whereas children tended to make multiple changes from one experiment to the next.

(c) *Pragmatics.* All but the youngest children (third graders) tended to create experiments that would not overwhelm their ability to observe and encode outcomes. The youngest children generated experiments with outcomes that, in principle, made clear distinctions between competing hypotheses, but that tended to be very hard to completely observe and/or recall, so that they were not effective. Thus, it seems that the youngest children had not yet developed sufficient metacognitive capacity to generate outcomes that were within their own perceptual and memory constraints (Brown, Bransford, Ferrara, & Campione, 1983).

These studies of children's problem-solving abilities involve problem solving in a "knowledge free" context. That is, while they reveal children's abstract reasoning abilities, they do not address questions about children's knowledge of the world. However, there is a rich literature on the development of children's concepts, and in the next section, we summarize studies that focus on specific concepts, and how they undergo change and development, rather than on problem-solving skills.

Conceptual Change in Children's Scientific Thought

One way in which children acquire new information is through conceptual change. Conceptual change refers to the process of reassigning a concept from one ontological category to another (Vosniadou, Vamvakoussi, & Skopeliti, 2008), such as when a child incorporates "plants" into his/her category of living things rather than inanimate objects. Conceptual change can also refer to the differentiation or the merging of concepts (Carey, 1991). Because conceptual change typically involves an ontological shift resembling the process of theory change, it can be distinguished from more incremental, additive, learning processes, in which gaps in students' knowledge are filled in or "enriched" by the addition of new information (Chi, 2008).

The extent to which children's concepts are organized is a highly contentious issue in conceptual change research. There is currently an active debate concerning whether children's conceptions are organized into coherent knowledge structures that have the consistency and predictive power, similar in many respects to scientific theories (Keil, 2011; Vosniadou & Brewer, 1992, 1994), or whether children represent knowledge in more context-dependent and fragmented ways (DiSessa, 2008; Smith, DiSessa, & Roschelle, 1994). Regardless of the nature of children's conceptual organization, it is clear that school-age children bring with them conceptions that differ from canonical scientific knowledge and that these preconceptions influence learning in a variety of scientific domains including physics (Clement, 1982), thermodynamics (Lewis & Linn, 1994), astronomy (Vosniadou & Brewer, 1994), biology (Inagaki & Hatano, 2008; Opfer & Seigler, 2004), geoscience (Gobert & Clement, 1999), and chemistry (Wiser & Smith, 2008). Moreover, conceptual change can occur for domain-general science concepts, such as learning the principles of experimental design (Siler, Klahr, & Matlen, in press) or understanding the purpose of scientific models (Grosslight, Unger, & Jay, 1991).

Preconceptions that are deeply rooted in children's everyday experiences are particularly resistant to instruction, and several studies have documented the continued existence of students' preconceptions even after semester-long courses in scientific domains (Clement, 1982; Lewis & Linn, 1994; Wiser & Smith, 2008). The underlying processes supporting conceptual change have been difficult to determine, although there have been a number of studies that have attempted to characterize the changes in knowledge representation that occur when switching from one way of representing scientific understanding to another (Carey, 1985; Chi, 1992; Chi & Roscoe, 2002; Clement, 1982; Thagard, 1992). One well-documented finding is that children construct new understanding on the basis of their prior knowledge, often by integrating teacher-provided information within the framework of their original skeletal beliefs (Vosniadou & Brewer, 1992, 1994). This process can lead to the adoption of "synthetic" understandings (Vosniadou & Brewer, 1994), characterized by a merging of information from instructional and noninstructional contexts. For instance, many early elementary-school children believe that the Earth is flat and that a person would eventually fall off the Earth if he or she were to walk straight for a long time. However, after being told that the Earth is in fact round, children can change their conception of the Earth to a round, disc-like object to account for what they are told (i.e., that the Earth is round) and what they observe (i.e., that the Earth is flat). Vosniadou and Brewer (1992) were able to identify a number of such synthetic understandings, including children's adoption of a dual earth model—where children believe there are two Earths: one in the sky, and another that comprises the flat ground where people live—or a hollow sphere model—where children believe that the Earth is a hollow sphere with a flat bottom on which we live. Facilitating conceptual change is thus a complex process that requires a detailed understanding of both the preconceptions children have and how those preconceptions are likely to change in the face of new information.

The idea that children undergo radical conceptual change in which old "theories" need to be overthrown or restructured has been a central topic of research on children's scientific thinking. Recent evidence suggests that rather than being replaced, preconceptions may continue to exist—and possibly compete with—scientifically accurate conceptions (Inagaki & Hatano, 2008). For example, adults are slower to categorize plants as living things than animals, even though they understand perfectly well that both are alive (Goldberg & Thompson-Schill, 2009), and under speeded conditions, adults sometimes will endorse teleological explanations in ways similar to children even though they reject such explanations under normal conditions (Kelemen & Rosset, 2009). These findings suggest that conceptual change may not be a process of "replacing" intuitive theories, but rather, that intuitive theories continue to exist alongside scientifically accurate ones (Schtulman & Valcarcel, 2011).

One strategy for inducing conceptual change is to introduce cognitive conflict—that is, to present children with information that conflicts with their existing conceptions. An example of a highly effective use of cognitive conflict comes from the domain of physics, where many students hold many misconceptions about forces. For instance, while a physicist understands that an object resting on a table has both downward (gravity) and upward (the table) forces acting upon it, many students believe that gravity alone is the only force acting upon the object. In a clever manipulation aimed at confronting students' misconceptions, Minstrell (1992) prompted his students to hold a book up with their hands and then asked them what forces were acting upon the object. When the students replied that gravity was the only force acting upon the object, Minstrell added an increasing number of books to the stack until students realized that their hand was exerting an upward force to counteract the gravitational force. Once students had revised their conceptual model of compensating forces, they were able to generalize this new knowledge to explain the forces acting on the object resting on a table (Minstrell, 1992). Posner, Strike, Hewson, and Gertzog (1982) note that inducing cognitive conflict is a particularly effective instructional method when (a) students have a dissatisfaction with their preconceptions, (b) the instructor introduces a novel, alternative conception that is intelligible to the student, (c) the novel conception is initially plausible, and (d) the novel conception can be generalized to explain other, related phenomena. Because children revise their knowledge by constructing new knowledge that is initially intelligible, analogies are useful tools for producing conceptual change (discussed in the following section).

The Role of Analogy in Scientific Thinking

Analogy is one of the most widely mentioned reasoning processes used in science. Analogical reasoning is the process of aligning two or more representations on the basis of their common relational structure (Gentner, 1983, 2010). When one of the representations is better understood than the other, information from the familiar case (i.e., by convention, termed the "base") can be used to inform the scientist's understanding of the unfamiliar case (i.e., by convention, termed the "target"). Another form of analogical reasoning is when learning proceeds by drawing comparisons between two partially understood cases. Many scientists have claimed that the making of certain analogies was instrumental in their process of scientific discovery, and several theories of analogical reasoning suggest how analogy can play a role in scientific thinking (see Gentner, Holyoak, & Kokinov, 2001). Moreover, real-world studies of contemporary scientific laboratories (i.e., in vivo studies) have revealed that scientists commonly incorporate analogies to generate hypotheses, explain scientific phenomena, and interpret and construct scientific models (Dunbar, 1995, 1997, 2001; Nercessian, 2009).

Early developmental researchers believed that children were incapable of reasoning by analogy until they had reached a stage of formal operational reasoning (i.e., at around age 11, Piaget, Montangero, & Billeter, 1977; Sternberg & Nigro, 1980). However, there is now extensive evidence to indicate that children have at least rudimentary analogical ability at very young ages, and that this ability develops gradually with the increase of domain knowledge (Bulloch & Opfer, 2008; Goswami, 1991, 2001), executive function (Thibault, French, & Vezneva, 2010), and development of the prefrontal cortex (Wright, Matlen, Baym, Ferrer, & Bunge, 2008).

Although children's analogical reasoning is present at early ages, however, it is often distorted by an overreliance on superficial features at the expense of recognizing deeper, relational structure (Gentner, 1988; Richland, Morrison, & Holyoak, 2006). For example, third-grade children who learn the principles of experimental design readily apply the newly learned strategy to a different feature in the training set of materials (e.g., a ball and ramp apparatus). But they are much less likely to apply the same strategy when asked to design experiments with a novel set of materials that are superficially dissimilar from the trained set (e.g., springs and weights), even though the basic process for simple experimental design is the same in both cases (Chen & Klahr, 1999; Matlen & Klahr, 2010). Instructional strategies that attempt to promote analogical reasoning by focusing children's attention on abstract schemas, away from superficial appearances, have been shown to significantly increase the use of analogical reasoning (Goldstone & Son, 2005; Sloutsky, Kaminski, & Heckler, 2005).

One way for students to abstract knowledge is to engage in explicit comparison of analogous cases. This process appears to help students encode important analogical relationships (Gentner, 2010). For example, Kurtz, Miao, and Gentner (2001) had students compare two instances of heat transfer with familiar objects: students who jointly interpreted each scenario and listed similarities between the cases were more likely to refer to the causal schema of heat transfer than students who studied each case separately. Comparing analogous cases (e.g., by being provided with diagrams or probe questions that highlight the common structure) has been shown to be an effective scaffold for analogical reasoning in a wide array of scientific domains and even relatively mild manipulations—such as side-by-side presentation of examples—can be an effective way to foster children's analogical comparison (Christie & Gentner, 2010; Camtrambone & Holyoak, 1989; Gentner, Loewenstein, & Hung, 2007).

Another effective way to evoke comparison is to use analogies from familiar domains: For example, a teacher might relate the workings of a factory to the functioning of a cell (Glynn & Takahashi, 1998). Instructional analogies are particularly effective scaffolds for conceptual change because they (a) help to form a bridge between students' prior knowledge and novel, unfamiliar information, (b) they make it easier for students to notice important links between the base and target representations, and (c) they help in the process of visualizing complex or unobservable concepts (Dagher, 1995; Iding, 1997; Jee et al., 2010). Because instructional analogies derive their power in part from being familiar, teachers must ensure that the base analog is well understood by students. To these ends, students and teachers can explicitly map analogies together, making sure to point out relevant similarities as well as "where the analogy breaks down" to ameliorate the effects of negative transfer (Glynn, 1991). Because no analogy is perfect, multiple analogies that target specific relations can be an effective way of inducing scientific understanding (Chiu & Lin, 2005), and analogies can be sequenced such that they progressively bridge students' understanding of the similarity between the base and target concepts (Clement, 1993). More recent studies suggest that students learn best when analogy-enhanced text is accompanied by visuals (Matlen, Vosniadou, Jee, & Ptouchkina, 2011), and when teachers use spatial cues—such as gesturing between base and target concepts—to facilitate students' comparisons (Richland, Zur, & Holyoak, 2007).

Another core component of science is the ability to construct, interpret, and revise scientific models (Duit, 1991; Harrison & Treagust, 1998). Reasoning with a scientific model inevitably relies on the back and forth process of mapping relationships from the model (i.e., the base) and phenomenon the model is trying to explain (i.e., the target), therefore relying on the process of inference projection and abstraction characteristic of analogical reasoning (Nersessian, 2002). While model-based reasoning is a relatively common practice in real-world science, it is rarely the focus of instruction in elementary education (Lehrer & Schauble, 2000). When a model's relational structure is supported by its superficial features, however, even young children can engage in relatively sophisticated reasoning with models. For example, Penner, Giles, Lehrer, & Schauble (1997) asked first-grade children to construct models of their elbow, and found that while children's models captured many superficial similarities (e.g., children insisted that their models include a hand with five fingers, represented by a foam ball and popsicle sticks), children were eventually able to construct models of their elbow that retained functional characteristics (e.g., incorporating the constraint that the elbow is unable to rotate 360°), and were also more likely than a nonmodeling peer group to ignore superficial distractors when identifying functional models. With sustained practice and scaffolding, children can overcome the tendency to attend to superficial similarities and can begin to reason with more abstract models that retain mostly relational structure, such as a graphing the relationship between plant growth and time (Lehrer & Schauble, 2004), or using a coin flip to model random variability in nature (Lehrer & Schauble, 2000). Often, superficial features can provide children the hook between perceiving relations between the model and the world, and as children gain experience and practice with modeling, superficial features can be progressively weaned away in favor of more abstract models.

The Role of Curiosity in Scientific Thinking

> *It is hard to satisfy the curiosity of a child, and even harder to satisfy the curiosity of a scientist.* (Bates, 1950)

Curiosity's role in scientific thinking is clear and unquestionably important. Curiosity involves recognizing when some information is missing or unknown. It motivates children to ask questions, make observations, and draw conclusions. Nevertheless, curiosity's role in these processes remains elusive, because there is no universally accepted definition of what it is (Jirout & Klahr, 2012). Defined in many different ways in the literature, operationalizations of curiosity can be categorized by level of specificity into three broad categories: curiosity as spontaneous exploration, curiosity as exploratory preference, and curiosity as preference for unknown and uncertainty.

When defining curiosity as exploratory behavior, researchers typically use observational measures of children playing, either with toys in a laboratory setting or in a highly stimulating environment, such as a museum. Any and all exploratory behaviors are considered to be curiosity, without consideration of children's differing levels of familiarity with the environment and materials, or the characteristics of the materials themselves. This method of studying curiosity has been used for studying children's maladaptive behavior (McReynolds, Acker, & Pietila, 1961), emotional and cognitive growth (Minuchin, 1971), and maternal behaviors (Endsley, Hutcherson, Garner, & Martin, 1979; Saxe & Stollak, 1971). Although the measure of curiosity as any exploratory behavior seems intuitively valid, the crucial element of stimuli characteristic is ignored. Additional factors are important to consider when using manipulations a child makes on an object as a measure of curiosity, such as the total opportunities or possibilities for manipulation on objects or the familiarity or novelty of an object. The use of inconsistent stimuli and a lack of consideration of both object familiarity and stimuli characteristics make it difficult to generalize any results beyond the limited scope of each study.

A more informative approach to studying curiosity—measuring children's exploratory preferences—involves determining specific factors that influence children's curiosity. In this approach, the total amount of exploratory behavior is not as important as the specific characteristics of objects or situations in which children choose to explore. Smock and Holt (1962) use preference for specific stimuli characteristics as a measure of curiosity. When given the opportunity to explore visual stimuli, preschool children were more likely to choose images higher in complexity, conflict, and incongruence, although there were wide individual differences in these choices. In addition, children preferred to play with an unknown toy to a known toy. Smock and Holt interpreted the characteristics of the preferred stimuli to be more general than complexity or incongruence, suggesting that the preferences were primarily driven by novelty, because children may have less experience with the type of complex, incongruent, and conflicting stimuli used in their study, and they suggest that this novelty is a more likely motivator of curiosity. A related approach to defining curiosity addresses the issue of children's curiosity as a novelty preference. The measurement of curiosity as a function of stimulus novelty assumes that more curious children prefer more novelty (Cantor & Cantor, 1964; Greene, 1964). There is empirical support for children's preference to explore novelty over familiarity, similar to the myriad studies on children's novelty preference unrelated to curiosity (Mendel, 1965). However, this work does not explain instances in which children prefer exploring familiar objects over novel ones, for example, children prefer exploring a known toy in which there is inconsistency, surprise, or ambiguity of its causal functions over a completely novel, unknown toy (Charlesworth, 1964; Schulz & Bonawitz, 2007). These examples are explained by the final approach to defining curiosity: curiosity as preference for the unknown, uncertainty, and ambiguity.

Instead of focusing on *stimuli* characteristics—that is, familiarity/novelty, complexity, and so on—this approach considers the relationship between the stimulus and the subject's knowledge, experience with, and understanding of the stimulus. These studies suggest that curiosity is a result of cognitive conflict or a gap in knowledge that is elicited by the stimuli or situation. For example, children are most curious when they see an outcome of an event that is inconsistent with their expectations (Charlesworth, 1964), when they don't understand how something works (Schulz & Bonawitz, 2007), and when they are aware of possible outcomes but aren't sure which will occur (Jirout & Klahr, 2009). With regard to the methods for measuring curiosity described above, measures of curiosity using uncertainty and ambiguity are the most specific and subsume the other methods used within their framework. Novelty, complexity, and the unknown can be interpreted as varying values on a continuum of uncertainty or ambiguity. The poles of this continuum—familiarity versus novelty, or known versus unknown—correspond to certain or unambiguous knowledge at one end of the spectrum, and total uncertainty and ambiguity at the other. After reviewing several theoretical perspectives on curiosity, Loewenstein (1994) arrived at the same conclusion and developed his Information-Gap Theory of curiosity, and Litman and Jimmerson (2004) developed their similar theory of curiosity as a feeling of deprivation. Both theories essentially defined curiosity in the same way, as the uncertainty/ambiguity measures discussed above, though Litman and Jimmerson also include a second type of curiosity similar to general interest.

Without a consistent operationalization of curiosity, it is difficult to discuss the role of curiosity in scientific thinking. Jirout and Klahr (2012) used the Information-Gap Theory to operationalize curiosity as the level of desired uncertainty in the environment most likely to lead to exploratory behavior. Studies of young children's curiosity, as defined here, suggest that curiosity is in fact related to children's ability to ask questions (Jirout & Klahr, 2011; Jirout, 2011). Children who are more curious—that is, those who prefer exploring greater uncertainty over less uncertainty—are also better at evaluating the effectiveness of questions and information in solving a mystery, and generate more questions overall (Jirout, 2011). When environments are experimentally manipulated to create varying amounts of uncertainty, both exploratory behavior (Litman, Hutchins, & Russon, 2005; Jirout & Klahr, 2009) and problem-solving accuracy (Mittman & Terrell, 1964) can be increased by creating an "optimal" level of uncertainty.

The curiosity literature supports the role of curiosity in scientific thinking as both a motivator of science and a crucial element of scientific reasoning, and the perceived importance of curiosity in science learning is evident by the inclusion of curiosity in all levels of science standards and goals (AAAS, 1993, 2008; Brenneman, Stevenson-Boyd, & Frede, 2009; Conezio & French, 2002; Kagan, Moore, & Bredekamp, 1995; National Education Goals Panel, 1995; NRC, 2000). Unfortunately, the body of knowledge about curiosity is limited. Although some will claim that "Real science begins with childhood curiosity..." (Conezio & French, 2002), further research is needed to better understand the role of curiosity in science learning.

SCIENTIFIC THINKING AND SCIENCE EDUCATION

Accounts of the nature of science and research on scientific thinking have had profound effects on science education at many levels, particularly in recent years. Up until the late 1970s, science education was primarily concerned with teaching students both the content of science (such as Newton's laws of motion), and the methods that scientists need to use in their research (such as using experimental and control groups). Beginning in the 1980s, a number of reports (e.g., AAAS, 1993; National Commission on Excellence in Education, 1983; Rutherford & Ahlgren, 1991) stressed the need for teaching children scientific thinking skills in addition to scientific procedures and content knowledge. This addition of scientific thinking skills to the science curriculum—from kindergarten through post-secondary levels—was a major shift in focus. Many of the particular scientific thinking skills that have been emphasized in this augmentation of the classical curriculum have been described in this chapter, such as teaching deductive and inductive thinking strategies. However, rather than focusing on any one specific skill, such as induction, researchers in science education have focused on ways to integrate the various components of scientific thinking, as well has how to better understand, and improve, collaborative scientific thinking.

What is the best way to teach and learn science? A clear, empirically supported, answer to this question has proven surprisingly elusive. For example, toward the end of the last century, influenced by several thinkers who advocated a constructivist approach to learning, ranging from Piaget (Beilin, 1994) to Papert (1978), many schools answered this question by adopting a philosophy dubbed "discovery learning." Although a clear operational definition of this approach has yet to be articulated, the general idea is that children are expected to learn science by reconstructing the processes of scientific discovery—in a range of areas from computer programming to chemistry to mathematics. The premise is that letting students discover principles on their own, set their own goals, and collaboratively explore the natural world, produces deeper knowledge that transfers widely. This approach sees learning as an active rather than a passive process, and suggests that students learn through constructing their scientific knowledge. We will first describe a few examples of the constructivist approach to science education. Following that, we will address several lines of work that challenge some of the assumptions of the constructivist approach to science education.

Often the goal of constructivist science education is to produce conceptual change through guided instruction where the teacher or professor acts as a guide to discovery, rather than the keeper of all the facts. One recent and influential approach to science education is the inquiry-based learning approach. Inquiry-based learning focuses on posing a problem or a puzzling event to students, and asking them to propose a hypothesis that could explain the event. Next, the student is asked to collect data that tests the hypothesis, make conclusions, and then reflect upon both the original problem and the thought processes that they used to solve the problem. Often students use computers that aid in their construction of new knowledge. The computers allow students to learn many of the different components of scientific thinking. For example, Reiser and his colleagues have developed a learning environment for biology, where students are encouraged to develop hypotheses in groups, codify the hypotheses, and search databases to test these hypotheses (Reiser, et al., 2001).

The research literature on science education is far from consistent in its use of terminology. However, our reading suggests that "Discovery learning" differs from "inquiry-based learning" in that few, if any, guidelines are given to students in discovery learning contexts, whereas in inquiry learning, students are given hypotheses, and specific goals to achieve. Although thousands of schools have adopted discovery learning as an alternative to more didactic approaches to teaching and learning, the evidence showing that it is more effective than traditional, direct, teacher-controlled instructional approaches is mixed, at best (Lorch et al., 2010; Minner, Levy, & Century, 2010). In several cases where the distinctions between direct instruction and more open-ended constructivist instruction have been clearly articulated, implemented, and assessed, direct instruction has proven to be superior to the alternatives (Chen & Klahr, 1999; Toth, Klahr, & Chen, 2000). For example, in a study of third- and fourth-grade children learning about experimental design, Klahr and Nigam (2004) found that many more children learned from direct instruction than from discovery learning. Furthermore, they found that among the few children who did manage to learn from a discovery method, there was no better performance on a far transfer test of scientific reasoning than that observed for the many children who learned from direct instruction.

The idea of children learning most of their science through a process of self-directed discovery has some romantic appeal, and it may accurately describe the personal experience of a handful of world-class scientists. However, the claim has generated some contentious disagreements (Kirschner, Sweller, & Clark, 2006; Klahr, 2009, 2010; Taber, 2010; Tobias & Duffy, 2009), and the jury remains out on the extent to which most children can learn science that way.

As noted above, scientific thinking involves thinking about both the *content* of science and the process of *"doing"* science. A recent NRC report describes a similar framework for science education (NRC, 2011). The overall framework of the report (see Table 9.1) includes three dimensions of science education: (a) scientific practices, which are general processes of science such as question asking, using models, and communicating information, (b) core ideas, including domain-specific topics like energy and evolution, and (c) cross-cutting concepts such as recognizing patterns.

On the basis of the assumption that understanding develops over time, the report emphasizes the importance of creating a framework with which to build on throughout K-12 science education, giving specific examples of learning progressions for different topics and concepts, and how to recognize understanding of the content taught at different grade levels. For example, in the domain "Organization for Matter and Energy Flow in Organisms," one set of goals, by grade, for students' scientific practices are: second grade students should be able to present information about why animals can be classified into simple groups, fifth grade students should be able to support claims using evidence, eighth grade students should be able to elaborate on arguments using explanations, and twelfth grade students should be able to present these arguments and explanations using sophisticated methods such as diagrams and models. The report also gives explicit suggestions on the depth of the standards that should be included in the more concrete, domain-specific content taught. To address the "mile wide and inch deep" issue, the report focuses on four main content areas to be covered: physical science, life science, Earth and space science, and engineering, technology, and the applications of science. The authors caution against including too many standards, and even suggest including what should *not* be taught as a way of keeping the standards manageable. The final dimension of science education addressed in the report is "cross-cutting concepts," another set

Table 9.1   *Framework for K-12 Science Education*

---

1. Scientific and Engineering Practices

1. Asking questions (for science) and defining problems (for engineering)
2. Developing and using models
3. Planning and carrying out investigations
4. Analyzing and interpreting data
5. Using mathematics and computational thinking
6. Constructing explanations (for science) and designing solutions (for engineering)
7. Engaging in argument from evidence
8. Obtaining, evaluating, and communicating information

---

2. Cross-Cutting Concepts

1. Patterns
2. Cause and effect: Mechanism and explanation
3. Scale, proportion, and quantity
4. Systems and systems models
5. Energy and matter: Flows, cycles, and conservation
6. Structure and function
7. Stability and change

---

3. Disciplinary Core Ideas

Physical Sciences

PS 1: Matter and its interactions
PS 2: Motion and stability: Forces and interactions
PS 3: Energy
PS 4: Waves and their applications in technologies for information transfer

---

Life Sciences

LS 1: From molecules to organisms: Structures and processes
LS 2: Ecosystems: Interactions, energy, and dynamics
LS 3: Heredity: Inheritance and variation of traits
LS 4: Biological evolution: Unity and diversity

Earth and Space Sciences

ESS 1: Earth's place in the universe
ESS 2: Earth's systems
ESS 3: Earth and human activity

Engineering, Technology, and the Applications of Science

ETS 1: Engineering design
ETS 2: Links among engineering, technology, science, and society

---

Source: Adapted from NRC, 2011.

of domain-general skills that are considered to be different from the scientific practices. These include patterns; cause and effect; scale, proportion, and quantity; systems and system models; energy and matter; flows, cycles, and conservation; structure and function; and stability and change. These concepts "provide students with an organizational framework for connecting knowledge from the various disciplines into a coherent and scientifically based view of the world." Although the report concedes that research on teaching these concepts is limited, it emphasizes the importance of developing these concepts as a way of instilling a common vocabulary and framework for learning across the core science domains. The report emphasizes the importance of integrating the three dimensions of science education, and provides examples of how this can be accomplished.

The authors of this framework acknowledge that it is only the beginning of a challenging process of improving science education, and they discusses issues of curriculum and instruction, teacher development, and assessments. An entire chapter is devoted to diversity and equity, and the importance of providing children a fair opportunity to learn. Recommendations for creating science standards are provided, and suggestions are made for research that is needed to effectively implement the provided framework and inform future revisions.

CONCLUSIONS AND FUTURE DIRECTIONS

In this chapter, we have argued that the basic cognitive processes used in everyday reasoning are the same processes used to make scientific discoveries. That is, the type of thinking that is deemed "scientific" and other traditional forms of thinking have more similarities than differences. Furthermore, we suggest that children are capable of such scientific thinking and that they employ it when learning scientific content, albeit at levels appropriate to their development. We have also offered suggestions for fostering scientific thinking in science education, including scaffolding children's scientific thinking through the use of explicit and guided-inquiry instruction.

At a course grain, the arguments we present in this chapter may seem somewhat paradoxical. On the one hand, we have suggested that children have many of the same basic reasoning abilities as scientists. This argument seems consistent with the constructivist notion that an optimal way to proceed with science education is to let children discover, on their own, scientific phenomena in much the same way that real-world scientists have throughout the course of history. On the other hand, we have suggested that an effective way to teach science—at least to domain novices—is to provide more directed guidance when a child engages in inquiry processes. Are these arguments necessarily inconsistent? We think not, for several reasons. First, while young children are indeed capable of employing many reasoning processes integral to scientific discovery, the developmental course of such reasoning processes is protracted, and extends even into adult education. Thus, children would likely benefit from some degree of scaffolding to avoid practicing incorrect strategies and more efficiently learn correct strategies. Second, guided and explicit instruction can be designed so as to use children's (often erroneous) preconceptions as starting points, an approach that is consistent with constructivism. Finally, the discoveries made by real-world scientists have sometimes taken centuries to achieve, and expecting young children to spontaneously recreate such discoveries is unrealistic. Withholding explicit guidance may indeed be beneficial for children once they have gained some expertise in science concepts (Kalyuga, 2007), however, expecting children to succeed in minimally guided settings is likely to lead to frustration and floundering (Koedinger & Aleven, 2007). Instead, we suggest that children should engage in sustained practice in scientific reasoning throughout their science careers and that providing guidance and feedback throughout this process can lead to optimal science learning outcomes.

REFERENCES

American Association for the Advancement of Science (AAAS). (1993). *Benchmarks for science literacy.* New York, NY: Oxford University Press.

Baillargeon, R. (2004). Infants' reasoning about hidden objects: Evidence for event-general and event-specific expectations. *Developmental Science, 7,* 391–424.

Bates, M. (1950). *The nature of natural history.* New York, NY: Charles Scribner's Sons.

Beilin, H. (1994). Jean Piaget's enduring contribution to developmental psychology. In R. Parke, P. A. Ornstein, & C. Zahn-Waxler (Eds.), *A century of developmental psychology* (pp. 257–290). Washington, DC: American Psychological Association.

Brenneman, K., Stevenson-Boyd, J. S., & Frede, E. (2009). Math and science in preschool: Policies and practice. *Preschool Policy Matters,* Issue 19. New Brunswick, NJ: National Institute for Early Education Research.

Brown, A. R., Bransford, R., Ferrara, R., Campione, J. (1983). Learning, remembering, and understanding. In P. Mussen (Ed.), *Handbook of child psychology: cognitive development.* New York, NY: John Wiley and Sons.

Bulloch, M. J., & Opfer, J. E. (2008). What makes relational reasoning smart? Revisiting the perceptual-to-relational shift in the development of generalization. *Developmental Science, 12,* 114–122.

Camtrambone, R., & Holyoak, K. J. (1989). Overcoming contextual limitations on problem-solving transfer. *Journal of Experimental Psychology, 15,* 1147–1156.

Cantor, J. H., & Cantor, G. N. (1964). Observing behavior in children as a function of stimulus novelty. *Child Development, 35,* 119–128.

Carey, S. (1985). *Conceptual change in childhood.* Cambridge, MA: MIT Press.

Carey, S. (1991). Knowledge acquisition: Enrichment or conceptual change? In S. Carey & R. Gelman (Eds.), *The epigenesis of mind* (pp. 257–291). Hillsdale, NJ: Erlbaum.

Charlesworth, W. R. (1964). Instigation and maintenance of curiosity behavior as a function of surprise versus novel and familiar stimuli. *Child Development, 35,* 1169–1186.

Chen, Z., & Klahr, D. (1999) All other things being equal: children's acquisition of the control of variables strategy. *Child Development, 70*(5), 1098–1120.

Chi, M. (2008). Three types of conceptual change: Belief revision, mental model transformation, and categorical shift. In S. Vosniadou (Ed.), *International handbook of research on conceptual change.*

Chi, M. T. H., & Roscoe, R. D. (2002). The processes and challenges of conceptual change. In M. Limon & L. Mason (Eds.), *Reframing the process of conceptual change: Integrating theory and practice.* (pp. 3–27). Dordrecht, the Netherlands: Kluwer Academic.

Chiu, M. H., & Lin, J. W. (2005). Promoting fourth graders' conceptual change of their understanding of electric current via multiple analogies. *Journal of Research in Science Teaching, 42,* 429–446.

Christie, S., & Gentner, D. (2010). Where hypotheses come from: Learning new relations by structural alignment. *Journal of Cognition and Development, 11,* 356–373.

Clement, J. (1982). Students' preconceptions in introductory mechanics. *American Journal of Physics, 50,* 66–71.

Clement, J. (1993). Using bridging analogies and anchoring intuitions to deal with students' preconceptions in physics. *Journal of Research in Science Teaching, 30,* 1241–1257.

Conezio, K., & French, L. (2002). Science in the preschool classroom: Capitalizing on children's fascination with the everyday world to foster language and literacy development. *Young Children, 57*(5), 12–18.

Crick, F. H. C. (1988). *What mad pursuit: A personal view of science.* New York, NY: Basic Books.

Dagher, Z. R. (1995). Review of studies on the effectiveness of instructional analogies in science education. *Science Education, 79,* 295–312.

Davenport, J. L., Yaron, D., Klahr, D., & Koedinger, K. (2008). Development of conceptual understanding and problem solving expertise in chemistry. In B. C. Love, K. McRae, & V. M. Sloutsky (Eds.), *Proceedings of the 30th annual conference of the cognitive science society* (pp. 751–756). Austin, TX: Cognitive Science Society.

Deloache, J. S.(1987). Rapid change in symbolic functioning in very young children. *Science, 238,* 1556–1557.

diSessa, A. (2008). A bird's-eye view of the "pieces" vs "coherence" controversy (from the "pieces" side of the fence). In S. Vosniadou (Ed.), *International handbook of research on conceptual change* (pp. 453–478). New York, NY: Routledge.

Duit, R. (1991). On the role of analogies and metaphors in learning science. *Science Education, 75,* 649–672.

Dunbar, K. (1995). How scientists really reason: Scientific reasoning in real-world laboratories. In R. J. Sternberg & J. Davidson (Eds.), *Mechanisms of insight* (pp. 365–395). Cambridge MA: MIT press.

Dunbar, K. (1997). How scientists think: Online creativity and conceptual change in science. In T. B. Ward, S. M. Smith, & S. Vaid (Eds.), *Conceptual structures and processes: Emergence, discovery and change* (pp. 461–493). Washington DC: APA Press.

Dunbar, K. (2001). The analogical paradox: Why analogy is so easy in naturalistic settings yet so difficult in the psychological laboratory. In D. Gentner, K. Holyoak, & N. Boicho (Eds.), *The analogical mind: Perspectives from cognitive science* (pp. 313–334). Cambridge, MA: MIT Press.

Duncker, K. (1945). On problem solving. *Psychological Monographs, 58,* i–113.

Duschl, R. A., Schweingruber, H. A., & Shouse, A. W. (Eds.). (2007) *Taking science to school: Learning and teaching science in grades K-8.* Washington, DC: National Academies Press.

Einstein, A. (1950). *Out of my later years.* New York, NY: Philosophical Library

Endsley, R. C., Hutcherson, M. A., Garner, A. P., & Martin, M. J. (1979). Interrelationships among selected maternal behaviors, authoritarianism, and preschool children's verbal and nonverbal curiosity. *Child Development, 50,* 331–339.

Fisher, A. V., Matlen, B., & Godwin, K. E. (2011). Semantic similarity of labels and inductive generalization: taking a second look. *Cognition, 118,* 432–438.

Gelman, S. A., & Coley, J. D. (1990). The importance of knowing a dodo is a bird: Categories and inferences in 2-year-old children. *Developmental Psychology, 26,* 796–804.

Gentner, D. (1983). Structure-mapping: A theoretical framework for analogy. *Cognitive Science, 7,* 155–170.

Gentner, D. (1988). Metaphor as structure mapping: The relational shift. *Child Development, 59,* 47–59.

Gentner, D. (2010). Bootstrapping the mind: Analogical processes and symbol systems. *Cognitive Science, 34,* 752–775.

Gentner, D., Loewenstein, J., & Hung, B. (2007) Comparison facilitates children's learning of names for parts. *Journal of Cognition and Development, 8,* 285–307.

Glynn, S. M. (1991). Explaining science concepts: A teaching-with-analogies model. In S. M. Glynn & R. H. Yeany (Eds.), *The psychology of learning science* (pp. 219–240). Hillside, NJ: Lawrence Erlbaum Associates.

Glynn, S. M., & Takahashi, T. (1998). Learning from analogy-enhanced science text. *Journal of Research in Science Teaching, 35,* 1129–1149.

Gobert, J. D., & Clement, J. J. (1999). Effects of student-generated diagrams versus student-generated summaries on conceptual understanding of causal and dynamic knowledge in plate tectonics. *Journal of Research in Science Teaching, 36,* 39–53.

Goldberg, R. F., & Thompson-Schill, S. L. (2009). Developmental "roots" in mature biological knowledge. *Psychological Science, 20,* 480–487.

Goldstone, R. L., & Son, J. (2005). The transfer of scientific principles using concrete and idealized simulations. *The Journal of the Learning Sciences, 14,* 69–110.

Gopnik, A., & Sobel, D. M. (2000). Detecting Blickets: How young children use information about novel causal powers in categorization and induction. *Child Development, 71,* 1205–1222.

Gopnik, A., Meltzoff, A., & Kuhl, P. (1999). *The scientist in the crib: Minds, brains, and how children learn.* New York, NY: William Morrow & Co.

Goswami, U. (1991). Analogical reasoning: What develops? A review of research and theory. *Child Development, 62,* 1–22.

Goswami, U. (2001) Analogical reasoning in children. In D. Gentner, K. J. Holyoak, B. N. Kokinov (Eds.), *The analogical mind: Perspectives from cognitive science* (pp. 437–470). Cambridge, MA: MIT Press.

Goswami, U., & Brown, A. L. (1990). Melting chocolate and melting snowmen: Analogical reasoning and causal relations. *Cognition, 35,* 69–95.

Greene, F. M. (1964). Effect of novelty on choices made by preschool children in a simple discrimination task. *Child Development, 35,* 1257–1264.

Grosslight, L., Unger, C., & Jay, E. (1991). Understanding models and their use in science: conceptions of middle and high school students and experts. *Journal of Research in Science Teaching, 28,* 799–822.

Harrison, A. G., & Treagust, D. F. (1998) Modeling in science lessons: are there better ways to learn with models? *School Science and Mathematics, 98,* 420–429.

Holton, G. (2003). Einstein's third paradise. *Daedalus, 132,* 26–34.

Iding, M. K. (1997). How analogies foster learning from science texts. *Instructional Science, 25,* 223–253.

Inagaki, K., & Hatano, G. (2008). Conceptual change in naïve biology. In S. Vosniadou (Ed.), *International handbook of research on conceptual change* (pp. 240–262). New York, NY: Routledge.

Jee, B. D., Uttal, D. H., Gentner, D., Manduca, C., Shipley, T., Sageman, B.,...Tikoff, B. (2010). Analogical thinking in geoscience education. *Journal of Geoscience Education, 58,* 2–13.

Jirout, J. (2011). Curiosity and the development of question generation skills. *Proceedings of the* aaai *2011 fall symposium on question generation.* Association for the Advancement of Artificial Intelligence (AAAI) Press, Menlo Park, CA.

Jirout, J., & Klahr, D. (2009). Children's recognition of uncertainty and exploratory curiosity. *Paper presented at the Society for Research in Child Development conference,* April 2–4, 2009, Denver, CO.

Jirout, J., & Klahr, D. (2011). Children's question asking and curiosity: A training study. *Paper presented at the Society for Research in Educational Effectiveness conference,* September 8–9, 2011, Washington, DC.

Jirout, J., & Klahr, D. (2012). Children's scientific curiosity: In search of an operational definition of an elusive concept. *Developmental Review, 32,* 125–160.

Kagan, S. L., Moore, E., & Bredekamp, S. (1995). *Reconsidering children's early development and learning: Toward shared beliefs and vocabulary.* Washington, DC: National Education Goals Panel.

Kalyuga, S. (2007). Expertise reversal effect and its implications for learner-tailored instruction. *Educational Psychology Review, 19,* 509–553.

Keil, F. C., Smith, W. C., Simons, D. J., & Levin, D. T. (1998). Two dogmas of conceptual empiricism: Implications for hybrid models of the structure of knowledge. *Cognition, 65,* 103–135.

Keil, F. C. (2011). Science starts early. *Science, 331,* 1022–1023.

Kelemen, D., & Rosset, E. (2009). The human function compunction: Teleological explanation in adults. *Cognition, 111*, 138–143.

Kirschner, P. A., Sweller, J., & Clark, R. E. (2006). Why minimal guidance during instruction does not work: An analysis of the failure of constructivist, discovery, problem-based, experiential, and inquiry-based teaching. *Educational Psychologist, 41*, 75–86.

Klahr, D. (2000). *Exploring science: The cognition and development of discovery processes.* Cambridge, MA: MIT Press.

Klahr, D. (2009). To every thing there is a season, and a time to every purpose under the heavens: What about direct instruction? In S. Tobias & T. M. Duffy (Eds.), *Constructivist theory applied to instruction: Success or failure?* New York, NY: Taylor and Francis.

Klahr, D. (2010). Coming up for air: but is it oxygen or phlogiston? a response to Taber's review of constructivist instruction: success or failure?. *Education Review, 13*(13), 1–6.

Klahr, D., & Dunbar, K. (1988). Dual space search during scientific reasoning. *Cognitive Science, 12*, 1–48.

Klahr, D., & Nigam, M. (2004). The equivalence of learning paths in early science instruction: Effects of direct instruction and discovery learning. *Psychological Science, 15*, 661–667.

Klahr, D., & Robinson, M. (1981). Formal assessment of problem solving and planning processes in preschool children. *Cognitive Psychology, 13*, 113–148.

Klahr, D., & Simon, H. (1999). Studies of scientific discovery: Complementary approaches and convergent findings. *Psychological Bulletin, 125*, 524–543.

Klahr, D., Dunbar, K., & Fay, A. L. (1990). Designing good experiments to test "bad" hypotheses. In J. Shrager & P. Langley (Eds.), *Computational models of discovery and theory formation* (pp 355–402). San Mateo, CA: Morgan-Kaufman.

Klahr, D., Fay, A. L., & Dunbar, K. (1993) Developmental differences in experimental heuristics. *Cognitive Psychology, 25*, 111–146.

Klayman, J., & Ha, Y (1987) Confirmation, disconfirmation and information in hypothesis testing. *Psychological Review, 94*, 211–228.

Koedinger, K. R., & Aleven V. (2007). Exploring the assistance dilemma in experiments with cognitive tutors. *Educational Psychology Review, 19*(3), 239–264.

Kuhn, D. (1989). Children and adults as intuitive scientists. *Psychological Review, 96*, 674–668.

Kurtz, K. J., Miao, C. H., & Gentner, D. (2001). Learning by analogical bootstrapping. *The Journal of the Learning Sciences, 10*, 417–446.

Lehrer, R., & Schauble, L. (2000). The development of model-based reasoning. *Journal of Applied Developmental Psychology, 21*(1), 39–48.

Lehrer, R., & Schauble, L. (2004). Modeling natural variation through distribution. *American Educational Research Journal, 41*, 635–679.

Lewis, E. L., & Linn, M. C. (1994). Heat energy and temperature concepts of adolescents, adults, and experts: Implications for curricular improvements. *Journal of Research in Science Teaching, 31*, 657–677.

Litman, J. A., & Jimmerson, T. L. (2004). The measurement of curiosity as a feeling of deprivation. *Journal of Personal Assessment, 82*, 147–157.

Litman, J. A., Hutchins, T. L., & Russon, R. K. (2005). Epistemic curiosity, feeling-of-knowing, and exploratory behaviour. *Cognition & Emotion, 19*(4), 559–582.

Loewenstein, G. (1994). The psychology of curiosity: A review and reinterpretation. *Psychological Bulletin, 116*(1), 75–98.

Lorch, R. F., Jr., Lorch, E. P., Calderhead, W. J., Dunlap, E. E., Hodell, E. C., Freer, B. D. (2010). Learning the control of variables strategy in higher- and lower-achieving classrooms: Contributions of explicit instruction and experimentation. *Journal of Educational Psychology, 102*, 90–101.

Matlen, B. J., & Klahr, D. (2010). Sequential effects of high and low guidance on children's early science learning. In K. Gomez, L. Lyons, & J. Radinsky, (Eds.), *Proceedings of the 9th International Conference of the Learning Sciences,* Chicago, IL.

Matlen, B.J., Vosniadou, S., Jee, B., & Ptouchkina, M. (2011). Enhancing science learning through visual analogies. In L. Carlson, C. Holscher, & T. Shipley (Eds.), *Proceedings of the XXXIV Annual Meeting of the Cognitive Science Society.*

McReynolds, P., Acker, M., & Pietila, C. (1961). Relation of object curiosity to psychological adjustment in children. *Child Development, 32*, 393–400.

Mendel, G. (1965). Children's preferences for differing degrees of novelty. *Child Development, 36*(2), 453–465.

Minner, D. D., Levy, A. J., & Century, J. (2010). Inquiry-based science instruction—what is it and does it matter? Results from a research synthesis years 1984 to 2002. *Journal of Research in Science Teaching, 47*, 474–496.

Minstrell, J. (1992). Facets of students' knowledge and relevant instruction. In R. Duit, F. Goldberg, & H. Niedderer (Eds.), *Research in physics learning: Theoretical issues and empirical studies* (pp. 110–128). Kiel, Germany: Institute for Science Education at the University of Kiel.

Minuchin, P. (1971). Correlates of curiosity and exploratory behavior in preschool disadvantaged children. *Child Development, 42*(3), 939–950.

Mittman, L. R., & Terrell, G. (1964). An experimental study of curiosity in children. *Child Development, 35*, 851–855.

National Commission of Excellence in Education (Ed.). (1983). *A nation at risk: The imperative for educational reform.*

National Education Goals Panel. (1995). *National education goals report: Building a nation of learners.* Washington, DC: U.S. Government Printing Office.

National Research Council (NRC). (2000) *Inquiry and the National Science Education Standards: A guide for teaching and learning.* Washington, DC: National Academies Press.

National Research Council (NRC). (2011). *A framework for K-12 science education: Practices, crosscutting concepts, and core ideas.* Committee on a Conceptual Framework for New K-12 Science Education Standards. Board on Science Education, Division of Behavioral and Social Sciences and Education. Washington, DC: The National Academies Press.

Nercessian, N. J. (2009). How do engineering scientists think? Model-based simulation in biomedical engineering research laboratories. *Topics in Cognitive Science, 1*, 730–757.

Nersessian, N. J. (2002). The cognitive basis of model-based reasoning in science. In P. Carruthers, S. Stich, & M. Siegal (Eds.), *The cognitive basis of science* (pp. 178–211). New York, NY: Cambridge University Press.

Newell, A., & Simon, H. (1972). *Human problem solving.* Englewood Cliffs, NJ: Prentice-Hall.

Opfer, J. E., & Seigler, R. S. (2004). Revisiting preschoolers' living things concept: A microgenetic analysis of conceptual change in basic biology. *Cognitive Psychology, 49*, 301–332.

Papert (1978). The mathematical unconscious

Penner, D. E., Giles, N. D., Lehrer, R., & Schauble, L. (1997). Building functional models: Designing an elbow. *Journal of Research in Science Teaching, 34*, 125–143.

Piaget, J., Montangero, J., & Billeter J. (1977). Les correlats. In J. Piaget (Ed.), *L'abstraction refleschissante* (pp. 115–129). Paris, France: Presses Universitaires de France.

Posner, G. J., Strike, K. A., Hewson, W., & Gertzog W. A. (1982). Accommodation of a scientific conception: Toward a theory of conceptual change. *Science Education, 66*, 211–227.

Rakison, D., & Lupyan, G. (2008). Developing object concepts in infancy: An associative learning perspective. *Monographs of the Society for Research in Child Development, 73*, 1–130.

Rattermann, M. J., & Gentner, D. (1998). More evidence for a relational shift in the development of analogy: Children's performance on a causal-mapping task. *Cognitive Development, 13*, 453–478.

Reiser, B. J., Tabak, I., Sandoval, W. A., Smith, B., Steinmuller, F., & Leone, T. J., (2001). BGuILE: Strategic and conceptual scaffolds for scientific inquiry in biology classrooms. In S. M. Carver & D. Klahr (Eds.), *Cognition and instruction: Twenty five years of progress* (pp. 263–305). Mahwah, NJ: Erlbaum.

Richland, L. E., Morrison, R. G., & Holyoak, K. J. (2006). Children's development of analogical reasoning: Insights from scene analogy problems. *Journal of Experimental Child Psychology, 94*, 249–273.

Richland, L. E., Zur, O., & Holyoak, K. J. (2007). Cognitive supports for analogies in the mathematics classroom. *Science, 316,* 1128–1129.

Roth, W. M. (2008). The nature of scientific conceptions: A discursive psychological perspective. *Education Research Review, 3,* 30–50.

Rutherford, F. J., & Ahlgren, A. (1991). *Science for all Americans.* Oxford, MA: Oxford University Press.

Saxe, R. M., & Stollak, G. E. (1971). Curiosity and the parent-child relationship. *Child Development, 42,* 373–384.

Schulz, L. E., & Bonawitz, E. B. (2007). Serious fun: Preschoolers engage in more exploratory play when evidence is confounded. *Development Psychology, 43*(4), 1045–1050.

Siegler, R. S. (1995). How does change occur: A microgenetic study of number conservation. *Cognitive Psychology, 28,* 225–273.

Siler, S., Klahr, D., & Matlen, B. J. (in press). Conceptual change when learning experimental design. In S. Vosniadau (Ed.), *International handbook of research on conceptual change* (2nd ed.). London, UK: Routledge

Simon, H. A., Langley, P., & Bradshaw, G. L. (1981). Scientific discovery as problem solving. *Synthese, 47,* 1–27.

Sloutsky, V. M., Kaminski, J. A., & Heckler, A. F. (2005). The advantage of simple symbols for learning and transfer. *Psychonomic Bulletin & Review, 12,* 508–513.

Sloutsky, V. M., Kloos, H., & Fisher, A. V. (2007). When looks are everything: Appearance similarity versus kind information in early induction. *Psychological Science, 18,* 179–185.

Smith, J. P. III, DiSessa, A. A., & Roschelle, J. (1994). Misconceptions reconceived: A constructivist analysis of knowledge in transition. *Journal of the Learning Sciences, 3,* 115–164.

Smith, L. B., Jones, S. S., & Landau, B. (1996). Naming in young children: A dumb attentional mechanism? *Cognition, 60,* 143–171.

Smock, C. D., & Holt, B. G. (1962). Children's reactions to novelty: An experimental study of "curiosity motivation". *Child Development, 33,* 631–642.

Sternberg, R. J., & Nigro, G. (1980). Developmental patterns in the solution of verbal analogies. *Child Development, 51,* 27–38.

Taber, K. S. (2010). Constructivism and direct instruction as competing instructional paradigms: An essay review of Tobias and Duffy's *Constructivist Instruction: Success or Failure? Education Review, 13*(8), 1–44.

Thagard, P. (1992). *Conceptual revolutions.* Cambridge, MA: MIT Press.

Thibault, J. P., French, R., & Vezneva, M. (2010). The development of analogy making in children: Cognitive load and executive functions. *Journal of Experimental Child Psychology, 106,* 1–19.

Tobias, S., & Duffy, T. M. (2009). *Constructivist instruction: Success Or failure?* New York, NY: Routledge.

Toth, E. E., Klahr, D., & Chen, Z. (2000) Bridging research and practice: a cognitively-based classroom intervention for teaching experimentation skills to elementary school children. *Cognition & Instruction, 18*(4), 423–459.

Vosniadou, S., & Brewer, W. F. (1992). Mental models of the earth: A study of conceptual change in childhood. *Cognitive Psychology, 24,* 535–585.

Vosniadou, S., & Brewer, W. F. (1994). Mental models of the day/night cycle. *Cognitive Science, 18,* 123–183.

Vosniadou, S., Vamvakoussi, X., & Skopeliti, I. (2008). The framework theory approach to the problem of conceptual change. In S. Vosnaidou, (Ed.), *International handbook of research on conceptual change* (pp. 3–33). New York, NY: Taylor & Francis.

Ward, S. L., & Overton, W. F. (1990). Semantic familiarity, relevance, and the development of deductive reasoning. *Developmental Psychology, 26,* 488–493.

Wiser, M., & Smith, C. L. (2008). Learning and teaching about matter in grades K-8: When should the atomic-molecular theory be introduced? In S. Vosniadou (Ed.), *International handbook of research on conceptual change* (pp. 205–239). New York, NY: Routledge.

Wright, S. B., Matlen, B. J., Baym, C. L., Ferrer, E., & Bunge, S. A. (2008). Neural correlates of fluid reasoning in children and adults. *Frontiers in Human Neuroscience.*

Special Topics

CHAPTER 10

Creative Genius in Science

Dean Keith Simonton

*T*his chapter concerns the conjunction of three concepts that overlap only partially: creativity, genius, and science. Not all creativity requires genius, as is evident in everyday forms of creativity. Coming up with this opening paragraph may have required some creativity, but I would be the last to claim that these sentences show any sparks of genius. Nor does all genius require creativity, a fact most apparent when genius is defined according to stellar score on an intelligent quotient (IQ) test. By the latter criterion, the greatest genius who ever lived was Marilyn vos Savant (McFarlan, 1989), a person who has yet to demonstrate genius-level creativity in any recognized domain of achievement. Finally, it is obvious that creativity and genius, both separately and together, can and do appear in domains that cannot be considered scientific by any stretch of the imagination. Certainly, creative geniuses appear in the arts, whether painting, sculpture, architecture, drama, poetry, fiction, music, opera, dance, or cinema. Yet we will ignore all of these luminaries, and instead concentrate on creative genius in science. Science is simply taken to indicate any of the disciplines making up the "sciences" at any contemporary research university. Mathematics, astronomy, physics, chemistry, biology, sociology, economics, and (of course) psychology come immediately to mind. I am writing this chapter because I believe that no psychology of science is complete without an empirical and theoretical treatment of this central phenomenon (Simonton, 1988c, 2004).

Why do I consider creative genius so central in science? The best answer I can give is to adopt the definition that the philosopher Kant offered in his 1790 *Critique of Judgement*. According to Kant, "Genius is the talent … for producing that for which no definite rule can be given" (Kant, 1790/1952, p. 525). This I take to mean that geniuses produce ideas that cannot be generated by any mere algorithm. Kant adds, *"originality* must be its primary quality" (p. 526). Yet given that "there may also be original nonsense, its products must at the same time be models, i.e., be *exemplary*" (p. 526). Genius generates works that provide models for others to follow, even imitate. Now I realize that Kant did not believe genius could not exist in science. He was under the mistaken assumption that scientific creativity was governed by a "method" or "logic" that precluded the need for genius.

Kant's definition has an obvious connection with the most favored definition of creativity (Simonton, 2000b). Creativity entails the conjunction of originality and usefulness. Hence, both genius and creativity produce originality. Moreover, both genius and creativity yield usefulness if we posit that nothing would be considered exemplary unless it had some utility. Still, the stipulation that the product be exemplary takes the second requirement a step further. The products of genius are not only accepted, but also admired by others in the field. Nonetheless, I would like to add a third component to the specification, namely, what the United States Patent Office uses when evaluating applications (http://www.uspto.gov/inventors/patents.jsp). To earn patent protection, the invention must be not only new and useful, but also nonobvious. That is, it cannot be an obvious extension or elaboration of a previous invention, where obviousness is gauged by someone with reasonable expertise in the field. In a sense, "no definite rule can be given" of how to connect the current invention with prior inventions. From a psychological rather than legal standpoint, we can say that the invention must be surprising. Generalizing the subject of this chapter, we will say that creative genius in science produces ideas that are novel, surprising, and exemplary. Ideas that score high on all three attributes can often be viewed as transformative, even revolutionary.

Such is the subject matter of this chapter. I start with a discussion of how to assess creative genius in science. I then turn to a treatment of two sets of factors associated with this phenomenon: individual differences and personal development (cf. Feist, 2006b). I then turn to a more brief discussion of some additional topics relevant to the subject. Where appropriate, I will mention when creative genius in science differs from that in other domains, especially the arts. In some respects, artistic geniuses provide a more telling comparison group than do scientific creators who do not attain genius levels.

MEASUREMENT

Although it is easy to define creative genius, it is much more difficult to measure it. So how do we go about determining who counts as an example? Given that scientific ideas may vary in the magnitude of their originality, utility, and surprise, how do we gauge the relative amount of creativity or genius displayed by a scientist? Empirical research has adopted four major assessment strategies:

Historiometric Eminence

Creative geniuses make history, and to make history indicates that such individuals leave an impression on the historical record. The latter consists of histories, encyclopedias, and biographical dictionaries. Moreover, the amount of space devoted to a given figure in these various reference works can be taken as an indicator of relative historical impact or eminence (e.g., Zusne & Dailey, 1982). Galton (1869) was the first to identify geniuses using this method, and his student, J. M. Cattell (1903), was the first to use this method to assess relative degrees of genius. Later, investigators applied historiometric eminence to the scrutiny of scientific genius (e.g., Simonton, 1984e, 1991a, 1992b; Sulloway, 1996). Using this technique, Murray (2003) identified the top five physicists as Newton (100), Einstein (100), Rutherford (88), Faraday (86), and Galileo (89), the numbers in parentheses indicating their quantitative level of distinction. At the bottom of the 218 physicists scored are John Canton, Marie Fabry, Aleksandr Popov, and Johann Schweigger (all 1 point each). Of the 218 "significant figures" in the sample, Murray singled out only 66 as "major figures"—physicists who could be considered creative geniuses in science.

Expert Evaluations

An alternative approach is to ask experts in a given domain to identify, and even rate, the leading figures in the field, both contemporary and historical. This technique was first introduced by J. M. Cattell (1906), and has since been extended to many different forms of creative genius. An example is the expert rating of 528 psychologists published by Annin, Boring, and Watson in 1968. At the top of the list are figures like Alfred Binet, Hermann Ebbinghaus, Sigmund Freud, Francis Galton, G. S. Hall, C. L. Hull, Wolfgang Köhler, C. E. Spearman, L. M. Terman, E. L. Thorndike, E. B. Titchener, E. C. Tolman, J. B. Watson, Max Wertheimer, Wilhelm Wundt, and R. M. Yerkes, whereas at the bottom are figures like C. E. Ferree, V. A. C. Henmon, H. G. Seashore, and Graham Wallas.

Journal Citations

The previous two methods can be applied to any form of creative genius, and even genius in general. In contrast, this third assessment method is most appropriate for the sciences, in which it has become the norm to cite the publications of others who have had an impact on the ideas reported in a given publication. In principle, scientists who have the biggest influence receive the most citations in the professional literature (Cole & Cole, 1973; Rushton, 1984). For instance, one study of 54 eminent psychologists placed Sigmund Freud and Jean Piaget at the top and Oscar Külpe and J. R. Angell at the bottom (Simonton, 2000c). Citations have even been used to predict who will receive the Nobel Prize (Ashton & Oppenheim, 1978), which takes us to the fourth and last criterion.

Scientific Awards

Galton (1874) was the first researcher to use awards, honors, or prizes as a criterion variable (cf. Candolle, 1873). In his particular case, the criterion was election as Fellow of the Royal Academy of London, which was then one of the most prestigious honors. Nonetheless, in the early 20th century, the Nobel Prize became the premiere honor in most major sciences. It should come as no surprise, then, that receipt of this award has long been taken as the criterion of scientific genius, creativity, or achievement (e.g., Berry, 1981; Clark & Rice, 1982; Manniche & Falk, 1957; Moulin, 1955; Rothenberg, 1983; Stephan & Levin, 1993; Zuckerman, 1977). In fact, it is difficult to conceive a single scientific genius in the last century who has not received this conspicuous honor, at least not in those disciplines recognized by the Nobel prize committees. In the physical sciences alone, the list includes Niels Bohr, Marie Curie, Albert Einstein, Enrico Fermi, Richard P. Feynman, Murray Gell-Mann, Werner Heisenberg, Linus Pauling, Max Planck, and Ernst Rutherford. Admittedly, unlike the previous three measures, receiving or not receiving this prestigious award is a dichotomous rather than quantitative variable. Not only that, but also it constitutes a criterion with an extremely low-base rate, a fact that undermines its utility if one seeks a large sample with greater variance. Fortunately, researchers can create an ordinal scale that places the Nobel at the top and lesser awards increasingly toward the bottom (Cole & Cole, 1973; cf. Vijh, 1987). This scale can then make finer differentiations regarding the relative impact of sampled scientists.

Each of the above four assessment methods has its own advantages and disadvantages. For this reason, the best measurement strategy is often to use two or more assessments, sometimes even combining them into a composite measure (e.g., Feist, 1993; Simonton, 1992a). The latter practice is justified because the alternative assessments tend to intercorrelate sufficiently to indicate a single latent variable that we might call a "creative genius factor" or "Galton's G" (Cole & Cole, 1973; Simonton, 1984c, 1991c). Such composites can claim to be more reliable and valid than their separate components. It is also worth mentioning that such measures tend to be highly stable over many decades, if not centuries (Over, 1982b; Simonton, 1984c). Fame is by no means fickle, a transhistorical stability that even holds in the arts (Ginsburgh & Weyers, 2006, 2010; Simonton, 1998).

INDIVIDUAL DIFFERENCES

Because scientists vary in the extent to which they exhibit creative genius, we can meaningfully ask what other variables correlate with that individual-difference variance. The research has examined the following three correlates: productivity, intellect, and personality.

Productivity

Scientists vary immensely in lifetime output. This exceptional variation has become especially conspicuous since the advent of professional journals, so that researchers can count articles rather whole books. How big is this variation? If a scientist is defined as someone who has earned a doctoral degree in a scientific specialty, then a large proportion of scientists publish nothing beyond their doctoral dissertations (Bloom, 1963), and "only about 1% of scientists publish more than about ten papers in their entire career" (Vijh, 1987,

p. 9). At the other extreme are those scientists who have hundreds of publications to their credit. The latter tend to be a rare minority, so much so that a small percentage of the scientists accounts for most of the research in their field (Lotka, 1926; Price, 1963; Shockley, 1957). In rough terms, the top 10% can be credited with nearly half of the total publications (Dennis, 1954a, 1954c, 1955). This prolific elite provides prime candidates for scientific genius.

Naturally, one could argue that quantity is not the same as quality. After all, close to half of all published research may not be cited at all (Redner, 1998). Perhaps the superbly prolific are mere mass producers, whereas the true geniuses are perfectionists who produce a relatively small number of high-impact works (cf. Cole & Cole, 1973; Feist, 1997). However, while there is no doubt that mass producers and perfectionists indeed exist, they are the exceptions rather than the rule: The most influential scientists tend to score high in both quantity and quality (Cole & Cole, 1973; Davis, 1987; Feist, 1993; Myers, 1970; Simonton, 1992a). Such scientists earn more awards, receive higher peer recognition, and attain higher posthumous eminence.

Indeed, I propose that we can speak of a continuous scale that begins from very low levels of scientific talent and extends to full-fledged scientific genius. This scale can be constructed by integrating two contrasting scales.

At the beginning is the "Scientific Discovery" subscale of the Creative Achievement Questionnaire (Carson, Peterson, & Higgins, 2005). The zero point is "I do not have training or recognized ability in this field," and then extends upwards through items like "I have received a scholarship based on my work in science or medicine" and "I have been the author or coauthor of a study published in a scientific journal" to items like "I have won a national prize in the field of science or medicine" and "I have received a grant to pursue my work in science or medicine" (p. 49).

At the higher end is the somewhat makeshift "Landau–Lotka" scale of research achievement, that begins with "one or no paper" and proceeds to "2 to 10 papers" and then "over 10 papers" until it reaches "local awards + prolific creativity," and "national and international awards + prolific creativity," and then culminates in "N.A.S., F.R.S., and near Nobel-prize quality, and so forth," ordinary "Nobel laureates," and finally the Nobel laureates of the highest class, such as "Dirac, Schrödinger, Heisenberg, and so forth"—scientific geniuses superceded by Einstein and Newton (Vijh, 1987, p. 11). This upper end clearly merges with historic levels of scientific eminence (e.g., Murray, 2003).

Because the above two scales clearly overlap, it should be possible to interweave them to create a single coherent dimension. This latent factor informs the remainder of this chapter: Scientific genius marks the upper limit of a creativity scale anchored in the complete absence of any special capacity for science.

Intellect

Because genius is often associated with extraordinary intelligence, it would seem probable that scientific genius would be very bright. That seems to be the case. In Cox's (1926) historiometric IQ estimates, the eminent scientists averaged scores between 135 and 152 (or between 155 and 175 if corrected for error). Using psychometric assessments, Roe (1953) found comparable verbal, mathematical, and spatial IQs in her sample of 64 eminent scientists (see also Cattell, 1963). Of course, even elite scientists will display some distribution around a mean so that those at the lower end of the distribution will not necessarily have genius-level IQs. Even so, the scores do not dip lower than an IQ around 120, the average for a U.S. college graduate at the time of the study (Roe, 1953). Just as important, variation in historiometric IQ correlates positively with achieved eminence (Cox, 1926; Simonton, 1991d; Simonton & Song, 2009; Walberg, Rasher, & Hase, 1978). Furthermore, if measured properly, the correspondence between intelligence and scientific achievement can appear even at the upper ends of the distribution, such as the top 1% (Lubinski, Webb, Morelock, & Benbow, 2001; Wai, Lubinski, & Benbow, 2005).

It is not only that eminent scientists are highly intelligent in general, but also that they tend to think in a complex manner about their works. This generalization stems from research on integrative complexity, a content analytical measure that can be applied to verbal material (Suedfeld, Tetlock, & Streufert, 1992). High scores on integrative complexity reflect the differentiation of separate points of view and the integration of those perspectives. When Suedfeld (1985) content analyzed the addresses of presidents of the American Psychological Association, he discovered that those scoring higher in integrative complexity also scored higher in eminence, as assessed by multiple criteria. Similarly, Feist's (1994) study of physicists, chemists, and biologists found that those scoring highest in integrative complexity when talking about their research also tended to receive higher peer ratings in eminence and higher rates of citation—yet high integrative complexity when talking about teaching was negatively associated with citations!

Personality

Feist (1998) conducted a comprehensive meta-analysis of the personality correlates of scientific (and artistic) creativity, including the variables that distinguish highly creative scientists from less creative scientists. Although most of the latter contrasts did not necessarily reach into scientific geniuses, the results are still useful: "creative scientists are more aesthetically oriented, ambitious, confident, deviant, dominant, expressive, flexible, intelligent, and open to new experiences than their less creative peers" (pp. 297, 298). Focusing just on the findings of research using the California Psychological Inventory, the more creative scientist tends to be "tolerant and open-minded, self-accepting, outgoing, confident, ambitious, persistent, and ... a good judge of character" (p. 298). Chiefly, scientific creativity seems associated with some rather positive personality traits.

However, other research results suggest that the scientific genius may not be quite so uniformly pleasant. Feist (1993) found that peer ratings of creativity and historical significance—which correlated with membership in the National Academy of Sciences—were linked directly or indirectly (via productivity and citations) with a hostile personality and an arrogant working style. Rushton (1990) obtained positive correlations between creativity measures and assessments on Psychoticism, one of the scales of the Eysenck Personality Questionnaire (EPQ) that describes an individual who is aggressive, cold, egocentric, impersonal, impulsive, antisocial, unempathetic, and tough-minded (Eysenck, 1995). R. B. Cattell, by applying the 16 Personality Factors Questionnaire (PFQ) to eminent scientists both historical and contemporary (Cattell, 1963; Cattell & Drevdahl, 1955), showed that they tend to score high in schizothymia (i.e., withdrawn, skeptical, internally preoccupied, precise, and critical) and desurgency (introspectiveness, restraint, brooding, and solemnity of manner). Finally, some studies indicate that scientific productivity and impact are positively correlated with a Type A "workaholic" personality (Helmreich, Spence, & Pred, 1988; Matthews, Helmreich, Beane, & Lucker, 1980; Taylor, Locke, Lee, & Gist, 1984; see also Roe, 1953). None of these studies suggests that the scientific genius would be a good spouse, friend, or even colleague.

These negative traits bring us to the controversial topic of the mad genius, or, more particularly, the mad scientific genius. Given that scientific creativity depends so heavily on rationality and objectivity, as constrained by fact and logic, one would expect great scientists who exhibit lower levels of psychopathology than would be the case for great artists. That expectation is confirmed in historiometric research (e.g., Ludwig, 1995; Post, 1994; Simonton & Song, 2009). Indeed, natural scientists display lower levels than do social scientists (Ludwig, 1995). Because most natural scientists create within paradigmatic disciplines (Kuhn, 1970), their creativity operates under greater constraints (Simonton, 2009). Nonetheless, although scientists who are motivated to preserve the paradigm achieve higher eminence if they exhibit less psychopathology, those revolutionary scientists who reject the paradigm attain higher eminence if they display more psychopathology (Ko & Kim, 2008). Because revolutionary scientists may enjoy a higher likelihood of being viewed as scientific geniuses (see Sulloway, 2009), the notion of the "mad scientist" has a grain of truth. But it is seldom strictly true.

PERSONAL DEVELOPMENT

Where do scientific geniuses originate? Are there any factors in their early biographies that would facilitate the identification and promotion of exceptional scientific talent? How does their creativity progress over the course of a lifetime? Below we will review two sets of developmental variables, the first concerning family background and the second, education and training. That review will be followed by a treatment of the adulthood career of the creative genius in science.

Family Background

Parents provide two kinds of developmental inputs to their offspring: genetic and environmental. Let us take each in turn.

Genetic Influences

Galton (1869) was the first to conduct a scientific study of whether genius was born or made. He did so by compiling extensive pedigrees for top creators and leaders in the hope of showing that exceptional achievement runs in family lineages (see also Bramwell, 1948). A whole chapter was devoted to making this case for scientific genius being the upshot of extraordinary endowment (see also Brimhall, 1922, 1923a, 1923b). These pedigrees have a curious connection with the psychopathology question raised earlier. Because mental illness has a strong genetic basis, psychopathology would also run in families, leading to the expectation that both genius and madness might be found in the same lineages. That is, the same families that produce a high proportion of mentally ill should also yield a higher rate of genius. There is some evidence that this is the case (Karlson, 1970). For instance, about one quarter of eminent mathematicians came from families whose members exhibit significant levels of psychopathology.

Recent researchers have often tried to minimize the role of genetic endowment, arguing that nurture is the primary agent (e.g., Howe, 1999). However, these arguments have gone too far. There are sound reasons for concluding that scientific genius has a partial genetic foundation (Simonton, 2008b). For example, all cognitive abilities and personality traits associated with scientific achievement exhibit substantial heritability coefficients. Nonetheless, to the extent that scientific talent is governed by emergenic inheritance, the genetic contribution can operate in a far more complex and elusive manner than usually assumed (Lykken, 1998a; Simonton, 1999). Among other implications, emergenesis implies that genius-level scientific talent would be much less common than would be expected from the population frequencies of the component genetic traits.

Environmental influences

The previous section notwithstanding, most researchers would agree that family contributions to the development of scientific genius are largely environmental rather than genetic. These environmental influences may be either nonshared or shared.

Nonshared environment—This set of influences entails those attributes of the family that are unique to each sibling. The nonshared environmental developmental factor that has received by far the most attention is birth order. Clearly, the first-born child does not grow up in the same home as a later or last-born. It was Galton (1874) who first indicated that first borns had a higher likelihood of achieving eminence in science, a finding replicated by many others (e.g., Eiduson, 1962; Helmreich, Spence, Beane, Lucker, & Matthews, 1980; Helson & Crutchfield, 1970; Roe, 1953; Terry 1989). Although the original research focused on male scientists, it is evident that this first-born effect is even stronger for female scientists (Simonton, 2008a). It is of interest that the same first-born overrepresentation appears in classical composers (Schubert, Wagner, & Schubert, 1977). Classical music features the kind of formal and logical constraints that render the domain more comparable to mathematical sciences. Nevertheless, first borns do not enjoy a universal advantage. Later borns are not only more prominent among creative writers (Bliss, 1970), but they are also more apparent among those scientists willing to accept progressive scientific revolutions (Sulloway, 1996). Indeed, some of the latter will introduce revolutionary ideas on their own (Sulloway, 2009). Charles Darwin was a latter born, as was Galton.

Shared environment—This set of influences include aspects of the family that a child and adolescent shares with his or her siblings. Research has shown that highly eminent scientists are more likely to come from families with specific socioeconomic, religious, and geographic characteristics (Berry, 1981; Candolle, 1873; Chambers, 1964; Helson & Crutchfield, 1970; Roe, 1953; Wispé, 1965; Feist, 1993). These characteristics include parents who value education and who subscribe to nondogmatic religious beliefs. Status as a first- or second-generation immigrant also appears important (e.g., Feist, 2006a; Levin & Stephan, 1999), a point also noted by Galton (1869). Although some researchers have suggested that early traumatic experiences—such as parental loss—might have a positive developmental impact (Eiduson, 1962; Roe, 1953; Silverman, 1974), others have questioned this factor (Woodward, 1974). Judging from the backgrounds of Nobel laureates, the impact may vary according to the domain of creative achievement (Berry, 1981). Physicists tend to come from much more stable and conventional homes than do other laureates, including chemists, whereas laureates in literature seem to hail from highly unstable and unconventional homes (see also Roe, 1953; Simonton, 1986). Nonetheless, some data suggest that revolutionary scientists may be more likely to grow up in the latter kind of family (Sulloway, 1996). This departure parallels what we observed before with respect to psychopathology.

Before leaving the home environment, it must be emphasized that shared environmental influences have a somewhat more ambivalent status than do nonshared influences. In particular, the shared influences can sometimes be conflated with underlying genetic effects (Scarr & McCartney, 1983). As a case in point, highly intelligent parents not only create home environments that will be intellectually stimulating, but they also contribute a genetic foundation for high intelligence in their children. So far, the research on scientific genius has not been able to tease out which of these pathways assume primary importance.

Education and Training

Galton (1874) was also the first investigator to examine the impact of education and training on exceptional scientific achievement. He noted that Fellows of the Royal Society tended to believe that their educational experiences made a positive development to their creative development. Subsequent historiometric work has shown that highly eminent scientists, compared to highly eminent artists, benefit more from formal education (Raskin, 1936; Simonton, 1983, 1986; cf. Hudson, 1958). That is, the scientific genius is more likely to display higher scholastic performance and attain a superior advanced education (Raskin, 1936; Rodgers & Maranto, 1989; Taylor & Ellison, 1967; Wispé, 1965). Furthermore, eminent scientists are much more likely than either eminent artists or noneminent scientists to have obtained their highest degree from a prestigious institution (Crane, 1965; Moulin, 1955; Wispé & Ritter, 1964; Zuckerman, 1977).

Much professional training takes place outside the lecture hall and seminar room. In particular, the development of creative scientific genius is positively affected by distinguished role models and mentors (Boring & Boring, 1948; Crane, 1965; Simonton, 1992a, 1992b; Wispé, 1965; Zuckerman, 1977). Indeed, Nobel laureates in the sciences have a strong likelihood of having studied under other recipients of the same honor (and not necessarily only after the master received that honor; Zuckerman, 1977). In this regard, the development of scientific genius parallels that of artistic genius, for the latter's development is also nurtured by early exposure to eminent role models and mentors (Simonton, 1984a). This developmental influence is probably important in all areas of incomparable achievement (Simonton, 1984a; Walberg, Rasher, & Parkerson, 1980).

Presumably, the purpose of both formal and informal training is to enable the young scientific talent to acquire the domain-specific expertise necessary for making world-class contributions to the field. This developmental necessity has often been expressed as the "10-year rule," meaning that a person must devote about a decade to extensive learning, training, and practice before he or she becomes a master rather than an apprentice (Ericsson, 1996). However, empirical investigations indicate that this principle is oversimplified (Simonton, 2000a). For example, although one study found that the 10-year rule held for classical composers as a rough statistical average (Hayes, 1989). Not only do composers vary immensely in the actual length of the apprentice period, but also the greatest composers in terms of both eminence and productivity take less time to acquire domain mastery than do lesser composers (Simonton, 1991b). This same result probably holds for great scientists as well, for not only do they tend to get their highest degrees at younger ages, but also they tend to start publishing at earlier ages than do lesser scientists (Feist, 2006a; Helson & Crutchfield, 1970; Poffenberger, 1930; Raskin, 1936; Simonton, 1991a, 1992a; Wispé, 1965). Thus, somehow, geniuses seem able to truncate the laborious process of expertise acquisition. An even more drastic truncation occurs when major breakthroughs come from those scientists whose training was in a different domain than the one in which they eventually made their mark (Hudson & Jacot, 1986; Jeppesen & Lakhani, 2010; Simonton, 1984d; cf. Gieryn & Hirsh, 1983). Yet, these events tend to be relatively rare, and seldom become the basis of a lifetime of creative discovery.

One last peculiarity is noteworthy: Creative geniuses are not inclined to narrow their interests and activities to an extremely specialized domain of achievement. On the contrary, such creators will often display voracious and omnivorous reading as well as an impressive versatility both avocational and vocational (Cassandro, 1998; Cassandro & Simonton, 2010; Raskin, 1936; Simonton, 1976, 1984c; White, 1931). The same breadth applies to outstanding creators in the sciences. For instance, distinguished scientists are prone to read widely (Blackburn, Behymer, & Hall, 1978; Smith, Albright, Glennon, & Owens, 1951; Van Zelst & Kerr, 1951), including wide-ranging reading in domains outside their main research specialty (Dennis & Girden, 1954; Manis, 1951; Simon, 1974). The same breadth can be found in the hobbies and other avocational activities engaged by outstanding scientists (e.g. Root-Bernstein et al., 2008; Root-Bernstein, Bernstein, & Garnier, 1995). For example, Nobel laureates in the sciences are more likely to have active artistic interests or hobbies than do Fellows of the Royal Society or members of the National Academy of Sciences, and the latter are more likely to have such avocations than do members of Sigma Xi or the U.S. public at large (Root-Bernstein et al., 2008). All told, scientific geniuses tend to be less engaged in disciplinary specialization than their less illustrious colleagues. This proclivity does not seem compatible with the idea that creative genius in science exclusively depends on domain-specific expertise.

To be sure, this openness and versatility might not be directly reflected in the work produced. Indeed, some empirical evidence implies that scientific eminence is positively correlated with research depth, that is, a focus on a small range of research topics (Feist, 1997; Simonton, 1992a; cf. Root-Bernstein, Bernstein, & Garnier, 1993). So, perhaps, the omnivorous reading and diverse hobbies influence creativity in a more indirect fashion, such as helping the scientist "think outside the box." Still, there is also some data showing that scientists who make contributions to more than a single scientific domain tend to become more famous than those who contribute to solely one (Sulloway, 1996). At this point, we can only hope that future research will tease out the complexities of these relationships. What we can certainly conclude is that scientific genius cannot be reduced to the acquisition of domain-specific expertise. They are not supreme experts but rather superlative creators.

Career Trajectories

Because scientists communicate their ideas largely through journal articles, it becomes natural to ask how these articles are distributed over the course of their careers. At what age does productivity begin, when does it peak, and when does output cease? It turns out that the relation between age and creative output is the oldest research topic in the behavioral sciences. The first empirical study of the subject was published in 1835 (Quételet, 1835/1968). Although this particular investigation was devoted to creative productivity in drama, the basic results have replicated multiple times in studies of scientific creators (e.g., Cole, 1979; Davis, 1987; Dennis, 1956, 1966; Horner, Rushton, & Vernon, 1986; Lehman, 1960; Simonton, 1985; Zuckerman, 1977). To be specific, productivity tends to begin about the early or mid-20s and then increase fairly rapidly to a peak somewhere in the late 30s or early 40s, and thereafter a gradual decline sets in, albeit the output rate in the 60s remains higher than at the beginning of the career. Hence, the overall curve is decidedly skewed, the career peak appearing about 10 to 20 years after the career onset. Although this longitudinal curve has been well established, the generalization must also be qualified various ways. These qualifications entail differential impact, interdisciplinary contrasts, and individual differences.

Differential Impact

As already mentioned, many publications are never cited, and even those that receive citations vary immensely in citation count, such that only a very small elite can be considered truly high-impact. Accordingly, it is useful to inquire whether the agewise distribution differs for high- versus low-impact publications. The answer is fairly easy: For the most part, the longitudinal trajectory of high-impact contributions closely tracks that for low-impact contributions (Lehman, 1953; Levin & Stephan, 1991; Simonton, 1988a, 1997). Most notably, the scientist's single most influential work will most often appear at roughly the same age at which he or she attains the peak output rate. Moreover, the scientist's first high-impact publication will appear shortly after his or her first low-impact publication and the last high-impact publication will appear somewhat before his or her last low-impact publication. Of course, because these statements only represent statistical averages, exceptions do take place, lessening the correspondence between the two age curves. Some scientists might have their most influential work occur earlier than expected, while others might have their most influential work occur later than expected.

Interdisciplinary Contrasts

A second qualification concerns the discipline to which a scientist contributes. Whether we look at total output or just high-impact output, different disciplines display contrasting career trajectories (Dennis, 1966; Lehman, 1953; Levin & Stephan, 1989, 1991). In highly abstract disciplines, like pure mathematics, the peak will appear somewhat earlier, whereas in more concrete disciplines, such as the earth sciences, the peak will occur noticeably later. These interdisciplinary contrasts hold not only for total output but also for the production of high-impact work (Simonton, 1991a). Consequently, a pure mathematician's most influential work will tend to come at a younger age than that of an earth scientist. In addition, the mathematician's first and last high-impact publications will come earlier, in comparison to those of the earth scientist. These contrasts have been given a theoretical explanation in terms of a mathematical model that considers the rates at which new ideas can be generated and elaborated (Simonton, 1984b, 1997). These rates tend to be faster in those disciplines in which the concepts are highly abstract and well defined (Simonton, 1989). This same principle likely distinguishes other scientific domains, such as theoretical physics versus evolutionary biology.

Individual Differences

I noted earlier that scientists vary immensely in total lifetime output with scientific genius most strongly associated with prolific productivity. This variation has powerful repercussions for the career trajectory (Simonton, 1997, 2004). At one extreme are those scientists whose productivity is so low that we cannot properly speak of a career trajectory. Instead, their minimal output is more or less distributed randomly across their career (Huber, 1998a, 1998b, 1999, 2000, 2001; Huber & Wagner-Döbler, 2001a, 2001b). At the other extreme, where we find scientific genius, are the scientists who begin their output at unusually young ages, produce at high annual rates, and end their output very late in life (Davis, 1987; Dennis, 1954b; Feist, 2006a; Simonton, 1992a,

1997; see also Chambers, 1964; Christensen & Jacomb, 1992; Clemente, 1973). These differences hold for both low- and high-impact works with only one curious exception: The scientist's single most influential work will probably appear at the same age, without respect to the level of total output. In this sense, all scientists working within the same discipline are expected to peak at the same age, regardless of whether they can be considered creative geniuses (Simonton, 1991a, 1997; but see Horner, Rushton, & Vernon, 1986). Even so, in the case of scientific geniuses, the first high-impact contribution will come very early and the last high-impact contribution will come very late. The former fact may be partly responsible for the impression that scientific breakthroughs are more likely to come from very young scientists (e.g., Kuhn, 1970). The creative genius in science will have his or her first major work at a young age, but the biggest breakthroughs are more likely to appear mid-career (Wray, 2003, 2004; cf. Dietrich & Srinivasan, 2007).

The above results are neither necessarily stable over time nor would we expect them to be (Wray, 2009; Zhao, 1984; Zhao & Jiang, 1986). Perhaps, the most provocative change has to do with the increasingly collaborative nature of scientific creativity (Over, 1982a; Smart & Bayer, 1986). The collaborations are seen not only in the research laboratories with their senior investigators, postdocs, graduate students, undergraduates, and technicians, but also in the ever increasing number of coauthors listed on a typical title page. One potential consequence of this shift could be the tendency for the age decrement in creative productivity to become far less pronounced for recent scientists (Feist, 2006a; Gingras, Lariviere, Macaluso, & Robitaille, 2008; Kyvik & Olsen, 2008; Stroebe, 2010). Creativity can be stimulated by the constant influx of new ideas from both younger collaborators and collaborators from different specialty areas or even distinct disciplines.

ADDITIONAL TOPICS

As I can personally testify, a whole book can be written on the psychology of scientific genius, something I have already done three times in my career (Simonton, 1988c, 2002, 2004). For that reason, this chapter has only scratched the surface of a very complex and significant phenomenon. Hence, I would like to give some indication of what has been omitted from the current treatment. Of the many omissions, perhaps the following three stand out:

Cognition

What psychological processes are involved in genius-level scientific creativity? How do these processes differ from those used either by artistic geniuses or by less illustrious scientific creators? Although some headway has been made in this regard, we are still far from a complete answer (see, e.g., Feist, 1994; Klahr & Simon, 1999; Roe, 1953; Rothenberg, 1983, 1996; Shavinina, 2003; Simonton, 2012). Although there have been attempts to write computer programs that simulate the discovery process of scientific geniuses, these have so far failed to live up to the expectations (e.g., Langley, Simon, Bradshaw, & Zythow, 1987).

Gender

Scientific geniuses are disproportionately male, exceptions like Marie Curie being truly exceptional. Although this gender gap has diminished over the years, it remains substantial in the mathematical and natural sciences. Fortunately, contemporary studies are providing some insights into the differential factors and developmental pathways responsible for this gender effect (e.g., Feist, 2006a; Lubinski, Benbow, Shea, Eftekhari-Sanjani, & Halvorson, 2001). Even so, we still have a long way to go before we have any definitive answer, if such an answer is even possible.

Zeitgeist

Scientific genius, like artistic genius, is not uniformly distributed across time and space (Schneider, 1937; Simonton, 1988b; Sorokin & Merton, 1935; Yuasa, 1974). On the contrary, major figures tend to be clustered in certain nations or civilizations in specific periods of their history. These "Golden" ages, punctuated by "Silver" and even "Dark" ages, imply that creative genius is highly contingent on the sociocultural milieu or zeitgeist (Candolle, 1873; Kroeber, 1944). This larger context can be either internal or external with respect to a given scientific discipline (Simonton, 2002). The internal zeitgeist concerns the current state of the domain (the prevalent paradigm, the number of active scientists, etc.), whereas the external zeitgeist involves contemporaneous events and conditions that impinge of scientists from outside the domain proper (peace and prosperity, ideology, political freedom, etc.). Although some psychologists might argue that this subject belongs more to the sociology of science, the topic can also be viewed as part of any comprehensive social psychology of science. Certainly, the sociocultural circumstances impose constraints and provide opportunities for the operation of individual-difference and developmental variables.

SUMMARY AND CONCLUSION

This chapter reviewed the scientific research on creative genius in science. After noting that scientific geniuses obtain that status by producing contributions that are novel, surprising, and exemplary, I then turned to the major approaches to measuring the phenomenon. There are four such assessment criteria: historiometric eminence, expert evaluations, journal citations, and scientific awards. Ideally, these standards can be integrated into a single composite scale that provides a reliable and valid indicator that covers the full range of scientific ability.

Having thus defined scientific genius, the next task was to review what we have learned about such notable figures. This review began with a treatment of individual differences, with a special focus on cross-sectional variation in productivity, intellect, and personality. Creative geniuses in science are not only highly productive and intelligent, but also display a distinctive dispositional profile. The review continued with a discussion of personal development including family background (including both genetic and environmental factors), education and training, and career trajectories (as moderated by differential impact, interdisciplinary contrasts, and individual differences). Finally, for the sake of completeness, I briefly touched upon three additional topics, namely, cognitive processes, gender differences, and the internal and external zeitgeist.

The last listed factor deserves special emphasis, for the impact of the zeitgeist might even affect psychological research on creative genius in science. It cannot simply be assumed that today's best scientists match the caliber of an Albert Einstein or Isaac Newton. Perhaps the most creative scientists active in the early 21st century are lower down in the Landau–Lotka scale discussed earlier in this chapter. If so, then psychologists who wish to study creative scientists of the highest caliber will have no other option but to use historiometric or other retrospective methods. Psychometric and experimental methods would then be restricted to sub-genius levels of scientific creativity. Admittedly, at present it cannot be decided whether we are still in the midst of a Golden Age—as represented by relativity and quantum theories, Neo-Darwinism, molecular biology, and so forth—or have already slipped into a Silver Age (or worse). That decision must await a future occasion when a later generation obtains sufficient perspective on present contributions to see whether scientific genius has become a phenomenon of the past, and thus, in a sense, obsolete. Maybe all scientists, even the greatest, now fit Kant's conception that scientific genius is an oxymoron.

REFERENCES

Annin, E. L., Boring, E. G., & Watson, R. I. (1968). Important psychologists, 1600–1967. *Journal of the History of the Behavioral Sciences, 4*, 303–315.

Ashton, S. V., & Oppenheim, C. (1978). A method of predicting Nobel prizewinners in chemistry. *Social Studies of Science, 8*, 341–348.

Berry, C. (1981). The Nobel scientists and the origins of scientific achievement. *British Journal of Sociology, 32*, 381–391.

Blackburn, R. T., Behymer, C. E., & Hall, D. E. (1978). Correlates of faculty publications. *Sociology of Education, 51*, 132–141.

Bliss, W. D. (1970). Birth order of creative writers. *Journal of Individual Psychology, 26*, 200–202.

Bloom, B. S. (1963). Report on creativity research by the examiner's office of the University of Chicago. In C. W. Taylor & F. X. Barron (Eds.), *Scientific creativity: Its recognition and development* (pp. 251–264). New York, NY: Wiley.

Boring, M. D., & Boring, E. G. (1948). Masters and pupils among the American psychologists. *American Journal of Psychology, 61*, 527–534.

Bramwell, B. S. (1948). Galton's "Hereditary Genius" and the three following generations since 1869. *Eugenics Review, 39*, 146–153.

Brimhall, D. R. (1922). Family resemblances among American Men of Science. *American Naturalist, 56,* 504–547.

Brimhall, D. R. (1923a). Family resemblances among American Men of Science. II. Degree of resemblance in comparison with the generality: Proportion of workers in each science and distribution of replies. *American Naturalist, 57,* 74–88.

Brimhall, D. R. (1923b). Family resemblances among American Men of Science. III. The influence of the nearness of kinship. *American Naturalist, 57,* 137–152.

Candolle, A. de. (1873). *Histoire des sciences et des savants depuis deux siècles.* Geneve: Georg.

Carson, S., Peterson, J. B., & Higgins, D. M. (2005). Reliability, validity, and factor structure of the Creative Achievement Questionnaire. *Creativity Research Journal, 17,* 37–50.

Cassandro, V. J. (1998). Explaining premature mortality across fields of creative endeavor. *Journal of Personality, 66,* 805–833.

Cassandro, V. J., & Simonton, D. K. (2010). Versatility, openness to experience, and topical diversity in creative products: An exploratory historiometric analysis of scientists, philosophers, and writers. *Journal of Creative Behavior, 44,* 1–18.

Cattell, J. M. (1903). A statistical study of eminent men. *Popular Science Monthly, 62,* 359–377.

Cattell, J. M. (1906). *American men of science: A biographical directory.* New York, NY: Science Press.

Cattell, R. B. (1963). The personality and motivation of the researcher from measurements of contemporaries and from biography. In C. W. Taylor & F. Barron (Eds.), *Scientific creativity: Its recognition and development* (pp. 119–131). New York, NY: Wiley.

Cattell, R. B., & Drevdahl, J. E. (1955). A comparison of the personality profile (16 P. F.) of eminent researchers with that of eminent teachers and administrators, and of the general population. *British Journal of Psychology, 46,* 248–261.

Chambers, J. A. (1964). Relating personality and biographical factors to scientific creativity. *Psychological Monographs: General and Applied, 78*(7, Whole No. 584).

Christensen, H., & Jacomb, P. A. (1992). The lifetime productivity of eminent Australian academics. *International Journal of Geriatric Psychiatry, 7,* 681–686.

Clark, R. D., & Rice, G. A. (1982). Family constellations and eminence: The birth orders of Nobel Prize winners. *Journal of Psychology, 110,* 281–287.

Clemente, F. (1973). Early career determinants of research productivity. *American Journal of Sociology, 79,* 409–419.

Cole, S. (1979). Age and scientific performance. *American Journal of Sociology, 84,* 958–977.

Cole, S., & Cole, J. R. (1973). *Social stratification in science.* Chicago, IL: University of Chicago Press.

Cox, C. (1926). *The early mental traits of three hundred geniuses.* Stanford, CA: Stanford University Press.

Crane, D. (1965). Scientists at major and minor universities: A study of productivity and recognition. *American Sociological Review, 30,* 699–714.

Davis, R. A. (1987). Creativity in neurological publications. *Neurosurgery, 20,* 652–663.

Dennis, W. (1954a, September). Bibliographies of eminent scientists. *Scientific Monthly, 79,* 180–183.

Dennis, W. (1954b). Predicting scientific productivity in later maturity from records of earlier decades. *Journal of Gerontology, 9,* 465–467.

Dennis, W. (1954c). Productivity among American psychologists. *American Psychologist, 9,* 191–194.

Dennis, W. (1955, April). Variations in productivity among creative workers. *Scientific Monthly, 80,* 277–278.

Dennis, W. (1956, July 5). Age and productivity among scientists. *Science, 123,* 724–725.

Dennis, W. (1966). Creative productivity between the ages of 20 and 80 years. *Journal of Gerontology, 21,* 1–8.

Dennis, W., & Girden, E. (1954). Current scientific activities of psychologists as a function of age. *Journal of Gerontology, 9,* 175–178.

Dietrich, A., & Srinivasan, N. (2007). The optimal age to start a revolution. *Journal of Creative Behavior, 41,* 54–74.

Eiduson, B. T. (1962). *Scientists: Their psychological world.* New York, NY: Basic Books.

Ericsson, K. A. (1996). The acquisition of expert performance: An introduction to some of the issues. In K. A. Ericsson (Ed.), *The road to expert performance: Empirical evidence from the arts and sciences, sports, and games* (pp. 1–50). Mahwah, NJ: Erlbaum.

Eysenck, H. J. (1995). *Genius: The natural history of creativity.* Cambridge, England: Cambridge University Press.

Feist, G. J. (1993). A structural model of scientific eminence. *Psychological Science, 4,* 366–371.

Feist, G. J. (1994). Personality and working style predictors of integrative complexity: A study of scientists' thinking about research and teaching. *Journal of Personality and Social Psychology, 67,* 474–484.

Feist, G. J. (1997). Quantity, quality, and depth of research as influences on scientific eminence: Is quantity most important? *Creativity Research Journal, 10,* 325–335.

Feist, G. J. (1998). A meta-analysis of personality in scientific and artistic creativity. *Personality and Social Psychology Review, 2,* 290–309.

Feist, G. J. (2006a). The development of scientific talent in Westinghouse Finalists and members of the National Academy of Sciences. *Journal of Adult Development, 13,* 23–35.

Feist, G. J. (2006b). How development and personality influence scientific thought, interest, and achievement. *Review of General Psychology, 10,* 163–182.

Galton, F. (1869). *Hereditary genius: An inquiry into its laws and consequences.* London, England: Macmillan.

Galton, F. (1874). *English men of science: Their nature and nurture.* London, England: Macmillan.

Gieryn, T. F., & Hirsh, R. F. (1983). Marginality and innovation in science. *Social Studies of Science, 13,* 87–106.

Gingras, Y., Lariviere, V., Macaluso, B., & Robitaille, J. P. (2008). The effect of aging on researchers' publication and citation patterns. *PLoS One, 3,* 1–8.

Ginsburgh, V., & Weyers, S. A. (2006). Persistence and fashion in art: Italian renaissance from Vasari to Berenson and beyond. *Poetics, 34,* 24–44.

Ginsburgh, V., & Weyers, S. A. (2010). On the formation of canons: The dynamics of narratives in art history. *Empirical Studies of the Arts, 26,* 37–72.

Hayes, J. R. (1989). *The complete problem solver* (2nd ed.). Hillsdale, NJ: Erlbaum.

Helmreich, R. L., Spence, J. T., Beane, W. E., Lucker, G. W., & Matthews, K. A. (1980). Making it in academic psychology: Demographic and personality correlates of attainment. *Journal of Personality and Social Psychology, 39,* 896–908.

Helmreich, R. L., Spence, J. T., & Pred, R. S. (1988). Making it without losing it: Type A, achievement motivation, and scientific attainment revisited. *Personality and Social Psychology Bulletin, 14,* 495–504.

Helson, R., & Crutchfield, R. S. (1970). Mathematicians: The creative researcher and the average Ph.D. *Journal of Consulting and Clinical Psychology, 34,* 250–257.

Horner, K. L., Rushton, J. P., & Vernon, P. A. (1986). Relation between aging and research productivity of academic psychologists. *Psychology and Aging, 1,* 319–324.

Howe, M. J. A. (1999). *Genius explained.* Cambridge, England: Cambridge University Press.

Huber, J. C. (1998a). Invention and inventivity as a special kind of creativity, with implications for general creativity. *Journal of Creative Behavior, 32,* 58–72.

Huber, J. C. (1998b). Invention and inventivity is a random, Poisson process: A potential guide to analysis of general creativity. *Creativity Research Journal, 11,* 231–241.

Huber, J. C. (1999). Inventive productivity and the statistics of exceedances. *Scientometrics, 45,* 33–53.

Huber, J. C. (2000). A statistical analysis of special cases of creativity. *Journal of Creative Behavior, 34,* 203–225.

Huber, J. C. (2001). A new method for analyzing scientific productivity. *Journal of the American Society for Information Science and Technology, 52,* 1089–1099.

Huber, J. C., & Wagner-Döbler, R. (2001a). Scientific production: A statistical analysis of authors in mathematical logic. *Scientometrics, 50,* 323–337.

Huber, J. C., & Wagner-Döbler, R. (2001b). Scientific production: A statistical analysis of authors in physics, 1800–1900. *Scientometrics, 50,* 437–453.

Hudson, L. (1958). Undergraduate academic record of Fellows of the Royal Society. *Nature, 182,* 1326.

Hudson, L., & Jacot, B. (1986). The outsider in science. In C. Bagley & G. K. Verma (Eds.), *Personality, cognition and values* (pp. 3–23). London, England: Macmillan.

Jeppesen, L. B., & Lakhani, K. R. (2010). Marginality and problem-solving effectiveness in broadcast search. *Organization Science, 21,* 1016–1033.

Kant, I. (1952). The critique of judgement. In R. M. Hutchins (Ed.), *Great books of the Western world* (Vol. 42, pp. 459–613). Chicago, IL: Encyclopaedia Britannica. (Original work published 1790).

Karlson, J. I. (1970). Genetic association of giftedness and creativity with schizophrenia. *Hereditas, 66,* 177–182.

Klahr, D., & Simon, H. A. (1999). Studies of scientific creativity: Complementary approaches and convergent findings. *Psychological Bulletin, 125,* 524–543.

Ko, Y., & Kim, J. (2008). Scientific geniuses' psychopathology as a moderator in the relation between creative contribution types and eminence. *Creativity Research Journal, 20,* 251–261.

Kroeber, A. L. (1944). *Configurations of culture growth.* Berkeley, CA: University of California Press.

Kuhn, T. S. (1970). *The structure of scientific revolutions* (2nd ed.). Chicago, IL: University of Chicago Press.

Kyvik, S., & Olsen, T. B. (2008). Does the aging of tenured academic staff affect the research performance of universities? *Scientometrics, 76,* 439–455.

Langley, P., Simon, H. A., Bradshaw, G. L., & Zythow, J. M. (1987). *Scientific discovery.* Cambridge, MA: MIT Press.

Lehman, H. C. (1953). *Age and achievement.* Princeton, NJ: Princeton University Press.

Lehman, H. C. (1960). The age decrement in outstanding scientific creativity. *American Psychologist, 15,* 128–134.

Levin, S. G., & Stephan, P. E. (1989). Age and research productivity of academic scientists. *Research in Higher Education, 30,* 531–549.

Levin, S. G., & Stephan, P. E. (1991). Research productivity over the life cycle: Evidence for academic scientists. *American Economic Review, 81,* 114–132.

Levin, S. G., & Stephan, P. E. (1999, August 20). Are the foreign born a source of strength for U.S. science? *Science, 285,* 1213–1214.

Lotka, A. J. (1926). The frequency distribution of scientific productivity. *Journal of the Washington Academy of Sciences, 16,* 317–323.

Lubinski, D., Benbow, C. P., Shea, D. L., Eftekhari-Sanjani, H., & Halvorson, M. B. J. (2001). Men and women at promise for scientific excellence: Similarity not dissimilarity. *Psychological Science, 12,* 309–317.

Lubinski, D., Webb, R. M., Morelock, M. J., & Benbow, C. P. (2001). Top 1 in 10,000: A 10-year follow-up of the profoundly gifted. *Journal of Applied Psychology, 86,* 718–729.

Ludwig, A. M. (1995). *The price of greatness: Resolving the creativity and madness controversy.* New York, NY: Guilford Press.

Lykken, D. T. (1998). The genetics of genius. In A. Steptoe (Ed.), *Genius and the mind: Studies of creativity and temperament in the historical record* (pp. 15–37). New York, NY: Oxford University Press.

Manis, J. G. (1951). Some academic influences upon publication productivity. *Social Forces, 29,* 267–272.

Manniche, E., & Falk, G. (1957). Age and the Nobel prize. *Behavioral Science, 2,* 301–307.

Matthews, K. A., Helmreich, R. L., Beane, W. E., & Lucker, G. W. (1980). Pattern A, achievement striving, and scientific merit: Does Pattern A help or hinder? *Journal of Personality and Social Psychology, 39,* 962–967.

McFarlan, D. (Ed.). (1989). *Guinness book of world records.* New York, NY: Bantum.

Moulin, L. (1955). The Nobel Prizes for the sciences from 1901–1950: An essay in sociological analysis. *British Journal of Sociology, 6*, 246–263.

Murray, C. (2003). *Human accomplishment: The pursuit of excellence in the arts and sciences, 800 B.C. to 1950.* New York, NY: HarperCollins.

Myers, C. R. (1970). Journal citations and scientific eminence in contemporary psychology. *American Psychologist, 25*, 1041–1048.

Over, R. (1982a). Collaborative research and publication in psychology. *American Psychologist, 37*, 996–1001.

Over, R. (1982b). The durability of scientific reputation. *Journal of the History of the Behavioral Sciences, 18*, 53–61.

Poffenberger, A. T. (1930). The development of men of science. *Journal of Social Psychology, 1*, 31–47.

Post, F. (1994). Creativity and psychopathology: A study of 291 world-famous men. *British Journal of Psychiatry, 165*, 22–34.

Price, D. (1963). *Little science, big science.* New York, NY: Columbia University Press.

Quételet, A. (1968). *A treatise on man and the development of his faculties.* New York, NY: Franklin. (Reprint of 1842 Edinburgh translation of 1835 French original).

Raskin, E. A. (1936). Comparison of scientific and literary ability: A biographical study of eminent scientists and men of letters of the nineteenth century. *Journal of Abnormal and Social Psychology, 31*, 20–35.

Redner, S. (1998). How popular is your paper? An empirical study of the citation distribution. *European Physical Journal B, 4*, 131–134.

Rodgers, R. C., & Maranto, C. L. (1989). Causal models of publishing productivity in psychology. *Journal of Applied Psychology, 74*, 636–649.

Roe, A. (1953). *The making of a scientist.* New York, NY: Dodd, Mead.

Root-Bernstein, R. S., Bernstein, M., & Garnier, H. (1993). Identification of scientists making long-term, high-impact contributions, with notes on their methods of working. *Creativity Research Journal, 6*, 329–343.

Root-Bernstein, R. S., Bernstein, M., & Garnier, H. (1995). Correlations between avocations, scientific style, work habits, and professional impact of scientists. *Creativity Research Journal, 8*, 115–137.

Root-Bernstein, R., Allen, L., Beach, L., Bhadula, R., Fast, J., Hosey, C., ... Weinlander, S. (2008). Arts foster scientific success: Avocations of Nobel, National Academy, Royal Society, and Sigma Xi members. *Journal of the Psychology of Science and Technology, 1*, 51–63.

Rothenberg, A. (1983). Psychopathology and creative cognition: A comparison of hospitalized patients, Nobel laureates, and controls. *Archives of General Psychiatry, 40*, 937–942.

Rothenberg, A. (1996). The Janusian process in scientific creativity. *Creativity Research Journal, 9*, 207–231.

Rushton, J. P. (1984). Evaluating research eminence in psychology: The construct validity of citation counts. *Bulletin of the British Psychological Society, 37*, 33–36.

Rushton, J. P. (1990). Creativity, intelligence, and psychoticism. *Personality and Individual Differences, 11*, 1291–1298.

Scarr, S., & McCartney, K. (1983). How people make their own environments: A theory of genotype environmental effects. *Child Development, 54*, 424–435.

Schneider, J. (1937). The cultural situation as a condition for the achievement of fame. *American Sociological Review, 2*, 480–491.

Schubert, D. S. P., Wagner, M. E., & Schubert, H. J. P. (1977). Family constellation and creativity: Firstborn predominance among classical music composers. *Journal of Psychology, 95*, 147–149.

Shavinina, L. V. (2003). Understanding scientific innovation: The case of Nobel Laureates. In L. V. Shavinina (Ed.), *International handbook of innovation* (pp. 445–457). Oxford, United Kingdom: Elsevier Science.

Shockley, W. (1957). On the statistics of individual variations of productivity in research laboratories. *Proceedings of the Institute of Radio Engineers, 45*, 279–290.

Silverman, S. M. (1974). Parental loss and scientists. *Science Studies, 4,* 259–264.

Simon, R. J. (1974). The work habits of eminent scientists. *Sociology of Work and Occupations, 1,* 327–335.

Simonton, D. K. (1976). Biographical determinants of achieved eminence: A multivariate approach to the Cox data. *Journal of Personality and Social Psychology, 33,* 218–226.

Simonton, D. K. (1983). Formal education, eminence, and dogmatism: The curvilinear relationship. *Journal of Creative Behavior, 17,* 149–162.

Simonton, D. K. (1984a). Artistic creativity and interpersonal relationships across and within generations. *Journal of Personality and Social Psychology, 46,* 1273–1286.

Simonton, D. K. (1984b). Creative productivity and age: A mathematical model based on a two-step cognitive process. *Developmental Review, 4,* 77–111.

Simonton, D. K. (1984c). *Genius, creativity, and leadership: Historiometric inquiries.* Cambridge, MA: Harvard University Press.

Simonton, D. K. (1984d). Is the marginality effect all that marginal? *Social Studies of Science, 14,* 621–622.

Simonton, D. K. (1984e). Scientific eminence historical and contemporary: A measurement assessment. *Scientometrics, 6,* 169–182.

Simonton, D. K. (1985). Quality, quantity, and age: The careers of 10 distinguished psychologists. *International Journal of Aging and Human Development, 21,* 241–254.

Simonton, D. K. (1986). Biographical typicality, eminence, and achievement style. *Journal of Creative Behavior, 20,* 14–22.

Simonton, D. K. (1988a). Age and outstanding achievement: What do we know after a century of research? *Psychological Bulletin, 104,* 251–267.

Simonton, D. K. (1988b). Galtonian genius, Kroeberian configurations, and emulation: A generational time-series analysis of Chinese civilization. *Journal of Personality and Social Psychology, 55,* 230–238.

Simonton, D. K. (1988c). *Scientific genius: A psychology of science.* Cambridge, England: Cambridge University Press.

Simonton, D. K. (1989). Age and creative productivity: Nonlinear estimation of an information-processing model. *International Journal of Aging and Human Development, 29,* 23–37.

Simonton, D. K. (1991a). Career landmarks in science: Individual differences and interdisciplinary contrasts. *Developmental Psychology, 27,* 119–130.

Simonton, D. K. (1991b). Emergence and realization of genius: The lives and works of 120 classical composers. *Journal of Personality and Social Psychology, 61,* 829–840.

Simonton, D. K. (1991c). Latent-variable models of posthumous reputation: A quest for Galton's *G. Journal of Personality and Social Psychology, 60,* 607–619.

Simonton, D. K. (1991d). Personality correlates of exceptional personal influence: A note on Thorndike's (1950) creators and leaders. *Creativity Research Journal, 4,* 67–78.

Simonton, D. K. (1992a). Leaders of American psychology, 1879–1967: Career development, creative output, and professional achievement. *Journal of Personality and Social Psychology, 62,* 5–17.

Simonton, D. K. (1992b). The social context of career success and course for 2,026 scientists and inventors. *Personality and Social Psychology Bulletin, 18,* 452–463.

Simonton, D. K. (1997). Creative productivity: A predictive and explanatory model of career trajectories and landmarks. *Psychological Review, 104,* 66–89.

Simonton, D. K. (1998). Fickle fashion versus immortal fame: Transhistorical assessments of creative products in the opera house. *Journal of Personality and Social Psychology, 75,* 198–210.

Simonton, D. K. (1999). Talent and its development: An emergenic and epigenetic model. *Psychological Review, 106,* 435–457.

Simonton, D. K. (2000a). Creative development as acquired expertise: Theoretical issues and an empirical test. *Developmental Review, 20,* 283–318.

Simonton, D. K. (2000b). Creativity: Cognitive, developmental, personal, and social aspects. *American Psychologist, 55,* 151–158.

Simonton, D. K. (2000c). Methodological and theoretical orientation and the long-term disciplinary impact of 54 eminent psychologists. *Review of General Psychology, 4,* 1–13.

Simonton, D. K. (2002). *Great psychologists and their times: Scientific insights into psychology's history*. Washington, DC: American Psychological Association.

Simonton, D. K. (2004). *Creativity in science: Chance, logic, genius, and zeitgeist*. Cambridge, UK: Cambridge University Press.

Simonton, D. K. (2008a). Gender differences in birth order and family size among 186 eminent psychologists. *Journal of Psychology of Science and Technology, 1*, 15–22.

Simonton, D. K. (2008b). Scientific talent, training, and performance: Intellect, personality, and genetic endowment. *Review of General Psychology, 12*, 28–46.

Simonton, D. K. (2009). Varieties of (scientific) creativity: A hierarchical model of disposition, development, and achievement. *Perspectives on Psychological Science, 4*, 441–452.

Simonton, D. K. (2012). Scientific creativity as blind variation: BVSR theory revisited. In R. Proctor & E. J. Capaldi (Eds.), *Psychology of science: Implicit and explicit reasoning* (pp. 363–388). New York, NY: Oxford University Press.

Simonton, D. K., & Song, A. V. (2009). Eminence, IQ, physical and mental health, and achievement domain: Cox's 282 geniuses revisited. *Psychological Science, 20*, 429–434.

Smart, J. C., & Bayer, A. E. (1986). Author collaboration and impact: A note on citation rates of single and multiple authored articles. *Scientometrics, 10*, 297–305.

Sorokin, P. A., & Merton, R. K. (1935). The course of Arabian intellectual development, 700–1300 A.D. *Isis, 22*, 516–524.

Stephan, P. E., & Levin, S. G. (1993). Age and the Nobel Prize revisited. *Scientometrics, 28*, 387–399.

Stroebe, W. (2010). The graying of academia: Will it reduce scientific productivity? *American Psychologist, 65*, 660–673.

Suedfeld, P. (1985). APA presidential addresses: The relation of integrative complexity to historical, professional, and personal factors. *Journal of Personality and Social Psychology, 47*, 848–852.

Suedfeld, P., Tetlock, P. E., & Streufert, S. (1992). Conceptual/integrative complexity. In C. P. Smith (Ed.), *Motivation and personality: Handbook of thematic content analysis* (pp. 393–400). Cambridge, UK: Cambridge University Press.

Sulloway, F. J. (1996). *Born to rebel: Birth order, family dynamics, and creative lives*. New York, NY: Pantheon.

Sulloway, F. J. (2009). Sources of scientific innovation: A meta-analytic approach. *Perspectives on Psychological Science, 4*, 455–459.

Taylor, C. W., & Ellison, R. L. (1967, March 3). Biographical predictors of scientific performance. *Science, 155*, 1075–1080.

Taylor, M. S., Locke, E. A., Lee, C., & Gist, M. E. (1984). Type A behavior and faculty research productivity: What are the mechanisms? *Organizational Behavior and Human Performance, 34*, 402–418.

Terry, W. S. (1989). Birth order and prominence in the history of psychology. *Psychological Record, 39*, 333–337.

Van Zelst, R. H., & Kerr, W. A. (1951). Some correlates of technical and scientific productivity. *Journal of Abnormal and Social Psychology, 46*, 470–475.

Vijh, A. K. (1987, January–March). Spectrum of creative output of scientists: Some psychosocial factors. *Physics in Canada, 43*(1), 9–13.

Wai, J., Lubinski, D., & Benbow, C. P. (2005). Creativity and occupational accomplishments among intellectually precocious youths: An age 13 to age 33 longitudinal study. *Journal of Educational Psychology, 97*, 484–492.

Walberg, H. J., Rasher, S. P., & Hase, K. (1978). IQ correlates with high eminence. *Gifted Child Quarterly, 22*, 196–200.

Walberg, H. J., Rasher, S. P., & Parkerson, J. (1980). Childhood and eminence. *Journal of Creative Behavior, 13*, 225–231.

White, R. K. (1931). The versatility of genius. *Journal of Social Psychology, 2*, 460–489.

Wispé, L. G. (1965). Some social and psychological correlates of eminence in psychology. *Journal of the History of the Behavioral Sciences, 7*, 88–98.

Wispé, L. G., & Ritter, J. H. (1964). Where America's recognized psychologists received their doctorates. *American Psychologist, 19*, 634–644.

Woodward, W. R. (1974). Scientific genius and loss of a parent. *Science Studies, 4*, 265–277.

Wray, K. B. (2003). Is science really a young man's game? *Social Studies of Science, 33,* 137–149.

Wray, K. B. (2004). An examination of the contributions of young scientists in new fields. *Scientometrics, 61,* 117–128.

Wray, K. B. (2009). Did professionalization afford better opportunities for young scientists? *Scientometrics,* doi: 10.1007/s11192–008-2254.

Yuasa, M. (1974). The shifting center of scientific activity in the West: From the sixteenth to the twentieth century. In N. Shigeru, D. L. Swain, & Y. Eri (Eds.), *Science and society in modern Japan* (pp. 81–103). Tokyo, Japan: University of Tokyo Press.

Zhao, H. (1984). An intelligence constant of scientific work. *Scientometrics, 6,* 9–17.

Zhao, H., & Jiang, G. (1986). Life-span and precocity of scientists. *Scientometrics, 9,* 27–36.

Zuckerman, H. (1977). *Scientific elite.* New York, NY: Free Press.

Zusne, L., & Dailey, D. P. (1982). History of psychology texts as measuring instruments of eminence in psychology. *Revista de Historia de la Psicología, 3,* 7–42.

CHAPTER 11

Gender and Science: Psychological Imperatives

Neelam Kumar

Theoretical debates around gender and science have primarily been of interest to the sociologists, historians, and philosophers of science. Psychologists have sporadically contributed to the gender theories in science, technology, and society (STS). STS has been getting rich contributions and regular theoretical inputs from the feminist philosophers, gender historians and theorists, anthropologists, and sociologists. The relative absence of psychological theories is easily noticeable and is a matter of great concern as well as surprise. Surprise is quite legitimate and intense if one looks at the two trends. First, there is an obsession in psychology to examine and test sex differences. Most of the research and experiments carried out to explain any psychological attributes and behaviors do take sex difference into account. Second, there is a fascination by psychologists to study scientists, their nature, and nurture as early as 1874 (Galton, 1874). To the question, why psychology of science is often missing from the gender and science literature, STS anthologies and handbooks thus need analysis and attention. Psychologists have discussed the personality traits, abilities, and interests of individual scientists for a long time. The psychology of science has been conspicuously absent as a subdiscipline—at least until the mid-1980s—from the science studies, also known as metascience (Feist, 2006a; Feist & Gorman, 1998). Psychological concepts and processes, such as imagination, creativity, thought processes, social influence, and motivation, have often been sources of theoretical inspiration to philosophers, historians, and sociologists doing social studies of science. Yet there is a striking absence of Psychology of Science as an established field within STS.

This chapter begins by examining the concepts and theories of sex differences used in psychology and the origin of the concept of gender and science. The concept of gender constitutes one aspect of the larger intellectual approach, a form of comparative and social history (Jordonova, 1993). Concerns of gender are thus essential to a social theory of science. The chapter then turns to the methods and perspectives advanced over the years to offer useful conceptual frameworks for exploring the gender–science relationship. Finally, gender–science is discussed in relation to some of the empirical researches and the leading theoretical perspectives, fields in psychology, and STS. What forms of engagement might be envisaged between these fields comprise the concluding section of the chapter.

SEX/GENDER AS A "VARIABLE" IN PSYCHOLOGY

Changing Perceptions

The history of psychology of gender is quite interesting. The nature and behavior of women had been an academic and social concern of philosopher psychologists throughout the ages. But formal psychology, which is said to have emerged in 1879 with the establishment of Wundt's laboratory, was relatively slow to study the subject. The "woman question" probably did not fall within the sharply defined limits of Wundt's psychology. The question of sexuality and personality formation were made prominent by psychoanalysts in the early 19th century but was restricted mainly to therapeutic context. Behaviorism, another important school of thought, was also not concerned with individual differences (such as gender differences) or personality. The result was that the psychology of women remained under the Freudian influence. It was actually the functionalist movement that fostered the study of sex differences within the academic psychology (Shields, 1975). Three topics were of special significance to the psychology of women during the functionalist era: (a) structural differences in the brains of males and females and the implications of these differences for intelligence and temperament, (b) the hypothesis of greater male variability and its relation to social and educational issues, and (c) maternal instinct and its meaning for a psychology of female "nature." Reviews of research on psychological sex differences began with Woolley's (1910, 1914) and Hollingworth (1914, 1916). Both of them conducted several studies and were convinced that the so-called sex differences were more a fabrication than reality. Female abilities were not hampered by their "unique" physical traits. In the United States, the entry of female academics in professional bodies made a big difference. Mary Calkins, for example, became the first female president of the American Psychological Association (APA) in 1905, and later of the American Philosophical Association in 1918. This was a huge recognition and must have served as an example to other academic bodies.

Yet the struggle was to continue. The problem remained as to how to find enduring traits that unambiguously distinguish one sex psychologically from the other. This gained momentum in the 1930s. The first masculinity–femininity (MF) scale was developed by Terman and Miles (1936), who were best known for research with high-intelligence quotient (IQ) children. Psychologists like Edwin Boring looked at the issue not solely as a problem "for and about" women, but as a problem of social dynamics (Boring, 1951). By the mid-20th century, psychologists used the phrases "psychological sex" and "sex-role identification" to point to a person's acquired sense of self as female or male. Henceforth "sex" turned out to be one of the most common variables for psychological explanation until the late 1960s and early 1970s. The Freudian version of psychology of women remained dominant until the feminist psychologists began to make challenges. Starting in the 1960s, feminist scholars argued that the discipline of psychology had neglected the study of women and often misrepresented women in its research and theories. Feminists also posed many questions worthy of being addressed by psychological science. As a result, a distinctively feminist psychology developed, which included a large and diverse research concentration on the psychology of women and gender. Gradually, these efforts led to the emergence of a new and influential body of research on gender and women (Eagly, Eaton, Rose, Riger, & McHugh, 2012). In the 1970s, feminist psychology in fact emerged as a movement that not only questioned, but, influenced theoretical as well as methodological approaches to psychology. In the case of social psychology, an article published in the United States in the 1970s (Unger, 1979) highlighted the distinction between sex and gender, and questioned the dominance of research on sex differences in psychology. However, Unger's distinction remained largely ignored by American mainstream social psychology. Feminist psychology, however, continued to figure in the epistemological debates and methodological developments of the 1980s and 1990s. The introduction of meta-analytical techniques in the 1980s was a significant point—it revolutionized the study of gender differences in psychology. In the 1980s many psychologists addressed women and female psychology and a number of classic books and articles were published during this decade (e.g., Cox, 1981; Gilligan, 1982; Hyde, 1985; Unger, 1979; Williams, 1983).

The women's movement played a significant role in drawing the attention of social scientists toward gender issues. It made clear to them that sex as a variable could not account for some of the gross discrepancies found between the lives of men and women. The concept of gender, thus, emerged in the studies of people and society. While the term sex was used to refer to biologically based distinctions, the term gender referred to the socially, often psychologically, derived or constructed differences between women and men. Joan Scott, a celebrated historian at Princeton, for example, defined gender as follows: "The core of the definition rests on an integral connection between two propositions; gender is a constitutive element of social relationships based on perceived differences between the sexes, and gender is a primary way of signifying relationships of power"(1986, p. 1067). Scott argued that gender is not merely an empirical "fact" attached to persons or even symbols, but an "analytical tool." Use of gender as an analytic tool in psychology, which is rather more recent, however mainly signified: (a) sex differences, such as sex-related differences in cognition; (b) within-sex variability; and (c) the gender-linked power relations in social institutions and interactions/and gender roles. The three approaches have generally been viewed as alternatives, and often are pursued in isolation from one another. As the psychology of gender matured as an area of study, changes in its terms, topics, methods, and its concepts occurred. Henceforth, there was no looking back; sex and gender researches in psychology advanced both theoretically as well as methodologically. The postmodern debates accelerated the pace of change. The social constructionist approach conceives of gender as a process. Psychology moved from discussing "sex and gender differences" to "doing gender" (Shields & Dicicco, 2011). Some scholars, however, argue that in psychology gender has mostly been used empirically, without proper understanding of its social or conceptual significance (Stewart & McDermott, 2004). We need to realize that the emergence and development of the feminist psychology has not been an esoteric intellectual exercise but has moved along the social and professional status and concerns of women over the course of the late 19th, 20th, and early 21st centuries (Rutherford, & Granek, 2010). Very recently two comparative volumes on gender, psychology, science, and culture have brought different geo-cultural locations together (Kumar, 2012; Rutherford et al., 2011). Debates about sex and gender differences will continue to attract attention.

HISTORICAL ASSOCIATIONS

Gender–science relation needs to be seen within larger historiographical issues. The initial history of science is, basically, that of exclusion of women. This exclusion can be understood from two perspectives, one on ideological grounds and the other through the development of science per se. Various ideological constructions of gender through different eras have served as barriers to women's access and progress in the sciences. Bacon saw science as a "chaste and lawful marriage between Mind and Nature"; mind as male and nature as female, to be subdued, dominated, and controlled. Ideology of masculinity and the "practice of the care of the self" led to the exclusion of women from the institution of new science in the seventeenth century (Golinski, 2002). Eighteenth century too saw intense debates over the different natures of female and male–whether women were mentally and socially inferior to men, or they were equal but different, or at least potentially equal! In the mid-19th century, social Darwinists invoked evolutionary biology to argue that a woman was a man whose evolution—both physical and mental—had been arrested in a primitive stage. Women's intellectual development, it was argued, would proceed only at great cost to reproductive development. There was prevalent a myth that claimed that *"as the brain developed, the ovaries shriveled"* (emphasis added) (Schiebinger, 1989, p. 2). Women were thus perceived intrinsically unsuited to natural philosophy (i.e., science) and those who did show any aptitude were made the butt of savage satires. This attitude however continued, with few revisions, throughout the history of Western science. In the 20th century, scientists gave new interpretations to the prejudices on women in science. For decades, whether boys' mathematical skills are superior to girls' has been a controversial topic among the social scientists.

The progress and development of science over the centuries had its own effects on women's access to it. There are, however, certain turning points in the history of science that led to exclusion of women from science. The institutionalization of science, for example, resulted in the women's marginalization. Universities became the home of science at the end of the 12th century when women were denied entry. The professionalization of science in the 17th century had further impact on the exclusion of women from this discipline (Whaley, 2003). The new scientific societies, which were becoming the prime disseminators of the latest scientific knowledge, like the university, were closed to women. Women were also excluded from the meetings of the scientific societies. Britain's Royal Society, established in 1662, for example, did not admit women until 1945. Similarly, the Académie Royale des Sciences (1666) in Paris refused to admit women until 1979. The developments within the disciplines also affected women. For example, during 1790 to 1830 women were particularly visible as writers of botany books in England, between 1760 and the 1820s, botany was perceived as feminized area, marked as especially suitable for women and girls. By 1830, botany began to make a modern profile as a science, and women's status as writers and cultural contributors began to be considered problematic (Shteir, 1996).

In the early years of the 19th century, when the production of science was still heavily dependent on the help of family members and private patrons, the situation of men and women engaged in scientific pursuits was in some ways comparable. However, over the course of the century, science's move, from the domestic and amateur context to the public and professionalized activity, resulted in women's disadvantage. As MacLeod and Moseley (1979) put it, science for women was at best useless, at most brutalizing. Those who tried to encourage women to study science had three common arguments; one, to make "better wives and mothers," second, to grant "equal opportunity," and finally, the Darwinist conviction that women were particularly "fitted" to pursuing certain kinds of knowledge. By the end of the 19th century, as research science began to replace classics, women's representation, particularly their involvement in the central activities of science, began to decline. Once scientific knowledge became culturally dominant at the turn of the 20th century, women were on its margins (Eisenhart & Finkel, 1998). Even the illustrious Marie Curie (1867–1934) was turned away (Schiebinger, 2003). The appearance of the scientific journals in the 1660s was a significant development, but women were largely absent from the pages of scientific journals until the early 20th century.

Women did not have access to the universities until the late 19th century (even until 20th century in a few countries), with the exception of Italy. In Italy, as early as the 13th century, a few exceptional women did study and teach at universities, and that too in such fields as physics and mathematics, which even today are thought especially resistant to women! The influence of Aristotelian philosophy in the universities' curricula provided a scientific basis for this exclusion, since Aristotle and his followers viewed women as intellectually inferior. During the 17th and 18th centuries, however, only a few outstanding women practiced and even lectured on natural philosophy (Fara, 2002).The 18th century, which constituted a distinct era in the organizational and institutional history of science, witnessed the transition from natural philosophy to the beginnings of an array of scientific disciplines and largely proscribed women. The world of organized science in the 18th century was almost exclusively male (McClellan, 2003). In the 19th century, *science* took a new meaning—science became devoid of its philosophical and theological concepts. Gradually the doors of universities, scientific societies, and research laboratories opened to women. Some like Mary Somerville made big names as science popularizers (Lightman, 2007). Women achieved access to institutions of higher education in the United States in 1833, Germany in 1908, and Japan in 1913. In India, the first graduate degrees were granted to women in 1883. One of the first university professions that women gained access into was medicine, in the face of tremendous resistance in most European countries. Women began to be admitted in recognized medical schools by the mid-19th century. In Philadelphia, Women's Medical College was founded and a complete medical course for women was offered in 1850. But around 1920 there was a reaction toward the growth of female education, saying that it "disadvantages every student" (Bonner, 2000, p. 339). Segregated education remained a hallmark of the medical practice (Bynum, 1994). Only a few top schools opened their doors to women medical students and it proved however easier to establish separate schools for women. In Germany medical schools remained closed to women until 1900. The leading British and American universities with a few exceptions barred women until World War I. (Harvard Medical School did not have a woman student until 1945!).

The 20th century has been described as a historic turnaround for women. Women were not only admitted to universities, but also found their place as professors. They became part, though in limited ways, of nurturing scientific subculture. Although in the 1880s there were only a few hundred female scientific workers in Europe and America, the number swelled to the thousands by 1940 (Outram, 1987). Women began to enter into all fields of science but mostly in subordinate positions. The best jobs remained a male preserve. During the 20th century, however, inequalities between female–male participation and recognition in the sciences began being questioned. Feminist theorists started emphasizing the biased nature of science. Feminist debates about science, however, made a fundamental change in the way in which science came to be understood and practiced. The concept of gender and science made its appearance in a psychoanalytic journal (Keller, 1978). But gender blindness in sociology of science could be noticed nearly a decade later (Delamont, 1987). By the 1980s, some feminist literature on science (such as *Science and Gender* by Ruth Bleiers) appeared along with the works on the history of women in science (e.g., Margaret Rossiter's book on women scientists in America, published in 1982). Most of these feminist historians and philosophers were from the United States. In most British history, even when the history of science appears at all, women were mostly absent from the "story" (Watts, 2007). The idea that the social structure and processes of science are gendered slowly emerged in diverse areas of feminist discourse and it was declared that science is "masculine," not only in practice, but also in its ethos and substance. The turn to gender as an analytic category (Connell, 1987; Harding, 1986; Scott 1986) introduced new ways of bringing gender into the discussions of power, culture, and the politics that seemed to have obvious implications for the study of science as well. Feminists started to elaborate gender as a concept to mean more than a socially constructed binary.

Gender as a concept developed further and contested the naturalization of sexual differences (Haraway, 1991). New and explosive questions were raised regarding gender and science, such as, *Has Feminism Changed Science?* (Schiebinger, 1999). Even visual and figural images have been used to study the workings of gender in science (Shteir & Lightman, 2006). Some have also argued that gender has moulded the very content of science (Schiebinger, 2003). Biological myths and assumptions were also challenged, and during the 1990s, feminist reconsideration of the sex/gender problem moved into full swing. Fausto-Sterling, for example, examined numerous scientific claims about biologically based sex differences between men and women. She argued that the sex–gender or nature–nurture accounts of difference fail to appreciate the degree to which culture is a partner in producing body systems commonly referred to as biology—something apart from the social. In her view, our bodies physically imbibe culture (Fausto-Sterling, 2005).

The development of feminist perspectives on the epistemology and methodology, history, philosophy, and sociology of science, thus became an ongoing process. Psychology somehow remained at bay! In dialogue with the history and sociology of technology, feminist technology studies (FTS) could also emerge and flourish. Interesting questions were raised by feminists on the gender and technology relations too. Technology is a social construction, so is gender. Technology is *doing* things, it is not mere *being*; so is gender. In the 1970s, radical feminists and ecofeminists initiated a critique of the inherently patriarchal nature of technology and of techno-science more generally. Women kept away from technology, as it was considered "masculine," and at the same time "masculinity" was being defined in terms of man's use of technology and its tools. It was wrong to define and confine technology to masculinity alone. Technology forms the kernel of material culture and what would material culture be without the role of women as both producer and consumer of certain techniques and technologies? Both technology and gender are thus mutually constitutive, though, empirically speaking, the influence of technology on gender is much more than that of gender on technology per se. Some feminists condemned all technology as intrinsically oppressive of women. Socialist feminists generally tried to be more contextual in their work, pushing Marxist analysis beyond class to ask why and how modern Western technology had become a male domain. In the 1980s science studies came to acknowledge the critical role of technology and its epistemologies in shaping the production of scientific knowledge (Bray, 2012). In the late 1980s, constructivist approaches emerged in technology studies. The new pioneers were Cockburn (1983), Cowan (1983), Haraway (1991), Oldenziel (1991), and Wajcman (1991). Like the feminist science studies, they interrogated the gendering of technology at virtually every level and from different disciplinary angles. There emerged a cross-fertilization from which both FTS and STS benefitted.

GENDER AND SCIENCE

Contemporary Realities

Science and technology remains gendered even in the contemporary period. A large body of social science research documents gender gap in science in various national contexts (Bosch, 2002; Costas, 2002; Guo, Tsang, & Ding, 2010; Hermann & Cyrot-Lackmann, 2002; Kumar, 2009; Long, 2001; Xie & Shauman, 2003). Overall, women account for a minority of the world's researchers (UIS, 2011). Race and ethnicity have erected additional barriers to the full participation of women in science. Although overt discrimination, observed in the earlier decades is out of style, covert and subtle forms of discrimination still prevail today. Underrepresentations, especially at higher echelons and prestigious postions, are observed all over the world, apart from the lower number of women in science. The scientific efforts and achievements of women do not receive the same recognition as do those of men. Even the awards in science, technology, engineering, and medical (STEM) fields are not free from gender biases (Lincoln, Pincus, Koster & Leboy, 2012). Women are underrepresented, though there has been an increase, on editorial boards of major journals in various disciplines, such as psychology (White, 1985), medicine (Kennedy et al., 2001; Morton & Sonnad 2007), management (Metz & Harzing, 2009), economics (Addis & Villa, 2003), and science (European Commission, 2004). Evidence of gender bias in the peer review of research grant applications are also reported (Wennerås & Wold, 1997). The issue of gender discrimination crosses national borders, thus, the question of gender inequality has become a universal question (Etzokowitz & Kemelger, 2001). Ironically, the universal socio-psychological conditions related to gendered science are often aggravated by the local circumstances. As a result, there has been increasing international concern about the more effective integration of women into science and technology. Numerous researches, using various disciplines, has been initiated, published, and reported. Studies dealing with cross-national similarities and differences in social contexts and systems to explain general and sex-specific STEM choice have been initiated (e.g., Poole, Bornholt, Summers, 1997; Van Langen & Dekkers, 2005; Kumar, 2012). Many international conferences have tried to determine the causes of women's underrepresentation in science, such as the one organized by the author on *Women in Science: Is the Glass Ceiling Disappearing?* at New Delhi in 2004. True, the ubiquitous presence of gender stratification in science is universally recognized, but what are the explanations? Why aren't more women pursuing careers in science, engineering, and math? Is the lack of women in these fields a consequence of societal discouragements, innate differences in ability between the sexes, or differences in aspirations? Top researchers debate over this (Ceci & Williams, 2007), and describe and dissect the evidence to know how biology and society conspire (Ceci & Williams, 2010). The challenge remains.

A Statistics of Scarcity and Segregation

Women remain underrepresented in the scientific and engineering workforce, although to a lesser degree than in the past. UNESCO figures reveal that globally, of 121 countries for which data are available, women represent slightly more than one-quarter of researchers (29%). There are some regional variations. Latin America and the Caribbean has the highest share (46%) of female researchers, which exceeds the world average of 29%, Commonwealth of Independent States (43%), Europe and Africa (33%), and Asia (18%) (UNESCO, 2010). In addition to simply being low in number, women are also segregated from men both vertically and horizontally. Vertical as well as horizontal, the two major types of gender segregation still characterize science all over the world. The number of women decreases at the upper hierarchy of the university, starting from PhD students, to Assistant Professor, to Associate Professor, to Full Professor. Women earned 41% of scientific and engineering (S&E) doctoral degrees awarded in the United States in 2008. Similar figures exist in Australia, Canada, the European Union, and Mexico. Women earned more than half of S&E doctoral degrees in Portugal and less than one-quarter of S&E doctoral degrees in the Netherlands, South Korea, and Taiwan (National Science Board, 2012). Yet, women continue to be low in number in the scientific careers and remain clustered in lower academic positions, and relatively few are able to be at the higher ranks. Women, for example, in the United States hold a larger share of Instructor and Assistant Professor positions (42%)

than of Associate (37%) or Full Professor (21%) positions in the disciplines of science, engineering, and health (National Science Board, 2012). This is more true for certain disciplines. Women obtain nearly 30% of the doctorates in chemistry, but as one goes up the ladder of prestige and seniority, the numbers are less encouraging (Cavallaro, Hansen, & Wenner, 2007 as cited in Ceci, Williams, & Barnett, 2009). Beyond the postdoctoral level, women scientists have slower rates of promotions, less recognition through awards, and hold fewer departmental chairs relative to the eligible pool (National Academy of Science, 2007). Attrition is higher among girls and, unfortunately, even those girls who are talented leave science. Feist (2006c) noticed that 43% of the female finalists of the prestigious Westinghouse Science Talent Search in the United States leave science, in contrast to only 11% of the male finalists. The situation in nonacademic venues is also not encouraging, where women leave science, engineering, and technology jobs at twice the rate of men, although this figure includes not only jobholders with PhD degrees but also those with bachelor's and master's degrees (Belkin, 2008). Segregation into certain disciplines has been another common problem worldwide. Women remain concentrated in certain fields and industries such as biology, health, agriculture, and pharmaceuticals, with low representation in physics and mathematics. Other obvious cross-national similarity is uniform underrepresentation of women in engineering and higher concentration in the humanities and social sciences. Everywhere the gender gap is most dramatic in engineering and technology. In the United States, compared with their male counterparts, women are more heavily concentrated in the fields of life sciences, social sciences, and psychology, with correspondingly lower shares in engineering, the physical sciences, mathematics, and computer sciences (NSB, 2012). Women are strongly underrepresented in science, engineering, and technical programs even in countries with high overall female enrolment rates (Jacobs, 2003; England & Li, 2006; England et al., 2007; Xie & Shauman, 2003). For example, in Portugal, the participation of women in science is exceptionally high, but segregation by scientific field is similar to that in other European countries.

The figures from Europe are no different! According to a new survey on Statistics and Indicators on Gender Equality in Science by the European Commission (She Figures, 2009), women in scientific research remain a minority, accounting for 30% of all researchers. Despite the number of female researchers growing faster than that of men and an increase in the proportion of female PhDs, the underrepresentation of women in scientific disciplines and careers remains a serious challenge also in Europe. During 2002 to 2006 there has been an increase in the overall number of female researchers in almost all fields of science in the European Union: the highest growth rates have been recorded in the fields of the medical sciences, the humanities, engineering and technology, and in the social sciences. Yet, She Figures (2009) shows that women still account for 37% of all researchers in the higher education sector, 39% in the government sector, and only 19% in the business enterprise sector. Career progression of women and men in the higher education sector confirms a pattern of "vertical segregation," whereby the majority of women in academia are to be found in the lower hierarchical positions. Women account for 59% of all graduates, but only 18% of full professors in Europe are women. The underrepresentation of women is even more striking in the field of science and engineering, where only 11% of professors are female. The report also reveals that on average only 13% of institutions in the higher education sector are headed by women.

What about the rest of the world? Asia has been discussed recently for its rising science and technology strength in a report by the United States's National Science Foundation (NSF, 2007). Can we assume that this reflects both female as well as male equally? The report does not however deal with gender dimension. The data from Asian countries are unfortunately not well documented and always as complete as one could wish. In Asia, women constituted only 18% of researchers, but there is considerable heterogeneity (UNESCO, 2009). South Asia had the lowest rate of 18%. Less than 30% of researchers were female in the Asian Arab states (21%), as well as in Japan (13%), and the Republic of Korea (15%). South East Asia reported a high share of female researchers at 40%. Most Central Asian countries reported gender parity (around 50%). Along with the expansion of higher education in China in recent years, the number of women staff and faculty also has shown significant increase. Here women represented 43% of total academic workforce in the year 2004, and 19.3% of the total professors were female (Yezhu, 2007). The situation in Asian context, however, is not so well researched as it is in the case of United States or Europe. A critique of a western bias in gender and technology studies that advocates more context sensitivity and focus on the cultural embeddedness of gender and technology relations deserves attention (see, for example, Mellström, 2009).

PSYCHOLOGICAL RELEVANCE AND RAMIFICATIONS

Theoretical Perspectives

One of the contested features of the late 19th-century psychological science was its gendered dimension. The generalized, "normal" adult mind that the experimentalists usually posited as their subject matter was at least implicitly the common property of both sexes, while the vocabulary and practices of objective science carried unmistakably masculine symbolism (Ash, 2005). The discipline of psychology itself became more open to women during the 1920s, but a gender hierarchy emerged in psychology, with industrial psychology remaining a male-dominated profession, while female "Binet testers" and social workers took on more people-oriented functions.

The ubiquitous presence of gender stratification in science is universally recognized and studied, yet certain psychological dimensions about its development remain unanswered. The psychology of science analyzes the cognitive, emotional, personal, social-psychological dimensions of science, and its subfields include general, social, cognitive, developmental, and organizational. Science has been investigated from each of the fundamental subdisciplines in psychology. There are as many psychologies of science as there are major subdisciplines in psychology (Feist, 2011). Psychology has witnessed multiple theories of gender too. The theories range from emphasizing the biological, individual, to sociocultural explanations. While some of these theories have strong social constructionist emphasis, others capture the interactions of situational and individual causes, and still others blend situational, personality, and biological causes (Eagly et al., 2012). Birth order such as first born predominance has been found to be important among eminent women psychologists (Simonton, 2008). Besides specific theories on gender, some of the general theories are also effectively applied to understand gender (Eagly et al., 2012). These can be also of great utility to understand gender–science relationships. For example, the social identity theory of Tajfel can be a useful framework for understanding sex discrimination in science too. System justification theory (Jost, Banaji, & Nosek, 2004) has provided insight into gender stereotyping and ideology and its effects on behavior (Jost & Kay, 2005). Can these explain the relationship between stereotypes–gender–science? Many theories are used in a specific manner to understand gender–science nexus, such as gender schema theory, attribution theory, and cognitive ability theory. Some of the feminist theories, such as theory of gendered organizations are of special importance in terms of contribution to understand the gender–science nexus. However, the process is not that of one way only. Feminist psychology has not only provided theories to explain the gender–science relationship, rather it has also been influenced by the feminist epistemological perspectives, particularly by the three perspectives of empiricism, standpoint theory, and postmodernism (Riger, 1992). The following three theoretical approaches within psychology provide some insight into socialization processes that might account for the gender stratification (Else-Quest, Hyde, & Linn, 2010):

Expectancy-Value Theory

This model or theory was first proposed by Eccles and her colleagues (Eccles, 1994; Jacobs, Davis-Kean, Bleeker, Eccles, & Malanchuk, 2005) and argues that people do not undertake a challenge unless they value it and have some expectation of success. Thus, if a girl believes that the career opportunities available to or appropriate for women do not require mathematics skills, she is less likely to invest in developing her mathematics skills by working hard in her required math courses. Perceptions of the value of the task are shaped by the cultural milieu and the person's short- and long-term goals. Expectations of success are shaped by the person's aptitude, relevant past events such as grades in the subject and scores on standardized tests, the person's interpretations of and attributions for these events, and the person's self-concept of ability. Sociocultural forces such as parents' and teachers' attitudes and expectations, including stereotypes, also shape self-concept and attitudes toward the subject.

Cognitive Social Learning Theory

Another psychological theory that is consistent with the gender stratification hypothesis is cognitive social learning theory (Bandura, 1986; Bussey & Bandura, 1999), which maintains that a number of social processes contribute to the development of gender-typed behavior, including reinforcements, modeling, and cognitive processes, such as self-efficacy. Social-cognitive theory of gender-role development and functioning integrates psychological and socio-structural determinants within a unified conceptual framework (Bandura, 1986, 1997; Bussey, & Bandura, 2004). Role models and socializing agents, as well as perceptions of gender-appropriate behavior, have an important influence on an individual's academic choices. This theory also emphasizes the role of self-efficacy in gender-typed behaviors, such as choosing to major in physics. It maintains that girls are attentive to the behaviors in which the women in their culture engage and thus feel efficacious in and model those behaviors. If girls observe that women in their culture do not become engineers or scientists, they may believe that such careers (and, by extension, STEM disciplines) are outside the realm of possibilities for girls and feel anxious about and/or avoid these subjects. In emphasizing the roles of observational learning and the internalization of cultural norms, cognitive social learning theory provides an individual-level explanation of why girls and women make gendered educational and vocational choices that recapitulate societal-level gender stratification.

Social Structural Theory

This theory, sometimes referred to as social role theory (Eagly, 1987; Eagly & Wood, 1999) is another relevant psychological theory that has been applied to explain access to mathematics and science education. It argues that psychological gender differences are rooted in sociocultural factors, such as gendered division of labor. A society's gendered division of labor fosters the development of gender differences in behavior by advancing different restrictions on female and opportunities to males on the basis of their social roles. If the cultural roles that women fulfil do not include math, girls may face both structural obstacles and social obstacles. This in turn impedes their access to mathematics and development of affect, interest, and ability in the discipline. According to social structural theory, across nations, gender equity in educational and employment opportunities should be associated with gender similarities in mathematics achievement, attitudes, and affect.

Empirical Findings

The underrepresentation of women in careers in science, technology, mathematics, and engineering points to an important and increasing role of psychological research in understanding gender–science relation. Psychologists of science need to be active in investigating empirically and explaining theoretically the questions and puzzles relating to gender and science (Feist, 2012). There are no single or simple answers to the complex questions about gender differences in science and mathematics (Halpern, Benbow, Geary, Hyde, & Gernsbacher, 2007). After all, the competing claims of biological and sociocultural causation still mark women's underrepresentation in science, and "sex" remains a variable in psychological studies, especially in relation to personality traits (Schmitt, Realo, Voracek, & Allik, 2008). A psychobiosocial model offers an alternative conceptualization to the debate (Halpern, &

LaMay, 2000). The age-old distinction between nature (i.e., biological factors) and nurture (i.e., environmental factors) has proved unproductive according to Halpern and LaMay. Their psychobiosocial model is based on the idea that some variables are both biological and social and, therefore, cannot be classified into one of these two dichotomous categories. One of the latest trends in psychology concerns the use of individual and sociocultural factors. For example, psychologists have investigated math-intensive STEM careers and found the relationship circular in nature (Ceci, Willimas, & Barnett, 2009).

Although psychologists have long been studying sex, gender, and science relationship, recent years have seen a remarkable upsurge in the literature. We need to keep in mind that the empirical literature on gender and science is very scattered and they cover issues overlapping each other. An exhaustive review of this literature may not be possible or even desirable, but this chapter will try to elucidate the literature dealing with those concepts and theories of psychology that may help explain gender and science relation (though it may be a difficult task, as there exist quite an overlap among psychological, socio-historical, and sociological, especially social-psychological dimensions). Psychologists generally use gender in empirical research in three different ways: to signify sex differences, within-sex variability, and the gender-linked power relations that structure many social institutions and interactions. Using the "sex differences approach," psychologists consider how and why average differences in personality, behavior, ability, or performance between the sexes might arise (see, e.g., Block 1984; Buss, 1995; Eagly, 1994; Levy & Heller, 1992; Maccoby, 1998; Maccoby & Jacklin, 1974). This approach reflects essentialism, that is, assumes that the differences arise from preexisting "essential" differences between male and female. Psychologists have been studying individual traits and personality characteristics of scientists. They have also investigated specific aspects of scientific processes and activities in relation to sex and gender differences, such as research and publication outputs. The second type of research looks at within-sex variability. It believes that in many cases a focus on sex differences ignores the large variance within gender on many characteristics, and therefore may tend to exaggerate sex differences, or even reinforce or create them in the mind of the public (Hare-Mustin & Marecek, 1990, Unger, 1989). Psychologists have also used gender for treating such topics as scientists' roles or science as an occupation, as well as scientists within organizations. The following section will deal with some of the typical themes and studies in the psychological studies dealing with gender–science relationships.

Gender and Cognition

The question of sex differences in cognitive processes has a long history in psychology. A common feature of late 19th-century psychological science was its gendered dimension (Ash, 2003). Since the early 1900s, modern intelligence researchers and theorists have documented extensively mean sex differences on many, if not most, cognitive ability measures. A common assumption that prevailed was that the male was more inclined to be intellectual and rational while female passionate and emotional (Burt & Moore, 1912). Cattell, the author of *American Men of Science,* and Thorndike, an influential educational

psychologist, commented on the lack of intellectually gifted women. In relation to the between sex differences, the notion of within-sex variability, especially among males, also became a hot topic. Cattell and Thorndike pointed
out that men and women had the same average level of intelligence, but men
were more variable (Fausto-Sterling, 1985). The debate on variability went
into the 1930s and gained momentum after Maccoby and Jacklin's (1974) four
fairly well-established gender differences: verbal, visual-spatial, mathematical ability, and aggressive behavior. One of the recent comprehensive studies
is by Deary, Thorpe, Wilson, Starr, and Whalley (2003) on 80,000+ Scottish
boys and girls, almost everyone born in Scotland between 1921 and tested
at the age 11 in 1932. They reported that there was no sex difference in mean
intelligent quotient (IQ) score. However, there was greater variability among
boys' scores, such that boys were overrepresented relative to girls at both
the highest and lowest extremes. In later research, Strand, Deary, and Smith
(2006), in an extremely large and nationally representative sample of school-
children, disaggregated ability in terms of verbal, quantitative, and nonverbal
reasoning tests, in contrast to previous UK studies focusing only on overall
IQ. Though for all three tests there were substantial sex differences in the
standard deviation of scores, with greater variance among boys, the mean
differences in the scores were small. Gallagher and Kaufman (2005) also propose that individual differences in ability and achievement *within* gender are
probably much larger than the differences *between* genders.

Gender differences in mathematical ability and performance is one of
the more contentious issues in the psychology of science (Feist, 2012). Many
of the earlier studies reported that males score higher than females (Astin,
1975; Backman, 1972; Benbow, 1988; Benbow & Stanley, 1980, 1983; Deaux,
1985; Holden, 1987; Keating, 1974). As a result, the stereotypes that girls and
women lack mathematical ability persisted for a long time. The stereotypes
are being challenged in recent years due to the mounting evidence of gender
similarities in math achievement (Hedges & Nowell, 1995; Hyde, Fennema,
& Lamon, 1990; Hyde, Lindberg, Linn, Ellis, & Williams, 2008; Lindberg,
Hyde, Linn, & Petersen, 2010). Recent research have found no systematic
sex differences in mean mathematics scores, although it has been observed
that male variances are greater, resulting in disproportionately more males
at both tails of the ability distribution (Ceci & Williams, 2010). The results
of meta-analysis also have provided evidence that, on average, males and
females differ very little in mathematical achievement, despite a more positive math attitude and affect among males (Else-Quest, Hyde, & Linn, 2010).
Furthermore, when analyses are restricted to those with high mathematical ability, fewer women than men choose these fields (Lubinski & Benbow,
2006). Even among mathematically gifted students, males were about twice
as likely as their female counterparts to attain a bachelor's degree in math
or in the physical sciences, and twice as likely to gain employment in such
fields (Benbow, Lubinski, Shea, & Eftekhari-Sanjani, 2000).

Numerous researches have been conducted to examine the causes of the
underrepresentation of women in science and particularly in math-intensive
STEM fields. Hypotheses span innate, biological differences to social factors
(e.g., effects of cultural beliefs, discrimination, and stereotypes). Studies using
sociocultural factors emphasize the primary, early socialization practices,
and biased teachers' and parents' attitudes toward girls and mathematics,

with a resultant stereotype threat and the discriminatory practices in science. These researchers cite the trend of decreasing gender differences and cross-cultural variations in favor of their support. Some of the latest research in psychology use individual and sociocultural factors and success in math-intensive STEM careers in a circular relationship framework (Ceci, Williams, & Barnett, 2009). In contrast, those who assign a greater role to biology stress hormonal, neural, and genetic factors that are alleged to result in a male advantage in spatial and mathematical abilities. Differences in brain development are thought to feed into potential differences in abilities. The brain organization hypothesis, termed brain organization theory, posits that male brain physiology is inherently "built" for efficient spatial and quantitative processing by these sex steroids. Sex steroids increase lateralization and a bias toward right hemispheric processing, the right hemisphere being the area preferentially involved in spatial and numerical processes, particularly those dealing with abstract numerical relations (Baron-Cohen, 2003).

Gender differences in mathematics are also explained in terms of spatial abilities. Benbow and Stanley (1980), for example, framed a hypothesis that sex differences in achievement in and attitude toward mathematics result from superior male mathematical ability, which may in turn be related to greater male ability in spatial tasks. Wai, Lubinski, and Benbow (2009) later identified spatial ability as a salient psychological characteristic to play a critical role in developing expertise in STEM. They, however, also suggested that individuals who are high in spatial ability, but not as exceptional in mathematical or verbal abilities, constitute an untapped pool of talent for STEM domains. A gender gap in mathematics achievement persists in some nations but not in others (Else-Quest et al., 2010).

Mathematics and science performance have for a long time been also associated with stereotypes—"math is male domain"; "math is hard." The stereotypes are known to influence competency beliefs or self-efficacy and may even influence the actual performance. Brown and Josephs (1999) found that women who believed that a math test would reveal weakness in math scored lower on math tests than did women who believed it would reveal whether they were exceptionally good at math. Men showed the reverse pattern. Stereotypes about female inferiority in mathematics, thus, pose a problem and anxiety for psychologists, and have thus led to numerous researches. Such negative stereotypes can have serious consequences for girls doing math (Nosek, Banaji, Greenwald, 2002). Studies reveal the negative effects of stereotyped identity and gender prime on the math test performance. In a study by Ambady, Shih, Kim, and Pittinsky (2001), Asian American girls were subtly reminded about their female identity, their Asian identity, or neither identity. Consistent with the earlier work by Shih, Pittinsky, and Ambady (1999), this study also revealed that when girls were reminded of their female identity, they performed worse on a math test than when their Asian identity or neither identity was primed. Associating the self with female and math with male made it difficult for women, even women who had selected math-intensive majors, to associate math with the self (Nosek, Banaji, & Greenwald, 2002). Steele and Ambady (2006) also observed that a subtle gender prime can shift women's implicit and explicit attitudes toward academic domains in a stereotype-consistent direction. Good, Aronson, and Harder (2008) found that stereotype threat suppresses test performance, even among the most highly qualified and persistent women, in college mathematics. Kawakami, Steele, Cifa, Phills, and Dovidio (2008), on the basis of results of their recent study suggest that shifting women's orientation may be one key strategy to address gender gap in mathematics-related careers. Gender differences in math anxiety and self-concept have received considerable research attention in recent years. Some of the earlier studies have shown that girls, whose competencies in math are regularly questioned by their proximal environment, develop a lower math self-concept, less confidence in their math aptitude, and are less motivated than boys in the math domain (Beyer & Bowden, 1997; Hyde, Fennema, Ryan, Frost, & Hopp, 1990; Parsons, Ruble, Hodges, & Small, 1976; Skaalvik & Rankin, 1994). Steele (1997) proposed that an important factor accounting for the different gender representation in mathematics is identification with the domain. Specifically, Steele suggested that many women have difficulty identifying with fields associated with mathematics because of inadequate support, few role models, and biased societal gender roles and stereotypes. He further argued that it is this dis-identification with math that subsequently impedes many women's motivation, which leads to their underperformance and, ultimately, to their underrepresentation in math-related fields. Recent study by Bonnot and Croizet (2007) have also revealed that women with low math self-evaluation displayed more errors and spent more time solving additions, that too on difficult items, than women with high math self-evaluation.

The issue of gender similarities/differences in mathematics achievement, however, has gained prominence and should be addressed as it has raised the important, rather unsettled, question of what is its cause. A gender gap in mathematics achievement persists in some nations but not in others (Else-Quest et al., 2010). Earlier also no consistent gender differences were found across countries in verbal, mathematical, or spatial abilities (Feingold, 1994). Moreover, some cognitive gender differences have shown a reduction over the years (Feingold, 1988). Many new studies are being conducted, which find realization not only as journal articles but specific research-based books as well. At the same time, various meta-analyses of the extant researches are the latest trend in the hope to find out causes behind women's relative lower representation in math-intensive science careers and to explain relation between gender, science, and mathematical abilities. Ceci and Williams (2010) draw on research in endocrinology, economics, sociology, education, genetics, and psychology to arrive at unique, evidence-based conclusion that the problem is due to certain choices that women (but not men) are compelled to make in our society; women tend not to favor math-intensive careers for certain reasons, and sex differences in math and spatial ability cannot adequately explain the scarcity of women in these fields.

MOTIVATION, ATTITUDES, AND INTEREST

A recurring theme in the psychology literature has been that men and women are motivated toward different goals and values. As a consequence, there have been widely held beliefs in sex-differentiated normative standards, as documented by psychological research on sex stereotypes (e.g., Banaji, Hardin, &

Rothman, 1993; Eagly & Mladinic, 1989; Swim, 1994). To psychologists studying sex differences, these normative beliefs are important because they structure many aspects of the everyday social interactions of men and women (Eagly, 1987; Eagly & Wood, 1991; Ridgeway & Diekema, 1992; Wood & Rhodes, 1992).

To explain the under-participation, which constitutes one of the most important issues of women in science, psychologists began looking at various motivational influences on learning and engagement with math and science. Differences in motivation/interests and activities are believed to lead directly to differences in the choices and, indirectly, to differences in mathematical and spatial abilities. This continues even in recent times. Baron-Cohen (2007) has, for example, argued that females are born with an innate motivation to orient toward people, whereas males have an orientation toward objects, which leads the sexes down differing paths of interests. Systemizing, as a cognitive style has also been identified, to predict greater male interests in science, technology, computers, and natural world (Nettle, 2007). Baron-Cohen (2003) argues that the empathizing–systemizing model has the explanatory power to explain most if not all psychological sex differences. Su, Armstrong, and Rounds (2009) provide a systematic review of the sex differences in the STEM interests and found that women prefer people-oriented careers over things-oriented careers. According to them, interests play a critical role in gendered occupational choices and gender disparity in the STEM fields. It has also been claimed that mean differences in the ways the sexes spend their time (e.g., playing with Legos vs. dolls) may contribute to differences in mathematical and spatial abilities. Some social-psychological research specifically targeting girls' interests in science and technology, however, have produced a few notable findings. For example, more feminine gender identities (Baker, 1987; Boswell, 1985; Handley & Morse, 1984; Hollinger, 1983) and perceptions of gender barriers (Barnett, 1975) decrease interest in scientific and technological curricula. Also, a scientific self-concept (e.g., thinking of oneself as a good science student) is related positively to success in science and mathematics classes (Peterson, Kauchak, & Yaakobi, 1980) and positive attitudes toward science and technology. A recent concern of psychologists has been on the effect of gender discrimination on the gender imbalance in scientific fields. Weisgram and Bigler (2007) carried a research on the effects of informing adolescent girls about gender discrimination and found that girls who learned about gender discrimination showed marked increase in science self-efficacy and belief in the value of science.

Sex differences in career preferences are often cited as among the most important underlying reasons for gender disparity in the STEM fields (e.g., Lubinski & Benbow, 1992, 2006, 2007). A recent meta-analysis revealed sex differences in the people-versus-things dimension of educational/vocational interests (Su, Rounds, & Armstrong, 2009). Ceci and Williams's (2010) synthesis of findings from psychology, endocrinology, sociology, economics, and education lead to the conclusion that, among a combination of interrelated factors, preferences and choices—both freely made and constrained—are the most significant cause of women's underrepresentation in math-intensive fields of studies. They conclude that differential gendered outcomes in the real world result from differences in resources attributable to choices, whether free or constrained (Ceci & Williams, 2011). Such choices could be influenced and better informed through education if resources were so directed. In their latest analyses, Williams and Ceci (2012) zeroed down to a single most important factor in explaining women's underrepresentation: a desire for children and family life. In their view, women's greater desire for lifestyle flexibility, reflecting differing ideas about work–life balance and different expectations regarding responsibility for raising children and working in the home, also plays a role. But how to explain the disparities in the careers within science? Kimberley, Lubinski, and Benbow (2009) opine that gender differences in *lifestyle preferences* and psychological orientation or *orientation toward life, which develop* during their emerging and young adulthood years, are responsible for women's underrepresentation in high-intensity STEM careers. Even among the gifted and top math/science graduate students, cognitive abilities, vocational interests, and lifestyle preferences matter for career choice, performance, and persistence (Kimberley, Smeets, Lubinski, & Benbow, 2010).

RESEARCH PRODUCTIVITY AND PERFORMANCE

Studies related to scientific processes and performance include those (a) focusing on psychological characteristics of scientists, such as creative ability of individual scientists, personality structure or characteristics associated with levels of performance, attitudes, interest, cognitive styles, age intelligence, and so on (e.g., Roe, 1954; Eiduson & Beckman, 1973); (b) involving an analysis of the interrelation between science as a social influence process and individual scientists' research behavior (these studies focus on the behavior of individual scientists in response to broad social influences); (c) focusing on structural context or environmental location and organizational variables; and (d) specific theories related to the process of science and scientific research.

Gender difference in scientific productivity has been an important point of discussion, especially for the academic rank. Various studies have attempted to probe the reasons for this difference (Leahey, 2006; Xie & Shauman, 2003). In particular, psycho-cognitive studies exploring verbal (Hyde & Linn, 1988), spatial (Linn & Peterson, 1985; Voyer, et al., 1995), and mathematical abilities (Hyde, et al., 1990) reveal that, for these abilities, sex differences are found only in few dimensions and, where they do occur, are very limited.

The gender gap in scientific productivity has also been examined in terms of the influence of certain demographic variables (Long & Fox, 1995; Xie & Shauman, 2003). But, in general, differences in background measures do not appear to explain gender and scientific publication relationships. Even the interactions of gender with family characteristics such as marriage and motherhood are found to be weak, inconclusive, or varied (Fox, 2005; Xie & Shauman, 2003). Cultural, social, and organizational factors play an important role. Psychological approaches have sometimes justified the existence of differences in scientific productivity based on individual characteristic, especially those related to personality, ability, and gender of the individual. Helmreich and Spence, for example, proposed a model to explain scientific productivity in which they hypothesized that the two main determinants of scientific productivity are motivation and gender of the researchers (Helmreich & Spence, 1982).

A commonly used measure of scientific productivity is publication rate (e.g., Fox, 1989; Joy, 2006; Leahey, 2006). Research during the earlier decades indicated that women published less than men (Cole, 1979; Cole & Zuckerman, 1984; Helmreich, Spence, Beane, Lucker, & Matthews, 1980; Long, 1992; Reskin, 1978). In contrast, recent studies have revealed small to no gender differences in publication rates (D'Amico & Di Giovanni, 2000; Joy, 2006; Long & Fox, 1995; Mauleón & Bordons, 2006; Sonnert & Holton, 1995; Ward & Grant, 1996; Xie & Shauman, 1998). Some of the latest research results are interesting and point to a more complex relationship between gender and publication patterns. Joy (2006), for example, observed that while male academic psychologists tend to publish more than females during the predoctoral and first eight postdoctoral years, which happen to be the periods for search for academic employment and initial push for tenure, but not thereafter. Females, unlike males, tend to increase their publication rates as they mature professionally. Vermigli and Canetto (2011) in a recent study found that the female–male differences in scientific publications disappeared when academic rank was considered—a finding consistent with publication trends by United States' academic psychologists (Joy, 2006) as well as publication trends by Spanish material scientists (Mauleón & Bordons, 2006). D'Amico, Patrizia, and Sara (2011) too observed that a strong female doctoral pipeline and scientific productivity are very slow at influencing the under representation of Italian women at the top ranks of academia.

To conclude, female and male patterns of scientific publication and academic advancement, which were described as a puzzle by Cole and Zuckerman in 1984, still remain unresolved. It is unclear what may be the best measures (Bell & Gordon, 1999; Ceci et al., 2009) and what are the best predictors of productivity (Long & Fox, 1995; Prpić, 2002).

ROLE OF ORGANIZATIONAL FACTORS

Despite the increasing representation of women in the workplace (glass ceilings and barriers withstanding) occupational psychology has progressed largely in a gender-blind way, marginalizing women as workers (Mills & Tancred, 1992). A few feminist theoretical critiques, however, shed light on the gendered nature of organizations, the power relations that exist within the wider society and are reproduced within the workplace. Although feminist psychologists have contributed to documenting problems women face in organizations with patriarchal structures and cultures (Cassell & Walsh, 1993) as well as the psychological implications for women of "gendered power and political worksites" (Nicolson, 1996), with a few exceptions, scientific organizations are largely absent from their purview. In fact, even within the psychology of science, organizational psychology remains the least discussed subdiscipline in comparison to other subfields. Studies of organizational factors are particularly important for the modern day science, which are being conducted in "big" organizations, mostly in teams and gropups, rather than by lone natural philosophers, like Newton or Galileo of earlier times! Some of the sociological studies do discuss the role of organizational issues, such as sex and location in the institution of science (Reskin, 1978), women in scientific community (Cole, 1979), and rank advancement in scientific careers (Long, Allison, & McGinnis, 1993).

The significance of hierarchies in terms of gender stratification within Indian scientific organizations was explored and established (Kumar, 2001, 2009). Later, Amâncio (2005) published on social-psychological dimensions of the history of women in science and articulates studies on gender and science with the theory of social representations. She utilizes organizational aspects as explanations for women's continued underrepresentation in the sciences. She argues that, in spite of structural changes in the participation and status of women in science, there is continuity in the gender representations that are embedded in the culture of scientific institutions. Amâncio sums up that there are two main reasons for the continued gendered disparity: gender representations are embedded in the organizational forms and dominant culture of science; and scientists and scientific institutions have contributed to the perpetuation of gender representations, in particular, the representation of women as a sexed category.

Virginia Valian (2005) sheds more light by focusing on a social-cognitive perspective. Her social-cognitive account relies on two key concepts: gender schemas and the accumulation of advantage. Gender schemas refer to cognitive structures of organized prior knowledge regarding the role expectations of individuals based on biological sex. The gender schemas, which are shared by all of us, result in our overrating men and underrating women in professional settings, in small, barely visible ways, and these small disparities accumulate over time to provide men with more advantages than women. In contrast to the accumulative advantage (metaphorically known as the Matthew Effect), Rossiter (1993) has identified the "Matilda Effect," by which women and their scientific contributions are credited to men or overlooked entirely. Contrary to the norms of universalism, women's scientific efforts are devalued compared with those of men (Long & Fox, 1995). One of the latest findings also confirms that the "Matilda Effect" persists—the scientific efforts and achievements of women do not receive the same recognition as do those of men. In contrast, men receive an outsized share of scholarly awards and prizes (Lincoln, Pincus, Koster, & Leboy, 2012).

Organizations bear an important responsibility through the way they reinforce or alleviate difficulties that women and men face in contributing to scientific research at all levels (Cheveigné, 2009). A qualitative study of perceptions of science careers in the Centre National de la Recherche Scientifique (CNRS), the main research institution in France, to understand the "glass ceiling" effect could identify factors such as tension between individual and collective dimensions of research activity, and long-term time-management problems. Cheveigné noticed that these affect both men and women, but in different ways. A number of mechanisms penalize women more than men within a research system, such as, poor internal communication and weak human resource management. Most fundamentally, women tend to mobilize a more collective model of scientific research than do men, but men fare better when the evaluation procedures focus on the individual. The actions necessary for improving the situation of women are not specific to them—rather, a systematic improvement in the conditions for practicing and evaluating scientific research, would improve the situation of all personnel, especially women. Reflection on the collective nature of the scientific enterprise and on its evaluation is needed to allow women and men to give their full measure while working together. Dasgupta (2011), using a model that integrates insights from social psychology and organizational behavior, reveals that ingroup experts and peers serve as social vaccines, increase social belonging, and inoculate fellow group members' self-concepts against stereotypes.

CONCLUSIONS

Historical experiences confirm beyond doubt the prevalence of gender discrimination in science. Women did not have access to universities, which was the center for science for almost 700 years, could not become members of scientific academies for a long time, and remained largely absent from historical writings. During the 20th century, the situation of women in science was not the only thing changed the historical, philosophical, and sociological studies and explanations also changed their focus and started including gender dimensions in their approach. Psychology, a discipline with a long history but a short past, has shown interest in sex as a variable for more than a century. It did not ignore science and Galton had looked at science as early as 1879. Later attention was paid to scientists' personality traits and their career choices and achievements and sex figured into these accounts. The debate still continues on gender similarities and differences in science, particularly in regard to abilities, interests, and achievements.

In recent years the gender–science relationship has changed, but is not totally free from biases. Women's enrolment into various science disciplines has shown remarkable and steady increase in most parts of the world, yet they remain underrepresented in the STEM workforce, although to a lesser degree than in the past. Overall, women account for a minority of the world's researchers (UIS, 2011). The most extreme gender disparities are within engineering. Women earned, for example, 41% of S&E doctoral degrees awarded in the United States in 2008, about the same percentage is earned by women in Australia, Canada, the European Union, and Mexico. Women earned more than half of S&E doctoral degrees in Portugal. Yet, women hold a larger share of junior faculty positions, that is, assistant professor than positions at senior level, that is professor rank. For example, in 2008, women constituted 21% of Full Professors, 37% of Associate Professors, and 42% of Junior Faculty in the United States (NSB, 2012). Female scientists and engineers are concentrated in certain fields, such as social, biological, and medical sciences and are relatively low in numbers in the engineering, computer, and mathematical sciences. The proportion of women in European countries among full professors, for example, is highest in the humanities and the social sciences (respectively 27.0% and 18.6%) and lowest in engineering and technology, at 7.2% (She Figures, 2009). Within social science also, some variations can be seen. Psychology, for example, has more women than economics. One goal of the psychology of science, therefore, is to unpack some of the factors behind this phenomenon gendered science.

In this chapter, I have demonstrated that the psychology of gender and science covers numerous topics and entails several methodological issues. Hundreds of published studies have discussed gender and science issues, yet, the gender difference in STEM raises the important, rather unsettled, question of what is its cause. Explanations include both biological differences in aptitude and more socially constructed differences, including the gendered socialization of boys and girls, gender-role stereotypes, negative expectancies and the biased social and organizational processes of science per se. Links between gender and science are of universal concern. Further research and theory on gender and science call for more rigorous testing of new concepts and greater interdisciplinarity. Theoretical sophistication is not only desirable but essential. The science and craft of psychology can play the role of a bridgehead here. Within the discipline of psychology, its subtexts can also benefit and enlighten science studies. New researches have enough to show how useful a psychological perspective can be for understanding gender–science relationships. In recent decades, science has made enormous advances; our economy is now entirely technology driven and people talk of knowledge economy and knowledge society. Even slow-paced societies in Asia and Africa are showing the signs of strain and structural changes. Yet, psychology is largely absent from gender-technology studies! We hope to make gender an important category of analysis, and indeed gender as a new and multidisciplinary configuration has recently emerged. Despite the mounting evidence of gender similarities, stereotypes that girls and women lack mathematical and scientific ability persist and gender discriminations are still ubiquitous. To understand the debates over the causes of gender differences in science and engineering careers, an integrative framework is needed, spanning different disciplines, methods, and approaches.

ACKNOWLEDGMENT

This paper was revised and finalized while I was a visiting scholar at the Centre for Feminist Research and the Department of Psychology, York University, Toronto.

REFERENCES

Addis, E., & Villa, P. (2003). The editorial boards of Italian economics journals: Women, gender, and social networking. *Feminist Economics, 9,* 75–91.

Amâncio, L. (2005). Reflections on science as a gendered endeavour: changes and continuities. *Social Science Information, 44,* 65–83.

Ambady, N., Shih, M., Kim, A., & Pittinsky, T. (2001). Stereotype susceptibility in children: Effects of identity activation on quantitative performance. *Psychological Science,12,* 385–390.

Ash, M. G. (2003). Psychology. In D. Ross & T. Porter (Eds.), *The modern social sciences, Cambridge history of science: Vol. 7,* (pp. 251–274). New York, NY: Cambridge University Press.

Ash, M. G. (2005). The uses and usefulness of psychology. *Annals of the American Academy of Political and Social Science, 600*, 99–114.

Astin, H. S. (1975). Sex differences in mathematical and scientific precocity. *Journal of Special Education, 9*(1), 79–91.

Backman, M. E. (1972). Patterns of mental abilities: Ethnic, socioeconomic, and sex differences. *American Educational Research Journal, 9*(1), 1–12.

Baker, D. (1987). The influence of role-specific self-concept and sex-role identity on career choices in science. *Journal of Research in Science Teaching, 24*, 739–756.

Banaji, M. R., Hardin, C., & Rothman, A. J. (1993). Implicit stereotyping in person judgment. *Journal of Personality and Social Psychology, 65*, 272–281.

Bandura, A. (1986). *Social foundations of thought and action: A social cognitive theory.* Englewood Cliffs, NJ: Prentice-Hall.

Bandura, A. (1997). *Self-efficacy: The exercise of control.* New York, NY: Freeman.

Barnett, R. (1975). Sex differences and age trends in occupational preference and occupational prestige. *Journal of Counseling Psychology, 22*, 35–38.

Baron-Cohen, S. (2003). *The essential difference: Men, women and the extreme male brain.* London, UK: Penguin.

Baron-Cohen, S. (2007). Sex differences in mind: Keeping science distinct from social policy In S. J. Ceci & W. M. Williams (Eds.), *Why aren't more women in science? Top researchers debate the evidence* (pp. 159–172). Washington, DC: American Psychological Association.

Bell, S., & Gordon, J. (1999). Scholarship: The new dimension to equity issues for academic women. *Women's Studies International Forum, 22*, 645–658.

Belkin, L. (2008). Diversity isn't rocket science, is it? *The New York Times*, p. A-2.

Benbow, C. P. (1988). Sex differences in mathematical reasoning ability in intellectually talented preadolescents: their nature, effects, and possible causes. *Behavioural and Brain Sciences, 11*(2), 169–183.

Benbow, C. P., & Stanley, J. C. (1980). Sex differences in mathematical ability: Fact or artifact? *Science, 210*, 1262–1264.

Benbow, C. P., Lubinski, D., Shea, D. L., & Eftekhari-Sanjani, H. E. (2000). Sex differences in mathematical reasoning ability at age 13: Their status 20 years later. *Psychological Science, 11*, 474–480.

Beyer, S., & Bowden, E. M. (1997). Gender differences in self-perceptions: convergent evidence from three measures of accuracy and bias. *Personality and Social Psychology Bulletin, 23*, 157–172.

Block, J. H. (1984). *Gender role identity and ego development.* San Francisco, CA: Jossey-Bass Publishers.

Bonner, T. N. (2000). *Becoming a physician: Medical education in Britain, France, Germany, and the United States, 1750–1945*, Baltimore, MD: The John Hopkins University Press.

Bonnot, V., & Croizet, J. (2007). Stereotype internalization and women's math performance: The role of interference in working memory. *Journal of Experimental Social Psychology, 43*, 857–866.

Boring, E. G. (1951). The woman problem. *American Psychologist, 6*, 679–682.

Bosch, M. (2002). Women and science in the Netherlands: A Dutch case? *Science in Context, 15*, 484–527.

Boswell, S. (1985). The influence of sex-role stereotyping on women's attitudes and achievement in mathematics. In S. F. Chipman & L. R. Brush, & D. M. Wilson (Eds.), *Women and mathematics: Balancing the equation* (pp. 275–328). Hillsdale, NJ: Erlbaum.

Bray, F. (2012) Gender and technology. In N. Kumar (Ed.), *Gender and science: Studies across cultures* (pp.37–60). New Delhi, India: Foundation Books, Cambridge University Press.

Brown, R. P., & Josephs, R. A. (1999). A burden of proof: Stereotype relevance and gender differences in math performance. *Journal of Personality and Social Psychology, 76*, 246–257.

Burt, C., & Moore, R. C. (1912). The mental differences between the sexes. *Journal of Experimental Pedagogy, 1*, 355–388.

Buss, D. M. (1995). Psychological sex differences: Origins through sexual selection. *American Psychologist, 50,* 164–168.

Bussey, K., & Bandura, A. (1999). Social cognitive theory of gender development and differentiation. *Psychological Review, 106,* 676–713.

Bussey, K., & Bandura, A. (2004). Social cognitive theory of gender development and functioning. In A. Eagly, A. Beall, & R. Sternberg (Eds.), *Psychology of gender* (pp. 92–119). New York, NY: Guilford Publications.

Bynum, W. F. (1994). *Science and the practice of medicine in nineteenth century.* New York, NY: Cambridge University Press.

Cassell, C., & Walsh, S. (1993). Being seen but not heard: barriers to women's equality in the workplace. *The Psychologist, 6,* 110–114.

Cattell, J. M. (1906). *American men of science.* New York, NY: Science Press.

Ceci, S. J., & Williams, W. M. (2007). *Why aren't more women in science? Top researchers debate the evidence.* Washington, DC: American Psychological Association.

Ceci, S. & Williams, W. (2010). *The mathematics of sex: how biology and society conspire to limit talented women and girls.* New York, NY: Oxford University Press.

Ceci, S. J., Williams, W. M., & Barnett, S. M. (2009). Women's underrepresentation in science: sociocultural and biological considerations. *Psychological Bulletin, 135,* 218–261.

Ceci, S. J., & Williams, W. M. (2010). Sex differences in math-intensive fields. *Current Directions in Psychological Science, 19,* 275–279.

Ceci, S. J., & Williams, W. M. (2011). Understanding current causes of women's underrepresentation in science. *Proceedings of the National Academy of Sciences, 108,* 3157–3162.

Cheveigné, S. de. (2009). The career paths of women (and men) in French research. *Social Studies of Science, 39,* 113–136.

Cockburn, C. (1983). *Brothers: male dominance and technological change.* London, England: Pluto.

Cole, J. R. (1979). *Fair science: Women in the scientific community.* New York, NY: Free Press.

Cole, J. R., & Cole, S. (1979). *Social stratification in science.* Chicago, IL: The University of Chicago Press.

Cole, J. R., & Zuckerman, H. (1984). The productivity puzzle: Persistence and change in patterns of publications of men and women scientists. In L. Martin, M. Maehr, & M. W. Steinkamp (Eds.), *Advances in Motivation and Achievement* (pp. 217–258). Greenwich, CN: JAI Press.

Connell, R. W. (1987). *Gender and power: Society, the person and sexual politics.* Cambridge, MA: Polity Press.

Costas, I. (2002). Women in science in Germany. *Science in Context, 15,* 557–576.

Cowan, R. S. (1983). *More work for mother: The ironies of household technology from the open hearth to the microwave.* New York, NY: Basic Books.

Cox, S. (Ed.). (1981). *Female psychology: The emerging self* (2nd ed.). New York, NY: St. Martin's Press.

D'Amico, R., & Di Giovanni, M. (2000). Pubblicazioni in riviste di psicologia: Un'analisi di genere. *Bollettino di Psicologia Applicata, 232,* 39–47.

D'Amico, R., Patrizia, V., & Sara, C. S. (2011). Publication productivity and career advancement by female and male psychology faculty: The case of Italy. *Journal of Diversity in Higher Education, 4,* 175–184.

Dasgupta, N. (2011). Ingroup experts and peers as social vaccines who inoculate the self-concept: The Stereotype Inoculation Model. *Psychological Inquiry, 22,* 231–246.

Deaux, K. (1985). Sex and gender. *Annual Review of Psychology, 36,* 49–81.

Deary, I. J., Thorpe, G., Wilson, V., Starr, J. M., & Whalley, L. J. (2003). Population sex differences in IQ at age 11: The Scottish mental survey 1932. *Intelligence, 31,* 533–542.

Delamont, S. (1987). Three blind spots? A comment on the sociology of science by a puzzled outsider. *Social Studies of Science, 17,* 163–170.

Eagly, A. H. (1987). *Sex differences in social behavior: A social-role interpretation.* Hillsdale, NJ: Erlbaum.

Eagly, A. H. (1994). On comparing women and men. *Feminism in Psychology, 4,* 513–522.

Eagly, A. H., & Mladinic, A. (1989). Gender stereotypes and attitudes toward women and men. *Personality and Social Psychology Bulletin, 15,* 543–558.

Eagly, A. H., & Wood, W. (1991). Explaining sex differences in social behavior: a meta-analytic perspective. *Personality and Social Psychology Bulletin, 17*, 306–315.

Eagly, A. H., & Wood, W. (1999). The origins of sex differences in human behavior. *American Psychologist, 54*, 408–423.

Eagly, A. H., Eaton, A., Rose, S. M., Riger, S., & McHugh, M. C. (2012, February 27). Feminism and psychology: analysis of a half-century of research on women and gender. *American Psychologist*, Advance online publication.

Eagly, A. H., Eaton, A., Rose, S. M., Riger, S., McHugh, M. C. (2012). Feminism and psychology: Analysis of a half-century of research on women and gender. *American Psychologist, 67*, 211–230.

Eccles, J. S. (1987). Gender roles and women's achievement-related decisions. *Psychology of Women Quarterly, 11*, 135–172.

Eccles, J. S. (1994). Understanding women's educational and occupational choices: Applying the Eccles et al. model of achievement-related choices. *Psychology of Women Quarterly, 18*, 585–610.

Eiduson, B. T., & Beckman, L. (Eds.). (1973). *Science as a career choice: Theoretical and empirical studies.* New York, NY: Russell Sage Foundation.

Eisenhart, M., & Finkel, E. (1998). *Women's science: Learning and succeeding from the margins.* Chicago, IL: University of Chicago Press.

Else-Quest, N. M., Hyde, J. S., & Linn, M. C. (2010). Cross-national patterns of gender differences in mathematics: A meta-analysis. *Psychological Bulletin, 136*, 103–127.

England, P., & Su Li, M. (2006). Desegregation stalled: The changing gender composition of college majors, 1971–2002. *Gender & Society, 20*, 657–677.

England, P., Allison, P., Su Li, M., Mark, N., Thompson, J., Michelle, B., & Sun, F. (2007). Why are some academic fields tipping toward female? The sex composition of U.S. fields of doctoral degree receipt, 1971–2002. *Sociology of Education, 80*, 23–42.

Etzokowitz, H., & Kemelgor, C. (2001). Gender inequality in science: A universal condition? *Minerva, 39*, 153–174.

European Commission. (2004). *Gender and excellence in the making.* Luxembourg: Office for the Official Publications of the European Communities.

Fara, P. (2002). Elizabeth Tollet: A new Newtonian women. *History of Science, 40*, 169–187.

Fausto-Sterling, A. (1985). *Myths of gender: Biological theories about men and women.* New York, NY: Basic Books.

Fausto-Sterling, A. (2005). The bare bones of sex: Part 1–Sex and gender. *Signs: Journal of Women in Culture and Society, 30*, 1491–1527.

Feingold, A. (1988). Cognitive gender differences are disappearing. *American Psychologist, 43*, 95–103.

Feingold, A. (1994a). Gender differences in variability in intellectual abilities: A cross-cultural perspective. *Sex Roles, 30*, 81–92.

Feingold, A. (1994b). Gender differences in personality: a meta-analysis. *Psychological Bulletin, 116*, 429–456.

Feist, G. J. (2006a). *The psychology of science and the origins of the scientific mind.* New Haven, CT: Yale University Press.

Feist, G. J. (2006b).The psychology of science: Development and personality in scientific thought, interest and achievement. *Review of General Psychology, 10*, 163–182.

Feist, G. J. (2006c). The development of scientific talent in Westinghouse finalists and members of the National Academy of Sciences. *Journal of Adult Development, 13*, 23–35.

Feist, G. J. (2011). Psychology of science as a new sub-discipline in psychology. *Current Directions in Psychological Science, 20*, 330–334.

Feist, G. J. (2012). Gender, science, and the psychology of science. In N. Kumar (Ed.), *Gender and science: Studies across cultures* (pp. 61–75). New Delhi, India: Foundation Books, Cambridge University Press.

Feist, G. J., & Gorman, M. E. (1998). Psychology of science: Review and integration of a nascent discipline. *Review of General Psychology, 2*, 3–47.

Fox, M. F. (1989). Disciplinary fragmentation, peer review, and the publication process. *American Sociologist, 20*, 188–191.

Fox, M. F. (2005). Gender, family characteristics, and publication productivity among scientists. *Social Studies of Science, 35*, 131–150.

Gallagher, A. M., & Kaufman, J. C. (2005). *Gender differences in mathematics: An integrative psychological approach.* Cambridge, England: Cambridge University Press.

Galton, F. (1874). *English men of science: Their nature and nurture.* London, England: Macmillan.

Gilligan, C. (1982). *In a different voice: Psychological theory and women's development.* Cambridge, MA: Harvard University Press.

Good, C., Aronson, J., & Harder, J. (2008). Problems in the pipeline: Stereotype threat and women's achievement in high-level mathematics courses. *Journal of Applied Developmental Psychology, 29*, 17–28.

Golinski, J. (2002). The care of the self and the masculine birth of science. *History of Science, 40*, 1–21.

Guo, C. Tsang, M. C. & Ding, X. (2010). Gender disparities in science and engineering in Chinese universities. *Economics of Education Review, 29*, 225–235.

Halpern, D. F. (2000). *Sex differences in cognitive abilities* (3rd ed.). Mahwah, NJ: Lawrence Erlbaum Associates.

Halpern, D. F., & LaMay, M. L. (2000). The smarter sex: A critical review of sex differences in intelligence. *Educational Psychology Review, 12*, 229–246.

Halpern, D. F., Benbow, C. P., Geary, D. C., Hyde, J. S., & Gernsbacher, M. A. (2007). The science of sex differences in science and mathematics. *Psychological Science in the Public Interest, 8*, 1–51.

Handley, H. M., & Morse, L. W. (1984). Two-year study relating adolescents' self-concept and gender role perceptions to achievement and attitudes toward science. *Journal of Research in Science Teaching, 21*, 599–607.

Haraway, D. (1991). Situated knowledges: The science question in feminism and the privilege of partial perspective. In D. Haraway (Ed.), *Simians, cyborgs, and women* (pp. 183–201). New York, NY: Routledge.

Haraway, D. (1991). *Simians, cyborgs, and women: The Reinvention of Nature.* New York, NY: Routledge.

Harding, S. (1986). *The science question in feminism.* Ithaca, NY: Cornell University Press.

Hare-Mustin, R., & Marecek, J. (Eds.). (1990). *Making a difference: Psychology and the construction of gender.* New Haven, CT: Yale University Press.

Hedges, L. V., & Nowell, A. (1995). Sex differences in mental test scores, variability, and numbers of high-scoring individuals. *Science, 269*, 41–45.

Helmreich, R. L., & Spence, J. T. (1982). Gender differences in productivity and impact. *American Psychologist, 36*, 1142.

Helmreich, R. L., Spence, J. T., Beane, W. E., Lucker, G. W., & Matthews, K. A. (1980). Making it in academic psychology: Demographic and personality correlates of attainment. *Journal of Personality and Social Psychology, 39*, 896–908.

Hermann, C., & Cyrot-Lackmann, F. (2002). Women in science in France. *Science in Context, 15*, 529–556.

Holden, C. (1987). Female math anxiety on the wane. *Science, 236*(4802), 600–601.

Hollinger, C. (1983). Self-perception and career aspirations of mathematically talented female adolescents. *Journal of Vocational Behavior, 22*, 49–62.

Hollingworth, L. (1916). Sex differences in mental traits. *Psychological Bulletin, 13*, 377–384.

Hyde, J. S. (1985). *Half the human experience* (3rd ed.). Lexington, MA: D. C. Heath & Co.

Hyde, J. S., Fennema, E., & Lamon, S. J. (1990). Gender differences in mathematics performance: A meta-analysis. *Psychological Bulletin, 107*, 139–155.

Hyde, J. S., Fennema, E., Ryan, M., Frost, L. A., & Hopp, C. (1990). Gender comparisons of mathematics studies and effects: A meta-analysis. *Psychology of Women Quarterly, 14*, 299–324.

Hyde, J. S., Lindberg, S. M., Linn, M. C., Ellis, A. B., & Williams, C. C. (2008). Gender similarities characterize math performance. *Science, 321*, 494–495.

Hyde, J. S., & Linn, M. C. (1988). Gender differences in verbal ability: A meta-analysis. *Psychological Bulletin, 104*, 53–69.

Jacobs, J. A. (2003). Detours on the road to equality: Women, work and higher education. *Contexts, 2*, 32–41.

Jacobs, J. E., Davis-Kean, P., Bleeker, M., Eccles, J. S. & Malanchuk, O. (2005). I can, but I don't want to: The impact of parents, interests, and activities on gender differences in math. In A. Gallagher & J. Kaufman (Eds.), *Gender differences in mathematics* (pp. 246–263). New York, NY: Cambridge University Press.

Jordonova, L. (1993). Gender and historiography of science. *The British Journal for the History of Science, 26,* 469–483.

Jost, J. T., Banaji, M. R., & Nosek, B. A. (2004). A decade of system justification theory: Accumulated evidence of conscious and unconscious bolstering of the status quo. *Political Psychology, 25,* 881–919.

Jost, J. T., & Kay, A. C. (2005). Exposure to benevolent sexism and complementary gender stereotypes: Consequences for specific and diffuse forms of system justification. *Journal of Personality and Social Psychology, 88,* 498–509.

Joy, S. (2006). What should I be doing, and where are they doing it? Scholarly productivity of academic psychologists. *Perspectives on Psychological Science, 1,* 346–364.

Kawakami, K., Steele, J. R., Cifa, C., Phills, C. E., & Dovidio, J. F. (2008). Approaching math increases math = me and math = pleasant. *Journal of Experimental Social Psychology, 44,* 818–825.

Keating, D. P. (1974). The study of mathematically precocious youth. In J. C. Stanley, D. P. Keating, & L. H. Fox (Eds.), *Mathematical talent: Discovery, description, and development* (pp. 23–46). Baltimore, MD: Johns Hopkins University Press.

Keller, E. F. (1978). Gender and science. *Psychoanalysis and Contemporary Thought, 1,* 409–433.

Kennedy, B., Lin, L. Y., & Dickstein, L. J. (2001). Women on the editorial boards of major journals. *Academic Medicine, 76,* 849–851.

Kimberley, F., Lubinski, D., & Benbow, C. P. (2009). Work preferences, life values, and personal views of top math/science graduate students and the profoundly gifted: Developmental changes and gender differences during emerging adulthood and parenthood. *Journal of Personality and Social Psychology, 97,* 517–532.

Kimberley, F., Smeets, S., Lubinski, D., & Benbow, C. P. (2010). Beyond the threshold hypothesis: Even among the gifted and top math/science graduate students, cognitive abilities, vocational interests, and lifestyle preferences matter for career choice, performance, and persistence. *Current Directions in Psychological Science, 19,* 346–351.

Kumar, N. (2001). Gender and stratification in science: an empirical study in Indian setting. *Indian Journal of Gender Studies, 8,* 51–67.

Kumar, N. (Ed.). (2009). *Women and science in India.* New Delhi, India: Oxford University Press.

Kumar, N. (Ed.). (2012). *Gender and science: Studies across cultures.* New Delhi, India: Foundation Books, Cambridge University Press.

Leahey, E. (2006). Gender differences in productivity: Research specialization as the missing link. *Gender & Society, 20,* 754–780.

Levy, J., & Heller, W. (1992). Gender differences in human neuropsychological function. In A. A. Gerall, H. Moltz, & I. L. Ward (Eds.), *Handbook of behavioral neurobiology.* New York, NY: Plenum.

Lightman, B. (2007). *Victorian popularizers of science: Designing nature for new audiences.* Chicago, IL: University of Chicago Press.

Lincoln, A. E., Pincus, S., Koster, J. B., & Leboy, P. S. (2012). The Matilda effect in science: Awards and prizes in the US, 1990s and 2000s. *Social Studies of Science, 42,* 307–320.

Linn, M. C., & Petersen, A. C. (1985). Emergence and characterization of sex differences in spatial ability: A meta-analysis. *Child Development, 56,* 1479–1498.

Long, J. S. (1992). Measures of sex differences in scientific productivity. *Social Forces, 70,* 159–178.

Long, J. S. (2001). *From scarcity to visibility: Gender differences in the careers of doctoral scientists and engineering.* Washington, DC: National Academy Press.

Long, J. S., & Fox, M. F. (1995). Scientific careers: Universalism and particularism. *Annual Review of Sociology, 21,* 45–71.

Long, J. S., Allison, P. D., & McGinnis, R. (1993). Rank advancement in academic careers: Sex differences and effect of productivity. *American Sociological Review, 58,* 703–723.

Lubinski, D., & Benbow, C. P. (1992). Gender differences in abilities and preferences among the gifted: Implications for the math/science pipeline. *Current Directions in Psychological Science, 1,* 61–66.

Lubinski, D., & Benbow, C. P. (2006). Study of mathematically precocious youth after 35 years: Uncovering antecedents for math science expertise. *Perspectives on Psychological Science, 1,* 316–345.

Lubinski, D., & Benbow, C. P. (2007). Sex differences in personal attributes for the development of scientific expertise. In S. J. Ceci & W. M. Williams (Eds.), *Why aren't more women in science?* (pp. 79–100). Washington, DC: American Psychological Association.

Maccoby, E. E. (1988). Gender as a social category. *Developmental Psychology, 26,* 755–765.

Maccoby, E. (1998). *The Two Sexes: Growing Up Apart, Coming Together.* Cambridge, MA: Belknap.

Maccoby, E., & Jacklin, C. (1974). *The psychology of sex differences.* Stanford, CA: Stanford University Press.

MacLeod, R., & Moseley, R. (1979). Fathers and daughters: Reflections on women, science and Victorian Cambridge. *History of Education, 8,* 321–333.

Mauleón, E., & Bordons, M. (2006). Productivity, impact and publication habits by gender in the area of materials science. *Scientometrics, 66,* 199–218.

McClellan, J. III. (2003). Scientific institutions and the organization of science. In R. Porter (Ed.), *Eighteenth century science* (pp. 87–106). (series in Cambridge history of science) Cambridge, MA: Cambridge University Press.

Mellström, U. (2009). The intersection of gender, race and cultural boundaries, or why is computer science in Malaysia dominated by women? *Social studies of Science, 39,* 885–907.

Metz, I., & Harzing, A. W. (2009). Gender diversity in editorial boards of management journals. *The Academy of Management Learning and Education, 8,* 540–557.

Mills, A. J., & Tancred, P. (1992). *Gendering organizational analysis.* Newbury Park, CA: Sage.

Morton, M. J., & Sonnad, S. S. (2007). Women on professional society and journal editorial boards. *Journal of the National Medical Association, 99,* 764–771.

National Science Board. (2012). *Science and engineering indicators 2012.* Arlington, VA: National Science Foundation (NSB 12–01).

Nettle, D. (2007). Empathizing and systemizing: What are they, and what do they contribute to our understanding of psychological sex differences? *British Journal of Psychology, 98,* 237–255.

Nicolson, P. (1996). *Gender, power and organization: A psychological perspective.* London, UK: Routledge.

Nosek, B. A., Banaji, M. R., & Greenwald, A. G. (2002). Math = male, me = female, therefore math ≠ me. *Journal of Personality and Social Psychology, 83,* 44–59.

Oldenziel, R. (1999). *Making technology masculine: Men, women, and modern machines in America, 1870–1945.* Amsterdam, Holland: Amsterdam University Press.

Outram, D. (1987). The most difficult career: Women's history in science. *International Journal of Science Education, 9,* 409–416.

Parsons, J. E., Ruble, D. N., Hodges, K. L., & Small, A. W. (1976). Cognitive developmental factors in emerging sex differences in achievement related expectancies. *Journal of Social Issues, 32,* 47–61.

Peterson, K., Kauchak, D., & Yaakobi, D. (1980). Science students' role-specific self-concept: Course, success, and gender. *Science Education, 64,* 169–174.

Poole, M., Bornholt, L., Summers, F. (1997). An international study of the gendered nature of academic work: Some cross cultural explorations. *Higher Education, 34:* 373–396.

Prpic, K. (2002) 'Gender and Productivity Differentials in Science', *Scientometrics 55:* 27–58.

Reskin, B. F. (1978). Scientific productivity, sex and location in the institutions of science. *American Journal of Sociology, 85,* 1235–1243.

Ridgeway, C., & Diekema, D. (1992). Are gender differences status differences? In C. Ridgeway (Ed.), *Gender, interaction, and inequality* (pp. 157–180). New York, NY: Springer-Verlag.

Riger, S. (1992). Epistemological debates, feminist voices: science, social values, and the story of women. *American Psychologist, 47*, 730–740.

Roe, A. (1954). A psychological study of eminent psychologists and anthropologists and a comparison with biological and physical scientists. *Psychological Monographs, 67*, 1–55.

Rossiter, M. W. (1997). Which science? Which women? *OSIRIS, 12*, 169–185.

Rutherford, A., & Granek, L. (2010). Emergence and development of the psychology of women. In J. Chrisler & D. McCreary (Eds.), *Handbook of gender research in psychology* (pp. 19–41). New York, NY: Springer.

Rutherford, A., Capdevila, R., Undurti, V., & Palmary, I. (Eds.). (2011). *Handbook of international feminisms: Perspectives on psychology, women, culture and rights.* New York, NY: Springer.

Schiebinger, L. (1989). *The mind has no sex? Women in the origin of modern science.* Cambridge, MA: Harvard University Press.

Schiebinger, L. (1999). *Has feminism changed science?* Cambridge, MA: Harvard University Press.

Schiebinger, L. (2003). Philosopher's beard: Women and gender in science. In R. Porter (Ed.), *Science in the eighteenth century,* (Vol. 4, pp. 184–210). (Series in the Cambridge history of science) Cambridge, England: Cambridge University Press.

Shields, S. A., & Dicicco, E. C. (2011). The social psychology of sex and gender: From gender differences to doing gender. *Psychology of Women Quarterly, 35*, 491–499.

Schmitt, D. P., Realo, A., Voracek, M., & Allik, J. (2008). Why can't a man be more like a woman? Sex differences in big five personality traits across 55 cultures. *Journal of Personality and Social Psychology, 94*, 168–182.

Scott, J. W. (1986). Gender—a useful category of historical analysis. *American Historical Review, 91*, 1053–1075.

She Figures. (2009). *Statistics and indicators on gender equality in science.* Luxembourg, Publications Office of the European Union.

Shields, S. A. (1975). Functionalism, Darwinism, and the psychology of women: A study in social myth. *American Psychologist, 30*, 739–754.

Shih, M., Pittinsky, T. L., & Ambady, N. (1999). Stereotype susceptibility: Identity salience and shifts in quantitative performance. *Psychological Science, 10*, 80–83.

Shteir, A. B. (1996). *Cultivating women, cultivating science: Flora's daughter and botany in England, 1760 to 1860.* Baltimore, MD: John Hopkins Press.

Shteir, A. B., & Lightman, B. (2006). *Figuring it out: science, gender and visual culture.* Hanover, NH: Dartmouth College Press.

Simonton, D. K. (2008). Gender differences in birth order and family size among 186 eminent psychologists. *Journal of Psychology of Science and Technology, 1*, 15–22.

Skaalvik, E. M., & Rankin, R. J. (1994). Gender differences in mathematics and verbal achievement, self-perception and motivation. *British Journal of Educational Psychology, 64*, 419–428.

Steele, C. M. (1997). A threat in the air: How stereotypes shape intellectual identity and performance. *American Psychologist, 52*, 613–629.

Steele, J. R., & Ambady, N. (2006). "Math is Hard!" The effect of gender priming on women's attitudes. *Journal of Experimental Social Psychology, 42*, 428–436.

Strand, S., Deary, I. J., & Smith, P. (2006). Sex differences in Cognitive Abilities Test scores: A UK national picture. *British Journal of Educational Psychology, 76*, 463–480.

Stewart, A. J., & McDermott, C. (2004). Gender in psychology. *Annual Review Psychology, 55*, 519–544.

Strand, S., Deary, I. J., & Smith, P. (2006). Sex differences in cognitive abilities test scores: A UK national picture. *British Journal of Educational Psychology, 76*, 463–480.

Su, R., Armstrong, P. I., & Rounds, J. (2009). Men and things, women and people: A meta-analysis of sex differences in interests. *Psychological Bulletin, 135*, 859–884.

Swim, J. K. (1994). Perceived versus meta-analytic effect sizes: An assessment of the accuracy of gender stereotypes. *Journal of Personality and Social Psychology, 66*, 21–36.

Terman, L. M., & Miles, C. C. (1936). *Sex and personality.* New York, NY: McGraw-Hill.

UNESCO. (2009). *Institute for statistics. A global perspective on research and development,* UIS fact sheet, No 2, UNESCO.

UNESCO. (2010). *Global education digest: Comparing education statistics across the world,* Author.

UIS. (2011). *Fact sheet: Women in science,* UNESCO Institute for Statistics.

Unger, R. K. (1979). *Female and male: Psychological perspectives,* New York, NY: Harper & Row.

Unger, R. K. (1989). Sex, gender and epistemology. In M. Crawford, & M. Gentry (Eds.), *Gender and thought: Psychological perspectives* (pp. 17–35). New York, NY: Springer-Verlag.

Valian, V. (2005). Beyond gender schemas: Improving the advancement of women in academia. *Hypatia, 20,* 198–213.

Van Langen, A., & Dekkers, H. (2005).Cross-national differences in participating in tertiary science, technology, engineering, and mathematics education. *Comparative Education, 41,* 329–350.

Voyer, D., Voyer, S., & Bryden, M. P. (1995). Magnitude of sex differences in spatial abilities: A meta-analysis and consideration of critical variables. *Psychological Bulletin, 117,* 250–270.

Wai, J., Lubinski, D., & Benbow, C. P. (2009). Spatial ability for STEM domains: Aligning over 50 years of cumulative psychological knowledge solidifies its importance. *Journal of Educational Psychology, 101,* 817–835.

Wajcman, J. (1991). *Feminism confronts technology.* Cambridge, UK: Polity.

Watts, R. (2007). Whose knowledge? Gender, education, science and history. *History of Education, 36,* 283–302.

Weisgram, E. S., & Bigler, R. S. (2007). Effects of learning about gender discrimination on adolescent girls' attitudes toward and interest in science. *Psychology of Women Quarterly, 31,* 262–269.

Wennerås, C., & Wold, A. (1997). Nepotism and sexism in peer-review. *Nature, 387,* 341–343.

Whaley, L. A. (2003). *Women's history as scientists: A guide to the debates.* Santa Barbara, CA: ABC Clio.

White, A. (1985). Women as authors and editors of psychological journals: A 10-year perspective. *American Psychologist, 40,* 527–530.

White, M. S. (1970). Psychological and social barriers to women in science. *Science, 170,* 413–416.

Williams, J. H. (1983). *Psychology of women: Behavior in a biosocial context* (2nd ed.). New York, NY: W. W. Norton & Co.

Williams, W. M. & Ceci, S. J. (2012). When women scientists choose motherhood. *American Scientist, 100,* 138–145.

Wood, W., & Rhodes, N. D. (1992). Sex differences in interaction style in task groups. In C. Ridgeway (Ed.), *Gender, interaction, and inequality* (pp. 97–121). New York, NY: Springer-Verlag.

Woolley, H. T. (1910). Psychological literature: A review of the recent literature on the psychology of sex. *Psychological Bulletin, 7,* 335–342.

Woolley, H. T. (1914). The psychology of sex. *Psychological Bulletin, 11,* 353–379.

Xie, Y., & Shauman, K. A. (2003). *Women in science: Career processes and outcomes.* Cambridge, MA: Harvard University Press.

Yezhu, Z. (2007). An analysis on the status of female faculty in Chinese higher education. *Frontiers of Education in China, 2,* 415–429.

CHAPTER 12

Conflicts, Cooperation, and Competition in the Field of Science and Technology

Anna Dorothea Schulze and Verena Seuffert

Some years ago, at an international conference in Seattle, Washington, a "full-colored thin-film electroluminescence display" was introduced by an American group, opening up new perspectives for telecommunication worldwide. The comments of the scientists then present (specialists from 15 countries) were very different: "Technically very advanced ..." (another American group); "The colors are pretty dark" (a German scientist); "The stability needs to be increased" (a Japanese scientist); "The physics of the whole thing needs to be better understood" (a French scientist). The majority of scientists and technical engineers considered this display a major progress. However, they are aware of the difficulties of developing a stable, bright, and full-colored display. Although so far only a single group (the American team) succeeded in developing the display, it is an example for the fascinating transformation of a physical phenomenon into a useful and high-quality product that combines the know-how and efforts of scientists from various countries.

The secret of this success is the communication between the specialists working in this field. Scientific and technical innovations are mainly the result of a complex communication process proceeding on several levels. The competencies of the various groups are materialized in the product. We will try to open "Pandora's Box" and describe controversies and conflicts playing an essential part in the genesis of this product. And now some characteristics of our approach. The topic of the chapter is the function of conflicts during the genesis of scientific innovations. After explaining the researchers' attitudes toward science and their self-image, we develop a concept characterizing the genesis of scientific innovations. By doing so, we focus on the function of social conflicts. Subsequently, the concept is demonstrated on the basis of empirical analyses of conflicts in the field of life sciences.

AMBIVALENCES IN THE ATTITUDE TOWARD SCIENCE AND THE SELF-IMAGE

When examining scientists' "self-image toward their work," we were frequently confronted by their idea that the recorded research results, or the course of their perception in its logical reconstruction, respectively, were considered as the "actual scientific work," whereas the real course was thought of as impure and, therefore, as one of secondary importance. According to this view, an examination of the real research process would offer no essential insights into its nature. Questions aiming at this were consequentially labeled as "disturbing," "irrelevant," or "unscientific."

It is expected of scientists to demonstrate a procedure as objective as possible and to rule out subjective factors. In our interviews with scientists, we could gain an insight into what, in their opinion, makes an activity scientific or not. Which criteria are decisive for the validity of reasons in the view of the scientists could be concluded from the frequently voiced opinion that the actual scientific work begins with a well-formulated problem only. When a scientist is explicitly asked to assess a new idea for which there is no clear information regarding its logical and experimental validity available yet, he often rejects the request because his judgment would be subjective and unscientific. However, we noticed while following informal discussions that scientists spontaneously make these kinds of assessments anyway, but do not like to provide information about it.

While analyzing scientific controversies, we asked which alternative hypotheses are considered probable. The answer was often: "I don't let myself in for probability statements; scientific evidence is all that matters to me." A young scientist reasoned as follows: "I'm only interested in experiments that plead for one model or another. During our education, it was drummed into us that only scientific evidence counts. Subjectivity, feelings, or personal opinions have to be excluded as much as possible, and now you want us to disobey this discipline."

Furthermore, it is remarkable that, while discussing these hypotheses with their colleagues, the same scientists spontaneously formulated: "I think, there is a high probability that colleague X's model can be proven." Here is shown how the scientist's personal attitude stands in contrast to the discipline demonstrated deliberately in public. This attitude only emerges in spontaneous conversations. Although personal convictions do exist on that score, one does not want to show them often, and does not want others to do so either.

In our opinion, this attitude is directly connected with the "criteria of scholarliness." Here it is clearly expressed that objectivity, rationalism, absence of contradictions, and emotions, as well as the rejection of mere opinions, value judgments, speculations, and convictions are meant by statements such as "Opinions do not matter to me, only arguments," "Science is not a bazaar at which articles of faith are traded," "Speculations without experimental findings are unscientific," "Scientific statements are only acceptable after their publication," "The way of research leads from experiment via interpretation to agreement," "It is taken for granted that every scientist sticks to a logical procedure," and "A model is not a speculation, but a suggestion that can be accepted or dismissed through experiments only." Objectivity and lack of prejudice, without which science is allegedly not possible, appear to be perfectly natural here.

This attitude now collides with one's own behavior patterns. The constantly effective subjective opinions and valuations by which the research process is always accompanied and which have an extremely important function are underestimated. In the justification context, the logical reconstruction of theories is focused on. Personal characteristics, individual opinions, and social relations of scientists are not considered as part of the research process, but as secondary or even disturbing. According to this concept, the research process follows general rules objectively, as it is also the case when research results are represented in magazines, articles, and research reports.

Here, the structure of the activity is reduced to its logical structure, and to empirical statements compatible with theoretical theorems. Thus, experimenting is not the practical analysis of dealing with the object, but the definition of starting conditions and the observation of results. We could observe that scientists take an objectivistic and detached view of their results when presenting them to a specialized public, whereas this understanding is neither cultivated in the research process itself nor in the direct communication. Latour and Woolgar (1980) produced evidence that researchers are relativists in the process of fact construction, who are aware of the possibility that they construct a reality, but that they turn into hard realists as soon as they make their ideas public (Latour & Woolgar, 1980; Callon & Latour, 1992; Knorr-Cetina, 1984). The researchers' reality concept in the process of fact production is rather constructivist and relativistic, which corresponds to the view of interpretative knowledge sociology. Researchers change both the nature and the content of their sentences depending on whether they are dealing with outsiders or insiders (Gilbert & Mulkay, 1984).

The researchers' ambivalent relation to the research process corresponds to their self-concept. Scientists' reports on their lives are usually stories about things, not people. Ideal behavior patterns, such as objectivity, rationalism, absence of contradictions and emotions, rejection of subjective opinions, value judgments, and convictions, are given special emphasis (Mitroff, 1974). This implicit understanding of his or her own personality gets shattered when the scientist gets involved in conflict situations and the ideal behavior repertoire alone does not lead to a solution. In these situations his or her emotions, likings, and prejudices become visible. The researcher reacts with "full subjectivity." However, in subsequent reflections on his or her behavior, this "subjective side" is repressed because it does not correspond to the standardized picture of an objective and unprejudiced scientist, and it is rather seen as a threat to the rational foundations of his or her behavior (Schulze & Wenzel, 1996). This implicit understanding is shaken when a researcher stumbles into a conflict situation with a behavioral repertoire oriented toward these ideals. Without adequate conflict resolution techniques, he or she reacts with "full subjectivity," that is, with heavy emotions, antipathies, and prejudices. But in reflections afterwards, this "subjective side" is repressed because such behaviors do not match the image of the rational, objective scientist and are therefore an eminent threat to his or her identity (Schulze & Wenzel, 1996). Yet, research is—just like any other organized activity—not free of conflicts, and researchers often have to deal with the resulting identity threats. In conflict situations, this dilemma provokes a strategy of either avoidance or imposing one's interests, though both strategies prove to have negative consequences for innovations. Leading scientists and engineers are successful with a strategy of forceful assertion and—because of their accepted status—become models worthy of imitation for younger scientists. In an investigation at high-ranking research institutes in Germany and the United States, leading scientists often said: "Conflicts in research? They go by, what remains are the scientific results!" Because of this downplaying by leading experts, students and young scientists are not prepared to handle conflicts productively. Some internalize such conflicts, blaming themselves for failure vis-à-vis their own aspiration level and they might react with psychic illness and/or complete retreat (Schulze, 2000). Science anthropologists and sociologists of science and technology name these phenomena "controversies" and "tensions" (e.g., Traweek, 1988e; Knorr-Cetina, 1984; Pickering, 1992; Pinch, 1986).

Gulbrandsen (2004) described the research process as "a process of discord: a process characterized by tension, ambiguities, conflicts, and resistance." Kuhn (1977, p. 342) reaches similar conclusions: "The ability to support a tension that can occasionally become almost unbearable is one of the prime requisites for the very best sort of scientific research." Kuhn names this phenomenon "creative tension."

So far a personality image that includes both professional as well as social competence is still pending for scientists and hardly discussed neither within academia nor the public.

As long as social-psychological and personality factors are declared secondary phenomena of scientific work, psychology can contribute very little to researching the nature of scientific work. Only after a "demythologization" of the views of scientific work, essential sides of the research process can be analyzed. For our further analyses, an understanding of science is necessary that overcomes the view that scientific work is only "... the relentless pursuit of logically secured conclusions from experimentally incontestable premises" (Holton, 1981, p. 51).

THE GENESIS OF SCIENTIFIC INNOVATIONS

The fabrication of scientific knowledge (Knorr-Cetina, 1984) is far from being completed by writing scientific papers about the knowledge constructed in a laboratory. Strictly speaking, scientific knowledge can only conditionally be considered a completed "structure," because ascriptions of meanings and interpretations of new knowledge should rather be described as a complicated and complex circulation process between research groups and scientific communities. The communication within a certain research group generates a group-specific context in which the constructed knowledge receives assigned meanings and interpretations.

The context generated in the laboratory is already considerably changed when formulating the corresponding scientific text (e.g., a research paper), because the authors orientate themselves on other readers/listeners who do not belong to the laboratory. The texts intended for the scientific community are written from the perspective of a "generalized other" (Mead, 1934), and their reception is only possible in a generalized context. In this generalized context the "laboratory cognitions" imparted by such texts get different meanings and interpretations in comparison to those that were assigned or given to them in the original laboratory context.

Historic examples often prove that only in a circulation process between research groups and scientific community comes to light whether these changed meanings and interpretations of the generalized context become orientation sizes for a more or less large circle of scientists of the community. In view of the complexity of the described circulation process, it seems useful to grasp the dynamics of scientific discourse with the terms "contextualization" and "context-generalization" to emphasize the diversity of the scientific discourse in research groups and the scientific community on one hand (generating a new scientific context), and by expressing a certain continuity (relative stabilization of new meanings and interpretations in the generalized context) at the chronological course of the circulation process between research groups and the scientific community on the other. The previous remarks clearly show that the analysis of modern science cannot stop with an examination of the direct communication at the laboratory. It must rather integrate the communication between the groups within the scientific community and the connection to other communication practices as well (Krüger, 1990).

On the basis of analysis of social conflicts during the genesis of scientific innovations, it can particularly be demonstrated that the acceptance of scientific results does not automatically emerge from their objective truth, assuming that a rational community of scientists exists, but that emotions, judgments of persons, processes of assertion, and conviction are working here as well and those social dimensions thus have to be analyzed.

SOCIAL CONFLICTS AND CONFLICT MANAGEMENT IN INNOVATION PROCESSES

Social conflicts have been studied in various vocational groups, such as teachers, politicians, managers, lawyers, or businessmen. However, there is a very obvious reserve to study social conflicts among scientists. This phenomenon is particularly remarkable because the innovativeness of research makes one expect conflict-ridden arguments par excellence. Although it is according to Coser (1957) out of the question that conflicts and their constructive resolution lead to emergent innovations within groups (by making separate elements visible and leading them to a new synthesis), the realization that conflicts must be considered as constituting features of a research process has hardly entered the scientific discourse.

The research process actually contains the basic contradiction that decisions for handling certain research problems, for selecting specific methods, or for promoting qualified scientists, have to be made at a time when the criteria for the decisions have not yet been acquired. This uncertainty in the decision-making process lies at the heart of the development of numerous conflicts.

Research resembles a constant selection and decision-making process in which certain alternatives are dismissed, while others are favored. The further away on a time-axis research is from its conclusion, the less standardized is this selection process by previous collective work of the participating scientists and, therefore, less limited and more decisive individual experiences of the participants become for the rejection or preference of certain alternatives. Apart from this, even in an advanced research process, new problems, new techniques, new effects are constantly generated, for whose integration or dismissal no common previous experiences (in the group) exist. Under these circumstances, constructive solutions to conflicts are an indispensable instrument for the qualification of both the decision-making and the research process.

Science anthropologists (e.g., Traweek, 1988) and representatives of laboratory studies (Knorr-Cetina, 1984) point to the passing toward conflicts during the genesis of scientific innovations. Turkle (1984) concedes more importance to conflicts in her studies, albeit without making them the focus of her analyses. The studies of scientific controversies (e.g., Pickering, 1992; Collins, 1981; Pinch, 1986) bear great resemblance to those of social controversies. However, the development of controversies into social conflicts and vice versa is largely ignored. The author's examinations of the scientific discourse suggest, however, that, when conflicts are viewed as processes, conflict-ridden phases are crucially important (Schulze & Wenzel, 1996).

In accordance with Latour (1980) we assume that scientific innovations have their origin in new scientific concepts that differ from the previously prevailing theories or methods. If they are really new, they are judged controversially. Among other things, this shows in the fact that scientists are involved in escalating conflict situations that, under certain circumstances, take place on different communication levels. During the genesis of scientific innovations, the communication is characterized by conflicts and controversies. When these are constructively dealt with, the course of scientific discourse might also become more constructive. That is why communication during the genesis of scientific innovations is characterized by conflicts and controversies. Through a model, we could demonstrate that basic steps of development of innovations go along with the development and the handling of conflicts (Rocholl & Schulze, 2008). Innovations bring discontinuity and hence breed uncertainty and unease, both of which are measures of conflict. Therefore, conflicts can have both detrimental as well as fruitful consequences depending on how they are managed. On the one hand, conflicts can give rise to and establish new ideas, products, and procedures, thus potentially enabling innovations (De Dreu, 1997). On the other hand, however, conflicts may cause enormous costs in terms of time, energy, or money, and they may lead to a destructive downward spiral in which much is lost and nothing gained (Sounder, 1987; De Dreu & Weingart, 2003). It has to be emphasized that the conflicts themselves are less important than the way in which they are managed (De Dreu, 1997; Scholl, 2004; Hauschildt, 1993). Despite their obvious importance for theory and practice, thorough analyses of conflict and conflict management in innovation processes are rare.

Here, at the end of this section, are a few definitions: The organizational literature provides no consensus on a precise definition of social conflict. In this research, we define social conflict in spirit with Thomas (1992a, p. 653) who views conflicts as a process "that begins when one party perceives that the other has negatively affected, or is about to negatively affect, something that he or she cares about." This definition is consistent with "the three general characteristics of conflict: interaction, interdependence, and incompatible goals" (Putnam, & Poole, 1987, p. 552), and it focuses on conflicts between individuals, groups, and social units rather than intrapersonal conflicts.

The Incidence of Conflicts in Innovation Processes

Studies in conflict research have consistently shown that conflicts can generally be classified into two main categories: task conflicts and relationship conflicts. Although relationship conflicts have been found to be dysfunctional for team performance, task-related conflicts—if handled with an integrative approach—seem to play a constructive role (Pelled, 1996; van de Vliert, 1997; Simons & Peterson, 2000).

De Dreu found conflicts to be related to innovation in a curvilinear fashion, so that moderate conflicts support innovation processes (2006, p. 105). Other authors stress the importance of considering the conditions under which task conflict turns into the detrimental form of a relationship conflict (e.g., De Church & Marks, 2001; Simons & Peterson, 2000; Turner & Pratkanis, 1997). Particularly in science and technology, researchers are confronted with situations characterized by high levels of ambiguity, extreme uncertainty, and involving high stakes. Under such conditions, task and relationship conflicts often merge and are highly correlated (De Dreu et al., 1999).

Therefore, it is often impossible to distinguish clearly between cognitive and relational conflicts during the innovation process, and these each demonstrate differing aspects in the course of the innovation process (Schulze, 1990).

Minority Dissent and Innovation

Because innovations often have their origin in new scientific concepts that differ from prevailing theories and methods, these concepts—if they are really new—are judged controversially. That is the reason why communication during the genesis of scientific innovations is characterized by conflicts and controversies (Schulze, 2000).

Scientists quickly find themselves in an escalating conflict situation, mostly involving pronounced feelings of injustice among the conflicting parties. "Conflict is an inherently emotional experience. It is tremendously difficult for individuals to remain objective about a situation when they feel that others are disagreeing, or even disapproving of their point of view" (Jehn, 1997). The more localized reaction to a conflict may then be to lash out at the team members in question and to consider the whole team to be functionally suboptimal, especially when it comes to perceptions of creative outcomes.

De Dreu stresses the importance of minority dissent for team innovations. "Results show more innovations and greater team effectiveness under high rather than low levels of minority dissent, but only when there was a high level of team reflexivity" (2002, p. 285). So far, the influence of minorities on innovation processes has received little attention in the research literature (for an exception see Rosenwein, 1994). In our investigations of the influence of minorities on innovation processes in science and technology, we were able to determine the following relationship: the Innovator who is in the minority faces a socially accepted majority. Innovations have to overcome existing theories and methods at a time when the innovations cannot yet be proved logically or experimentally (Schulze, 1990, p. 59). During this phase, behavior characteristics, such as perseverance, a high measure of assertiveness, self-confidence, self-assurance, hardiness, and stability are particularly useful. A dominating behavior pattern further adds to the reservations of opponents. "One has cold admiration for a person when one admires them but does not like them" (Moscovici, 1971, p. 241).

Innovators are often not nice in the sense of being able to get along with other people with a minimum amount of friction (Gulbrandsen, 2004; Hemlin, 2006; Kwang, 2001). In this struggle for domination, successful innovators will not give in or lower their demands, instead certain behavior patterns will become more effective.

The level of reflexivity is very low in the phase of the occurrence of a new idea when a minority in a group publicly opposes the beliefs, attitudes, ideas, procedures, and policies adopted by the majority of the group (De Dreu, 2002). Laboratory experiments suggest minority dissent prevents defective decision making, as dissent encourages other group members to resist conformity pressures (Nemeth, 1986) and it reduces the individual's tendency to become more extremist over time.

Temporal Dimensions and Phases of Conflicts in Innovation Processes

In recent decades, the issue of innovation has taken on an ever-increasing importance. "Innovation is a scintillating and fashionable term" (Hauschildt, 1993, p. 3). However, there is no definition or understanding shared by all researchers, even within the same discipline. Many researchers relate innovation to novelty, complexity, uncertainty, and to the emergence of conflicts (e.g., Hauschildt, 1993; Pleschak & Sabisch, 1996). As the focus in this chapter lies on social conflicts arising in processes of innovation, we specifically look at its process-related dimensions.

For the most part, previous conflict research has treated conflict as a rather static phenomenon in the course of a team's activity by using participants' post hoc summaries of the conflicts experienced in their team (e.g., Amason, 1996; Jehn, 1995, 1997). It is only recently, that researchers (e.g., Jehn & Mannix, 2001) have begun to look at the temporal dimensions of conflict by means of longitudinal research designs. Findings along those lines indicate that conflicts differ and therefore have different effects on performance depending on their temporal location, that is, at the beginning, in the middle, or at the end of a process. The strength of such investigations lies in the use of longitudinal data, making it possible to examine the influence of timing on the relationship between conflict and innovation.

Aspects of an innovation process are distinguished in this chapter on the basis of suggestions by Hauschildt (1993, pp. 17–19). His phase model (ibid., p. 3) describes the dynamic processes of innovations in detail. The process begins with what he calls a *"pet idea"*—the fascination with a certain, but still vague idea in a specific field. This is followed by *"start,"* when the decision is taken to work on the idea, successively making it more concrete. After a number of other phases, Hauschildt's model concludes with *"market entrance,"* that is, the introduction of a new product into the relevant market. Two models describe the psychological micro processes that occur: the "Competence Model" by Amabile (1996) and the "Phase Model of Innovation Processes" by Schulze (1990). The Competence Model describes creativity (creative processing) as an iterative process whereby a creative idea is generated. This process includes several steps, such as problem identification, preparation, response generation, response validation, and communication. These are not thought of as occurring in any particular linear order. In contrast, the Phase Model of innovation processes suggests two phases of innovation: first, the conceptualization phase, in which new ideas and concepts are developed, and, second, the program phase characterized by articulated problems and the planning and undertaking of the research project. Particularly in basic research, the conceptualization phase is characterized by a high frequency of personal conflicts and interrelations between cognitive and personal conflict aspects. The program phase (especially in applied research) is marked by a separation between cognitive and personal conflict aspects along with a high frequency of task and process-related conflicts. The discussion of a range of ideas lies at the center of the conceptualization phase. Consequently, a problem-solving style of conflict management seems most appropriate here. In contrast, the implementation of the research programme and its successful commercialization are crucial for the program phase, especially in applied research, and hence a conflict-management style of dominance seems most appropriate to lead to desired results.

It is particularly vital for innovation processes to return from a program phase to a new conceptualization phase in which new ideas are discussed for the first time. Our longitudinal studies on the conflict behavior of researchers in molecular biology, solid-state physics, and computer science have repeatedly shown that leading researchers make use of different conflict-management styles depending on the phase of the innovation process. In the stable phase of the research process when the research program is already established and being worked on by the research group, leading researchers show behavioral features of an integrative conflict management style. At a later stage of the program phase, however, contrary forces tend to come to the fore, diminishing consensus and identifying weak points in the research program. Ensuing expressions of dissenting opinions that substantially criticize the theoretical and methodological assumptions of the group thus mark the emergence of a phase of group opening and the starting point for a new conceptual phase. In this phase of group opening, the behavior of leading researchers in the scientific community is characterized by dominant conflict-management behavior, which tends to block the development of innovations. Success and failure in innovation processes thus seem to depend to a significant degree on whether and when a group opening is made possible (Gersick, 1989; Schulze & Wenzel, 1996). As a consequence, it was crucial for our investigation to identify this transitional situation from a program, back to a conceptualization phase, and to measure the corresponding conflict management styles. This suggests that innovations result from two component processes: "For instance, conflict that may promote creativity or new perspectives on a problem may undermine the ability of a group to implement that new idea" (O'Reilly, Williams, & Barsade, 1989, p. 186, 187; Ziegler, 1968; Gebert, 2004).

Despite their obvious importance for theory and practice, thorough analyses of conflict and conflict management in innovation processes are rare. On the basis of the assumption that conflicts and conflict management decisively influence processes of innovation, that is, they are crucial variables for the success or failure of innovations, this chapter examines their role in greater detail. Trying to fill an important research gap, this study contributes to current research on innovations by focusing on the role of conflicts and their management in processes of innovation.

It has to be emphasized that conflicts themselves are thereby not nearly as important as the way in which they are managed (De Dreu, 1997; Scholl, 2004; Hauschildt, 1993).

Models of Conflict-Management Styles

The Dual-Concern Model

Blake and Mouton (1964, 1970) first presented a two-dimensional conceptual scheme—the conflict-management grid—classifying styles of handling interpersonal conflicts into five distinct types: dominating, avoiding, obliging, compromising, and problem solving (Figure 12.1). Rahim (1983, 2002) reinterpreted this research by outlining conflict management styles on a range of dimensions that either emphasize concern for personal needs or the needs of others. Rahim distinguishes five conflict-handling strategies: (a) a problem-solving conflict-management strategy demonstrates a high concern both for the self and for others, (b) an obliging strategy demonstrates a low concern for the self and a high concern for others, (c) a compromise strategy demonstrates moderate concern both for the self and for others, (d) a dominating strategy demonstrates a high concern for the self and a low concern for others, and (e) an avoiding strategy demonstrates low concern both for the self and for others. This classical model of interpersonal conflict management has given rise to several theoretical approaches, further conceptualizing different conflict-management styles (Pruitt & Rubin, 1986; Rahim, 1992; Thomas, 1992a) and empirically evaluating these styles (Rahim, 1983; van de Vliert & Hordijk, 1989).

Undercover Conflict-Management Strategy

Another relevant strategy of conflict management in innovation processes is the "undercover conflict management strategy," which is not explicitly included in classic Dual-Concern Models. Undercover conflict management is characterized by external compliance and the apparent relinquishment of personal interests. However, these are continued surreptitiously as a high priority (Rocholl & Schulze, 2008). "Some innovations could only be successfully concluded conspiratorially, that is, against the plans and

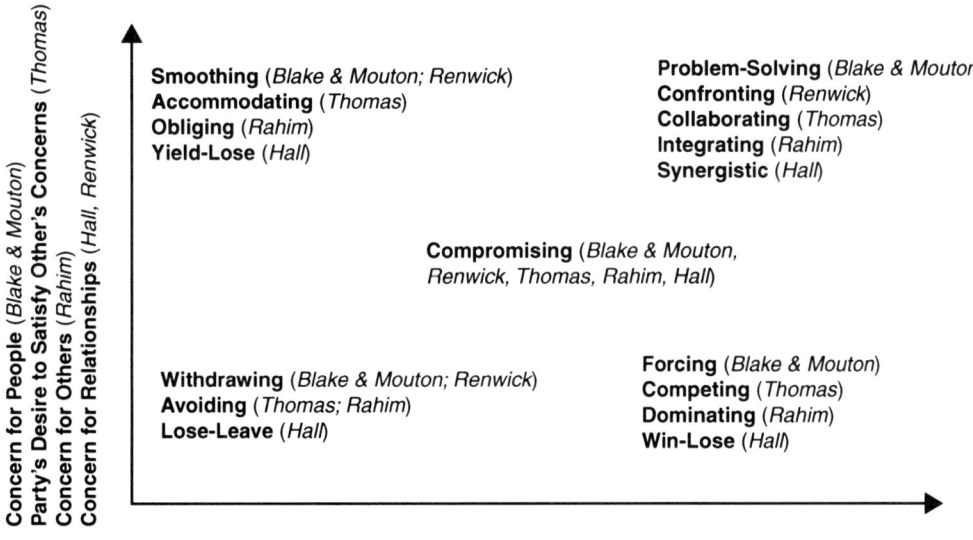

Figure 12.1   *Overlay of conflict-management styles and authors derived from the dual-concern theory.*

decisions of the higher levels of management" (Scholl, 2004, p. 5; Eglau et al., 2000, p. 41). In the following, an example for this strategy will be presented.

Example: Submarine research in pharmaceutical development. In Bayer's Leverkusen synthesis laboratory, chemist Klaus Grohe developed a new method to produce drugs against bacterial infections. The substances he sent to the company's Wuppertal pharmaceutical research centre for testing, showed steadily increasing bacteriological effects. The scientist was then all the more disappointed when pharmaceutical management decided to close down his project in 1980. Despite this decision, Grohe continued with his research and his perseverance paid off. In 1981, he was able to increase the effectiveness of his substance ciprofloxacin so spectacularly that Bayer reversed its decision to kill off the project and decided instead to make it a priority. After only 5 years, a very short time for the development of a new drug, the first medical doctors were able to prescribe Ciprobay as a broad-band antibiotic. Since then this has become one of Bayer's most successful drugs (Eglau et al., 2000, p. 45).

The Conglomerate Conflict Behavior Model

Van de Vliert et al. (1995) and van de Vliert (1997) recently developed a new perspective on conflict management that they termed conglomerated "conflict behavior." Instead of just one single style, this theory states that the components of conflict behavior should be understood as a configuration of various behavioral styles. The underlying assumption in this model is that mixed motives result in complex rather than simple behavior, which is then best analyzed as a mixture of components. Interpersonal conflicts are thus complex situations in which behavior is directed by various motives and concerns about one's own goals and the other's goals, the relationship between oneself and the other person as well as short- and long-term objectives (Thomas, 1992b).

Perspectives on Effectiveness

The analysis of the effectiveness of different conflict-management behavior has traditionally been approached from three perspectives: the "one-best-way perspective" (e.g., Blake & Mouton, 1970, 1981; Fisher & Ury, 1981; Lewicki et al., 1992; Tjosvold, 1991), "the contingency perspective" (e.g., Axelrod, 1984; Rahim, 1992), and the "complexity perspective" (van de Vliert, 1997; Munduate et al., 1999). While the focus in this chapter is on the "contingency perspective," we actually aimed for synthesis of all three. The findings on combinations of different styles could then be interpreted as further evidence that conflict behavior is a complex form of behavior, just as the complexity perspective predicts.

To be truly effective, Amason and Schweiger (1994) propose that teams must produce high-quality decisions and team members have to share information rather than suppress opposing positions (Driskell & Salas, 1991; Gruenfeld et al., 1996; Scholl, 2004). Further they have to cultivate an understanding of and commitment to those decisions. Finally, they must maintain a level of affective acceptance sufficient to sustain their ability to work together in the future.

CONFLICT AND CONFLICT MANAGEMENT: A STUDY OF LIFE SCIENCES

Objectives

Thorough analyses of conflicts and conflict management in innovation processes are rare; this insight was the starting point of our research proposal. Aiming at expanding research on conflict management, this study examined processes of innovation along with variables of innovation success in the field of life sciences. One hundred and fifty-two basic and applied researchers were surveyed regarding their conflict-management styles.

Although most innovation studies are retrospective in nature, we analyzed the evolution of ongoing as well as of previous processes. Thereby a multimethod approach incorporating both survey and qualitative methods was used to test the model and to examine the impact of both conflict type and conflict-management style on innovation.

Single conflict-management styles were compared in terms of aspects of innovation performance.

Characterization of Conflicts in the Life Sciences Field

The field of life sciences with its core areas of gene theory and technology/genetic engineering was chosen for the investigation, because it is a highly innovative area where significant work has been done over the last few years and will continue to be done in the future. It is a field of groundbreaking discoveries, innovations, and conflicts. It is our hypothesis that the genesis, the stakes, forms, and outcomes of innovation-related conflicts vary significantly with the "radicalism" of innovations in a field. The life sciences are thereby not simply a field of more radical and new technology. What is more is that they fundamentally affect the very foundations of nature and humankind.

Methods

For a deeper exploration and understanding of conflicts in innovation processes and as a preparation for the questionnaire survey, interviews were conducted with researchers in gene science and technology ($N = 50$). On the basis of suggestions from these interviews and on our theoretical frame of reference we developed questionnaires and sent them to about 800 researchers. In any case, 220 researchers responded either by mail or via a web-based version. About 30% of them were approached and notified as especially high ranking in science (52 or more publications in high-ranking journals in the last 5 years) or in the economy (six or more patents). Complete or almost complete questionnaires were obtained from 152 researchers with which we made our analyses. We asked gene researchers in the interviews to describe concrete conflict situations. Three important conflict situations were chosen from the interviews and presented in the questionnaire: (a) A conflict about a new idea that implied a deviation from the previous project plan: this was experienced by 74% of the respondents, either as proponent of the new idea or as proponent of the established project plan. (b) A conflict about the carefulness of testing before a scientific publication or a market introduction in firms: this was experienced by 58% of the respondents on one side or the other. (c) A conflict about the termination of a project after disappointing results: this was experienced by 68% of the respondents on either side. If the respondents had not experienced at least one of the three mentioned situations they could describe a personally experienced conflict situation, but this possibility was rarely chosen. Taken together, 90% of the respondents had experienced and answered at least one of these concrete conflict situations. Thus, they could be asked to answer further questions on the most important conflict situation they had experienced.

For further analyses of these conflicts, we could reliably and validly measure the styles of conflict management of the researchers with our questionnaire and we could extract them from the interviews. As a unique feature of conflict-management research, we also asked for the style of the other main person involved in the conflict, the opponent. The styles of conflict management were the commonly accepted (e.g., Pruitt et al., 2004; Thomas, 1992b): *problem solving* (trying together to find better ways for serving all interests involved), *dominant* (trying to push through one's own interests against those of the other), *obliging* (giving up one's own interests and accepting the position of the other), and, finally, *avoiding* (trying to avoid open conflict). The four styles of the respondent and the four styles of his or her opponent were analyzed.

Results

Occurrence of Conflicts

The main objective of the present study was to explore conflicts in innovation processes in the life sciences. Generally, social conflicts were familiar to researchers participating in this study: 90% had experienced at least one of the specific conflict situations described in the questionnaire. This once again shows that conflicts are a part of a researcher's everyday life, underlining the importance of research on this topic.

The conflict situation most frequently experienced by researchers was the following: "Researchers in the same unit had developed a new idea that differed significantly from the current project. This researcher is desperate to work on the new idea but runs against the rest of the team because working on the new idea would consume much time and resources from the current project." This transitional situation from an implementation to a new conceptualization phase was therefore chosen for the measurement of its corresponding conflict-management styles.

Adding to the literature, which has so far paid insufficient attention to potentially different antecedents of specific innovation phases, we could establish the following phase-specific results: With respect to the frequency and the intensity of conflicts, researchers in applied research, experienced conflicts more often than those in basic research. Another difference between applied and basic research relates to conflict causes. While conflicts concerning process issues arise more frequently in applied research, conflicts about personal differences play a greater role in basic research. This may be explained by the assumption that work in basic research is more deeply tied to the researcher's self-concept and self-esteem than work in enterprises. In other words, a broader role-person merger may exist in basic research, resulting in significantly more person-role conflicts, which then become more personal.

In addition to differences in the frequency and the issues of conflicts between basic and applied research, we could—very importantly—establish that conflicts, escalate more often in applied research.

Conflict Management and Innovation Success

Moreover, the problem-solving, conflict-management style was used to a lesser extent and dominance to a greater extent by researchers in organizations of applied science. This holds true both for the self-assessed conflict-management style as well as for the description of the other's style. Furthermore, a higher frequency of a problem-solving conflict-management style could be observed for the subsample of basic research, again regarding the assessment of one's own style as well as of that of the other. A possible explanation for these results might be that higher levels of complexity along with more rapid developments require fast decision processes in the commercial sphere. In addition, research in applied research is more orientated toward practical questions, with an early release often being crucial, that is, for the survival of a small start-up company. Finally, financial and market aspects play an important role in applied research, in addition to ordinary research difficulties. Our results also indicate that a dominating conflict management style was significant correlated with the factor Project Newness ($rs = 0.32$, $p < 0.01$). This style was the significant predictor of Project Newness with a moderate effect size of $rpart = 0.28$ ($p < 0.01$).

Conflict Management and Process Characteristics of Innovations

Communication. In terms of self-assessment, our study finds a positive relationship between a conflict management style of dominance and the level of external communication, that is, researchers successful ($rs = 0.21$, $p < 0.05$). The dominating conflict management style had a moderate effect effect on external communication ($rpart = 0.21$, $p < 0.05$). Communicate much more with other research groups of their community, thereby making frequent use of a dominant conflict management style. Together with this results, we could find out that there is a significant association between external communication and the Project Newness Factor, which leads to the conflict-management style of dominance being used frequently. The importance of a dominant conflict-management style for the communication in innovation processes is further underlined by studies on cooperation and competition in the life sciences field. Those show that researchers often compete with other groups working in their field (Seuffert, 2007) and suggest a close association between a dominant conflict-management style and competitive relationships. The development of gene science and genetic engineering is characterized by scientific personalities with the vision to make outstanding contributions to human well-being (Gardner et al., 2001). At the same time, however, the history of science and recent studies show that the key motivation of researchers is to be the first to publish results or to launch new products on the market. Striving for personal recognition, individualistic work is on the foreground.

Conflict Management and Performance

The results of this study suggest that the relationship between conflict management and performance needs to be viewed in a differentiated manner. The finding that a problem-solving style was used more often in the early stages of innovation processes, that is, in basic research, than later on in applied research, where dominance prevailed, can be explained by a contingency perspective. This holds that different stages of an innovation process pose different demands. For example, the discussion of various views on the basis of a problem-solving approach in the early innovation phases proves crucial for the development of new ideas. Consequently and based on a contingency perspective, further research should examine the set of conditions constituted by conflict-management style, type of conflict, and task to be accomplished (cf. De Dreu & Weingart, 2003, p. 151). The results regarding a dominant conflict management style could, however, point to conflict management as being a rather complex process, as suggested by the complexity perspective (van de Vliert et al., 1997). When further elaborating this perspective, it might be worthwhile to keep the following two aspects in mind. First, Scholl (2009) establishes in his studies on conflict management styles and growth of knowledge that the latter is clearly fostered by a problem-solving style. Second, Thomas (1992) argues that the problem-solving style should be considered in terms of a long-term perspective such as a final goal.

In conclusion, the crucial question is how researchers can move from a dominant to a problem-solving style. To this end, a synthesis of the different perspectives would obviously be ideal: "The synthesis would be perfect if it could be shown that, in the long run, problem solving determines the personal and dyadic effectiveness more than other components, even when the latter components are more appropriate and occur in more particular situations" (van de Vliert, 1997, p. 121). Possibly termed a "constructive controversy" (Johnson et al., 1989; Tjosvold, 1985), the most productive conflict-management behavior appears to be a combination of integration and then, if no conclusion has been reached, the use of dominance to resolve the conflict (van de Vliert et al., 1997; Munduate et al., 1999).

COOPERATION AND COMPETITION IN SCIENCE AND TECHNOLOGY

Effects of Cooperation and Competition on Conflicts

Gebert (2004) claims that strong competition leads to more conflicts. The close relationship between competition and conflicts becomes obvious in the fact that both concepts are frequently used interchangeably (cf. Deutsch, 2003). Grunwald (1982) tries to clarify the distinction between both concepts and argues that competition can be seen as a subcategory of conflict, indicating that competition is always conflict-laden, whereas not every conflict is competitive. He suggests that conflict is antagonist-oriented, whereas competition is goal-oriented. These assumptions fit the distinction by Deutsch (1973, 2003), who differentiates between conflicts and competition in the following way: Competition is characterized by incompatible goals, whereas conflicts are characterized by incompatible actions. This means that a conflict can arise without experienced or actual goal incompatibility. However, there is still no agreement about the distinction between conflict and competition in the literature. De Dreu (1997) lists a number of possible issues that might cause conflicts: The division of scarce resources, policies, what to consider in the decision-making process, how to approach the task, what humor is funny, what norms and values are valid and appreciated, and which beliefs are to be respected (ibid., p.16). In this enumeration it can be seen that, in accordance with Grunwald (1982), not all of these possible causes of conflict do include competition. Jehn (1997) argues that conflicts are more likely to occur and to escalate in competitive environments than in cooperative ones. However, despite the assumption that conflicts are more likely to occur in competitive environments, it is possible and likely that conflicts also occur in cooperative environments (Deutsch, 2003). However, it is argued that conflicts are more likely to escalate and to have negative consequences under competitive circumstances, while they may even release synergies under cooperative circumstances (e.g., De Dreu & Weingart, 2003; Deutsch, 2003; Johnson & Johnson, 1989). Therefore, it is assumed that there will be more conflicts in competitive environments, and existing conflicts will be of higher intensity in competitive environments than in cooperative environments.

Concepts and Definitions of Cooperation and Competition

The definition of cooperation and competition used in the present chapter is based on the theoretical framework of Deutsch (1949), who defined cooperative and competitive situations. Accordingly, a cooperative social situation is one "in which the goals of separate individuals are so linked together that there is a positive correlation among their goal attainments. An individual can attain his or her goal if and only if the other participants can attain their goals. Thus a person seeks an outcome that is beneficial to all those with whom he or she is cooperatively linked" (Johnson, Maruyama, Johnson, & Nelson, 1981, p. 47). In contrast, a competitive situation is defined as "one in which the goals of the separate participants are linked in a way so that there is a negative correlation among their goal attainments. An individual can attain his or her goal if and only if the other participants cannot attain their goals. Thus a person seeks an outcome that is personally beneficial but is detrimental to the others with whom he or she is competitively linked" (Johnson et al., 1981, p. 47). All in all, Deutsch's theory is based on two continua: the first leading from promotive to contrariant, describing the type of interdependence among the goals of the participants, the second leading from effective to bungling, describing the type of actions taken by participants (Deutsch, 1949, 1962). Summarized by Johnson and Johnson (1992): "Essentially, in cooperative situations the actions of participants substitute for each other, participants positively react to each other's effective actions, and there is a high inducebility among participants. In competitive situations, the actions of participants do not substitute for each other, participants react negatively to each other's effective actions, and inducebility is low" (Johnson & Johnson, 1992, p. 175). Another concept that has to be considered in the context of cooperation and competition is individualistic behavior. Grunwald (1982) argues that the absence of cooperation is not competition, but individualistic behavior. Accordingly, the absence of competition does not automatically result in cooperation, but again in individualistic behavior (cf. Johnson & Johnson, 1989). Situations that lead to this individualistic behavior can be described by individualistic goal structures, without any interdependence between the actors: "Whether an individual accomplishes his or her goal has no influence on whether other individuals achieve their goals. Thus a person seeks an outcome that is personally beneficial, ignoring as irrelevant the goal achievement efforts of other participants in the situation" (Johnson et al., 1981, p. 48). Apart from this, the possibility that cooperation and competition can occur simultaneously has to be taken into account. Deutsch (1949) states that there are very few situations in real life that are purely cooperative or competitive, but rather "involve a complex set of goals and subgoals. Consequently, it is possible for individuals to be promotively interdependent with respect to one goal and contrary interdependent with respect to another goal" (Deutsch, 1949, p.132; cf. also Deutsch, 2003). He gives the example of a basketball team, where members "may be co-operatively interrelated with respect to winning the game, but competitively interrelated with respect to being the 'star' of the team" (Deutsch, 1949, p. 132). Especially in a very complex environment like that of scientific research, the simple distinction between cooperation and competition is insufficient to describe the situation properly. In this highly innovative field scientists have to cooperate with each other because of the high complexity of the problems with which they are dealing. Every scientist becomes a specialist for his or her topic and needs to cooperate with other specialists, often from other disciplines, in order to be successful. Still, besides cooperating with other scientists, each person tries to stand out from the group to work on his or her own career (cf. Young, 2003). In this way the situation within a research group is comparable to the one Deutsch (1949) describes for the basketball team.

The assumption that situations are hardly ever purely cooperative or purely competitive leads to the general question of why people cooperate or compete in specific situations. This question will be explored in the following paragraph. Under what circumstances do individuals cooperate or compete?

Under What Circumstances Do Individuals Cooperate or Compete?

As stated above, the main difference between cooperative and competitive situations lies in the kind of (social) interdependence between participants (e.g., Deutsch, 1949; Johnson & Johnson, 1989, 1992). Social interdependence is given when the outcomes of individuals in a situation are affected by the actions of each other (Johnson & Johnson, 1989, 1992). In the case of positive interdependence, individuals are likely to cooperate, whereas negative interdependence fosters competition; besides, the absence of social interdependence leads to individualistic behavior (e.g., Deutsch, 1949, 1962; Johnson & Johnson, 1989, 1992). Therefore, the type of social interdependence present in a given situation should explain whether individuals will cooperate or compete with each other. However, obstacles for deciding whether a situation is best characterized by positive or negative interdependence are: first, there are various possible aspects in which individuals can be interdependent, for example, goals, rewards, resources, roles, and divisions of labor (cf. Johnson & Johnson, 1989, p.175). Second, the type of interdependence can differ with respect to subgoals and goals: individuals can be positively interdependent with respect to their subgoals and simultaneously negatively interdependent with respect to their goals or vice versa (Deutsch, 1949). Third, according to Deutsch's theory (1949, 1973, 2003) it is central how people perceive the interdependence of their goals and not how they are objectively interrelated. Thus, "how individuals behave in a situation is largely determined by their perceptions of the outcomes desired and the means by which the desired outcomes may be reached" (Johnson & Johnson, 1992, p. 181). Bierhoff and Muller (1993) identified various structural factors that foster cooperation within an organization such as reward systems, organizational promotion policy, matrix organization, and compatibility of interests between team members and departments. In this list it already becomes obvious that working relationships do not stop at the level of the working group but are embedded in a wider range of working relationships within and outside the organization which can all influence the process of innovation (e.g., Hemlin et al., 2004; Schulze, 2000; Schulze & Wenzel, 1996). The importance of differentiation between various levels also becomes clear in the review of research on innovation by Anderson, De Dreu, and Nijstad (2004) who impressively stress the importance of multi-level analyses in innovation research. Therefore, the following abstracts will target under what circumstances individuals tend to cooperate or compete with other individuals within their group, and with different groups inside or outside their organization.

COOPERATION AND COMPETITION: A STUDY IN THE FIELD OF LIFE SCIENCES

Objectives

The study aimed to examine cooperation and competition in the process of innovation. The field of interest was life sciences, which has been identified as a knowledge-intensive area with a strong pressure to be innovative. Thereby it was targeted to account for the complexity of working environments by including various environmental layers and different forms of organizations.

Methodology

It was the same sample that we described in Chapter 11. In a first phase, 45 scientists from different forms of organizations were interviewed, using half-structured problem-centered interviews that were based on an interview guideline. Two independent raters coded the description of working relationships on various levels. In a second phase, 146 scientists were surveyed with a questionnaire. Their working relationships in terms of cooperation and competition on various levels were assessed on a 7-point Likert scale ranging from "purely competitive" over "both equally" to "purely cooperative." Scientists' orientation in terms of cooperation and competition were assessed on three indices: internal competitive, external competitive, and cooperative orientation. It was explored how these aspects of cooperation and competition were associated with each other.

Additionally, it was examined how they were associated with three indicators for innovative success (high performance vs. normal performance group, success within the last 5 years, and newness of projects) and with the frequency and intensity of conflicts. Both approaches compared scientists from different forms of organizations, as well as superiors and subordinates.

Results

Differences Between Scientists Working in Different Forms of Organizations

Both approaches showed that working relationships on the group level are described as more cooperative than those on higher levels.

Various differences between scientists working in different forms of organizations were revealed: In the interviews, scientists working in basic research reported more individualistic behavior on the group level than those working in applied research. Scientists in basic research reported more competitive relationships with external basic research than those in other forms of organizations. In the interviews, scientists in research and development (R&D) companies reported more cooperation with external start-ups than those in start-ups. Scientists working in universities were externally less competitively oriented than those in R&D companies or start-ups, while scientists working in start-ups were less cooperatively oriented than those in R&D companies or basic research institutes. External working relationships were the more competitive the higher the external competitive orientation, whereas internal working relationships were the more cooperative the higher the cooperative orientation. Cooperatively oriented scientists reported that the more internal their cooperation, the more cooperative their external working relationships would be.

Comparisons between superiors and subordinates revealed that subordinates were more likely to report cooperation within the department (interview data), whereas superiors were more likely to have external working relationships with applied research and described their existing working relationships with external applied research as more cooperative (questionnaire data).

Working Relationships and Innovative Success

Success within the last 5 years was positively associated with cooperation within the group and newness of projects with the existence of external working relationships. Interactions between external working relationships and organizational forms in predicting success could be revealed in terms of membership in the high performance or normal performance group. None of the scientists working in a university who reported no working relationships with external basic research were high performers, whereas for scientists working in another form of organization there was no association between the two factors. Scientists working in basic research institutes were more likely to be high performers the more cooperatively they described their working relationships with external basic research, or their external working relationships in general. Scientists in universities, on the other hand, were less likely to be high performers the more cooperatively they described their external working relationships.

Internal competitive orientation was positively associated with all three indicators for success. External competitive orientation was negatively associated with being a high performer and with success within the last 5 years.

Working Relationships, Orientations, and Conflicts

Considering the association between cooperation and competition and the frequency and intensity of conflicts, results supported the assumption that the more intense the conflicts, the more competitive the working relationships. In regard to orientation in terms of cooperation and competition, results revealed a positive correlation between external competitive orientation and frequency of conflicts, and between internal competitive orientation and intensity of conflicts. However, it remains to be explained why an externally more competitive orientation was associated with a higher frequency of conflicts within the direct working environment.

Altogether, the results go in line with the assumption that conflicts are more likely to occur under competitive circumstances (e.g., Gebert, 2004; Jehn, 1997) and that conflicts are more likely to escalate under competitive than under cooperative circumstances (e.g., De Dreu & Weingart, 2003; Deutsch, 2003; Jehn, 1997; Johnson & Johnson, 1989). Knowing that the occurrence and intensity of conflicts within the process of innovations is associated with working relationships and orientation in terms of cooperation and competition, leads to the question of how frequency and intensity of conflicts are associated with innovative success. Deutsch (2003) argues that conflicts can have positive effects if they take a constructive course, and that the course of a conflict heavily depends upon the context in terms of cooperation and competition. In line with this assumption, other authors argue that the way in which conflicts are handled strongly influences the consequences of conflicts (e.g., West & Hirst, 2003).

Results indicate that scientists in the field of life sciences are aware of both the competitiveness of their field and the need to cooperate to be innovative. The study revealed a number of positive associations between cooperation and innovative success. Scientists who were internally more competitively oriented were more successful, but none of the results revealed a positive association between competitive working relationships and innovative success.

Practical Implications

The findings have important practical implications for conflict-management training. Researchers in science and engineering often underestimate or downplay the importance of conflicts in innovation processes. In particular, if they are poorly managed, conflicts can have quite severe consequences, such as delays, terminations, and loss of time and money.

It is the objective of conflict-management training to familiarize managers and researchers with basic methodological principles of analysis and constructive conflict management in innovation processes.

Because experts underestimate the importance, students and young scientists are not prepared to handle conflicts productively. Some internalize the conflicts, blaming themselves for failure vis-à-vis their own aspiration level and may react with psychic illness and/or complete retreat (Schulze, 2000). Special courses for PhD students should include the detailed analysis of conflict situations in processes of innovation, which will then enable them to identify critical incidents. Simulation exercises can demonstrate where the key opportunities are and what steps the students could take going forward.

DIRECTIONS FOR FUTURE STUDIES AND CONCLUSIONS

First, future studies should include possible moderators for the association between cooperation and competition and innovative success. For example, Scott and Bruce (1994) argue that task interdependence may mediate the relationship between cooperation and performance (cf. also Beersma et al., 2003; Slavin, 1977; Stanne et al., 1999). In the present study, participants were only asked how they experience their working relationships in terms of cooperation and competition, but not how interdependent they were with others in accomplishing their tasks. Considering the assumption that the perception of interdependence is also crucial for behavior (cf. Deutsch, 1949, 1973, 2003; Jehn, Northcraft, & Neale, 1999; Johnson & Johnson, 1989, 1992; Young, 2003), future studies should also include the assessment of task interdependence and/or reward structures (cf. Wageman, 1995), which might help to explain the occurrence of cooperation as well as the relationship between cooperation and innovative success. Another example for a possible moderator in the association between cooperation and competition and innovative success is the sex of participants. Gardner and colleagues (2001) suggest that competitiveness is harder for women than for men. This possible sex difference in (a) the description of working relationships and (b) the effect of these working relationships should be targeted in future studies. A third concept that should be considered in future studies is trust.

Various authors (e.g., Tjosvold et al., 2003; Young, 2003) argue that cooperation is more likely to occur if individuals trust each other, whereas mistrust fosters competition. In the interviews conducted in the present study, it was revealed that scientists mistrust each other, especially in their external working relationships. Mistrust partly resulted in the nondisclosure of information and results even from the so-called cooperation partners. This stresses the importance of exploring the association between trust and working relationships in terms of cooperation and competition, as well as the question whether trust functions as a moderator in the association between working relationships and innovative success. Second, future studies should account for the fact that aspects of cooperation and competition may not only influence innovative success but also a variety of other aspects, for example, quality of communication (e.g., Deutsch, 1973; Qin et al., 1995; Wetzel, 1994), functioning of interdisciplinary collaboration (e.g., Jehn et al., 1999; Scholl, 2004, 2005), perceived support (e.g., Deutsch, 1973; Johnson & Johnson, 1989), intrinsic motivation (e.g., Tauer & Harackiewicz, 2004; West & Hirst, 2003), and psychological health and well-being (e.g., Johnson & Johnson, 1989; Standage, Duda, & Pensgaard, 2005; Tjosvold et al., 2003; West et al., 2003). These and other factors should be included in future studies to examine the role of cooperation and competition in its whole complexity.

The present study indicates that scientists in the field of Life Sciences are aware of both the competitiveness of their field and the need to cooperate to be innovative. Several significant positive associations between cooperation and innovative success could be revealed. However, aside from the finding that scientists who are internally more competitively oriented were more successful in the present study, there was no evidence of positive effects of competition. Together with the results from previous research showing positive associations between cooperation and psychological health and well-being (cf. Johnson & Johnson, 1989; Standage et al., 2005; Tjosvold et al., 2003; West et al., 2003), these results stress the requirement to foster cooperation in science. Gardner and colleagues (2001) express strong concern that the need to make profit is a great obstacle for science: first, as it influences the contents of scientific research and, second, as it intensifies competition between scientists. They complain that science becomes increasingly more conditioned by market forces and argue that "we must acknowledge that markets are human inventions, and that humans have the power to limit them or to change the rules by which they operate" (Gardner et al., 2001, p. 92).

Bearing in mind these considerations and the fact that life sciences target examining harmful diseases to invent possible treatments, it should be considered an ultimate ambition to help scientists focus on the superordinate goal, namely helping people, and to pull together to reach this goal. If psychological research can help to reach this by examining the conditions that enable cooperation and by developing interventions that foster cooperation, it can make an important contribution for the good of scientists and mankind as a whole.

ACKNOWLEDGMENT

We would like to thank the Volkswagenstiftung for supporting the research for this chapter. In addition, we very much appreciated the advice and support of W. Scholl. Thanks also to our collaborators, K. Meischner and C. Walther.

REFERENCES

Amabile, T. M. (1996). *Creativity in context*. Bouldner, CO: Westview Press.

Amason, A. C. (1996). Distinguishing the effects of functional and dysfunctional conflict on strategic decision making: Resolving a paradox for top management teams. *Academy of Management Journal, 39*, 123–148.

Amason, A. C., & Schweiger, D. M. (1994). Resolving the paradox of conflict, strategic decision making, and organizational performance. *International Journal of Conflict Management, 5*(3), 239–253.

Anderson, N., De Dreu, C. K. W, & Nijstad, B. A. (2004). The routinization of innovation research: A constructively critical review of the state-of-the science. *Journal of Organizational Behavior, 25,* 147–173.

Axelrod, R. A. (1984). *The evolution of cooperation.* New York, NY: Basic Books.

Beersma, B., Hollenbeck, J. R., Moon, H., Conlon, D. E., & Ilgen, D. R. (2003). Cooperation, competition, and team performance: Towards a contingency approach. *Academy of Management Journal, 46*(5), 572–590.

Bierhoff, H. W., & Muller, G. F. (1993). Kooperation in Organisationen [Cooperation in Organizations]. *Zeitschrift für Arbeits- und Organisationspsychologie, 37*(2), 42–51.

Blake, R. R., & Mouton, J. S. (1964). *The managerial grid,* Houston, TX: Gulf Publishing Co.

Blake, R. R., & Mouton, J. S. (1970). The fifth achievement. *Journal of Applied Behavioral Science, 6,* 413–426.

Blake, R. R., & Mouton, J. S. (1981). Management by grid principles or situationalism: Which? *Group and organizations studies, 6,* 439–445.

Callon, M., & Latour, B. (1992). Don't throw the baby out with the Bath school—a reply to Collins and Yearley. In A. Pickering (Ed.), *Science as practice and culture* (pp. 1–16). Chicago, IL: University of Chicago Press.

Collins, H. M. (1981). Stages on the empirical programme of relativismus. *Social Studies of Science. 11,* 3–10.

Coser, L. A. (1956). *The function of social conflicts.* Glencoe, IL: Free Press.

De Church, L. A., & Marks, M. A. (2001). Maximizing the benefits of task conflict: The role of conflict management. *The International Journal of Conflict Management, 12,* 4–22.

De Dreu, C. K. W. (1997). Productive conflict: The importance of conflict management and conflict issue. In C. K. W. De Dreu & E. Van de Vliert (Eds.), *Using conflict in organizations* (pp. 9–22). London, UK: Sage.

De Dreu, C. K. W. (2002). Team innovation and team effectiveness: The importance of minority dissent and reflexivity. *European Journal of Work and Organizational Psychology, 11,* 285–298.

De Dreu, C. K. W. (2006). When too little or too much hurts: Evidence for a curvilinear relationship between task conflict and innovation in teams. *Journal of Management, 32,* 83–107.

De Dreu, C. K. W., & Weingart, L. R. (2003). Task versus relationship conflict, team performance, and team member satisfaction: A meta-analysis. *Journal of Applied Psychology, 88*(4), 741–749.

De Dreu, C. K. W., Harinck, F., & van Vianen, A. E. M. (1999). Conflict and performance in groups and organizations. In C. L. Cooper & I. T. Robertson (Eds.), *International review of industrial and organizational psychology* (Vol. 14, pp. 369–414). New York, NY: John Wiley & Sons.

Deutsch, M. (1949). A theory of co-operation and competition. *Human Relations, 2,* 129–152.

Deutsch, M. (1962). Cooperation and trust: Some theoretical notes. In M. R. Jones (Ed.), *Nebraska symposium on motivation* (pp. 275–319). Lincoln, NE: University of Nebraska Press.

Deutsch, M. (1973). *The resolution of conflict.* New Haven, CT: Yale University Press.

Deutsch, M. (2003). Cooperation and conflict: A personal perspective on the history of the social psychological study of conflict resolution. In M. A. West, D. Tjosvold, & K. G. Smith (Eds.), *International handbook of organizational teamwork and cooperative working* (pp. 9–43). Chichester, UK: Wiley & Sons.

Driskell, J. E., & Salas, E. (1991). Group decision making under stress. *Journal of Applied Psychology, 76*(3), 473–478.

Eglau, H., Kluge, J., Meffert, J., & Stein, L. (2000). *Durchstarten zur Spitze: McKinseys Strategien für mehr Innovation* [Take off to the top: McKinsey's strategies for more innovation] (2nd ed.). Frankfurt am Main. Germany: Campus.

Fisher, R., & Ury, B. (1981). *Getting to yes: Negotiating agreement without giving in.* Boston, MA: Houghton Mifflin.

Gardner, H., Csikszentmihalyi, M., & Damon, W. (2001). *Good work. When excellence and ethics meet.* New York, NY: Basic Books.

Gebert, D. (2004). *Innovation durch Teamarbeit. Eine kritische Bestandsaufnahme* [Innovation through teamwork. A critical status quo assessment]. Stuttgart, Germany: Kohlhammer.

Gersick, C. J. (1989). Marking time: Predictable transitions in task groups. *Academy of Management Journal, 32,* 274–309.

Gilbert, G., & Mulkay, M. (1984). *Opening Pandora's Box.* Cambridge, MA: Cambridge University Press.

Gruenfeld, D. H., Mannix, E. A., Williams, K. Y., & Neale, M. A. (1996). Group composition and decision making: How member familiarity and information distribution affect process and performance. *Organizational Behavior and Human Decision Processes, 67*(1), 1–15.

Grunwald, W. (1982). Konflikt-Konkurrenz-Kooperation: Eine theoretisch empirische Konzeptanalyse [Conflict–competition–cooperation. A theoretical empirical concept analysis]. In W. Grunwald & H. G. Lilge (Eds.), *Kooperation und Konkurrenz in Organisationen* (pp. 50–96). Bern, Switzerland: UTB.

Gulbrandsen, M. (2004). Accord or discord? Tensions and creativity in research. In S. Hemlin, C. M. Allwood, & B. R. Martin (Eds.), *Creative knowledge environments. The influences on creativity in research and innovation* (pp. 31–57). Cheltenham, UK: Elgar.

Gulbrandsen, M. (2004). Accord or discord? Tensions and creativity in research. In S. Hemlin (Ed.), *Creative knowledge environments. The influences on creativity in research and innovation.* Cheltenham, UK: Elgar.

Hauschildt, J. (Ed.). (1993). *Innovationsmanagement* [Innovation Management]. München, Germany: Vahlen.

Hemlin, S., Allwood, C. M., & Martin, B. R. (2004). What is a creative knowledge environment? In S. Hemlin, C. M. Allwood, & B. R. Martin (Eds.), *Creative knowledge environments. The influences on creativity in research and innovation* (pp. 1–28). Cheltenham, UK: Elgar.

Hemlin, S. (2006). Creative knowledge environments in research groups in biotechnology. *Scientometrics, 67,* 121–142.

Holton, G. (1981). *Thematische Analyse der Wissenschaft* [Thematic analysis of science]. Frankfurt, Germany: Suhrkamp.

Jehn, K. A. (1995). A multimethod examination of the benefits and detriments of intragroup conflict. *Administrative Science Quarterly, 40,* 256–282.

Jehn, K. A. (1997). Affective and cognitive conflict in work groups: Increasing performance through value-based intragroup conflict. In C. K. W. De Dreu & E. Van de Vliert (Eds.), *Using conflict in organizations* (pp. 72–86). London, UK: Sage.

Jehn, K. A., & Mannix, E. A. (2001). The dynamic nature of conflict: A longitudinal study of intragroup conflict and group performance. *Academy of Management Journal, 44,* 238–251.

Jehn, K. A., Northcraft, G. B., & Neale, M. A. (1999). Why differences make a difference: A field study of diversity, conflict, and performance in workgroups. *Administrative Science Quarterly, 44,* 741–763.

Johnson, D. W., Johnson, R., & Smith, K. G. (1989). Controversies in decision making situations. In M. A. Rahim (Ed.), *Managing conflict: An interdisciplinary approach* (pp. 251–264). New York, NY: Praeger.

Johnson, D. W., & Johnson R. T. (1989). *Cooperation and competition: Theory and research.* Minnesota, MN: Interaction Book Company.

Johnson, D. W., & Johnson, R. T. (1992). Positive interdependence: Key to effective cooperation. In R. Hertz-Lazarowitz & N. Miller (Eds.), *Interaction in cooperative groups* (pp. 174–199). Cambridge, MA: University Press.

Johnson, D. W., Maruyama, G., Johnson, R., & Nelson, D. (1981). Effects of cooperative, competitive, and individualistic goal structures on achievement: A meta-analysis. *Psychological Bulletin, 89*(1), 47–62.

Knorr-Cetina, K. (1984). *Die Fabrikation von Erkenntnis* [The fabrication of knowledge]. Frankfurt, Germany: Suhrkamp.

Krüger, H. P. (1990). *Kritik der kommunikativen Vernunft.* Berlin, Germany: Akademieverlag.

Kuhn, T. S. (1977). *The essential tension.* Chicago, IL: University of Chicago Press.

Kwang, N. A. (2001). Why creators are dogmatic people: "nice" people are not creative, and creative people are not "nice". *International Journal of Group Tensions, 30*(4), 293–324.

Latour, B., & Woolgar, S. (1980). *Laboratory Live. The social construction of scientific facts.* London, UK: Sage.

Lewicki, R. J., Weiss, S. E., & Lewin, D. (1992). Models of conflict, negotiation and third party intervention: A review and synthesis. *Journal of Organizational Behavior, 13,* 209–252.

Mead, G. H. (1934): *Mind, self and society.* Chicago, IL: Chicago University Press.

Mitroff, I. (1974). *The subjective side of science.* Amsterdam, Holland: Elsevier.

Moscovici, S. (1971). *Sozialer Wandel durch Minoritäten* [Social Change through minorities]. München, Germany.

Munduate, L., Ganaza, J., Peiró, J. M., & Euwema, M. (1999). Patterns of styles in conflict management and effectiveness. *The International Journal of Conflict Management, 10,* 5–24.

Nemeth, C. J. (1986). Differential contributions of majority and minority influence, *Psychology Rewiew, 93,* 23–32.

O'Reilly, C. A. III, Williams, K. Y., & Barsade, S. (1998). Group demography and innovation: Does diversity help? In D. H. Gruenfeld (Ed.), *Composition: research on managing groups and teams* (pp. 183–207). New York, NY: Elsevier Science/JAI Press.

Pelled, L. H. (1996). Relational demography and perceptions of group conflict and performance: A field investigation. *International Journal of Conflict Management, 7,* 230–246.

Pickering, A. (Ed.) (1992). *Science as practice and culture.* Chicago, IL: University of Chicago Press.

Pinch, T. (1986). *Confronting nature: The sociology of neutrino detection.* Dordrecht, Holland: Reidel.

Pleschak, F., & Sabisch, H. (1996). Innovationsmanagement. [Innovation Management]. Stuttgart, Germany: Schäffer–Poeschel.

Pruitt, D., & Rubin, J. (1986). *Social conflict: Escalation, stalemate, and settlement.* New York, NY: Random House.

Pruitt, D., Rubin, J., & Kim, S. H. (2004). *Social conflict: escalation, stalemate, and settlement* (3rd ed.). New York, NY: McGraw-Hill.

Putnam, L. L., & Poole, M. S. (1987). Conflict and negotiation. In F. M. Jablin, L. L. Putnam, K. H. Roberts, & L. W. Porter (Eds.), *Handbook of organizational communication: An interdisciplinary perspective* (pp. 549–599). Thousand Oaks, CA: Sage Publications.

Qin, Z., Johnson, D. W., & Johnson, R. T. (1995). Cooperative versus competitive efforts and problem solving. *Review of Educational Research, 65*(2), 129–143.

Rahim, M. A. (1983). *Rahim Organizational Conflict Inventories.* Palo Alto, CA: Consulting Psychologist Press.

Rahim, M. A. (1992). *Managing conflict in organizations* (2nd ed.). New York, NY: Praeger.

Rahim, M. A. (2002). Toward a theory of managing organizational conflict. *International Journal of Conflict Management, 13*(3), 206–236.

Rocholl, A., & Schulze, A. (2008). *The genesis of innovations as a process of conflict emergence and conflict management.* Paper presented at the International Congress of Psychology, July 20–25th, 2008, Berlin, Germany.

Rosenwein, R. E. (1994). Social influence in science: Agreement and dissent in achieving scientific consensus. In W. R. Shadish & S. Fuller (Eds.), *The social psychology of science* (pp. 262–285). New York, NY: Guilford Press.

Scholl, W. (2004). *Innovation und Information. Wie in Unternehmen neues Wissen produziert wird* [Innovation and information. How new knowledge is produced in enterprises]. Goettingen, Germany: Hogrefe.

Scholl, W. (2005). Grundprobleme der Teamarbeit und ihre Bewältigung–Ein Kausalmodell [Fundamental problems of team work and their handling—A causal model]. In

H. G. Gemunden & M. Hogl (Eds.), *Management in teams. Theoretische Konzepte und empirische Befunde* (3rd ed., pp. 33–66). Wiesbaden, Germany: Gabler.

Scholl, W. (2009). Konflikte und Konflikthandhabung bei Innovationen [Conflicts and conflict management in innovation processes]. In E. Witte & C. Kahl (Eds.), *Sozialpsychologie der Kreativität und Innovation* (pp. 67–86). Lengerich, Germany: Pabst.

Schulze, A. (1990). On the rise of scientific innovations and their acceptance in research groups: A socio-psychological study. *Social Studies of Science, 20*, 53–64.

Schulze, A. (2000). Mediation bei asymmetrischen Konflikten in der Forschung [Mediation in asymmetric conflicts in research]. In A. Dieter, L. Montada, & A. Schulze (Eds.), *Gerechtigkeit im Konfliktmanagement und in der Mediation* (pp. 181–196). Frankfurt, Germany: Campus.

Schulze, A., & Wenzel, V. (1996). *Faszination Licht-Portrait einer wissenschaftlichen Gemeinschaft.* [Fascination light – portrait of a scientific community]. Muenster, Germany: Waxmann.

Scott, S. G., & Bruce, R. A. (1994). Determinants of innovative behaviour: A path model of individual innovation in the workplace. *Academy of Management Journal, 37*(3), 580–607.

Seuffert, V. (2007). *Cooperation and competition in the process of innovation. A study in the field of Life Science.* unpublished Master's thesis, Humboldt-Universität zu Berlin, Berlin, Germany.

Simons, T. L., & Peterson, R. S. (2000). Task conflict and relationship conflict in top management teams: The pivotal role of intragroup trust. *Journal of Applied Psychology, 85*, 102–111.

Slavin, R. E. (1977). Classroom reward structure: An analytic and practical review. *Review of Educational Research, 47*, 633–650.

Sounder, W. (1987). *Managing new product innovation.* Lexington, MA: Lexington Books.

Standage, M., Duda, J. L., & Pensgaard, A. M. (2005). The effect of competitive outcome and task-involving, ego-involving, and cooperative structures on the psychological well-being of individuals engaged in a co-ordination task: A self-determination approach. *Motivation and Emotion, 29*(1), 41–68.

Stanne, M. B., Johnson, D. W., & Johnson, R. T. (1999). Does competition enhance or inhibit motor performance: A meta-analysis. *Psychological Bulletin, 125*(1), 133–154.

Tauer, J. M., & Harackiewicz, J. M. (2004). The effects of cooperation and competition on intrinsic motivation and performance. *Journal of Personality and Social Psychology, 86*(6), 849–861.

Thomas, K. W. (1992a). Conflict and conflict management: Reflections and update. *Journal of Organizational Behavior, 13*, 265–274.

Thomas, K. W. (1992b). Conflict and negotiation processes in organizations. In M. D. Dunnette & L. M. Hough (Eds.), *Handbook of industrial and organizational psychology* (2nd ed., pp. 651–717). Palo Alto, CA: Consulting Psychologists Press.

Tjosvold, D., West, M. A., & Smith, K. G. (2003). Teamwork and cooperation: Fundamentals of organizational effectiveness. In M. A. West, D. Tjosvold, & K. G. Smith (Eds.), *International handbook of organizational teamwork and cooperative working* (pp. 3–8). Chichester, UK: Wiley & Sons.

Tjosvold, D. (1985). Implications of controversy. *Journal of Management, 11*, 221–238.

Tjosvold, D. (1991). Rights and responsibilities of dissent: Cooperative conflict. *Employee Responsibilities and Rights Journal, 4*, 13–23.

Traweek, S. (1988). *Beamtime and livetime.* Cambridge, MA: Harvard University Press.

Turkle, S. (1984). *The second self.* New York, NY: Simon and Schuster.

Turner, M. E., & Pratkanis, A. R. (1997). Mitigating groupthink by stimulating constructive conflict. In C. K. W. De Dreu & E. van de Vliert (Eds.), *Using conflict in organizations* (pp. 53–71). Thousand Oaks, CA: Sage Publications.

van de Vliert, E. (1997): *Complex interpersonal conflict behaviour.* Hove, UK: Psychology Press.

van de Vliert, E., & Hordijk, J. W. (1989). A theoretical position of compromising among other styles of conflict management. *The Journal of Social Psychology, 129*, 681–690.

van de Vliert, E., Euwema, M. C., & Huismans, S. E. (1995). Managing conflict with a subordinate or a superior: Effectiveness of conglomerated behavior. *Journal of Applied Psychology, 80,* 271–281.

van de Vliert, E., Nauta, A., Euwema, M. C., & Janssen, O. (1997). The effectiveness of mixing problem solving and forcing. In C. K. W. De Dreu & E. van de Vliert (Eds.), *Using conflict in organizations* (pp. 38–52). Thousand Oaks, CA: Sage Publications.

Wageman, R. (1995). Interdependence and group effectiveness. *Administrative Science Quarterly, 40,* 145–180.

West, M. A., & Hirst, G. (2003). Cooperation and teamwork for innovation. In M. A. West, D. Tjosvold, & K. G. Smith (Eds.), *International handbook of organizational teamwork and cooperative working* (pp. 297–319). Chichester, UK: Wiley & Sons.

West, M. A., Smith, K. G., & Tjosvold, D. (2003). Past, present, and future perspectives on organizational cooperation. In M. A. West, D. Tjosvold, & K. G. Smith (Eds.), *International handbook of organizational teamwork and cooperative working* (pp. 575–597). Chichester, UK: Wiley & Sons.

Wetzel, J. (1994). Problemlösen in Gruppen: Auswirkungen von psychologischen Trainingsmaßnahmen und Expertenbeteiligung unter kooperativen und kompetitiven Arbeitsbedingungen [Problem solving in groups: effects of psychological training interventions and of the participation of experts under cooperative and competitive working conditions]. Dissertation. Braunschweig, Germany.

Young, G. (2003). Contextualizing cooperation. In M. A. West, D. Tjosvold, & K. G. Smith (Eds.), *International handbook of organizational teamwork and cooperative working* (pp. 77–109). Chichester, UK: Wiley & Sons.

Ziegler, R. (1968). *Kommunikationsstruktur und Leistung sozialer Systeme.* [Communication structure and performance of social systems]. Hain, Germany: Meisenheim am Glan.

CHAPTER 13

Postmodernism and the Development of the Psychology of Science

E. J. Capaldi and Robert W. Proctor

The psychology of science is a new field of study with many needs not yet fully met (Feist, 2006, 2011; Simonton, 2009). If the field is to prosper, it must be aware of how to meet those needs and how to counter negative views of science that may undermine achieving this goal. One means of countering views that would undermine the psychology of science is to be aware of them and their potential for harm. The purpose of this chapter is to describe core beliefs associated with postmodernism and its variants (social constructionism and contextualism) that are inimical to progress in the psychology of science. We note at the outset, though, that not all beliefs of postmodernism and its variants are inimical to the psychology of science. For example, an emphasis on agreement among scientists for theory acceptance can be embraced if it is intended to mean agreement on the basis of empirical as well as theoretical considerations.

The naturalistic worldview, which is regarded as the basis of science as conventionally understood, embraces empiricism while rejecting supernatural and a priori propositions. Naturalism is compatible with a variety of empirical methods, ranging from observational to correlational to experimental, and it bases its conclusions on the results of such empirical methods. Naturalism underlies, and in fact is essential to, all of contemporary science, and it also has been extended to philosophy in general and philosophy of science in particular (Callebaut, 1993; Mayo & Spanos, 2010).

Prior to naturalism prevailing in the philosophy of science, the accepted approach was that of foundationism, in which one's conception of science was based on logic and intuition (Curd & Cover, 1998). This approach was largely replaced due to the influence of Kuhn (1962), who, in his monumental book, *The Structure of Scientific Revolutions*, suggested that philosophy of science should be based on empirical data, that is, it should be naturalized. The specific data employed by Kuhn were historical examinations of past theoretical accomplishments, which led him to a variety of novel conclusions, among them being that scientists accept a new paradigm only when they are prepared to abandon a prior paradigm.

Since Kuhn, a variety of methods have been introduced to study how science is accomplished (see, e.g., Donovan, Laudan, & Laudan, 1992; Proctor & Capaldi, 2006). These include, for example, experiments in which scientists and nonscientists solve problems in controlled environments, statistical analyses of social/personality data of scientists for differences from nonscientists, personal reports of scientists obtained from laboratory notes and by other means, and studies of groups of scientists working in laboratory settings (e.g., Dunbar & Fugelsang, 2005; Simonton, 2009). According to a philosopher of science Laudan (1996), naturalism demands that the philosophy of science not be incompatible with the findings of science. The same would be true of the psychology of science. A fundamental difference between foundationism and naturalism, can be expressed as follows: Foundationism approaches science from the top down, whereas naturalism approaches science from the bottom up.

Clearly, naturalism is the view that can profitably be embraced by the psychology of science. As reasonable as naturalism may seem to most working scientists, however, there are many who reject it in favor of a wholly different, nonscientific approach to knowledge (see Gross & Levitt, 1994 and Laudan, 1996, for discussions of this point). Those who embrace these nonscientific views may be in a position to thwart the psychology of science, not only in an intellectual sense, which we consider later, but also in a practical one, which we deal with now. By practical, we mean stymieing efforts by the psychology of science to enter the academy through denying access to what is necessary: courses, degree programs, support for students, and so on. For these reasons, the psychology of science must be aware of viewpoints in the academy that might act to thwart its development.

As indicated, the views to which we refer are those of postmodernism and its variants, which include social constructionism and contextualism, among others. The views are held by a considerable number of individuals, albeit a minority, in the academy (Gross & Levitt, 1994), in the government (Devine, 2004), and in the American Psychological Association (APA). Examination of the *American Psychologist*, the flagship journal of APA, shows that it regularly publishes articles that are highly positive of postmodernism (Gergen, 2001), social constructionism (Gergen, 1985; Gergen, Gulerce, Lock, & Misra, 1996), and contextualism (Rogler, 2002). Also, there are forces seeking to guide the APA away from an emphasis on experimentation and quantitative methods and in the direction of variants of qualitative methods that are at bottom subjective and relativistic (e.g., Camic, Rhodes, & Yardley, 2003). An example is a recent move to incorporate a section called the *Society for Qualitative Inquiry in Psychology* (SQI) into Division 5 of the APA, "Evaluation, Measurement, and Statistics," which has a long history of advocating quantitative methods in psychological research. The task force studying the matter said, "On the task force we have viewed both approaches (quantitative and qualitative) as complementary and compatible, with both methods needed to fully examine questions of interest" (Boodoo, Schmitt, & Weiner, 2009, p. 4). All of this sounds reasonable until one realizes that qualitative *inquiry*, as will be discussed later, is not referring to use of qualitative *methods* in general, something with which few researchers would disagree, but to a particular approach that is linked to postmodernism and its variants.

In agreement with our assessment, Shrout and Shadish (2011) opposed the initiative for the new section on qualitative inquiry in Division 5 for the following reason:

> It seeks to welcome into Division 5 *all* the members of SQI, many of whom have little or no interest in quantitative approaches, some of whom intellectually oppose assigning numbers to people in the process we call measurement, and some of whom even find problematic the very word "method" as a description of what we might share in common. The fact that the section is to be called qualitative *inquiry* rather than qualitative *methodology* is no accident. (p. 7)

Shrout and Shadish emphasized that views associated with qualitative inquiry are so different from those of most members of Division 5 that "the risks [associated with the new section] are more substantial than have been stated" (p. 7). Despite the concerns voiced by Shrout and Shadish, as we were completing this chapter, members voted to make SQI an official section in Division 5 of the APA. Gergen, Josselson, Freeman, and Futch (2012) announced this decision in *The Qualitative Report* of March 5, 2012, describing it as "a time for collective jubilation" and a "historic occasion." Given that the first three of those individuals have expressed views in the past that are antithetical to those of most members of Division 5, we do not see this as a very propitious event.

It is easy to imagine that people who oppose measurement, experimentation, and the like, or quantitative methods in general, may be in a position to obstruct developments both in psychological science and the psychology of science. This can be done by their assuming various positions of considerable influence in organizations, as in the case of Division 5 of APA, and in the academy, such as those of department head, deans, and so on. Once in place, such individuals are in a position to exercise their biases by opposing programs that would advance scientific psychology and the psychology of science.

CORE BELIEFS OF POSTMODERNISM

Modernism is of the view that there is a reality that exists independent of our selves. This reality can be known rationally and, hence, modernism places considerable emphasis on science. The modernist view is optimistic in terms of what rational inquiry can achieve in relation to furthering knowledge and improving the lot of humankind (Wilson, 1998). In short, the outlook of modernism is rosy and optimistic, and it emphasizes that progress is possible.

In contrast, postmodernism can be defined as: "A worldview characterized by the belief that truth doesn't exist in any objective sense but is created rather than discovered" (McDowell & Hostetler, 1998, p. 208). In the postmodern view, truth is "created by the specific culture and exists only in that culture. Therefore, any system or statement that tries to communicate truth is a power play, an effort to dominate other cultures" (McDowell & Hostetler, 1998, p. 208). As these quotes imply, a core belief of postmodernism, which it shares with social constructionism and contextualism, is that objective reality does not exist, that truth merely constitutes an agreement among individuals and exists only at a specific time and in a specific context. Any attempt to go beyond this view of truth is considered to be illegitimate.

Postmodernism sees modernism as having overconfidence in the rational process to improve the lot of humankind (Devine, 2004). Also, from this view, there is in modernism overconfidence in science as being the ultimate form of knowledge. Postmodernism also rejects the view that understandings arising from rational scientific processes are truer than other understandings. That is, science is merely one form of knowledge, no better than any other form. For example, Gergen, perhaps the leading postmodernist in psychology and one of the aforementioned leaders of the movement to incorporate the Society for Qualitative Inquiry within Division 5 of the APA, indicated that understandings from scientific and spiritual perspectives are equally valid forms of knowledge. According to Gergen (2001),

> This is so in the case of both natural sciences and spiritual practices. They both constitute traditions of understanding; among their major differences are the rules of agreement...that they embrace and the kinds of outcomes that they provide for the culture. (p. 806)

We quote and reference Gergen's views extensively in the remainder of the chapter because he is the most prominent advocate of postmodernism and social constructionism in psychology and his views are representative of those of postmodernists in general.

Postmodernism is skeptical of progress, suggesting that modernism has an overconfidence in progress, whether economic, social, or technical. Postmodernism views science as playing what Gergen (2001) describes as "the truth game." Postmodernists also see science as oppressive, arbitrary, and authoritarian because science sees itself as producing the most reliable and valid knowledge for dealing with the empirical world. Contrary to this view, postmodernists believe that different approaches such as science and witchcraft simply employ and play by different sets of rules. According to the postmodern worldview, theories do not get better, they only change by adopting new rules. Postmodernists also believe that there is no way to adjudicate among knowledge claims. All that can be done is to determine whether the knowledge claim is consistent with the rules of the particular system one employs.

Because postmodernists believe that knowledge claims cannot be evaluated outside of the particular system that produces them, they suggest that when some knowledge claim is put forth as true its basis is not truth but power. According to this view, science occupies a prominent position in the university and society in general not by being true but by having the power to advance its claims. Not only does power play a positive role in advancing these claims, but it also plays a negative role in silencing any potential and actual critics. Here we see the authoritarian, nonhumanistic side of science, according to the postmodernists.

Another view of postmodernists that should be troublesome to conventional scientists is that methods such as experimentation, which are important in modernism, are seen as irrelevant in postmodernism and its variants. At one extreme, Gergen and Gergen (1991) suggest that conducting experiments is useless because they merely produce the conclusion wished for in the first place. In their words,

> We argue that not only are experimental attempts at 'testing' general theoretical ideas wasteful of effort and resources, but that by engaging in reflexive techniques a scientist can learn more about a theoretical position in a brief period than research can ever "demonstrate." (p. 79)

They also say, "The confirmations (or disconfirmations) of hypotheses through research findings are achieved through social consensus, not through observation of the 'facts'" (p. 81). Indeed, Gergen and Gergen advocate a research procedure that they call *hypothetical data rotation*, which involves considering all hypotheses about what outcomes might occur in certain situations and then not bothering to test those hypotheses because the researcher's preconceived biases would determine the alternative eventually selected. In their words:

> In the first step of this procedure the investigator undertakes the traditional preparatory steps in generating an idea for a laboratory study. Theoretical preferences are singled out, hypotheses are formed, a research design is elaborated, and procedures (setting, subject, etc.) are envisioned. However, at this juncture the standard research procedure is terminated. No funds are spent on equipment, subjects, data collection, data analysis and the like. Rather, the researchers may lay out the research design and arrange the predicted pattern of results in a matrix. At this point they are poised for the expansion of understanding. (pp. 82, 83)

Postmodernists suggest that things are not what they seem to be, that appearances are deceiving, because there are always hidden meanings. These hidden meanings are the result of hidden power claims. For example, "woman" can mean a mother, a victim, a sexual object, a member of a superior or inferior gender, or many other things depending on the context in which the utterance occurs. Postmodernists favor *deconstruction*, which means reading a text to detect meanings other than the one that the text intends to suggest. One result of deconstruction is to indicate that the text contains contradictions between the intent of the text and the surface of the work. Essentially, deconstruction suggests that a text has multiple and contradictory meanings, and therefore can be employed in a variety of incompatible ways.

Surprisingly, postmodernists find some justification for these views in the writings of the philosopher/historian of science, Kuhn. At one point, Kuhn (1962) pointed out that the criteria for judging whether a theory is or is not worthwhile may be at odds with each other. For example, a theory that is fruitful may not be stated in a quantitative, and therefore unambiguous, manner. Alternatively, a quantitative, unambiguous theory need not contain implications of particular theoretical importance. Postmodernists see relativism in this same view as that of Kuhn's. They also see relativism in Kuhn's idea that individuals holding different theories may not be able to fully communicate with each other, a situation he describes as incommensurability. Also, there is some ambiguity in Kuhn's writings as to when one theory can be rejected and another accepted. Kuhn is vague as to when this may occur, as perhaps can be appreciated from the fact that he sees the criteria for accepting or rejecting theories as often being at odds with each other. This certainly applies to anomalies, and in Kuhn's view, it also applies to evaluation of theories in general.

However, Kuhn vigorously denied that he was a relativist. At one point, Kuhn (1970) said, "For me, therefore, scientific development is, like biological evolution, unidirectional and irreversible. One scientific theory is not as good as another for doing what scientists normally do. In that sense, I am not a relativist" (p. 264). Later, even more emphatically, he stated,

> The most extreme form of the [negotiation] movement, called by its proponents "the strong program," has been widely understood as claiming that power and interest are all there are. Nature itself, whatever that may be, has seemed to have no part in the development of beliefs about it. Talk of evidence, of the rationality of claims drawn from it, and of the truth or probability of those claims has been seen as simply the rhetoric behind which the victorious party cloaks its power. What passes for scientific knowledge becomes, then, simply the belief of the winners.
>
> I am among those who have found the claims of the strong program absurd: an example of deconstruction gone mad. (Kuhn, 1992, pp. 8, 9)

In truth, postmodernists who suggest that Kuhn provides support for their position overlook that there are indeed many aspects of Kuhn's view that are by no means relativistic. For example, Kuhn accepts experimentation and empirical-based arguments, two things that are rejected by postmodernists. Moreover, Kuhn's book of 50 years ago was the starting point of much subsequent philosophy of science that is by no means relativistic (e.g., Laudan, 1996; Mayo & Spanos, 2010), which postmodernists tend to ignore. Also, Kuhn suggested that theory evaluation involved *relative* theory evaluation, a view incompatible with the relativistic notion that a theory should be evaluated according to its consistency with its own criteria.

CORE BELIEFS OF SOCIAL CONSTRUCTIONISM

Social constructionism is based on the idea that knowledge is a social construction. Gergen (2009a) sums up the main implication of knowledge being socially constructed as follows: "The basic idea of social construction may seem simple enough. But consider the consequences: if everything we consider real is socially constructed, then nothing is real unless people agree that it is" (p. 4). Gergen's social constructionist views are viewed by some psychologists as being of considerable significance to the field. For example, in 2011, a special issue of the *Journal of Constructivist Psychology* was devoted to Gergen's (2009b) book, *Relational Being*, with the issue's organizer stating, "It is safe to say that those of us who read and responded to this book consider it a benchmark contribution to social psychology, if not psychology generally. At the very least, it is a vital counterweight to individualism" (Slife, 2011, p. 277).

Social constructionism shares many of its beliefs with postmodernism, placing an emphasis on the social construction of knowledge, which amounts to agreement among individuals. From this perspective, social constructs arise from countless human choices rather than exclusively from laws and external reality. Social constructionism holds that all knowledge, including the most basic, is derived from and maintained by social interaction. A major concern of social constructionism is to discover the ways in which individuals and groups participate in the creation of their perceived social reality.

Although social constructionism shares many views with postmodernism, it is worth treating the two separately because it has had a great impact on a field that is closely related to the psychology of science, namely on the sociology of science (Latour, 1999). The philosopher Friedman (1998) states, the strong program of the sociology of scientific knowledge (SSK):

> has been framed by an explicitly philosophical agenda—an agenda that aims to reject the traditional philosophical ideal of universal standards of rationality, objectivity, and truth (which ideal has, of course, been traditionally taken to be paradigmatically exemplified in modern science itself) in favor of a relativistic conception of scientific rationality, objectivity, and truth that grounds these concepts, in the end, in local and particular social and cultural circumstances. According to what I will call the philosophical agenda of SSK, that is, all there ultimately is to the notions of rationality, objectivity, and truth are local socio-cultural norms conventionally adopted and enforced by particular sociocultural groups. (pp. 239, 240)

A characteristic of the strong program, whose claims Kuhn (1992), as quoted previously, found "absurd," is that it has almost no regard for the philosophy of science, and only a slight regard for the role of the psychology of science. Because the sociology of science is concerned with socially produced knowledge, not individually produced knowledge, the individual scientist is of limited concern. The philosopher of science, Laudan (1996) describes the position of Bloor, who espouses the strong program, as follows, "Bloor sets out to redefine the disciplinary boundaries for the study of science, giving sociology pride of place, leaving a limited scope for psychology and dealing philosophers, on the strength of their prior track record, largely out of the game altogether" (p. 184). The limited scope for psychology is illustrated by the following statement of Bloor's (1991): "The knowledge of our culture, as it is represented in our science, is not knowledge of a reality that any individual can experience or learn about for himself.... Knowledge, then, is better equated with Culture than Experience" (p. 16). The strong program of the sociology of science is a threat to the psychology of science because, as Laudan has indicated, the view of advocates of the strong program is: "Science is a social activity, therefore it is best understood and explained in sociological terms" (p. 201).

A number of social constructionists who have advocated the strong program of the sociology of science, including Bloor, Barnes, and Collins, use social constructionism to suggest that what science typically regards as objective reality is really subjectivity imposed on those facts by groups of individuals with similar beliefs. In their view, these are subjective but taken to be objective. This can be interpreted as a form of relativism, as acknowledged by Collins (1983), who said, "Barnes, Bloor and I maintained an unambiguous concern with the sociology of knowledge examined from a relativist perspective ..." (pp. 267, 268). More recently, Collins has departed from this view, which he calls the second wave of scientific development, by suggesting what he calls the third wave, which consists of consulting not all individuals but only those persons who are knowledgeable on a particular topic (Collins & Evans, 2002).

According to social constructionists, science is not privileged in any way. All scientific constructs, from physical laws to theories, are arrived at by consensus among similarly minded people, and all are social constructs. In agreement with postmodernism, social constructionism does not see science as having a special position among other approaches to knowledge. This is evident in what Bloor (1973) identifies as two of the four requirements of the strong program: "The program must be impartial with respect to truth and falsity," and "Not only must true and false beliefs be explained, but the same sort of causes must generate both classes of belief" (pp. 173, 174), the latter of which is known as the symmetry requirement. From this view, knowledge that is sociologically interesting does not arise from the physical world but is a result of group/social processes. For example, Fosnot (1996) says of constructionist theory, "Based on work in psychology, philosophy, and anthropology, the theory describes knowledge as temporary, developmental, nonobjective, internally constructed, and socially and culturally mediated" (p. ix).

We have reached the conclusion that the psychology of science can be usefully informed by the philosophy of science, and vice versa, whereas the sociology of science, at least in its strong form that emphasizes social constructionism, could not be a part of advancing knowledge in the psychology or philosophy of science. It should be clear that the strong program of the sociology of science is as unfriendly to social psychology as it is to the psychology of science. For example, Slife (2011) states, "Indeed, several scholars consider social psychology to be so individualistic that it rarely merits the adjective 'social'" (p. 277). We do not mean by our arguments that sociological factors are irrelevant to determining how an individual approaches science (see, e.g., Nersessian, 2005). But, they must be evaluated according to accepted scientific criteria, which are eschewed by the strong program.

CORE BELIEFS OF CONTEXTUALISM

Contextualism is a view that has been adopted by a variety of psychologists in the past two decades. The main proponents of contextualism are to be found in developmental psychology (Ford & Lerner, 1992) and behaviorism (Gifford & Hayes, 1999), but they can be found in other areas including cognitive psychology (Gillespie, 1992) and clinical psychology (Sarbin, 1993). Probably the main reason individuals are attracted to contextualism is that it eschews theory because theory constrains empirical outcomes.

Contextualism is one of four worldviews, each separate and distinct from the other, identified by Pepper (1942) to be formism, organicism, mechanism, and contextualism. Pepper assigned to each of these worldviews a distinct root metaphor that characterizes it. For formism, the root metaphor is similarity: similarity exists insofar as all members of the class conform to the norm, or as Aristotle would put it, to the essence. For organicism, the root metaphor is the growing organism: development occurs in certain stages, directed toward some end. For mechanism, the root metaphor is the machine: the function of the entire machine is produced by its interacting parts. For contextualism, the root metaphor is the act in progress: because of the emphasis on the act in progress, contextualists tend to see novelty and change as constantly occurring. In Pepper's words, "The ineradicable contextualistic categories may thus be said to be *change* and *novelty*" (p. 232). The emphasis in contextualism on novelty and change has many implications for understanding science, which cannot be overemphasized.

Contextualism is a branch of pragmatism, and as such, its concept of truth is what works in practice (successful working). Pepper says, "Truth is utility or successful functioning, and that is the end of it" (p. 43). Hypothesis testing is shunned. Avoiding hypothesis testing has made contextualism attractive to behavior analysts who follow Skinner (see, e.g., Hayes, 2010; Hayes & Hayes, 1989), a variety called functional contextualism. Skinner, of course, was a radical behaviorist, which is a subset of radical empiricism. Radical empiricism emphasizes the view that what we see is what we get, and there is no durable entity behind the empirical phenomena. James (1912) was a prominent exponent of the idea. A consequence of radical empiricism is to shun theory construction, a position strongly adopted by the behavior analysts.

Many developmental psychologists have also embraced contextualism for slightly different reasons. They see development as occurring as a result of many interacting processes, at different levels, which modify each other. Lerner and Kauffman (1985) concluded, "A 'pure' contextualism, in being completely dispersive would not be suitable for use as a philosophical model from which to derive a concept of development" (p. 319). Pure, or philosophic, contextualism sees development as terminating in many different ends, whereas a view called modified contextualism wants to see development terminating in a specific, predetermined end. Capaldi and Proctor (1999) concluded that modified contextualism is not a true instance of contextualism due its desire to see development as culminating in a predetermined end (see also Overton, 2007). We are not concerned here with the functional contextualism of the behavior analysts, nor with the developmental contextualism of the developmental psychologists. The reason is that philosophic contextualism has many more important implications for conventional science than do either of these modified forms (see Capaldi & Proctor, 1999).

In philosophic contextualism, the role of analysis can be seen from three different points of view (Capaldi & Proctor, 1999, p. 44):

> First, any attempt to analyze a complex event into its elements is presumed to distort the event and possibly may be misleading. Second, all analyses are tentative in the sense that no event can be completely analyzed. In dealing with particular events, we may go from one event to another, and there is no stopping place. Third, an event may be analyzed from many points of view, depending on one's purposes. For a contextualist, there is no correct or incorrect analysis; analysis always proceeds from some point of view and serves some practical purpose. (p. 44)

Capaldi and Proctor (1999) identified various fundamental ways in which philosophic contextualism differs from the naturalism of the mechanistic approach. Contextualists, unlike naturalists, reject any procedure that would restrict the range of variables examined. For example, when experiments are run, variables are selected to be examined, necessitating that other variables are held constant or controlled in some other manner. Contextualists also reject the idea that laws can be isolated. They accept that novelty always occurs. Unlike naturalism, which embraces conventional science, contextualism rejects it. Finally, contextualism accepts radical empiricism, rejecting the idea, common in science, that phenomena are the result of underlying processes.

To get the flavor of contextualists' characterizations of conventional science, we quote Gillespie (1992): "Contextualist thinkers have been especially active in showing the inadequacy of objectivist and reductionist methodologies for understanding human behavior ... but they have also been actively producing different conceptions of science and theories of cognition" (p. 39). Gillespie is in favor of subjective methods in psychology that provide holistic descriptions or interpretations.

Perhaps the most basic belief of contextualism is that adding or subtracting something, even a minor something, to an ongoing event, however simple or complicated, may affect that event substantially. As Pepper (1942) has remarked, in contextualism, even a sneeze may take on cosmic significance. Contextualists believe that novelty is always to be expected because changes in ongoing events will transform them into some form of novelty. It is perhaps obvious that controlled experiments would be anathema to contextualists because in the act of controlling variables, the phenomenon is modified and differs from its original form. Philosophic contextualism therefore rejects the view that controlled observations are useful. In short, experimentation is rejected. The contextualists' emphasis on novelty ensures that they would reject lawfulness, and therefore they reject science as it is conducted.

IMPLICATIONS FOR THE PSYCHOLOGY OF SCIENCE

Earlier, we discussed the practical implications of the postmodernist, social constructionist, and contextualist views for the psychology of science, namely that to the extent that advocates of those views achieve prominence in the academy, they may be in a position to impede the development of the psychology of science. Here, we examine some additional practical implications of postmodernism for the psychology of science, but the main focus is on intellectual implications. Most generally, according to the views of postmodernism, social constructionism, and contextualism, the psychology of science would not be considered to be a valid discipline. This assessment is based on a number of considerations, each of which is described below.

Social Constructionism and the Individual

According to social constructionism, truth is the province of the group and not the individual. Social constructionism emphasizes the social to the exclusion of the individual because truth is social consensus. This view rules out the psychology of science based on the contributions of individual scientists. In this connection, Gergen (2009a) has said,

> Likewise, to extend the logic, objectivity and truth are not byproducts of individual minds but of community traditions. And also, science cannot make claims to universal truth, as all truth claims are specific to particular traditions—lodged in culture and history. (p. 8)

Gergen (2009a) is explicit in indicating that even scientific reality is entirely constructed, saying, "Once you enter the halls of social construction, there is no material world in itself. That is, what we call the material world is itself a construction" (p. 54). In another place, Gergen says, "What we assume to be scientific knowledge is therefore a byproduct of a social process" (p. 23). On this basis, he suggests, "No longer was it possible to justify science as a quest for *the* truth." (p. 24)

Theory by Agreement

According to postmodernists and social constructionists, because all theories come about by agreement, there is nothing unique about science, including psychological science. On this view, all theories are on equal par, and none is privileged. Agreement is limited to a particular time, a particular context, and a particular set of individuals. That is, at some particular time in the development of a science, a group of scientists may agree that there are entities such as atoms or genes, whereas at a later time they may not. Any actions to create a psychology of science, therefore, would be only an exercise in power, according to this view.

Modernism, Postmodernism, and Methods

Modernism stressed methods, as does the psychology of science, which these views tend to downplay (see Proctor & Capaldi, 2006). All three of the views examined—postmodernism, social constructionism, and contextualism—oppose the use of scientific methods such as experimentation. Postmodernism rejects scientific methods because these cannot possibly establish a truth independent of the investigator's belief. Social constructionists oppose scientific methods because arriving at the truth is achieved through social agreement and not on the basis of the results of experiments. Contextualists reject experimentation because, in the process of achieving experimental control, the phenomenon that we wish to study is modified and changed. According to all three views, perhaps, bias is why scientific methods should be eschewed. Bias is so great that it cannot be modified or changed by data: It is inevitable that we would interpret the data in terms of our biases. Because all views have equal validity, acceptance of views promulgated by the psychology of science would be no exception and would be only another oppressive voice.

Animal Investigations

According to all three of the views under examination, animal investigations such as those used in learning, memory, and neuroscience (e.g., Pearce, 2008) are incapable of providing any data that are useful. There might be some use of animal data from a social constructionist perspective if researchers agree on what the animal's behavior implies. However, this interpretation would stem from agreement among the researchers and nothing more. The almost exclusive focus on human interactions would deprive the psychology of science from utilizing knowledge gained from the behavior of nonhumans. Arbitrary criteria for evaluating the worth of a psychological theory, such as excluding the study of animals, are eschewed in the psychology of science.

Theory Evaluation

The normal criteria for evaluating scientific theories such as agreement between data and theory, the fruitfulness of a theory, or lack of ambiguity in a theory, to mention a few, are rejected by postmodernists, social constructionists, and contextualists alike. For the social constructionist, as we have seen, truth is arrived at on the basis of agreement and not on the basis of predictive accuracy and so on. For the contextualist, truth varies with a variety of factors and would differ if any of these factors were modified or changed. For the postmodernist, theory is not something that is in agreement with objective reality because there is no such thing as objective reality. As an example, Gergen (2009a) has said about science,

> There is no attempt to deny that "something is happening" when the scientific community is at work. However, the significant question is whether the scientist's terms for naming or describing this "something" can reflect what is actually the case. To be sure, we have a language of atomic properties, chemical elements, and neuro-transmission. The danger is in concluding that these words are somehow privileged as "reflecting" or "mapping" what exists,

that these terms tell or inform us about the nature of the real world. From a constructionist perspective, for example, all that we call "chemical elements" could be given the names of Greek gods. Or, with no loss in accuracy, in physics, we could substitute the term Neptune for the neutron and Zeuss for the proton. Let us not mistake the word for the "world." (p. 172)

What Gergen intends here is not that we could without loss substitute some scientific word with another word, but rather that the referent of the scientific word does not exist.

CRITIQUE OF THE POSTMODERNIST DEPICTION OF PSYCHOLOGICAL SCIENCE

It should be clear from our depiction of postmodernism and its variants that they are possessed of a very negative view of much of psychology because of its scientific approach. This negative view is a necessary consequence of their assumptions and beliefs because the scientific approach seeks to arrive at the single best theory by evaluating and eliminating alternative interpretations. Since the psychology of science seeks to provide explanations using the methods and theories of psychology, it is important to evaluate the negative depiction of psychological research provided by postmodernists.

The negative view of psychological science has been expressed in many places by Gergen, who, as indicated, is the leading exponent of postmodernism and its variants among psychologists. We focus on Gergen's view of psychological science because, unlike many other postmodernists, he specifically singles out psychology for explicit and extended criticism. The analysis provided here is a summary of arguments made in more detail in our 2009 article, *Two Radically Different Worldviews of Psychological Science: Implications for the Psychology of Science*.

Gergen's critique of psychological science is to see it as antiquated, out of date, out of touch, and misinformed. We suggest that Gergen's analysis contains two kinds of errors in his critique of psychological science. First, he is often factually incorrect. Second, his characterizations of psychological science are often highly oversimplified.

The Truth Game

Gergen suggests that psychological science is concerned with truth and thus plays what he calls the "truth game." Yet most psychologists in evaluating theory would stress prediction and not claim that their theories necessarily express truth. More important, the evaluation of scientific theories is often said to occur on the basis of inference to the best explanation (Capaldi & Proctor, 2008; Hanson, 1958; Thagard, 1978). The best explanation available is not necessarily considered to be true, but rather, the best of available explanations. An obvious example from biology is the theory of evolution, which is accepted because there are no reasonable scientific alternatives. In psychology, Logan (2004) has identified two types of mathematical models of attention that have been shown to be superior to alternatives and are currently in competition as being the best of the available explanations.

Oppressive Discourse

Gergan and Zielke (2006) state, "To propound a theory is to join the ranks of the potentially oppressive" (p. 302) because "theory is nothing more or less than a form of human discourse" (p. 303). No theory is better than any other theory and, therefore, has no right to suppress its rivals. As noted, from a scientific perspective, it is reasonable to accept the best alternative. Moreover, theories, contrary to Gergan and Zielke, are used for a number of nonoppressive purposes, as, for example, explaining phenomena, making predictions, and showing relations between phenomena that appear initially to be unrelated. A well-known quote attributed to the social psychologist Lewin (1951, p. 169) is, "There is nothing so practical as a good theory." This stresses not the oppressive side of theories but their practical, useful side.

Theories Only Change

Gergen and Thatchenkerry (2004) have the following view of theory change: "As research operates to replace one scientific theory with another,...we are—as many would say—simply replacing one way of putting things with another" (p. 237). Many philosophers of science (e.g., Lakatos, 1970; Laudan, 1996) suggest that theories not only change but may be improved. They may be improved by explaining more phenomena than previously or by becoming simplified, more quantitative and therefore less ambiguous, or more fruitful, and in still other ways.

Psychology's Conception of Science Is Historically Frozen

Gergen (2001) suggests that psychology's conception of science is historically frozen by "its isolation from the major intellectual and global transformations of the past half century" (p. 803). Such isolation of psychology, as Gergen views it, is from postmodernism and its variants. Gergen is mistaken because it is clear that psychology is not isolated from the major developments in the philosophy of science. For example, when logical positivism was at its peak, psychology was clearly influenced by it in emphasizing behaviorism (see, e.g., Bergmann & Spence, 1941). That day is long past. Beginning in the 1960s, a number of new scientific ideas were incorporated into psychology (see Driver-Linn, 2003, and Gholson & Barker, 1985) that compare and evaluate the nonpositivistic views of Kuhn, Lakatos, and Laudan.

Passé Research

Gergen (2001) says, "Research conducted even a decade ago is virtually confined to a casket" (p. 809). It takes little in the way of scholarship to show that this view is egregiously wrong. For example, in Capaldi and Proctor (2009), we compared the content of four introductory psychology textbooks selected more or less at random. We found in all four favorable coverage of research that dated from as far back as 1898. None of the research cited occurred later than 1983. The influence of prior research is not restricted to textbooks. As one example, Paul Fitts' pioneering research on stimulus-response compatibility (Fitts & Seeger, 1953) and the time for aimed movements (Fitts, 1954) continues to be influential in contemporary research on human action and motor control. As another, research on ideomotor activity traces itself to its roots in the 1800s, up through James (1890), with contemporary research being derived from Greenwald's (1970) article (see Shin, Proctor, & Capaldi, 2010, for a review).

Transcendent Adjudication

According to Iversen, Gergen, and Fairbanks (2005), different paradigms (or theories) have different sets of agreements, and thus there is no way of evaluating a particular paradigm other than from its own point of view. This is a rather strange view since much of the activity of scientists consists in evaluating rival paradigms. For example, as indicated, Gholson and Barker (1985) compared the paradigms of Kuhn, Lakatos, and Laudan, suggesting that Laudan's was superior. As another example, Donovan et al. (1992) examined many implications of Kuhn's position from a rival standpoint, in some cases supporting Kuhn and in some cases not supporting him. Theories, which are below the level of paradigm, are tested all of the time, by all sorts of scientists, including psychologists. This activity is so commonplace that no example is needed.

Bias in Science

According to Gergen (1999), scientists see no place in science for passion and emotion because these necessarily lead to bias, which scientists seek to avoid. Many examples could be provided of scientists admitting to great passion for their work. It is not uncommon for scientists to suggest that they have a passion for their work that gives rise to great joy (see, e.g., Kahneman, 2007; Loftus, 2007). What they also say is that they employ methodological devices to protect their work from any bias that their passions may engender.

POSTMODERNISM AND ITS VARIANTS: ALTERNATIVE RESEARCH METHODS

To the extent that postmodernism and its variants stress methodology, they propose methods fundamentally different from those used in conventional science, including psychology. The proposed methods are not those that have been found to be valid in practice, but rather methods that stem from their ontological convictions. They often discourage the use of quantitative methods and, to the extent that they favor qualitative methods, emphasize description, understanding, interpretation, and induction rather than prediction, control, and hypothesis testing (Smith, 2008). For example, Freeman (1993), states, with regard to a project studying creativity, "when the research project began, I did flirt with the notion of doing a 'content analysis' of some sort, mainly to comply with my elders, who were much more given to quantification" (p. 28), but he chose instead to do "a full-blown interpretive study" because it would provide "a more valid piece of work" (p. 29). When postmodernists refer, as they often do, to qualitative methods, they are really referring to something quite different from just using the methods as part of the scientists' toolkit. Instead, they are advocating an approach that is sometimes called qualitative inquiry, as in the new section of APA's Division 5. This is made clear in the introduction to a recent book by Wertz, Charmaz, McMullen, Josselson, Anderson, and McSpadden (2011), where the authors say, "Qualitative analyses are not the mere application of technical procedures; they are not simply additional tools for the researcher's toolbox. When properly practiced, *such analyses require a unique qualitative stance and worldview*" (p. 4, emphasis added).

Previously, we argued that postmodernism and its allies often tend to mislead in fundamental ways individuals who do not share their views (Capaldi & Proctor, 2009). We offer examples of this below. As mentioned earlier, the Society for Qualitative Inquiry was recently incorporated into Division 5 of the APA, which historically has been concerned with quantitative methods. The ostensible reason for incorporating qualitative inquiry into a division concerned with quantitative methods is that qualitative methods have been and are frequently used in science as well as in psychology. This rationale assumes that qualitative inquiry is concerned with the qualitative methods as these are normally understood. But, this is far from the truth. Qualitative inquiry, as indicated, has the goal of providing interpretations of behavior in specific environments and rejects the objective standards of science. In this regard, note that both Freeman in the second-to-last quote and Josselson in the last quote, who along with Gergen spearheaded the move to get the Society for Qualitative Inquiry incorporated into Division 5, are not advocating qualitative *methods* as tools to complement quantitative methods but qualitative *inquiry* as a distinct worldview. The main point we are making was stated forthrightly and succinctly by Shweder (1996), a leading proponent of qualitative inquiry, several years ago: "I actually think that the well-publicized tension between quantitative and qualitative approaches has a greater ring of truth when formulated as a problem in ontology rather than as a problem in method" (p. 177). In other words, the difference is one of the worldviews of naturalism versus postmodernism and not which specific methods one brings to bear on research issues.

Willig (2008), also a strong exponent of qualitative inquiry, similarly makes clear that use of qualitative methods is not what is meant by qualitative inquiry. She distinguishes qualitative inquiry from conventional qualitative methods. The former is "concerned with…the exploration of lived experience and participant-defined meanings" (Willig, 2008, p. 9); a characteristic of this approach is that the participant and researcher are regarded as equals in the research process. Conventional qualitative researchers, according to Willig, tend to impose their own meaning during data collection, analysis, and interpretation. They begin an investigation with predetermined categories for the coding and analysis of data. In contrast, she says that qualitative inquiry is more fluid and inductive, and the researcher develops interpretive categories as the research progresses, with input from the participant. Conventional researchers are often concerned with cause–effect relationships, but advocates of qualitative inquiry are concerned with developing a deeper understanding of the participant's experience and not with prediction and control. Willig says that the use of qualitative methods in the conventional manner in psychology "is, in my view, not compatible with the spirit of qualitative 'methodology' [inquiry]" (p. 9), a position consistent with that taken by Wertz et al. (2011), above, in their more recent advocacy of qualitative inquiry.

If one were not aware of the distinction that Willig (2001) and Wertz et al., (2011) are making, one could easily mistake qualitative inquiry for a conventional qualitative approach, which most researchers would see as complementary to a quantitative approach. Yet, that would be an error. It is not just more widespread use of qualitative methods that is being advocated, but rather a quite different approach to psychological research. One of the aims of qualitative inquiry is to get its methods and procedures incorporated into graduate curricula (Camic et al., 2003), which would in fundamental ways change the ways that graduate students are trained.

SUMMARY AND CONCLUSIONS

We examined three approaches to psychology that can be expected to be unfriendly to the psychology of science: postmodernism, social constructionism, and contextualism. These views may be characterized as being relativistic and anti-science. Among other things, they reject the use of standard methodologies in science, such as experimentation. We suggest that persons interested in the psychology of science may profit by being familiar with these relativistic, unscientific views so as to recognize and oppose their negative influences.

We demonstrated that postmodernism and its variants are based not on empirical considerations, but on a priori ideas derived from a preconceived approach. Postmodernism and social constructionism deny that we can know external reality. Contextualism does not go as far in this regard, but it suggests that normal scientific methods are not useful empirically because, when employed, they distort and modify phenomena. Each of these views is concerned with described experience, which is limited to a particular time and context. In postmodernism and social construction-ism, such descriptions arise solely from agreement among interested par-ties. A scientific theory does not occupy a privileged place among these descriptions because, like all other theories, it arises from shared agree-ment. Given this view, the psychology of science would simply constitute another approach arrived at by agreement and so would not have any par-ticular claim to being valid or truthful.

REFERENCES

Bergmann, G., & Spence, K. W. (1941). Operationism and theory in psychology. *Psychological Review, 48,* 1–14.

Bloor, D. (1973). Wittgenstein and Mannheim on the sociology of mathematics. *Studies in History and Philosophy of Science, 4,* 173–191.

Bloor, D. (1991). *Knowledge and social imagery* (2nd ed.). Chicago, IL: University of Chicago Press.

Boodoo, G. M., Schmitt, N., & Weiner, I. (2009). Results of the straw vote on adding a new section on qualitative inquiry. *The Score Newsletter,* XXXI (4), 3–5.

Callebaut, W. (1993). *Taking the naturalistic turn or how the real philosophy of science is done.* Chicago, IL: University of Chicago Press.

Camic, P. M., Rhodes, J. E., & Yardley, L. (Eds.). (2003). *Qualitative research in psychology: Expanding perspectives in methodology and design.* Washington, DC: American Psychological Association.

Capaldi, E. J., & Proctor, R. W. (1999). *Contextualism in psychological research? A critical review.* Thousand Oaks, CA: Sage.

Capaldi, E. J., & Proctor, R. W. (2008). Are theories to be evaluated in isolation or relative to alternatives? An abductive view. *American Journal of Psychology, 121,* 617–641.

Capaldi, E. J., & Proctor, R. W. (2009). Two radically different worldviews of psychological science: Implications for the psychology of science. *Journal of Psychology of Science and Technology, 2,* 44–58.

Collins, H. M. (1983). The sociology of scientific knowledge: Studies of contemporary sci-ence. *Annual Review of Sociology, 9,* 265–285.

Collins, H. M., & Evans, R. J. (2002). The third wave of science studies: studies of expertise and experience. *Social Studies of Science, 32,* 235–296.

Curd, M., & Cover, J. A. (Eds.) (1998). *Philosophy of science: The central issues.* New York, NY: W. W. Norton.

Devine, S. (2004). Postmodernism and science. *New Zealand Science Review, 61*(1), 2–6.

Donovan, A., Laudan, L., & Laudan, R. (Eds.). (1992). *Scrutinizing science: Empirical studies of scientific change.* Baltimore, MD: Johns Hopkins University Press.

Driver-Linn, E. (2003). Where is the psychology going? Structural fault lines revealed by psychologists' use of Kuhn. *American Psychologist, 58,* 269–278.

Dunbar, K. N., & Fugelsang, J. A. (2005). Causal thinking in science: How scientists and students interpret the unexpected. In M. E. Gorman, R. D. Tweney, D. C. Gooding, & A. P. Kincannon (Eds.), *Scientific and technological thinking* (pp. 57–97). Mahwah, NJ: Lawrence Erlbaum.

Feist, G. J. (2006). The psychology of science and the origins of the scientific mind. New Haven, CT: Yale University Press.

Feist, G. J. (2011). Psychology of science as a new subdiscipline in psychology. *Current Directions in Psychological Science, 20*(5), 330–334.

Fitts, P. M. (1954). The information capacity of the human motor system in controlling the amplitude of movement. *Journal of Experimental Psychology, 47,* 381–391.

Fitts, P. M., & Seeger, C. M. (1953). S-R compatibility: Spatial characteristics of stimulus and response codes. *Journal of Experimental Psychology, 46,* 199–210.

Ford, D. H., & Lerner, R. M. (1992). *Developmental systems theory: An integrative approach.* Thousand Oaks, CA: Sage Publications, Inc.

Fosnot, C. T. (1996). Preface. In C. T. Fosnot (Ed.), *Constructivism: Theory, perspectives and practice* (pp. ix–xi). New York, NY: Teachers College Press.

Freeman, M. (1993). *Finding the muse: A sociopsychological inquiry into the conditions of artistic creativity.* New York, NY: Cambridge University Press.

Friedman, M. (1998). On the sociology of scientific knowledge and its philosophical agenda. *Studies in the History and Philosophy of Science, 29,* 239–271.

Gergen, K. J. (1985). The social constructionist movement in modern psychology. *American Psychologist, 40,* 266–275.

Gergen, K. J. (1999). *An invitation to social construction.* Thousand Oaks, CA: Sage.

Gergen, K. J. (2001). Psychological science in a postmodern context. *American Psychologist, 56,* 803–813.

Gergen, K. J. (2009a). *An invitation to social construction* (2nd ed.). Thousand Oaks, CA: Sage.

Gergen, K. J. (2009b). *Relational being: Beyond self and community.* New York, NY: Oxford University Press.

Gergen, K. J., & Gergen, M. M. (1991). Toward reflexive methodologies. In F. Steir (Ed.), *Research and reflexivity* (pp. 76–95). London, UK: Sage.

Gergen, K. J., Gulerce, A., Lock, A., & Misra, G. (1996). Psychological science in cultural context. *American Psychologist, 51,* 496–503.

Gergen, K., Josselson, R., Freeman, M., & Futch, V. (2012). The Society for Qualitative Inquiry in Psychology, now an official section in Division 5 of the American Psychological Association, Calls for Members. *The Qualitative Report, 17,* March 5, 2012. Retrieved March 11, 2012, from www.nova.edu/ssss/QR/QR17/17_10.html

Gergen, K. J., & Thatchenkery, T. J. (2004). Organization science and social construction: Postmodern potentials. *Journal of Applied Behavioral Science, 40,* 228–249.

Gergen, K. J., & Zielke, B. (2006). Theory in action. *Theory & Psychology, 16,* 299–309.

Gholson, B., & Barker, P. (1985). Kuhn, Lakatos, and Laudan: Applications in the history of physics and psychology. *American Psychologist, 40,* 755–769.

Gifford, E. V., & Hayes, S. C. (1999). Functional contextualism: A pragmatic philosophy for behavioral science. In W. O'Donohue, R. Kitchener, W. O'Donohue, & R. Kitchener (Eds.), *Handbook of behaviorism* (pp. 285–327). San Diego, CA: Academic Press.

Gillespie, D. (1992). *The mind's we: Contextualism in cognitive psychology.* Carbondale, IL: Southern Illinois University Press.

Greenwald, A. G. (1970). Sensory feedback mechanisms in performance control: With special reference to the ideo-motor mechanism. *Psychological Review, 77,* 73–99.

Gross, P. R., & Levitt, N. (1994). Higher superstition: The academic left and its quarrels with science. Baltimore, MD: Johns Hopkins University Press.

Hanson, N. R. (1958). *Patterns of discovery: An inquiry into the conceptual foundations of science.* Cambridge, UK: Cambridge University Press.

Hayes, S. C. (2010). Contextualism. In I. B. Weiner & W. E. Craighead (Eds.), *The Corsini Encyclopedia of Psychology* (4th ed., Vol. 1., pp. 402–404). Hoboken, NJ: John Wiley.

Hayes, S. C., & Hayes, L. J. (1989). Is behavior analysis contextualistic? *Theoretical & Philosophical Psychology, 9,* 37–40.

Iversen, R. R., Gergen, K. J., & Fairbanks, R. P. II. (2005). Assessment and social construction: Conflict or co-creation? *British Journal of Social Work, 35,* 689–708.

James, W. (1890). *The principles of psychology.* New York, NY: Henry Holt.

James, W. (1912). *Essays in radical empiricism.* New York, NY: Longman, Green, & Co.

Kahneman, D. (2007). Daniel Kahneman. In G. Linzey & W. M. Runyan (Eds.), *A history of psychology in autobiography* (Vol. IX., pp. 154–197). Washington, DC: American Psychological Association.

Kuhn, T. S. (1962). *The structure of scientific revolutions.* Chicago, IL: University of Chicago Press.

Kuhn, T. S. (1970). Reflections on my critics. In I. Lakatos & A. Musgrave (Eds.), *Criticism and the growth of knowledge* (pp. 231–278). Cambridge, UK: Cambridge University Press.

Kuhn, T. S. (1992). *The trouble with the historical philosophy of science.* Robert and Maurine Rothschild distinguished lecture; Occasional publication of the Department of the History of Science, Harvard University. Cambridge, MA.

Lakatos, I. (1970). Falsification and the methodology of scientific research programmes. In. I. Lakatos & A. Musgrave (Eds.), *Criticism and the growth of knowledge* (pp. 91–196). New York, NY: Cambridge University Press.

Laudan, L. (1996). *Beyond positivism and relativism: Theory, method, and evidence.* Boulder, CO: Westview Press.

Latour, B. (1999). For David Bloor…and beyond: A reply to David Bloor's "Anti-Latour". *Studies in the History and Philosophy of Science, 30,* 113–129.

Lerner, R. M., & Kauffman, M. B. (1985). The concept of development in contextualism. *Developmental Review, 5,* 309–333.

Lewin, K. (1951) *Field theory in social science: selected theoretical papers.* D. Cartwright (Ed.), New York, NY: Harper & Row.

Loftus, E. F. (2007). Elizabeth F. Loftus. In G. Linzey & W. M. Runyan (Eds.), *A history of psychology in autobiography* (Vol. IX., pp. 198–227). Washington, DC: American Psychological Association.

Logan, G. D. (2004). Cumulative progress in formal theories of attention. *Annual Review of Psychology, 55,* 207–234.

Mayo, D. G., & Spanos, A. (Eds.) (2010). *Error and inference: Recent exchanges on experimental reasoning, reliability, and the objectivity and rationality of science.* New York, NY: Cambridge University Press.

McDowell, J., & Hostetler, B. (1998). *The new tolerance.* Carol Stream, IL: Tyndale House.

Nersessian, N. J. (2005). Interpreting scientific and engineering practices: Integrating the cognitive, social, and cultural dimensions. In M. E. Gorman, R. D. Tweney, D. C. Gooding, & A. P. Kincannon (Eds.), *Scientific and technological thinking* (pp. 17–56). Mahwah, NJ: Lawrence Erlbaum.

Overton, W. F. (2007). A coherent meta-theory for dynamic systems: Relational organicism-contextualism. *Human Development, 50,* 154–159.

Pearce, J. M. (2008). *Animal learning and cognition* (3rd ed.). Hove, UK: Psychology Press.

Pepper, S. C. (1942). *World hypotheses.* Berkeley, CA: University of California Press.

Proctor, R. W., & Capaldi, E. J. (2006). *Why science matters: Understanding the methods of psychological research.* Malden, MA: Blackwell Publishing.

Rogler, L. H. (2002). Historical generations and psychology: The case of the Great Depression and World War II. *American Psychologist, 57,* 1013–1023.

Sarbin, T. R. (1993). The narrative as the root metaphor for contextualism. In S. C. Hayes, L. J. Hayes, H. W. Reese, & T. R. Sarbin (Eds.), *Varieties of scientific contextualism* (pp. 51–65). Reno, NV: Context Press.

Shin, Y. K., Proctor, R. W., & Capaldi, E. J. (2010). A review of contemporary ideomotor theory. *Psychological Bulletin, 136,* 943–974.

Shrout, P. E., & Shadish, W. R. (2011). Why we oppose a new section on qualitative inquiry in Division 5. *The Score, XXXIII* (4), 7, 13.

Shweder, R. A. (1996). *Quanta* and *qualia:* What is the "object" of ethnographic method? In R. Jessor, A. Colby, & R. A. Shweder (Eds.), *Ethnography and human development: Contextual meaning in social inquiry* (pp. 175–182). Chicago, IL: University of Chicago Press.

Simonton, D. K. (2009). Applying the science of psychology to the psychology of science: Can psychologists use psychological science to enhance psychology as a science? *Perspectives on Psychological Science, 4,* 2–4.

Slife, B. S. (2011). Introduction: The special issue on Ken Gergen's book *Relational Being. Journal of Constructivist Psychology, 24,* 277–279.

Smith, J. A. (2008). *Qualitative psychology: A practical guide to research methods* (2nd ed.). Thousand Oaks, CA: Sage.

Thagard, P. (1978). The best explanation: Criteria for theory choice. *The Journal of Philosophy, 75,* 76–92.

Wertz, F. J., Charmaz, K., McMullen, L., Josselson, R., Anderson, R., & McSpadden. E. (2011). *Five ways of doing qualitative analysis.* New York, NY: Guilford Press.

Willig, C. (2008). *Introducing qualitative research in psychology* (2nd ed.). Maidenhead, UK: McGraw Hill Open University Press.

Wilson, E. O. (1998). *Consilience: The unity of knowledge.* New York, NY: Vintage.

CHAPTER 14

Psychobiography and the Psychology of Science: Encounters With Psychology, Philosophy, and Statistics

William McKinley Runyan

The discipline of psychology is concerned with at least three different levels of generality: Learning what is true about people in general, about groups of people, and about individual lives (Kluckhohn & Murray, 1948; Runyan, 1982). Similarly, the psychology of science is concerned with learning what is true about scientists in general (Simonton, 1988, 2002), about groups of scientists (Feist & Gorman, 1998; Maslow, 1966; Roe, 1953a, 1953b), and about the work and lives of individual scientists (Gardner, 1993; Gruber, 1974).

Lee Cronbach argued in his presidential address to the American Psychological Association that there are "Two Disciplines of Scientific Psychology," correlational and experimental (1957). He later wrote about the interaction of personal and social factors in "Beyond the Two Disciplines of Scientific Psychology" (1975) and about the importance of historical accounts of individual cases (1982). There are at least "Three Disciplines of Scientific Psychology:" quantitative, experimental, and historical-interpretive (Runyan, 2005). This chapter explores examples of historical-interpretive analyses of single cases in psychology, philosophy, and statistics.

Previously, I argued for the relationship of personal experience to psychological theorizing in a paper on "Psychobiography and the Psychology of Science: Understanding Relations Between the Life and Work of Individual Psychologists" (Runyan, 2006). Examples discussed included Sigmund Freud, Karen Horney, B. F. Skinner, Henry A. Murray, Paul Meehl, Edwin G. Boring, and Michel Foucault. This chapter builds on that paper, and argues that personal experience can be relevant not only in psychology, but can also be influential (perhaps in somewhat different ways) in philosophy and statistics.

Understanding relations between life and work can help in understanding the sources and meanings of a theory. I should make clear at the beginning, however, my own view that personal experience can be a source of great insights, or of great errors, and that identifying personal, social, or cultural sources of a theory does not answer questions about its more general validity.

Several powerful traditions in the history of science deny or minimize the role played by personal factors (Popper, 1959). An internalist tradition in history of science focusing on the interplay of scientific theory and research might see personal-experiential factors as little more than distractions from rigorous scientific inquiry. Externalist traditions analyze science in its social and cultural contexts, ranging from Marxist analyses of science in society, through the sociology of science, to postmodern social and cultural constructivist views, each of which sometimes slights or ignores the personal-psychological dimensions of science.

The psychology of science analyzes the cognitive, emotional, experiential, personal, social, and other psychological dimensions of science. These are not minor issues: "Individuality is found in feeling; and the recesses of feeling, the darker, blinder strata of character, are the only place in the world in which we catch real fact in the making, and directly perceive how events happen and how work is actually done" (William James, 1902, in Murray, 1967, p. 293). The "blinder strata of character" is at least one of the places in the world in which we can see facts in the making, along with social, cultural, and historical levels of analysis.

This chapter begins with two examples from philosophy, Bertrand Russell and Ludwig Wittgenstein. I start there because the life stories are dramatic and because the work has great conceptual scope, with implications for how we think about science in the world. The next section discusses relationships between the work and life of several eminent psychologists including Freud, B.F. Skinner, Karen Horney, and others (drawing from Runyan, 2006). The third section is on statisticians, with examples from Karl Pearson, R. A. Fisher, and Jerzy Neyman. The argument, in brief, is that personal experience as studied in psychological biography can be relevant to understanding work not only in psychology, but also in philosophy, and statistics.

WHAT PSYCHOLOGY TO INCLUDE IN THE PSYCHOLOGY OF SCIENCE?

If psychology of science is going to be included in science studies, how can this integration be achieved? What kinds of psychological theory, research, and research methods are available for developing the psychology of science? One valuable resource is *Psychology of Science: Contributions to Metascience* (Gholson, Shadish, Neimeyer, & Houts, 1989). This edited collection includes a guide to the literature on psychological epistemology by Donald Campbell. Campbell expresses hope that this volume and the 1985 conference from which it originated "will catalyze the critical mass needed to establish psychology of science as a discipline with its own journals, organizations, courses and doctoral programs"(Campbell, 1989, p. 21). As this critical mass may currently be forming, the present chapter argues that psychobiographical inquiry into relations between the life and work of individual scientists can be a valuable part of an evolving psychology of science.

Campbell says that origins of his chapter, "Fragments of the fragile history of psychological epistemology and theory of science" result from 45 years of "back burner" attention to these issues (as early as a 1950 lecture on "The psychology of knowledge" at the University of Chicago), and he hopes that the thread will be picked up by younger scholars. One step in this direction is the *Psychology of Science* volume itself (1989), developing from a 1985 conference at Memphis State (now the University of Memphis). The volume includes chapters by a number of major contributors to the psychology of science, including Dean Simonton, Howard Gruber, William J. McGuire, Ryan Tweney, the four editors of the volume (Gholson, Shadish, Neimeyer, Houts), Donald Campbell, and others.

A further step in the institutionalization of the psychology of science was *The Social Psychology of Science* (Shadish & Fuller, 1994). This book was intended to counter the view that the psychology of science consists solely of the cognitive psychology of science. The book emphasizes contributions of social psychology to the psychology of science. It anthologizes contributions to the social psychology of science, including both psychological and social perspectives, examines conceptual underpinnings, and suggests future directions for the social psychology of science.

A later contribution to the psychology of science by Greg Feist and Michael Gorman (1998) reviewed work in five different areas of psychology contributing to the psychology of science. This discussion was extended in Feist's *The Psychology of Science and the Origins of the Scientific Mind* (2006). Here there were individual chapters reviewing work in Biological Psychology of Science (Chap. 2), Developmental Psychology of Science (Chap. 3), Cognitive Psychology of Science (Chap. 4), Personality Psychology of Science (Chap. 5), and Social Psychology of Science (Chap. 6). In Chapter 1, Feist situated the psychology of science in relation to the three more established disciplines of the history of science, the philosophy of science, and the sociology of science. Drawing on earlier work by Nicholas Mullins (1973), Feist argues that disciplines can go through three distinct stages of development: isolation, identification, and institutionalization. In the first stage of isolation, scholars work on problems in isolation, yet without the social organization of training centers, conferences, or professional organizations. In the second stage of identification, after intellectual achievements by the founders outline a field of inquiry, students and other scholars can identify themselves with the field and may begin to meet with each other and establish journals. Third, in the stage of institutionalization, professional societies are more formally organized, annual conferences are established, and training centers develop.

Feist suggests that the history, philosophy, and sociology of science are each well into formal institutionalization, whereas psychology of science is slowly emerging out of the isolation stage. Individual isolated workers are increasingly identifying and communicating with each other, as in the edited volumes referred to above in the psychology of science (Gholson et al., 1989), in the social psychology of science (Shadish & Fuller, 1994), or in "The Psychology of Science" Special Issue of *The Review of General Psychology* (June, 2006). Indeed, more recently, a journal devoted to the psychology of science was started as well as a society for the same (*International Society for the Psychology of Science and Technology, ISPT*).

MY PERSONAL JOURNEY THROUGH THESE QUESTIONS

Like any intellectual project, this inquiry into the biographical sources of psychological theory, philosophy, and statistics has unfolded in changing social, cultural, and personal contexts, several of which are discussed here. I had long been interested in the study of lives within the social sciences, writing a dissertation on "Life Histories: A Field of Inquiry and a Framework for Intervention" in a program in Clinical Psychology and Public Practice at Harvard in 1975. This was followed by a book on *Life Histories and Psychobiography: Explorations in Theory and Method* (Runyan, 1982), analyzing alternative accounts of lives, the case study method, idiographic methods, and the psychobiography debate.

In 1988 I started teaching a course on "Personality Theory." To better understand the theories, I attended not only to the interaction of theory and empirical research, but also discussed their biographical, social, and cultural contexts. It dawned on me that this was at least partly a project in the history of science. To do a more rigorous job, I tried to learn about recent developments in the history and philosophy of science. On a sabbatical in the spring of 1994, and a leave in 1995–1997, I spent time learning about developments in the history of science at Harvard's History of Science Department, M.I.T.'s Dibner Institute for the History of Science and Technology, and Boston University's Colloquium series in the philosophy and history of science. These experiences were tremendously thought-provoking, challenging many of my assumptions about what science is and how it fits into the world.

Yet, they were also tremendously stressful, in that much recent literature in history and social studies of science explicitly discounted the role of personal, psychological, or experiential factors in science, topics that were of primary interest to me. For example, in the first-year graduate seminar on Methods of Research in the History of Science at Harvard in the Fall of 1995, one of the instructors said that "Last year's seminar decided that biography is not a useful or appropriate method in the history of science." After some initial shock, I raised my hand, and asked, "What is the argument here?" As far as I could tell, there was not much of an argument, but that social, cultural, and material studies of science were valued, and were seen as the cutting edge. Within this view, talk of biography was lumped with a discredited "Great Man Theory of History." From this perspective, talk of individuals and their psychology was seen as intellectually or politically regressive for overemphasizing individuals and neglecting the extent to which science is socially constructed.

An obvious response is that, although it may not be easy, one can pay attention to social and cultural dimensions of science, along with studying individuals, groups, and populations. I will argue that analyzing relations between the life and work of individual scientists is a valuable component of the psychology of science, a place where the "rubber meets the road," with scientific tasks being performed by particular individuals and groups in particular social, cultural, and historical contexts.

I had gone to the history of science looking for more powerful intellectual instruments and found an approach to understanding science that was more detailed and sophisticated than what I had previously been exposed to. Yet, at the same time, I felt I had found a severely flawed telescope, bringing the social and cultural dimensions of science into the foreground, yet blurring, or sometimes ignoring the personal-psychological dimensions. The psychology of science can help to bring the personal-psychological dimensions of science back into focus. The following sections discuss relations between work and life in philosophy, psychology, and statistics.

THE PERSONAL SIDE OF PHILOSOPHY

On Bertrand Russell and Ludwig Wittgenstein

Bertrand Russell and Ludwig Wittgenstein are sometimes seen as two of the 20th century's most influential philosophers. They both had influences on the development of logical positivism in the Vienna Circle in the 1920s and early 1930s. However, they arrived at dramatically different conceptions of philosophy in relation to science, with Russell favoring a more "scientific philosophy" and Wittgenstein opposing such a view.

Bertrand Russell (1872–1970) was coauthor of *Principia Mathematica* with Alfred North Whitehead (3 volumes, 1910, 1912, 1913), analyzing the logical bases of mathematics. This book was drawn upon in the development of logical positivism in the Vienna Circle of the late 1920s and early 1930s. The Vienna Circle also drew upon Wittgenstein's *Tractatus Logico-Philosophicus* (1921). Wittgenstein (1889–1951) started as a potential protégé to Russell, but they ended up with dramatically different views of philosophy in relation to science, Russell favoring a more "scientific philosophy" and Wittgenstein opposing such a view in *Philosophical Investigations* (1953). The relations of Russell and Wittgenstein illustrate ways in which philosophical beliefs can be related to personal psychology and interpersonal relationships.

What was the personal context of Russell's work in philosophy? Russell's three-volume autobiography opens with an inspiring Prologue, "What I Have Lived For":

> Three passions, simple but overwhelmingly strong, have governed my life: the longing for love, the search for knowledge, and unbearable pity for the suffering of mankind.
>
> These passions, like great winds, have blown me hither and thither, in a wayward course, over a deep ocean of anguish, reaching to the very depths of despair.
>
> I have sought love, first because it brings ecstasy—ecstasy so great that I would have sacrificed all the rest of life for a few hours of this joy. I have sought it, next, because it relieves loneliness—that terrible loneliness in which one shivering consciousness looks over the rim of the world into the cold unfathomable lifeless abyss....
>
> With equal passion I have sought knowledge. I have wished to understand the hearts of men. I have wished to know why the stars shine. And I have tried to apprehend the Pythagorean power

by which number holds sway above the flux. A little of this, but not much, I have achieved.

Love and knowledge, so far as they were possible, led up to the heavens. But always pity brought me back to earth. Echoes of cries of pain reverberate in my heart. Children in famine, victims tortured by oppressors, helpless old people, a hated burden to their sons, and the whole world of loneliness, poverty, and pain make a mockery of what human life should be. I long to alleviate the evil, but I cannot, and I too suffer. (Russell, 1967, pp. 3,4)

This prologue is an eloquent statement of major human values. It impressed me when I first read it in the summer of 1969, as I was beginning graduate school that Fall. Russell wrote so much and knew so many eminent people in philosophy, literature, and politics. Were there things I could learn from him then? Are there things that we can learn from his work and life now?

Is Russell a model of intellectual productivity in many different fields? Or, is the lesson that great intellectual achievement can come at a high personal cost to those around one, as in Russell's relations with his first three wives and with his two children, Kate and John?

A two-volume psychological biography of Russell by Ray Monk (1996, 2000) provides a different interpretation of Russell's life, more critical than his autobiography or than three prior biographies. Monk says that three earlier biographies of Russell failed to adequately relate his work and life. The first by Alan Wood (1957) had Russell's cooperation, but failed to explore his inner life. The latter two by Ronald Clark (1975) and Carolyn Moorehead (1992) had more on Russell's inner life, but did not seriously relate his life to his work. Monk tries to relate Russell's three passions for love, knowledge, and politics to each other in his two-volume *Bertrand Russell: The Spirit of Solitude, 1872–1921* (1996) and *Bertrand Russell: The Ghost of Madness, 1921–1970* (2000).

Monk provides a detailed interpretation of Russell's work and its relationship to his inner life. It may change your perception of Russell's work and life; it changed mine. Monk argues that his three great passions were attempts by Russell "to overcome his solitariness through contact with something outside himself: another individual, humanity at large, or the external world" (p. xviii). To an extent greater than I had realized in reading Russell's work or in reading his autobiography, he feared the depths of his emotions, felt cut off from others, and was afraid of going mad. Monk may be too critical of Russell, but provides a level of detail about Russell's work and life that requires reexamination of both. Russell lived from 1872 to 1970, and was enormously productive, having published 70 books and more than 2,000 articles.

Ludwig Wittgenstein, however, lived from 1889 to 1951 and during his lifetime published a total of one book review (1912), one book, *Tractatus Logico-Philosophicus* (1921), and one article (1929), his third and final publication. After his death in 1951, *Philosophical Investigations* (1953) was published, along with many other volumes based on Wittgenstein's lectures, conversations, and notebooks.

Russell continued to argue for making philosophy more scientific, while Wittgenstein criticized such a view, as in *Philosophical Investigations* (1953).

I won't try to summarize their whole history, but rather I focus on Russell's relationship with Wittgenstein from 1911 to 1914. The younger Wittgenstein, born in 1889 to one of the wealthiest families in Vienna, then a student in aeronautical engineering at Manchester University, age 22, wanted to know if he had the talent to make a significant contribution to philosophy. He first visited the logician Gottlob Frege in Jena, who advised him to consult Russell.

Without a prior appointment, Wittgenstein arrived at Russell's rooms in Trinity College, Cambridge on October 18, 1911, to introduce himself. Wittgenstein actively participated in Russell's seminar through the term and argued with him afterwards. Before Christmas, Wittgenstein asked Russell whether he had the ability to make a contribution to philosophy. Russell said he didn't know, and asked to see a piece of his writing. Wittgenstein returned to Cambridge in January 1912 with a manuscript he had written over vacation. Russell was impressed, and believed that Wittgenstein might do great things. (Unfortunately the manuscript has not survived.) Wittgenstein later told a friend that Russell's encouragement "had proved his salvation, and had ended nine years of loneliness and suffering, during which he had continually thought of suicide" (Monk, 1990, p. 41).

Over the next term, Wittgenstein worked so intently in mathematical logic that Russell felt he had learned what Russell had to teach and maybe gone beyond him. Russell felt that Wittgenstein might be the protégé he had been looking for. However, by June 1913, Wittgenstein became severely critical of Russell's work. Russell was devastated by the criticisms. He wrote to his lover Lady Ottoline Morrell, that after Wittgenstein's severe criticism of his work, he "felt ready for suicide" (June 19, 1913). In a letter to her several years later Russell wrote that he didn't think she realized this at the time, but Wittgenstein's criticism in 1913 "was an event of first-rate importance in my life, and affected everything I have done since. I saw he was right and I saw that I could not hope ever again to do fundamental work in philosophy. My impulse was shattered like a wave dashed to pieces against a breakwater" (March 4, 1916; in Monk, 1996, pp. 301,302). In Russell's autobiography he wrote that Wittgenstein was "perhaps the most perfect example I have ever known of genius as traditionally conceived, passionate, profound, intense, and dominating" (p. 46).

In Wittgenstein's later work he became strongly critical of the view that scientific knowledge is the model of all knowledge. An overly scientific view gets in the way "not just of philosophical clarity, but of a full understanding of art, music, literature, and, above all, ourselves" (Monk, 2005, p. 106).

THE PERSONAL SIDE OF PSYCHOLOGICAL THEORISTS

Case Studies

Sigmund Freud

There is enormous literature on the relations between Freud's personal biography and his intellectual development, concentrating on his self-analysis, interpretations of his dreams, or his identification with historical figures such as Leonardo daVinci or Moses, starting with Wittels in 1923, through Jones (1953–1957), Ellenberger (1970), Roazen (1975), Sulloway (1979), Gay (1988), Breger (2000), Elms (2005), and many others.

I focus on two brief examples, each controversial or contested in its own way. Part of the story of psychoanalysis is how the theory drew upon Freud's self-analysis, as well as from his clinical work and cultural resources. In two key letters to his friend Wilhelm Fliess, Freud wrote on September 21, 1897, "And now I want to confide in you immediately the great secret that has been slowly dawning on me in the last few months. I no longer believe in my *neurotica* [theory of the neuroses]" (i.e., no longer believing in childhood sexual seduction as the cause of neuroses). And on October 15, 1897:

> Dear Wilhelm,
>   My self-analysis is in fact the most essential thing I have at present and promises to become of the greatest value to me if it reaches its end.... Being totally honest with oneself is a good exercise. A single idea of general value dawned on me. I have found, in my own case too, [the phenomenon of] being in love with my mother and jealous of my father, and I now consider it a universal event in early childhood, even if not so early as in children who have been made hysterical.... If this is so, we can understand the gripping power of Oedipus Rex...the Greek legend seizes upon a compulsion which everyone recognizes because he senses its existence within himself. (Freud, quoted by Masson, 1984, p. 272)

This certainly sounds as if Freud's personal experience is being used to support his belief in the Oedipal theory. (The cautious methodologist may be concerned about overgeneralization as Freud moves from his own case to a "universal event in early childhood.") There are, of course, controversies about the extent to which this abandonment of the seduction theory and conception of the Oedipus complex was shaped by his self-analysis, his clinical patients, assumptions about the prevalence of childhood sexual abuse, and/ or political expediency (e.g., Breger, 2000; Malcolm, 1984; Masson, 1984).

A second example from Freud's work illustrates some of the difficulties in linking personal experience to the development of theory and also suggests something about the possibilities of critically examining such claims. Freud's first biographer, Fritz Wittels (1880–1950), had suggested in 1923 that Freud's idea of the "death instinct," introduced in *Beyond the Pleasure Principle* (1959/1920) occurred to Freud while "under the impress" of the death of his daughter, Sophie (Wittels, 1923). Freud read the biography and wrote to Wittels on December 18, 1923:

> That seems to me most interesting, and I regard it as a warning. Beyond question, if I had myself been analyzing another person in such circumstances, I should have presumed the existence of a connection between my daughter's death and the train of thought present in *Beyond the Pleasure Principle*. But the inference that such a sequence exists would have been false. The book was written in 1919, when my daughter was still in excellent health. She died in

> January, 1920. In September, 1919, I had sent the manuscript of the
> little book to be read by some friends in Berlin.... What seems
> true is not always the truth. (Vol. 19, p. 187)

This last sentence may be a useful motto for work in this area: "What seems true is not always the truth." In this case, what seems a personal connection may not actually be one. However, Freud's disclaimer may itself not be entirely true in that he originally sent out the manuscript in 1919, but he also worked on the manuscript for several additional months in 1920, after Sophie had died. Freud was correct in that the death of his daughter could not have started this line of thought, but it is possible that her death influenced his later revisions to the manuscript.

Another personal factor proposed as related to his origin of the death instinct was that of Freud's cancer of the jaw. This, however, was not diagnosed until 1923, so it is clearly after the introduction of the concept in 1920. Others have suggested that Freud was influenced by the traumas of the Great War and by anxiety about his two sons serving in the military. Another explanation is that the concept of a death instinct played a significant role in the structure of Freud's theorizing, with intimations of it going as far back as his unpublished *Project for a Scientific Psychology* in 1895. I will not attempt to resolve all these issues here, but it is clear that a whole field of personal factors can be proposed as sources of a concept. However, as Freud argued, apparent connections are not always true and it is necessary to critically assess them.

Karen Horney

Karen Horney (1885–1952), the distinguished neoanalytic or social psychoanalyst, is best known for works such as *The Neurotic Personality of Our Time* (1937), *New Ways in Psychoanalysis* (1939), *Self-Analysis* (1942), and *Neurosis and Human Growth* (1950). She was an early advocate for understanding the cultural contexts of psychopathology, and a critic of Freud's misunderstanding of women's psychology with a posthumous collection of papers titled *Feminine Psychology* (1967).

A recent biography of Horney is by Bernard Paris, a Horneyan literary critic, professor of English at the University of Florida, and founder and director of the International Karen Horney Society. Paris says that working on the biography *Karen Horney: A Psychoanalyst's Search for Self-Understanding* (1994) changed his perception of her, and his sense of how the person was related to her work. Reading her books over the years, Paris had "formed an image of her as a wise, benign, supportive woman who, having worked through her own problems, was now free to help others" (p. 175). However, earlier biographies of Horney by Jack Rubins and Susan Quinn, and his own research led to revisions in his understanding of her. He now sees her as a "tormented woman with many compulsions and conflicts who violated professional ethics and had difficulties in her relationships" (1994, p. 175).

In particular, she had compulsive affairs with colleagues and with students in training or in supervision with her for many years. She had a relationship with Erich Fromm from approximately 1934 to 1939, while also having affairs during this time with Paul Tillich and Erich Maria Remarque. She also had several affairs with analysands of hers including Harold Kelman in the 1940s, who was a major figure in the Association for the Advancement of Psychoanalysis, which she had cofounded in 1941.

In Horney's *Self-Analysis* (1942), she writes about a patient named Clare, who is struggling to sort out problems in her relationship with a man named Peter. Paris speculates that Horney is really writing about her relationship with Erich Fromm, which romantically ended around 1939 and continued professionally for a few years beyond that. Paris suggests that the Clare–Peter relationship was similar to the Horney–Fromm relationship with an "unworkable combination of a dependent woman and a man hypersensitive to any demands upon him" (p. 146). Paris also suggests that Fromm's *Escape from Freedom* (1941) also indirectly discusses their relationship, and that perhaps "Fromm and Horney were writing in part for each other, each trying to show the other how much he or she understood" (Paris, 1994, p. 147).

Those disturbed by Horney's character might "wish to discard her ideas" (p. 175). In contrast, Paris argues that being disturbed by her behavior, or even considering it pathological, need not lead to rejecting her ideas. His view is that although Horney had significant character flaws, she was "also a rather heroic figure whose courage in seeking the truth about herself enabled her to make a major contribution to human thought" (p. 176). Her difficulties may well have been the sources of her ideas, leading to continuing self-analysis and to continuing theoretical creativity: "We do not achieve profound psychological understanding without having had the need to look deeply into ourselves. Where would Horney's insights have come from had she not experienced her difficulties?" (p. 176).

To this last question, I would respond that insights can come not only from personal difficulties and experience, but also from clinical work, empirical research, cultural sources, from integrative reading and thinking, or various combinations of these (a point with which Paris may well agree). There is no need to weaken the claim for the relevance of personal experience to theoretical creativity by exaggerating it. An interesting set of questions is raised: To what extent does profound psychological understanding require deep introspection, and to what extent is such self-understanding a precondition for other kinds of learning and creativity?

Henry A. Murray

Henry A. Murray (1893–1988) was a founder of personality psychology, author of *Explorations in Personality* (1938), coinventor of the T.A.T. (Thematic Apperception Test) in 1935, editor with Clyde Kluckhohn of *Personality in Nature, Society and Culture* (1948), and director of the Harvard Psychological Clinic from 1928. He was admired by many, including myself, as a critic of sterile scientism, a champion in linking psychodynamic and academic psychology, and a personally compelling advocate of the study of whole persons and the deepest human experiences (Runyan, 2008).

Two incidents from his life will be presented as illustrations of the connections between life and work. When Forrest Robinson first proposed doing a biography of Murray in 1970, Murray replied that a central theme was a 40-year secret love affair that had revolutionized his life (Robinson, 1992). The object of his affections was Christiana Morgan, born in 1897, daughter of a professor at Harvard Medical School, and coinventor of the Thematic Apperception Test in 1935. Murray and Morgan, both married, first met each other in 1923. By Easter vacation, 1925, Murray, with an MD and a PhD in biochemistry near completion, was talking with Jung about his growing attachment to Christiana Morgan, and Jung told Murray about his own relationship with his wife Emma Jung and his "inspiratrice" Toni Wolff.

Jung advised Murray against going into psychology and was not encouraging about the relationship with Christiana, but Murray ended up following Jung's example more than his advice. Murray and Morgan told their spouses of their relationship, yet remained married, and pursued a passionate, emotionally involved relationship until the end of her life in 1967. They saw each other as paths to the study of the unconscious and to their own deepest selves. Morgan saw Jung in therapy in 1926, and Jung taught a series of Vision Seminars on her visions from 1930 to 1934, which have recently been published in two volumes.

In 1959, Murray published a chapter on "Vicissitudes of Creativity" in which he describes the experience of a couple he called Adam and Eve, both of them coming out of dead marriages:

> The hypothesis that is suggested by the history of this particular dyad is that periodic complete emotional expression within the compass of an envisaged creative enterprise—not unlike the orgiastic Dionysian rites of early Greek religion in which all participated—is a highly enjoyable and effective manner of eliminating maleficent...tendencies as well as of bringing into play beneficent modes of thought and action.... In sharp contrast to this is both the traditional Christian doctrine of repression of primitive impulses and the psychoanalytic notion of the replacement of the id by the ego (rationality), which results so often in a half-gelded, cautious, guarded, conformist, uncreative, and dogmatic way of coping with the world. (Shneidman, 1981, p. 327)

Murray elaborates on the power of dyads for regenerating culture, but without knowing something of Murray's relationship with Christiana, it is sometimes hard to see what he is talking about.

A second moment in Murray's life is his tenure meeting in 1936, chaired by Harvard President James Bryant Conant. As an illustration of the passions aroused by debates about the place of psychoanalysis in the university, Karl Lashley, a neuropsychologist recently hired by Harvard as supposedly the most distinguished psychologist in the country, said that he would resign if Murray received tenure. A major supporter, social and personality psychologist Gordon Allport said that he would resign if Murray did not receive tenure. Edwin G. Boring, an experimental psychologist who was chair of the psychology department, and who will be discussed later, also opposed tenure. They later reached a compromise in which Murray was given two 5-year appointments, but not tenure, and to mollify Lashley, he was made a research professor, with no teaching responsibilities.

As an indication of Lashley's hostility to psychoanalysis, there is a story that Lashley had briefly been in psychoanalysis with Franz Alexander at the University of Chicago, had left in a rage, and then unsuccessfully tried to get Alexander fired from the university. This story needs additional evidence to support or refute it, to move it from the penumbra of possibly true to the categories of probably true or probably false. In the meantime, what is more certain is that even Lashley's friends, like Boring, said Lashley was irrationally hostile to psychoanalysis.

Examples we have considered so far are from the psychodynamic, experiential side of psychology, such as Freud, Karen Homey, and Henry Murray. Are personal-experiential factors operative only in such "soft" traditions, but not in "hard" natural science traditions? I will argue that personal–psychological–experiential factors can also be important within quantitative or experimental natural science traditions, although perhaps in somewhat different ways. Examples will be drawn from the life and work of B.F. Skinner on behaviorism and Paul Meehl in psychological measurement.

B. F. Skinner

In an excellent book on psychobiography, *Uncovering Lives: The Uneasy Alliance of Biography and Psychology* (1994), Alan Elms argues that even though B. F. Skinner (1904–1990) was the preeminent behaviorist of his time and, in the view of some, the preeminent psychologist, the personal sources of his ideas may be somewhat obscure.

Elms argues that Skinner's *Walden Two* (1948), his best-selling book with more than two million copies sold, provides some insight into Skinner's changing self-conceptions and his relations with behaviorism. Skinner indicates that he usually wrote slowly and in longhand, but that *"Walden Two* was an entirely different experience. I wrote it on the typewriter in seven weeks." Parts of it were written "with an emotional intensity that I have never experienced at any other time" (Elms, 1994, p. 86).

Walden Two is partly a dialogue between Burris, "a pedestrian college teacher," and Frazier, "a self-proclaimed genius who has deserted academic psychology for behavioral engineering." B. F. Skinner, whose full name was Burris Frederic Skinner, says the novel was "pretty obviously a venture in self-therapy, in which I was struggling to reconcile two aspects of my own behavior represented by Burris and Frazier"(Elms, 1994, p. 87). As Skinner told Elms in an interview in 1977, when he wrote *Walden Two,* he was not really a Frazierian, a social engineer. However, writing the book convinced him: "I'm now a thoroughgoing Frazierian as a result and I'm no longer Burris" (Elms, 1994, p. 99). In other words, Skinner was no longer the pedestrian college teacher, but more a brilliant maverick applying behavioral principles to the redesign of society.

Elms argued that writing *Walden Two* was Skinner's response to a midlife crisis at age 41. This may have reactivated an earlier identity crisis Skinner had during his "Dark Year" at age 22, when he concluded that he could not be a fiction writer as he had nothing to say, which led to confusion and disastrous consequence for his self-respect. "The crisis (at age 22) was finally resolved, as such intense identity crises often are through the wholehearted acceptance of an ideology indeed, an extreme ideology. In Skinner's case, the ideology was radical behaviorism" (Elms, 1994, p. 90).

Skinner's identity crisis and formulation of a new identity as a more scientific psychologist advocating radical behaviorism, may be related to a wider field of social and cultural issues. Skinner's major supporter in graduate school was not a professor in the psychology department, but an experimental biologist, W. J. Crozier who established a Laboratory of General Physiology at Harvard in 1925. Crozier's stance toward biology was strongly influenced by Jacques Loeb (1859–1924) and was dedicated to the experimental study of whole behaving organisms, as in tropisms, as contrasted with biochemical experiments in physiology. Crozier's world view was one that resonated with Skinner's experimental study of behavior, and Crozier was a major supporter in getting Skinner National Science Foundation (NSF) fellowships and getting him elected to the first class of Junior Fellows at Harvard in 1933.

One theme found over and over again among some of the more eminent experimental psychologists was insecurity that gets converted to conceit and arrogance. Pauly (1987), for instance, argues that there was a shared social background in many in this aggressively experimental research tradition. Many in this tradition felt like social outsiders, were not psychologically well adjusted, "lived with feelings of insecurity and inferiority, and compensated with exaggerated displays of conceit and self-assertion" (Pauly, 1987). When I first read this, it struck me that some of this may apply to Skinner's relations with other psychologists.

Paul E. Meehl

Paul E. Meehl (1920–2003) was a major contributor to psychological measurement, taxonomy, and philosophical psychology. He received his BA in 1941 and PhD in 1945 from the University of Minnesota, where he spent his entire career. He is author of the classic *Clinical versus Statistical Prediction* (1954), *Psychodiagnosis: Selected Papers* (1973a), and *Selected Philosophical and Methodological Papers* (1991). Most recently, *A Paul Meehl Reader: Essays on the Practice of Scientific Psychology* (2005) has been published. He has a reputation among many as one of the most brilliant psychologists in the history of the discipline and was elected President of the American Psychological Association in 1962.

In his 1973 book, *Psychodiagnosis: Selected Papers*, my favorite piece is a 75-page paper, "Why I Do Not Attend Case Conferences." Meehl describes this as a diatribe, a polemic against the kind of faulty reasoning he sees as endemic in clinical case conferences because of inadequate training of most clinicians in logic, statistics, diagnosis, psychometrics, and biology. He says this paper is intended as destructive criticism in that you have to shake people up before you can get them to do something different.

Meehl wants to change both the quality of reasoning and the "buddy–buddy" norms in case conferences, in which everything, "gold and garbage alike" is positively received: "The most inane remark is received with joy and open arms as part of the groupthink process" (Meehl, 1973b, p. 228). Negative feedback is heard with horror and disbelief, and if it is delivered, one is seen as an ogre. In clinical case conferences and other academic groups, he says, people seem to undergo a kind of intellectual deterioration when they gather around a table in one room. Meehl decries what he sees as the "groupy" attitude, in which all evidence is seen as equally good, and a "mush-headed approach which says that everybody in the room has something to contribute (absurd on the face of it, since most persons don't usually have anything worthwhile to contribute about anything, especially if it's the least bit complicated)" (Meehl, 1973b, p. 227). In a similar tone, he goes on to identify and make fun of common fallacies in clinical reasoning.

Personally, I love this paper and find its aggressive polemics amusing. I have used it in classes, with students split on it, some loving it, finding it one of the most illuminating things they have ever read, as well as funny; while others find it threatening, or intimidating, and get so upset they do not finish reading it. I once wrote Meehl a letter about the piece saying that these strong criticisms may make clinicians feel anxious, defensive, or misunderstood, and perhaps angry at the critic, but will not necessarily lead to significant change. Would it not be more effective to also provide models of more rigorous clinical reasoning, which practitioners could draw from? He wrote back, "We're not quite communicating. You assume I hope to cure the slobs by attack. But when did I ever assert such?" (personal communication, Sept. 16, 1974). In another letter, "I agree entirely with your view that clinicians are largely unaffected by tough, incisive, aggressive argument—I spend more of my time with lawyers and philosophers, and so have fallen into 'nontherapeutic' habits.... On the subjective side, you should remember that I have been in this field for over 30 years, and one becomes impatient after the tenth time he has to hear the same dumb errors made by PhD's. (That's no excuse, it's by way of personal explanation.)" (personal communication, Aug. 10, 1974). Meehl's letter led me to write a paper trying to follow my own advice, outlining average, optimal, and the best feasible approaches to clinical decision making in "How Should Treatment Recommendations Be Made? Three Studies in the Logical and Empirical Bases of Clinical Decision-Making" (Runyan, 1977).

Paul Meehl published an autobiographical chapter in 1989, and I want to raise here the question of whether a few of these biographical facts contribute anything to understanding the content or tone of his writing.

> My father was a bank clerk, who, despite extraordinary intelligence quit high school to help support a widowed mother and unmarried sister. He was fond of me in a cool way, and I knew it. Fortunately, I got his "brain" genes, because he held Admiral Rickover's view that if a man is dumb he might just as well be dead. I identified strongly with him.... In 1931 my father, who had embezzled money to play the stock market, committed suicide. (Meehl, 1989, p. 337)

Meehl's mother had been misdiagnosed for over a year as having Meniere's disease, a disturbance of the semicircular canal in the ear. Finally, a neurologist was called in, who correctly diagnosed a brain tumor. When Meehl was 16, his mother died after surgery for this brain tumor: "This episode of gross medical bungling permanently immunized me from the child-like faith in physician's omniscience that one finds among most persons, including educated ones" (Meehl, 1989, p. 340).

A question for psychologists of science arises here: Is there any connection of this event to his interests in correct diagnosis with the strong affect and anger associated with it? The answer is not, as I see it, absolutely certain, although at first glance, it seems there might well be a connection. Even if there is, other factors may also be at work, including his cyclothymic temperament, and his social and cultural contexts, such as his association with Herbert Feigl and other philosophers in the Minnesota Center for the Philosophy of Science, which Meehl helped create in 1953. Meehl also spent time in the medical school and with lawyers who may each have different cultures and styles of argument than in the clinical case conferences of which he was so critical.

CHANGING PERSPECTIVES IN THE HISTORY OF PSYCHOLOGY

What are the different ways that the personal or biographical dimension has been included or not in different histories of psychology? Historians of psychology may focus on the internal interplay of theory and research, on external social-political or cultural factors, and/or on the personal-biographical contexts of psychology (Runyan, 1988; Smith, 1997).

I will not attempt a comprehensive review here, but rather discuss the views of two individuals who exemplified the two ends of the continuum: first, a sophisticated advocate of biography in the history of psychology, Edwin Boring, and second, a major postmodernist critic of personal-experiential approaches to the history of science, Michel Foucault.

Edwin G. Boring

Edwin G. Boring (1886–1968) was a professor at Harvard from 1922, director of the Psychological Laboratory from 1924, President of the American Psychological Association in 1927, and author of the dominant history of academic psychology, *A History of Experimental Psychology* (1929/1950). Boring's lineage may be traced back to the founding of experimental psychology, with Wundt's establishment of his laboratory in 1879 in Leipzig. Boring was the favorite student of E. B. Titchener (1867–1927), an Englishman who had studied with Wundt in Leipzig, and then came to Cornell University in 1892, where he became a major figure in translating Wundt's work (at least the experimental and physiological parts of it), and in organizing experimental psychologists in the United States. After Titchener's death in 1927, Boring, as long-term chair of the Harvard Psychology Department, may have been the most influential experimental psychologist in the United States, at least institutionally, if not intellectually, and a recognized founder of the history of psychology.

Boring's *A History of Experimental Psychology* (1929/1950) is a massively informed history of the work and lives of experimental psychologists, which became standard reading as psychology attempted to stake out its territory as a natural science. Boring's text included a tremendous amount of biographical information on experimental psychologists and was an indispensable resource: "Perhaps I should say also why there is so much biographical material in this book, why I have centered the exposition more upon the personalities of men than upon the genesis of the traditional chapters of psychology. My reason is that the history of experimental psychology seems to me to have been so intensely personal. Men have mattered much" (1950/1929, p. viii). The authority of particular individuals was sometimes influential "quite independently of the weight of experimental evidence" for their views. Personalities were important in shaping schools and "the systematic traditions of the schools have colored the research" (1950/1929, p. viii).

Boring's interest in more biographical information led him to write a letter to Carl Murchison at Clark proposing a series of autobiographical essays in psychology, which began in 1930 as *The History of Psychology in Autobiography* (Murchison, 1930) and continued, after a break (Boring & Lindzey, 1967), up to the present (Lindzey, 1989; Lindzey & Runyan, 2007). In 1929, Boring emphasized the importance of individual great psychologists in shaping the field, but by the 1950 edition, he was also attending to the "zeitgeist" or cultural factors of the age.

Boring was a leading advocate of experimental psychology, so it may be somewhat surprising to see him try his hand at psychobiography in explaining the divisions between different types of psychologists. In a 1942 essay on William James, on the centennial of James's birth, Boring explores the differences between phenomenologists, like William James, and experimentalists, like himself. He speculates that "the phenomenologist must have faith in himself and his own observations, whereas the experimentalist mistrusts himself and is forever looking to controls... to correct his own errors" (as quoted in Boring, 1961, p. 203). How are these two stances generated?

> Perhaps some future empiricist will, indeed, solve the problem, will show that a phenomenologist must have had a happy childhood with love and security to spare, a childhood in which it was natural to accept the givens without demanding accounts of their origins. The empiricists and reductionists would then turn out to be the insecure children, who learned early to look beyond the given, suspecting a catch in what is free.... Sensed insecurity is nevertheless the sanction for science itself. (Boring, 1961, p. 208)

This seems to me too monolithic an interpretation of the personal motives for experimentation. It may well be consistent with Boring's self-understanding, as he saw himself as insecure and not attaining "maturity" until in his fifties, but like Freud, he may well have overgeneralized from his own experience. One could also argue the converse, that experimentalists are more secure adults, who are willing to have their ideas tested experimentally. One can think of examples like Edward Tolman of the University of

California, Berkeley, who seemed self-confident and secure, at least in some ways, and a dedicated experimental psychologist working primarily with rats, whose bookplate contains an image of a rat in a maze. There need not be any one-to-one relation of personality to theoretical or methodological preferences, although in some contexts there may be aggregate group differences (Simonton, 2000; Stolorow & Atwood, 1979).

Like many psychologists, Boring's view of psychology changed by his experience in World War II. Boring became more open to applied psychology, seeing its value in the war effort, and made efforts to be more eclectic. In the 1961 introduction to his William James essay of 1942, Boring writes "the progress of thought and discovery depends to some extent upon the personalities of the thinkers and the discoverers...Psychology's great scientific divide needs not only division of labor but also the division of personality that makes complementary and even incompatible activities essential for progress" (Boring, 1961, p. 194).

Michel Foucault

It is sometimes charged that biographical approaches to the history of science have been overemphasized while the social and cultural sides have been neglected. Sometimes the personal–psychological–experiential side of the human sciences is downplayed or denied, whether by Marxists, sociologists of scientific knowledge, or by some postmodernists. An extreme case of this is in the work of Michel Foucault (1926–1984) who has been enormously influential in the history and social studies of science.

He and many others emphasize the ways in which science is socially, politically, economically, culturally, materially, and historically constructed. These are important perspectives, sometimes supported with exquisitely detailed social analysis of topics in the history of science (Galison, 1997; Shapin & Shaffer, 1989). They can open one's eyes to processes previously not seen or attended to.

Foucault often denied the relevance of the personal or psychological and said that what counts is the political aspect of his work. This view was expressed through most of his career with an unexpected change at the end. I will discuss a few elements of his work because he is one of the most influential postmodern historians and critics of the human sciences. In a 1969 interview about his book, *The Archaeology of Knowledge* (1969), Foucault said he absolutely refuses the psychological and wants to focus on discourse itself without "looking underneath discourse for the thought of man" (Foucault, 1996, p. 58).

The denial of the psychological can be done for intellectual, political, and/or personal reasons. I would guess that all three are operative in Foucault. To mention just one of his political and intellectual objections to the psychological, he says in an interview in 1974 on the Attica prison uprising that does not "everything that is a psychological or individual solution for the problem, mask the profoundly political character both of society's elimination of these people and of those people's attack on society. All of that profound struggle is, I believe, political. Crime is a *'coup d' etat* from below" (1996, p. 121).

My response to Foucault is: Yes, psychological analysis can mask the political. However, the converse can also happen, in which the political masks the personal and the psychological. Sometimes personal hurt or rage is projected onto wider political arenas. Often, the personal-experiential, the political, and the intellectual–cultural are interwoven in complex and reciprocally influencing ways. And there are few better examples of this than Foucault himself.

What are the sources of Foucault's desire to critique modernist culture, to critique the human sciences, or to dismantle extant power relations? Does this come from disinterested intellectual reflection, from social-political contexts, and/or from personal experience? It seems possible that aspects of Foucault's critical stance can be related to his personal experience of feeling persecuted as a homosexual in France, attempting suicide in 1948, threatening or attempting suicide a number of other times, and feeling mistreated by the mental health establishment. A doctor at the "Ecole Normale Superieure," citing confidentiality, would say only that "these troubles resulted from an extreme difficulty in experiencing and accepting his homosexuality" (Eribon, 1991, p. 21). According to Eribon, after homosexual encounters, "Foucault would be prostate for hours, ill, overwhelmed with shame" (p. 27), and a doctor was called on frequently to keep him from committing suicide. These personal experiences, in a particular social and cultural context, may well be a source of his antipathy to the mental health establishment and of his perceptions of the human sciences as invasive and harmful rather than beneficent. These personal experiences and others may be interwoven with the formation of political stances and changing intellectual programs throughout Foucault's career.

Foucault maintained what I would describe as a heavily political yet underpsychologized approach to the human sciences through his early archeology of knowledge phase and to his middle genealogical or power/ knowledge period. However, after the transformative experience of participating in the gay community in San Francisco in 1975 and of taking LSD in 1975, his intellectual position changed, with attention turned toward the history of sexuality, history and technologies of the self, and ethics. After 1975 and 1976, the style of his writing also changed to a more clear, lucid style.

At the end of his life, in what is said to be his last interview on May 19, 1984, Foucault says that in his earlier books *Madness and Civilization, The Order of Things,* and *Discipline and Punish,* "I tried to mark out three types of problems: that of the truth, that of power, and that of individual conduct. These three domains of experience can be understood only in relation to each other, and only with each other. What hampered me in the preceding books was to have considered the first two experiences without taking into account the third" (Foucault, 1996, p. 466). In other words, these early works were concerned first with discourse itself, then with the relations of truth and power, but neglected individual conduct, which he tried to address somewhat more in his last books on the history of sexuality, ethics, and techniques of the self. In adding individual conduct, he said "I had a guiding thread which didn't need to be justified by resorting to RHETORICAL methods (capitalization added) by which one could avoid one of the three fundamental domains of experience"(Foucault, 1996, p. 466). Foucault acknowledges, more so in his later life, that all of his work had origins in fragments of his personal experience, including his writings on madness, prisons, and the history of sexuality.

PSYCHOBIOGRAPHIES OF PSYCHOLOGISTS

There is some excellent recent work on the biographical side of psychological theory and research. At its best, it includes discussions of individual psychobiography with relevant social, cultural, and historical contexts. I will mention only a few selected books. A strong advocacy of the importance of the personal side of psychological theory came with Stolorow and Atwood's (1979) *Faces in a Cloud: Subjectivity in Personality Theory*, inspired in part by Silvan Tomkins' work on the psychology of knowledge. They argued that the subjective experiential worlds of Freud, Jung, Rank, and Reich all powerfully influenced their theories of personality. More recent interpretations of Freud, Skinner, and Carl Rogers are provided in Demorest (2005). Erik Erikson's life and work have been reinterpreted by Friedman (1999) and by Erikson's daughter, Sue Erikson Bloland (2005).

In *Pioneers of Psychology* (2012) Raymond Fancher and Alexandra Rutherford demonstrate the advantages of a biographical approach to psychological theory in 15 chapters, beginning with Rene Descartes, and including Wundt, Darwin, Galton, William James, Pavlov, Watson and Skinner, Freud, Binet, and Piaget, with the next-to-last chapter organized not around a single person but around a machine, the computer with the last chapter on a variety of applied psychologies.

Irving Alexander provides psychobiographical interpretations of Freud, Jung, and most intriguingly, a hypothesis about the missing years in young adulthood of Harry Stack Sullivan (Alexander, 1990). In addition to his study of B. F. Skinner discussed above, Elms also has published studies of Freud, Jung, Allport, and others (Elms, 1994, 2005). Gordon Allport has been the subject of a complex analysis of the social, cultural, and psychological sources of his thought (Nicholson, 2003) with additional studies of Allport by Barenbaum (2005).

The *Handbook of Psychobiography* (Schultz, 2005) contains a section on the psychobiography of psychologists, including chapters on the life and work of Freud, Gordon Allport, Erik Erikson, and S. S. Stevens. The handbook also has sections on "Psychobiographies of Artists" (including Elvis Presley, Sylvia Plath, J. M. Barrie, and Edith Wharton) as well of others such as Truman Capote and Diane Arbus. Scholarly interest remains strong in the lives of both Charles Darwin and William James. Their lives have been studied from social, cultural, and psychological perspectives. Both Darwin and James each have good biographies, standard editions of their works, and published volumes of their correspondence, year by year, providing valuable resources for later biographers, psychobiographers, and historians. Excellent examples are the biographies of Darwin by Desmond and Moore (1991) and the two volumes by Janet Browne (1995, 2002). A major recent biography on William James is by Richardson (2006) and on James and his early associates in *The Metaphysical Club* by Menand (2001).

The psychological interpretation of psychologists is also engaged in by psychologists themselves. Between 1930 until the present, the series *A History of Psychology in Autobiography* has produced nine published volumes. Personally, I first became aware of this series in 1967, which contained autobiographies by Gordon Allport, Henry Murray, Carl Rogers, and B. F. Skinner (Boring &

Lindzey, 1967). Volume 7 (Lindzey, 1989) includes interesting autobiographies by Roger Brown, Lee Cronbach, Eleanor Maccoby, Paul Meehl, George Miller, and others, whereas Volume 9 has illuminating autobiographies by Elliot Aronson, Gordon Bower, Jerome Kagan, Daniel Kahneman, Elizabeth Loftus, Ulrich Neisser, Walter Mischel, and others (Lindzey & Runyan, 2007).

Statisticians: R. A. Fisher and Jerzy Neyman

There are complex relations between thinking statistically, thinking histori-cally, and thinking personally or experientially. What might be learned about these issues from looking at the lives and interpersonal relationships of stat-isticians? This discussion will focus on Sir Ronald A. Fisher (1890–1962) and Jerzy Neyman (1894–1981), two of the most influential statisticians of the 20th century, with an introductory note on Karl Pearson, a founder of the field.

Statistics was developed and instutionalized in part by Karl Pearson (1857–1936). Pearson's book, *The Grammar of Science* (1892/1900) presented a view of the importance of statistics in relation to science which inspired many over the years, including both Fisher and Neyman. *The Grammar of Science* made a claim for the unlimited scope of science, and "a moral vision of scientific method as the very basis of modern citizenship, because it pro-vides standards of knowing that are independent of all individual interests and biases" (Porter, 2004, p. 7.) The second edition of *The Grammar of Science* in 1900 included a philosophical rationale for statistics, which was not in the first edition. Until the end of his life, Pearson saw it as his mission to "reshape science using the tools of statistical mathematics" (Porter, 2004, p. 8). Pearson was a follower of Francis Galton (1822–1911), author of *Hereditary Genius* (1869) and other works. In collaboration with Galton and W. F. R. Weldon, Pearson founded the journal *Biometrika* in 1901, which Pearson edited with a heavy editorial hand until his retirement in 1934.

An excellent biography of Pearson by Theodore Porter (2004) analyzes the complex personal meanings that statistics had for Pearson, who earlier planned to be a poet, and wrote a romantic novel (*The New Werther*) modeled after Goethe's *The Sorrows of Young Werther* (1774). Pearson had earlier been a German Scholar, a socialist, and an advocate for women's causes before turning to science by 1892 and to statistics by 1900.

Pearson was one of the individuals who helped to define what it meant to be a statistically sophisticated scientist. In *Karl Pearson: Scientific Life in a Statistical Age* (Porter, 2004) Porter writes about how Pearson's statistical and scientific work is related to his personal life, religious anxieties, and family dynamics. In 1901, Pearson began editing *Biometrika*, which he edited with an iron hand, advocating his own views of statistics until almost the end of his life in 1936.

R. A. Fisher (1890–1962) had more mathematical training than Pearson and became critical of Pearson's work. Pearson would not allow Fisher's work to appear in *Biometrika*, the journal he had founded and edited. Fisher had manuscripts refused at *Biometrika* in 1916, 1918, and 1920. As editor, Pearson wrote "I am regretfully compelled to exclude all that I think is erroneous in my own judgment, because I cannot afford controversy" (Box, 1978, p. 83).

In 1919, Fisher was offered an appointment in the Galton Laboratory at University College, London, which he declined as it seemed he would not be able to publish his ideas without Pearson's approval (Box, 1978, p. 82). Instead, Fisher joined the staff of the Rothamsted Experimental Station in October 1919. They had collected many years of data on agricultural experiments that had not been adequately analyzed and Fisher drew on this data in his classic books *Statistical Methods for Research Workers* (1925) and *The Design of Experiments* (1935), each of which went through many later editions. In 1922, Fisher published an article severely criticizing Pearson's chi-square test in the *Journal of the Royal Statistical Society*. Pearson counterattacked in *Biometrika* and the two spent the rest of their life in conflict.

Who replaced Karl Pearson in the Galton Chair at University College, London, after he retired in 1933? Pearson asked that it be anyone but Fisher. The chair was eventually divided into two positions, one in statistics and one in eugenics. The Head of the Department of Applied Statistics, went to Egon Pearson (Karl's son), and directorship of the Galton Laboratory went to R. A. Fisher.

The University of California at Berkeley was interested in hiring a distinguished statistician, and invited R. A. Fisher to give the Hitchcock Lectures in the Fall of 1936. Fisher came and gave the Hitchcock Lectures at UC Berkeley on *The Design of Experiments*. This was supposed to include a 3- or 4-week stay at the campus so that faculty and students could meet informally with him. Raymond Birge, Chair of the Physics department at UC Berkeley who was involved with the recruitment, felt that the visit did not go well. Fisher spent the first week with a friend in San Francisco, rather than at the Berkeley campus. Fisher was seen as so arrogant that the department did not offer him a position. Birge wrote that Fisher was the most conceited man he had ever met, and "that is saying a lot with such competitors as Millikan et al." (Reid, 1998, p. 144).

On November 10, 1937 a letter was sent to Jerzy Neyman (1894–1981) inviting him to teach statistics in the math department at UC Berkeley, and Neyman arrived in the Fall of 1938. He was able to turn the statistics laboratory into a separate Statistics Department by 1955 (Reid, p. 148). Neyman built Berkeley into "the largest and most important statistics center in the world" in the years after World War II (McGrayne, 2011, p. 98). Neyman was a frequentist and Berkeley was an "anti-Bayesian powerhouse" (p. 51). There is a fascinating story of two centuries of controversy between frequentist and Bayesian views of statistics. The complex story is well told in McGanahan (2011), and I won't try to summarize it all here. In her view, Bayesian views were frequently rejected or attacked, but eventually came to triumph (McGanahan, 2011).

Fisher and Neyman also developed different views of statistics, Fisher concentrating on inductive inferences, and Neyman on inductive behavior. It seems that each could not or would not recognize any merit in the other's viewpoint (Lehmann, 2008, p. 168). Lehmann argues that the two approaches are not as incompatible as Fisher and Neyman seemed to believe (p. 168). Important elements are integrated in decision theory as developed by Abraham Wald (1902–1950), in his *Statistical Decision Functions* (1950), which many saw as a magnificent new integrative framework for the field. Wald saw himself as a follower of Neyman, and Neyman was enthusiastic about the new decision-theory approach, but Fisher strongly objected to it (Lehmann, 2008, pp. 166,167).

Karl Pearson's idea of statistics as a site of intellectual consensus has not yet been reached, with several alternative views of statistics, such as frequentist or subjectivist (Bayesian) views still influential. Historians of statistics have argued that statistics textbooks often paper-over these unresolved theoretical differences (Gigerenzer et al., 1987).

Jerzy Neyman, a leading frequentist, held the view that the theory of probability with which a statistician works is a matter of taste. When asked in 1979 about his view of the alternative Bayesian view, Neyman said "it does not interest me. I am interested in frequencies" (Reid, 1982, 1998, p. 274).

People's viewpoints are not set in stone, and can be shaped by culture, temperament, intelligences, and experience. When approached about being the subject of a biography, Neyman initially said he was not interested, and refused to read a biographical sketch written at his 80th birthday, which he referred to as his "obituary." However, "It's a free country...and if people want to write about me, I can't stop them" (Reid, 1998, p. 1). Constance Reid came to talk with him about his life on Saturdays in 1978. Near his 85th birthday in 1979, she offered him a ride home from his 85th birthday party celebration, and said she needed to begin writing, and would not be coming to talk with him next week, which he seemed to regret. A valuable history of the Berkeley Statistics Department, with biographical sketches, is provided by Erich Lehmann, graduate student at Berkeley since 1941, since 1942 a student of Neyman's, and later chair of the department (Lehmann, 2008).

Does biography and psychological biography alone determine the history of statistics? No. Are there biographical, psychological, and interpersonal relations that serve as *strands* of the whole history? Yes. My argument is that psychological biography is one important strand of the history of statistics, as it is also a strand of the history of psychology and the history of philosophy.

HOLISM AND PSYCHOBIOGRAPHY

Psychobiography can be one dimension of providing a wider holistic context for the psychology of science. A valuable review of multiple meanings of holism is Holism in *Reenchanted science: Holism in German culture from Wilhelm II to Hitler* (1996) by Anne Harrington. In Germany, holism was linked to high humanistic ideals, yet also used by Nazi theorists to claim that Aryan holistic thinking was to be prized over mechanistic, atomistic "Jewish thinking." Ironically, several Jewish immigrants from Germany, including Max Wertheimer (1880–1943) and Kurt Goldstein (1878–1965) were major contributors to gestalt psychology and to holistic neuroscience in the United States. Wertheimer and Goldstein were both important influences on Abraham Maslow (1908–1970) in his development of humanistic or "third force" psychology in the United States. Harrington uses a "multiple biography" approach in unraveling different strands of holistic and analytic research in relation to each other. This multiple biography approach provides a valuable way of analyzing the different ways in which "the personal, the scientific, and the sociopolitical continually co-construct each other over time" (Runyan, 1998, p. 390).

CONCLUSION

This chapter has touched on psychobiographical examples from the three fields of philosophy, psychology, and statistics. In studies of Bertrand Russell and Ludwig Wittgenstein in philosophy, the work and interpersonal relationships sometimes have surprising personal meanings. In psychology, the works of Henry Murray, Karen Horney, or Paul Meehl have meanings to them, which can be discovered only in individual biography. With Foucault, the denial of the personal is discovered to have not only political meanings, but also personal ones. Even though statistics can be thought of as an objective search for illuminating quantitative analysis, there has been a surprising amount of disagreement and interpersonal conflict running through the history of the field. What are we to conclude? My own view is that psychobiography is a valuable addition to the uses of cognitive and social psychology in the psychology of science. We may need multiple psychobiographical studies of individuals, as well as experimental and statistical analyses at the aggregate level. Along with social, cultural, and political analysis, psychological biography can be part of the answer to the question of: What is really going on here?

ACKNOWLEDGMENTS

I thank Jim Anderson, Nicole Barenbaum, Mary Coombs, Alan Elms, William Todd Schultz, and members of the Society for Personology and of the San Francisco Bay Area Psychobiography Group for their comments on earlier drafts of this chapter.

REFERENCES

Alexander, I. (1990). *Personology: Method and content in personality assessment and psychobiography.* Durham, NC: Duke University Press.

Barenbaum, N. B. (2005). Four, two, or one? Gordon Allport and the unique personality. In W. T. Schultz (Ed.), *Handbook of psychobiography* (pp. 223–239). New York, NY: Oxford University Press.

Bloland, S. E. (2005). *In the shadow of fame: A memoir by the daughter of Erik H. Erikson.* New York, NY: Viking.

Boring, E. G. (1950). *A history of experimental psychology.* New York, NY: Appleton Century. (Original work published 1929)

Boring, E. G. (1961). *Psychologist at large: An autobiography and selected essays.* New York, NY: Basic Books.

376   SECTION IV. SPECIAL TOPICS

Boring, E. G., & Lindzey, G. (Eds.). (1967). *A history of psychology in autobiography, Vol. 5.* New York, NY: Appleton-Century-Crofts.

Box, J. F. (1978). *R. A. Fisher: The life of a scientist.* New York, NY: Wiley.

Breger, L. (2000). *Freud: Darkness in the midst of vision.* New York, NY: Wiley.

Browne, J. (1995). *Charles Darwin: Vol 1. Voyaging.* Princeton, NJ: Princeton University Press.

Browne, J. (2002). *Charles Darwin: Vol 2. The Power of Place.* Princeton, NJ: Princeton University Press.

Campbell, D. T. (1989). Fragments of the fragile history of psychological epistemology and theory of science. In B. Gholson et al., (Eds.), *Psychology of science: Contributions to metascience* (pp. 21–46). Cambridge, England: Cambridge University Press.

Cronbach, L. J. (1957). The two disciplines of scientific psychology. *American Psychologist, 12,* 671–684.

Cronbach, L. J. (1975). Beyond the two disciplines of scientific psychology. *American Psychologist, 30,* 116–127.

Cronbach, L. J. (1982). *Designing evaluations of educational and social programs.* San Francisco, CA: Jossey-Bass.

Clark, R. A. (1975). *The life of Bertrand Russell.* London, UK: Cape.

Demorest, A. (2005). *Psychology's grand theorists: How personal experiences shaped professional ideas.* Mahwah, NJ: Lawrence Erlbaum Associates.

Desmond, A., & Moore, J. (1991). *Darwin: The life of a tormented evolutionist.* New York, NY: Norton.

Ellenberger, H. E. (1970). *The discovery of the unconscious.* New York, NY: Basic Books.

Elms, A. (1994). *Uncovering lives: The uneasy alliance of biography and psychology.* New York, NY: Oxford University Press.

Elms, A. (2005). Freud as Leonardo: Why the first psychobiography went wrong. In W. T. Schultz (Ed.), *Handbook of psychobiography* (pp. 210–222). New York, NY: Oxford University Press.

Eribon, D. (1991). *Michel Foucault.* Cambridge, MA: Harvard University Press.

Feist, G. J. (2006). *The psychology of science and the origins of the scientific mind.* New Haven, CT: Yale University Press.

Feist, G. J., & Gorman, M. E. (1998). The psychology of science: Review and integration of a nascent discipline. *Review of General Psychology, 2,* 3–47.

Fisher, R. A. (1922). On the mathematical foundations of theoretical statistics. *Philosophical Transactions of the Royal Society of London, 222,* 309–368.

Fisher, R. A. (1925). *Statistical methods for research workers.* Edinburgh, UK: Oliver & Boyd.

Fisher, R. A. (1935). *The design of experiments.* Edinburgh, UK: Oliver and Boyd.

Foucault, M. (1969). *The archaeology of knowledge.* New York, NY: Pantheon.

Foucault, M. (1996). *Foucault live (interviews, 1961–1984).* New York, NY: Serniotext(e).

Friedman, L. (1999). *Identity's architect: A biography of Erik H. Erikson.* New York, NY: Scribner.

Fromm, E. (1941). *Escape from freedom.* New York, NY: Holt, Rinehart & Winston.

Galison, P. (1997). *Image and logic: A material culture of microphysics.* Chicago, IL: University of Chicago Press.

Galton, F. (1869). *Hereditary genius.* London, UK: Macmillan.

Gardner, H. (1993). *Creating minds: An anatomy of creativity seen through the lives of Freud, Einstein, Picasso, Stravinsky, Eliot, Graham, and Gandhi.* New York, NY: Basic Books.

Gay, P. (1988). *Freud: A life for our times.* New York, NY: Norton.

Gholson, B., Shadish, W. R., Neimeyer, R. A., & Houts, A. C. (Eds.). (1989). *Psychology of science: Contributions to metascience.* Cambridge, England: Cambridge University Press.

Gigerenzer, G., et al. (1989). *The empire of chance: How probability changed science and everyday life.* Cambridge, New York, NY: Cambridge University Press.

Gross, P., Levitt, N., & Lewis, M. (Eds.). (1996). *The flight from science and reason.* New York, NY: The New York Academy of Sciences.

Gruber, H. E. (1974). *Darwin on man: A psychological study of scientific creativity.* New York, NY: E. P. Dutton.

Harrington, A. (1996). *Reenchanted science: Holism in German culture from Wilhelm II to Hitler.* Princeton, NJ: Princeton University Press.

Harrington, A. (1996). Reenchanted science: Holism in German culture from Wilhelm II to Hitler. *Contemporary Psychology, 43*(6), 389–392.

Horney, K. (1937). *The neurotic personality of our time.* New York, NY: Norton.

Horney, K. (1939). *New ways in psychoanalysis.* New York, NY: Norton.

Horney, K. (1942). *Self-analysis.* New York, NY: Norton.

Horney, K. (1950). *Neurosis and human growth: The struggle toward self-realization.* New York, NY: Norton.

Horney, K. (1967). *Feminine psychology* (H. Kelman, Ed.). New York, NY: Norton.

Horney, K. (1994). *A psychoanalyst's search for self-understanding.* New Haven, CT: Yale University Press.

Jones, E. (1953–1957). *The life and work of Sigmund Freud* (3 volumes). New York, NY: Basic Books.

Kluckhohn, C., & Murray, H. A. (Eds.). (1948). *Personality in nature, society and culture.* New York, NY: Knopf.

Lehmann, E. L. (2008). *Reminiscences of a statistician: The company I kept.* New York, NY: Springer.

Lindzey, G. (Ed.). (1989). *A history of psychology in autobiography, Vol. VIII.* Stanford, CA: Stanford University Press.

Lindzey, G., & Runyan, W. M. (Eds.). (2007). *A history of psychology in autobiography,* Vol. IX. Washington, DC: American Psychological Association.

Malcolm, J. (1984). *In the Freud archives.* New York, NY: Knopf.

Maslow, A. (1966). *The psychology of science.* New York, NY: Harper & Row.

Masson, J. (1984). *The assault on truth: Freud's suppression of the seduction theory.* New York, NY: Farrar, Straus & Giroux.

McGrayne, S. (2011). *The theory that would not die.* New Haven, CT: Yale University Press.

Meehl, P. E. (1954). *Clinical versus statistical prediction.* Minneapolis, MN: University of Minnesota Press.

Meehl, P. E. (1973a). *Psychodiagnosis: Selected papers.* Minneapolis, MN: University of Minnesota Press.

Meehl, P. E. (1973b). Why I do not attend case conferences. In P. E. Meehl (Ed.), *Psychodiagnosis: Selected papers* (pp. 225–302). Minneapolis, MN: University of Minnesota Press.

Meehl, P. E. (1989). Paul E. Meehl. In G. Lindzey (Ed.), *History of psychology in autobiography* (Vol. VIII, pp. 337–389). Stanford, CA: Stanford University Press.

Meehl, P. E. (1991). *Selected philosophical and methodological papers.* (C. A. Anderson & K. Gunderson, Eds.). Minneapolis, MN: University of Minnesota Press.

Meehl, P. E. (2005). *A Paul Meehl reader: Essays on the practice of scientific psychology.* Mahwah, NJ: Erlbaum.

Menand, L. (2001). *The metaphysical club.* New York, NY: Farrar, Straus and Giroux.

Monk, R. (1990). *Ludwig Wittgenstein: The Duty of Genius.* New York, NY: Free Press.

Monk, R. (1996). *Bertrand Russell: 1872–1920, The Spirit of Solitude.* New York, NY: Free Press.

Monk, R. (2000). *Bertrand Russell: 1921–1970, The Ghost of Madness.* New York, NY: Free Press

Monk, R. (2005). *How to Read Wittgenstein?* New York, NY: Norton.

Moorehead, C. (1993). *Bertrand Russell: A life.* New York, NY: Viking.

Mullins, N. (1973). *Theories and contemporary theory groups in contemporary American sociology.* New York, NY: Harper and Row.

Murchison, C. (Ed.). (1930). *A history of psychology in autobiography, Vol. 1.* Worcester, MA: Clark University Press.

Murray, H. A. (1938). *Explorations in personality.* New York, NY: Oxford University Press.

Murray, H. A. (1959). Vicissitudes of creativity. In E. S. Shneidman (Ed.), *Endeavors in psychology: Selections from the personology of Henry A. Murray* (pp. 312–330). New York, NY: Oxford University Press.

Murray, H. A. (1967). Henry A. Murray. In E. G. Boring & G. Lindzey (Eds.), *A history of psychology in autobiography,* (Vol. V, pp. 283–310). New York, NY: Appleton-Century-Crofts.

Nicholson. I. A. M. (2003). *Inventing personality: Gordon Allport and the science of selfhood.* Washington, DC: American Psychological Association.

Pauly, P. J. (1987). *Controlling life: Jacques Loeb & the engineering ideal in biology.* New York, NY: Oxford University Press.

Pearson, K. (1892/1900). *The grammar of science* (2nd ed.). London, UK: Adam and Charles Black.

Popper, K. (1959). *The logic of scientific discovery.* New York, NY: Science Editions.

Porter, T. (2004). *Karl Pearson: The scientific life in a statistical age.* Princeton, NJ: Princeton University Press.

Reid, C. (1998). *Neyman.* New York, NY: Springer.

Richardson, R.D. (2006) *William James: In the maelstrom of American modernism.* Boston, MA: Houghton Mifflin.

Roazen, P. (1975). *Freud and his followers.* New York, NY: Knopf.

Robinson, F. (1992). *Love's story told: A life of Henry A. Murray.* Cambridge, MA: Harvard University Press.

Roe, A. (1953a). *The making of a scientist.* New York, NY: Dodd, Mead.

Roe, A. (1953b). A psychological study of eminent psychologists and anthropologists, and a comparison with biological and physical scientists. *Psychological Monographs, 67*(2, Whole No. 352).

Runyan, W. M. (1977). How should treatment recommendations be made? Three studies in the logical and empirical bases of clinical decision making. *Journal of Consulting and Clinical Psychology, 45,* 552–558.

Runyan, W. M. (1982). *Life histories and psychobiography: Explorations in theory and method.* New York, NY: Oxford University Press.

Runyan, W. M. (Ed.). (1988). *Psychology and historical interpretation.* New York, NY: Oxford University Press.

Runyan, W. M. (2005). Evolving conceptions of psychobiography and the study of lives: Encounters with psychoanalysis, personality psychology and historical science. In W. T. Schultz (Ed.), *Handbook of psychobiography* (pp. 19–41). New York, NY: Oxford University Press.

Runyan, W. M. (2006). Psychobiography and the psychology of science: Understanding relations between the life and work of individual psychologists. *Review of General Psychology, 10*(2), 147–162.

Runyan, W. M. (2008). Henry Alexander Murray. In N. Koertge (Ed.), *The new dictionary of scientific biography* (Vol. 1, pp. 214–219). Detroit, MI: Thomson Gale.

Russell, B. (1967). *The autobiography of Bertrand Russell.* (Vol. 1). Boston, MA: Little Brown.

Schultz, W. T. (Ed.). (2005). *Handbook of psychobiography.* New York, NY: Oxford University Press.

Shadish, W. R., & Fuller. S. (Eds.). (1994). *The social psychology of science.* New York, NY: Guilford Press.

Shapin, S., & Schaffer, S. (1989). *Leviathan and the air-pump: Hobbes, Boyle, and the experimental life.* Princeton, NJ: Princeton University Press.

Shneidman, E. S. (Ed.). (1981). *Endeavors in psychology: Selections from the personology of Henry A. Murray.* New York, NY: Oxford University Press.

Simonton, D. K. (1988). *Scientific genius: A psychology of science.* Cambridge, England: Cambridge University Press.

Simonton, D. K. (2000). Methodological and theoretical orientation and the long-term disciplinary impact of 54 eminent psychologists. *Review of General Psychology, 4,* 13–21.

Simonton, D. K. (2002). *Great psychologists and their times: Scientific insights into psychology's history.* Washington, DC: American Psychological Association.

Skinner, B. F. (1948). *Walden two.* New York, NY: Macmillan.

Smith, R. (1997). *The Norton history of the human sciences.* New York, NY: Norton.

Stolorow, R. D., & Atwood, G. E. (1979). *Faces in a cloud: Subjectivity in personality theory.* New York, NY: Jason Aronson.

Sulloway, F. (1979). *Freud, biologist of the mind: Beyond the psychoanalytic legend.* New York, NY: Basic Books.

Wald, A. (1950). Statistical decision functions. New York, NY: Wiley.

Whitehead, A. N., & Russell, B. (1910, 1912, 1913). *Principia mathematica,* (3 vols.). Cambridge, MA: Cambridge University Press.

Wittels, F. (1923). Sigmund Freud: His personality, his teaching, and his school. New York, NY: Dodd, Mead.

Wittgenstein, L. (1913). Review of P. Coffey, The Science of Logic, *The Cambridge Review,* *34,* 853.

Wittgenstein, L. (1922). *Tractatus logico-philosophicus, with an introduction by Bertrand Russell.* New York, NY: Harcourt, Brace & Company.

Wittgenstein, L. (1953). Philosophical investigations; translated by G.E.M. Anscombe. Oxford, MA: Blackwell.

Wood, A. (1957). *Bertrand Russell the passionate skeptic; a biography.* New York, NY: Simon and Schuster.

Applied Psychologies of Science

CHAPTER 15

The Psychology of Technological Invention

Michael E. Gorman

My first job out of graduate school was at Michigan Tech, an engineering school in Michigan's Upper Peninsula. I straddled three departments: humanities, social sciences, and education, and for several years I was the only psychologist. The hill right across the university was dominated by the Quincy mine's hoist building (now a national park) and there was a large collection of buildings right across the Portage Waterway—an abandoned stamp mill. Signs of both technological ingenuity and environmental irresponsibility were visible everywhere.

The social science department suggested I team-teach a course on invention and innovation with Bernard Carlson, a newly hired historian of technology. As he lectured about inventors, I saw how psychology of science could be applied to studying them. We applied for a National Science Foundation (NSF) grant, got it on the second try, and began an extended study of the invention of the telephone where we tried to use a cognitive–historical approach to comparing three inventors: Alexander Graham Bell, Elisha Gray, and Thomas Edison.

The point here is to show how I fell serendipitously into the study of inventors. If there was little psychology literature on discovery, there was less on this kind of technological invention. Robert Weber, who was conducting experimental studies on the invention process, invited Bernie and me to a conference at Oklahoma State he organized with creativity researcher David Perkins from Harvard. The purpose of the conference was to bring together those doing work on the cognitive psychology of technological invention from multiple fields.

This chapter will begin with a brief discussion of the work Weber did with creativity researcher David Perkins, then will shift to three studies that have used the cognitive–historical approach. Using the framework described in Gorman (Gorman, 2008), we will refer to the experimental work as in vitro and the cognitive–historical work as *Sub Species Historiae*. Next, I review recent in vivo observations of design and invention teams using different cognitive methods. The focus on teams inevitably links to social psychology of science. The chapter will conclude with reflections on the ethical responsibilities of inventors and suggestions for future research.

HEURISTICS AND INVENTION

Weber and Perkins (1989) explored heuristics that have been used in the invention of new technologies, going back to prehistory, and identified both frames and heuristics that were used in the past and could be used in the future. According to Weber and Perkins, "a frame is an entity with slots in which particular values, relations, procedures, or even other frames reside; as such the frame is a framework or skeletal structure with places in which to put things"(p. 51).

The findings have important practical implications for conflict management training. Researchers in science and engineering often underestimate or downplay the importance of conflicts in innovation processes. In particular, if they are poorly managed, conflicts can have quite severe consequences, such as delays, terminations and loss of time and money.

It is the objective of conflict management training to familiarize managers and researchers with basic methodological principles of analysis and constructive conflict management in innovation processes.

Because experts underestimate the importance, students and young scientists are not prepared to handle conflicts productively. Some internalize the conflicts, blaming themselves for failure and may react with psychological problems and/or complete retreat from the conflict situation. Special courses for PhD students should include the detailed analysis of conflict situations in processes of innovation, which will then enable them to identify critical incidents. Simulation exercises can demonstrate where the key opportunities are and what steps the students could take going forward. Consider the knife: it can have slots corresponding to different designs for its edge, for its grip, for the materials it is made of, and the Swiss Army knife opens up a whole new set of slots for other devices on the knife that can then be exploited by other manufacturers.

Heuristics are strategies that can be applied to search for new options to include in a slot, or new slots, or even to switch frames. One heuristic is to represent an invention idea as a frame with slots, allowing the inventor to focus on different aspects of the idea that could be altered. Consider the fork—dividing it into slots like grip and tines permits a designer to think about alternatives—what about a better grip? What about a different combination of tines? Changing the tines slot to a more functional representation like grasp food allows for even more innovation. Why not combine spoon and fork and have one device instead of two—very handy for backpackers, especially when one opens a material slot and substitutes titanium, thereby reducing weight.

Weber and Dixon (1989) tried experiments to test their problem-solving model of invention. One of their heuristics was "reduce the time in switching between operations." They hypothesized that this heuristic had driven the development of the sewing needle; they tested it by asking both expert and novice sewers to use several classic forms of sewing needle; for both, the modern form led to significantly fewer switching operations, providing proof-of-concept that the heuristic could have been a driver for innovation.

To establish ecological validity, this sort of proof-of-concept could be tested against accounts of actual use and evolution of design—in effect, combining psychology and history of technology. At the time Weber and Perkins were publishing their work, I had already begun my NSF-supported collaboration with Bernard Carlson. It was he who selected the problem domain based on the available records and the fact that three inventors were involved: Alexander Graham Bell, Elisha Gray, and Thomas Edison.

A COGNITIVE PSYCHOLOGICAL COMPARISON OF THREE TELEPHONE INVENTORS

Our method was to take advantage of Carlson's knowledge about, and access to, the relevant archival records, including the Bell notebooks at the Library of Congress, Gray documents and artifacts at the Smithsonian, Edison sketches and artifacts from the reconstructed Menlo Park in Dearborn, early Bell telephones from an AT&T archive, and many other sources. We were particularly inspired by the work of David Gooding, who had not only studied Faraday's detailed notebooks, but also studied whatever apparatus was still preserved and even redid some of Faraday's experiments to understand his discovery path (Gooding, 1990a, 1990b).

My job was to develop a framework that would be useful in comparing these inventors. I combined Weber's ideas with those of Ryan Tweney, who, like Gooding, did extensive, detailed studies of Faraday (Tweney, 1985, 1989). Both emphasized heuristics, but whereas Weber liked the concept of a frame to describe problem representation, Tweney preferred schema, which are sets of expectations about what would happen next.

I preferred the notion of a mental model. To demonstrate why, consider the following sketch from Bell's notebooks that I "discovered" in the Library of Congress. I say discovered because it had been seen before, but no one had grasped its significance. It shows the bones of the ear attached to a diaphragm and a speaking tube, vibrating close to two different arrangements of electromagnets. Over this crude sketch, Bell announces his approach: that he would follow the analogy of nature by using the ear as a model for the telephone, which means that he would need an armature that would function like the bones of the ear, translating the vibrations of the diaphragm into a current that precisely mimicked the form of the sound wave (see Figure 15.1).

This seemed to me a perfect example of a mental model, which is a better form of representation than frame or schema for describing what inventors do.[1] Bell's sketch shows the bones of the ear acting as an armature vibrating as someone speaks into the tube and diaphragm on the left. The bones vibrate in front of two possible arrangements of an electromagnet to translate motion into sound. The sketch comes after goal statements: follow the analogy of nature by using the ear as a model for a device to transmit speech and specifically finding an armature that does what the bones of the middle ear do.

So this sketch provides a model of Bell's strategy, a model that includes alternative designs, that he could run mentally, imagining how it might work. Like a schema, this mental model embodies expectations. Bell had

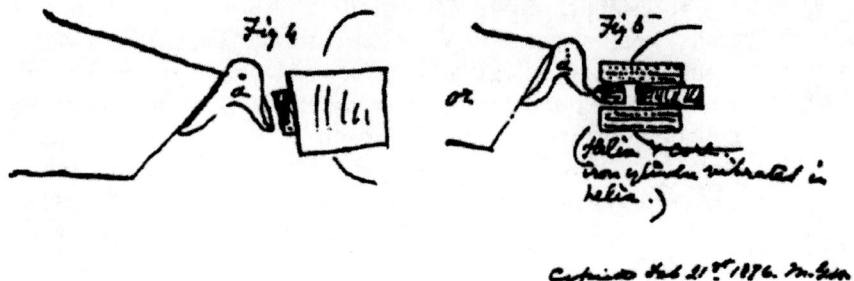

Figure 15.1   Bell's drawing of the bones of the middle ear ("a") between a speaking tube and two arrangements of electromagnets from his "Experimental Notebook."

Note: The text under "Fig 5" reads "(helix & core, iron cylinder vibrated in helix)" and at the bottom right Mabel Gardiner, Bell's future wife, notes that she copied the figure on February 21st.

previously found a way to mount the bones of the ear on a speaking tube and trace the motion on a piece of smoked glass. His goal was to reproduce the sound wave in what he referred to as an undulating electric current; naturally, he started by imagining the bones themselves in a circuit, expecting that a device like this would perfectly reproduce the sound wave mechanically.

The two electromagnetic arrangements suggest that mental models can have slots, in this case one that might be functionally represented as "translated sound vibration into electric current." A slot designates a part of a representation that functions like a variable—an inventor can try different arrangements to accomplish the function. Bell illustrates by showing two, but others might be possible. Bell seems to be following one of Weber and Perkin's heuristics—he is representing the problem of transmitting speech in a way that allows him to target particular functional slots for improvement.

Another slot is the one corresponding to the bones of the ear. Bell knew he could not build speaking telegraphs out of bones, but the bones served as a mental model of what he needed in this slot—an armature that functioned like the ossicles. It was particularly significant that Bell drew this sketch about the same time as he submitted one of the most valuable—and hotly contested—patents in American history.

To see if my hypothesis about Bell's mental model and slots was correct, I had to study and understand both his previous and later work in telegraphy, working at as fine a level of detail as records permitted. Bell did not keep a notebook throughout his invention process, though he did send letters to his mother and father to deliberately record his ideas. I put all these materials on a website where others could see my data and draw their own conclusions.

Where the records were sufficiently detailed, I used a technique called the problem-behavior graph that Tweney (1989) applied to Faraday's incredibly detailed notebook. I adapted it to Bell's notebook. It helped me understand a sequence of experiments Bell began after he submitted his patent. The great Bell's biographer Robert Bruce referred to this sequence of experiments

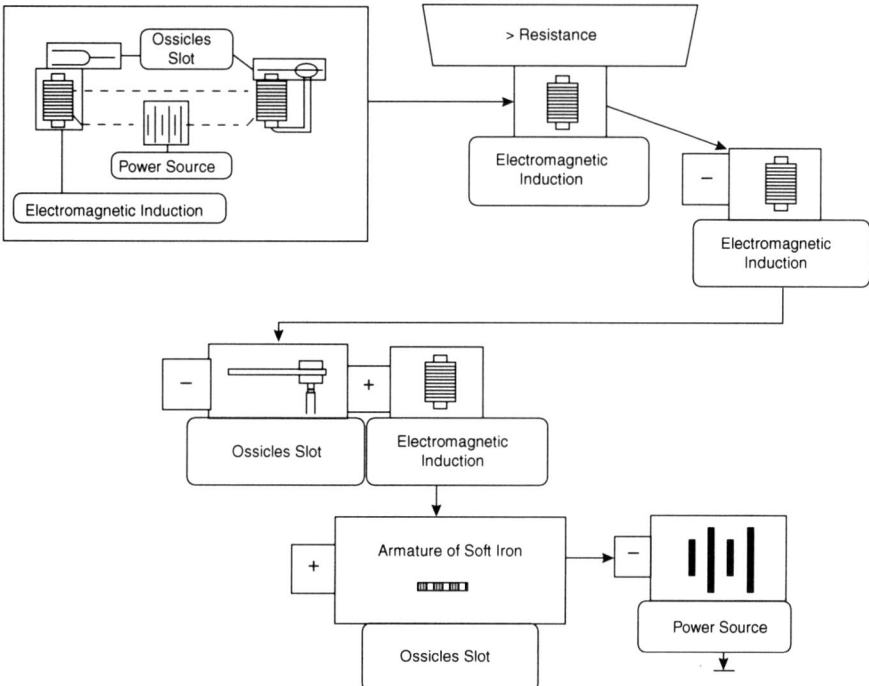

Figure 15.2   Problem behavior graph showing Bell's initial experiments after submitting his patent proposal.

Note: Horizontal arrows indicate Bell heard sound; vertical indicates no sound; diagonal indicates partial success at transmitting sound.

as random (Bruce, 1973), but a cognitive psychological analysis showed they were anything but.

Figure 15.2 begins with a slot diagram based on Bell's mental model. The electromagnet and ossicles slots come straight out of the ear diagram. Note that in the ossicles slot Bell's initial set-up includes a tuning fork and a reed, the fork to send a tone and the reed to receive it. This is a standard set-up Bell had used often in his multiple telegraph experiments. What he was doing was testing his experimental set-up to make sure it would work before making variations. Baker and Kevin Dunbar refers to this as a known standard control Baker & Dunbar (2000).

The straight horizontal line in a problem-behavior graph indicates a result consistent with the goal: Bell heard the expected tone. Bell then increased the resistance in the electromagnet. The diagonal line indicates that the transmission of the tone was not as good, which is what Bell expected. Removing the electromagnet altogether stopped transmission—again, as Bell expected. Put the electromagnet back in and remove the tuned metal reed on the receiving end also produces no tone. Substitute a soft iron armature for the reed and the tone comes back. Remove the power source and there is no sound.

Bell now knew his set-up worked as expected, and he moved to the set of experiments that result in the first transmission of speech. He substituted a needle going into water for the tuning fork, likely in response to what a patent examiner let slip about his rival Elisha Gray. Gray had submitted a caveat for a speaking telegraph to the patent office on the same day that Bell submitted his patent application.[2] Bell found out his patent was held up, and went to the patent office to find out why. The patent examiner pointed to a line in Bell's patent concerning variable resistance to indicate that this was the source of the conflict between him and Gray. After establishing his known standard control, Bell went back to his laboratory and started a sequence of variable resistance experiments using water as the medium.

The end result was the famous line, "Watson, come here, I want you" to which Watson responded by coming from the receiving room to where Bell was transmitting.[3] Bell and Watson arrived at this result by a systematic series of experiments in which they varied the distance between and relative sizes of the contacts separated by water with acid or salt.

Who Invented the Telephone?

The battle over who invented the telephone continues to this day, and Elisha Gray is frequently put forward as the real inventor. He was a highly accomplished inventor of telegraph systems. Cognitive psychology does not tell us who deserves to be called the first inventor; that is a societal label, determined through the courts, through arguments among scholars and backers, and so on. But cognitive psychology can tell us based on the records whether these two inventors had the same mental models.

Gray's caveat featured a device in which a needle handing from a diaphragm vibrated near the bottom of a column of water as someone spoke into the diaphragm.

Gray's design depended on the depth of the needle in the water; Gray never built it, but was sure it would work. In experiments diagrammed in a problem-behavior graph, Bell compared sticking a bell in the water to having the point of a needle barely touching the surface, and concluded that the latter arrangement produced a better effect. Instead of putting a needle deep into the water, as Gray proposed, Bell's needle barely touched the surface. Therefore, Bell and Gray were operating from distinct mental models concerning the relationship between the contact and the water.

Bell's and Gray's overall mental models were also distinct. Gray's mental model for a speaking telegraph was a device called the "lover's telegraph," the old two tin cans connected by a string most of us built in childhood. Gray at one point briefly considered that the ear might be a useful analogy for the speaking telegraph, but he did not have Bell's detailed knowledge of the workings of the ear. Bell in fact had built a device that used the bones of the middle ear to trace sound waves, therefore his mental model was based on hands-on expertise.

The strength and weakness of Bell's mental model are immediately apparent. The ear is a great receiver for sound waves, but not a transmitter. Other inventors like Edison realized that Bell had solved the problem of receiving sound waves, but that his transmitter was far from optimal. Edison, in particular, developed a greatly improved transmitter.

Bell admitted earlier that he would have to be a theoretical inventor because his hands-on skills and resources were limited. Bell followed a vary one thing at a time (VOTAT) strategy in his experiments—vary one thing at a time. Edison, in contrast, was running the first independent research and department (R&D) laboratory at Menlo Park and could have had assistants to help him try multiple variations. When asked to describe his invention process in litigation over who deserved credit for inventing the telephone, he provided sketches that are still available for study. Some of the sketches were of relatively complete apparati, but most were little more than scribbled diagrams with occasional notes—the kind of thing Edison used to think about alternate designs and variations. As he said, "My sketches were rough ideas of how to carry out that which was necessary in my mind…" (Carlson, 2000, p. 11). This quote suggests that the sketches served as a space where Edison could work out variations on his mental models, including creating slots and imagining the kinds of devices or materials that might be used in them.

My analysis of Bell's invention process highlighted the utility of cognitive methods like problem-behavior graphs and concepts like mental models for analyzing and comparing historical records of inventors. But the comparison part was hindered by the fact that the Edison, and Gray, records did not include the kind of detail that was available for Bell. Edison's records were detailed, but very hard to decipher and interpret. Gray's were enough to infer his mental model, but not sufficient to construct more than a partial graph of his invention process.

Both Gray and Edison pursued more than one problem at a time— especially Edison, who was running the first R&D laboratory, and therefore could pursue multiple experiments. As Carlson wrote, "In the early months of 1877, [Edison] investigated several different classes of telephone (dragging, rubbing, switching, induction, and squeezing); in the summer months he broke his research on the carbon pressure telephone into several different areas; and in the autumn he supplemented work on the rubber tube telephone with studies of induction receivers, battery telephones and a few alternative lines" (2000, p. 155). Edison could select promising designs from among this network of related alternatives and mobilize the team to focus on a single problem, like the carbon pressure telephone. After extensive experimentation and mental modeling, he settled on a button that contained a mixture of carbon and rubber for the transmitter, resulting in greatly improved transmission over Bell's crude device.

The Invention of the Airplane

Two cognitive psychologists, Gary Bradshaw and Philip Johnson-Laird, conducted studies of the Wright brothers using different conceptual frameworks. Like Faraday and Bell, the Wrights left extensive records, including notebooks, patent applications, letters, and observations from those who saw or participated in their experiments. But what makes the Wright brothers a particularly important case study is that they were far ahead of any of their contemporaries. Once the Wrights achieved their historic short flight in Kitty

Hawk in 1903, they improved both the plane and their piloting abilities to the point where they could fly long distances, including loops. They stayed away from air shows because they wanted to patent their invention and sell it to the military. Gary Bradshaw argues that the Wright brothers' flying capabilities were not matched until 1909, after their patents had been available for 3 years (Bradshaw, 1992).

The telephone, in contrast, had several inventors who were close and could have invented a speaking telegraph—indeed, the argument about who deserves credit continues to this day. Not so with the airplane. Samuel Langley, supported by the Smithsonian, tried to argue that his plane could have flown at about the same time as the Wrights launched theirs, but Langley's plane crashed immediately on take-off into the Potomac and nearly killed the pilot.

Mental Models in the Invention of the Airplane and Telephone

Johnson-Laird attributes the success of the Wright brothers to mental models, specifically their ability to animate such representations to work out the flow of wind over an aircraft or to design a transmission system. They could use models in imaginative play constrained by their knowledge to come up with a novel way to truss wings. They could manipulate models in their reasoning to check the consequences of an assumption, to derive a counterexample to a claim, to find a set of possible explanations for inferior performance, or to diagnose a malfunction. And they were most adroit in using a model of one thing, such as bicycle, as an analogy for another, such as an aircraft (Johnson-Laird, 2005, p. 21).

Bell was a mental modeler par excellence, and it was the key to his patent application. Bell's ear mental model was grounded in the experience of actually building a device that used the bones of the ear to translate sound into waves. Similarly, the Wrights used mental models grounded in their experience and their mechanical skills were vastly superior to Bell's. They were bicycle builders and repairers. One unique element they gained from their bicycle experience was the knowledge that stability was not the key to aircraft design. Bicycles are not stable, yet they can be managed very effectively by a rider. Similarly, a pilot of an aircraft could learn to turn by leaning, as on a bicycle. Another aspect of their mental model came from the observation of birds. Wilbur noted that to turn, a vulture twists the rear edge of its right wing upward and the rear edge of its left downward. But how could one do this with wings? Wilbur added to the mental model when he picked up a box used to hold bicycle tires and noted that he could twist the ends in opposite directions—suggesting that a biplane design might allow a pilot to twist or warp the wings so as to turn.

But having a mental model is not enough. The Wrights were experimentalists par excellence. They found the coefficients of lift calculated by Otto Lilienthal were not consistent with their results, and recalculated them using a wind tunnel they built. To create a propeller, they made an analogy between its blades and wings, concluding that the blades should not be flat but cambered (Crouch, 1989).

Mental models can be both a strength and a weakness. Bell's ear was a great model for the receiver, but a poor model for a transmitter. The Wright brothers' wing-warping relied heavily on pilot skill; their airplane was very difficult to fly, and several of its pilots had fatal crashes. Orville Wright was himself almost killed in one; the passenger in front of him died. Wilbur alone could reliably fly the plane (Crouch, 1989). While the Wrights focused on patenting their design, others studied it and were able to make improvements, like adding ailerons to replace wing warping. Therefore, the Wrights fell behind the invention curve. But their airplane design, and the recalculation of curves of lift, was a breakthrough that enabled all others working on the problem.

Invention as Dual Space Search

Gary Bradshaw attributes the success of the Wright brothers to another cognitive capability: the ability to search two problem spaces. Klahr and Dunbar (1988) established the value of this approach for studying how people learned to program a remote-controlled vehicle called a Big Trak by searching in two spaces, one focused on possible hypotheses, another on experiments. Some participants preferred to work more in the hypothesis space, others in the experiment space; Klahr and Dunbar called the former "theorists" and the latter "experimenters" (Klahr & Dunbar, 1988). A theorist style was preferred by students with prior programming experience, which helped them come up with the right frame or mental model to guide their experiments.

Bradshaw extended this dual space approach to invention by proposing that the Wright brothers worked in design and function spaces, whereas many of their rivals preferred iterations in a design space. According to Bradshaw, the Wrights did a functional decomposition of the functions involved in flight, and conducted experiments to determine design parameters that would fit these functions, including building a wind tunnel and recalculating the existing tables of lift. This kind of systematic iteration between function and design reduces the need for testing multiple prototypes—the first prototype is likely to be close to the goal. Indeed, the Wrights built and tested only three gliders before they constructed the first airplane. Rival inventors like Samuel Langley, Octavio Chanute, and Otto Lilienthal tended to build gliders and, in Langley's case, an aircraft, and see how far and how long they flew, varying design parameters like the number and placement of wings without systematic functional analysis. When the Wrights' second glider failed in 1901, systematic experimentation revealed that the problem was with current measurements of the coefficient of lift. The brothers constructed a wind tunnel and used it to correct the coefficient of lift. While Bradshaw focused on design and function spaces, it is clear the brothers also worked in specific experimental spaces related to problems like lift—applying results to new designs.

Weber might have taken Bradshaw's approach and turned it into a heuristic: decompose an invention problem into design and function spaces. Bradshaw employed his dual space analysis again in a study of the Rocket

Boys (Bradshaw, 2005). In order to build their rocket, these boys could have decomposed the design space into a variety of more specific problem spaces, including alternatives for propellants, fins, casements, and nozzle geometries. Instead, they focused mainly on propellant alternatives and nozzle design, and for the other spaces they just launched different rockets. Each rocket was cheap to make, so in effect, they could conduct experiments by firing rockets and studying the remains to see what had happened. The exception was propellants, which they test-fired in a coal furnace. Bradshaw's study of the Rocket Boys is a reminder that inventors and scientists make choices about which problem spaces they focus on. These choices, in turn, affect the final outcome of the invention.

Was the Wrights' success due to mental models or problem decomposition? The answer is both. The Wrights were efficient, in part, because they could imagine and run different designs mentally. As Wilbur noted, "My imagination pictures things more vividly than my eyes" (Johnson-Laird, 2005, p. 8). In particular, the Wrights could concentrate their research on those places where their mental models were incomplete or were at variance with the data—as in their discovery that the tables of lift in use at the turn of the century were incorrect.

The problem with *Sub Species Historiae* studies is the unevenness of records that would correspond to a protocol taken at the time. Bell kept notebooks for only a part of his invention process, and Gray and Edison did not keep any notebooks. The Wright brothers fortunately left detailed records, but even those had gaps. So mental models have to be inferred from partial records, and problem-behavior graphs will have gaps.

In Vivo Studies of Working Inventors

Studies of modern working inventors potentially could be done without gaps. It is possible to observe and record the invention process, if an inventor or invention team is willing and restrictions on intellectual property are allowed. Christensen and Schunn were able to use this kind of in vivo observation of engineering designers at a medical plastics company, organized into teams focusing on different aspects of design (Christensen & Schunn, 2009). The authors protected the company's IP by not disclosing the name of the company or details regarding the products, which meant no problem-behavior graph was possible. The researchers observed team meetings and videotaped the interactions. The major focus was on mental simulations, which is the running of a mental model of a component or aspect of a design. One third of the task-related conversation segment were mental simulations, which occurred at a rate of about one every 2 minutes. Two thirds of these mental simulations included an initial representation, a simulation run and a result that could involve a change in the aspect of the design under consideration or a confirmation. Half of these initial representations were derived from sketches or prototypes, half were not based on any external representation, suggesting they were derived from mental models.

There were two main types of mental models observed in these in vivo meetings, those having to do with technical and functional aspects of a design, which is the space in which Bell and the Wright brothers mostly worked, and those having to do with end use, which the Wrights especially thought about and experimented on to make an aircraft capable of being flown. About two thirds of the mental simulations were technical and functional, and one third involved end use. Christensen and Schunn concluded that mental simulations did significantly reduce the uncertainty of aspects of the design, moving the team they selected for study toward solutions. But because the design details were proprietary, it is not possible to show the design paths taken by the team.

Christensen and Schunn used a similar in vivo methodology to do a study of analogical reasoning in design teams (Christensen & Christensen, 2007). Bell's mental model was based on an analogy with the human ear, and the Wrights got the idea of wing warping in part from watching the flight of birds. These were what are referred to as cross-domain analogies—from biological systems to mechanical systems. Christensen and Schunn found that 45% of the analogies used in one of the medical design teams they followed were between domain and 55% were within. This finding is in contrast to Dunbar's observation that molecular biology laboratories used few between-domain analogies (Dunbar, 1995). Perhaps engineering designers use more between-domain analogies? One cannot make that generalization from two in vivo studies done in different problem domains, but it is an interesting hypothesis for further work.

While Dunbar focused on laboratory meetings in his study of molecular biology laboratories, Nancy Nersessian and her colleagues adopted the methods of cognitive anthropology and became participant-observers of ongoing research and interactions among researchers in biomedical engineering laboratories, developing a coding scheme iteratively (Nersessian, 2005). They also studied the texts generated by the researchers. One major phenomenon Nersessian investigated parallels a theme of this chapter: the role of mental models and mental simulations. In biology and psychology, many questions that could be answered by in vitro research involve studies that are unethical. One possible solution is to rely more on mental models and thought experiments. But just like Bell and the Wrights, modern researchers have to complement mental simulation with actual prototyping and testing. Furthermore, unlike these early inventors, modern researchers have powerful computational modeling tools.

The biomedical engineering scientists in Nersessian's research constructed systems that were hybrids of computational models and physical devices, which allowed for experimentation and exploration. Nersessian's researchers characterized what they were doing as "putting a thought to the bench top and seeing if it will work" (Nersessian, 2005, p. 749). In effect, they created model systems that allow for control and manipulation of biological systems. For example, one researcher worked with colleagues over a 3-year period to develop a system that would allow her to test the effect of shear forces on endothelial cells drawn from a baboon outside of the organism. The development of this model system involved constant iteration that was a combination of mental simulation and physical experimentation. The end result had an organism, a baboon, connected to mechanical and electrical systems that measured characteristics of blood flowing out of the baboon and through these devices, then back into the baboon. The goal was to find out precisely how shear stress could be used to reduce platelet clotting in a system simpler than the actual organism, but generalizable to the in vivo setting.

These model systems allow for close interplay between mental, physical, and computational simulations to the point where all function as a distributed cognitive system. As Nersessian notes, "our data provide evidence that researcher representations are themselves model-like in structure—traditionally called 'mental models'. From the perspective of distributed cognition, researcher mental models and device model systems constitute distributed inferential systems through which candidate problem solutions are achieved" (Nersessian, 2005, p. 745). It would be possible to construct problem-behavior graphs that show at least some of the processes involved in model creation. This kind of distributed analysis merits fine-grained analysis. Nersessian also emphasizes shared cognition in her work. These model systems emerge in laboratories and involve interactions within and between teams. Here it might be worth comparing the kind of shared cognition in Christensen's teams with those in Nersessian's.

The model systems also serve as "hubs" linking cognitive and social practices in the laboratory: they represent the current understanding of a problem, link practices of multiple disciplines together in a system, and serve as sites of learning for new generations of scientists and for those from other laboratories who want more details on how the science is done.

CONCLUSIONS AND FUTURE RESEARCH

This brief survey of seminal work in the cognitive psychology of technological invention has focused particularly on mental models, heuristics, and techniques like problem-behavior graphs that allow one to track changes in problem representations. To this analysis was added the idea of dividing the search for solutions into two or more problem spaces containing different functions that had to be performed and different designs to match to the functions. This dual-space search is a classic example of a heuristic— the division into separate spaces can remind an inventor to iterate between designs and functions.

Models are not only mental. Nersessian's work reminds us that cognition is distributed among devices like experimental apparati and computational tools. Dunbar's work reminds us that cognition is also shared among those working on problems, leading sometimes to shared mental models, other times to important divergences that can fuel new experiments and designs.

Methods for studying invention could potentially include experiments, observational studies, computational simulations, and other approaches (Gorman, 2006). This handbook chapter particularly focused on two:

1. *Sub Species Historiae* work, which is limited by the available records— those inventors who kept detailed notebooks are the best objects of study. This was particularly evident in comparative studies of Bell, Gray, and

Edison, where much more detail was available on Bell's simple processes than the more complex experiment and design processes of Gray and Edison.

2. In vivo studies of current inventors and innovators. Work by Christensen and Schunn shows how design teams can be studied while protecting identities and intellectual property, but this approach has clear limitations—one cannot, for example, do a problem-behavior graph. Nersessian's deep engagement method holds greater promise because she was able to talk about specific experiments and projects. Distributed cognition can be mapped over time as Hutchins and others have shown.

What is needed is more fine-grained detail and analysis. It is possible to protocol an inventor or invention team as they work, if willing volunteers could be found. These protocols could be used as the basis for problem-behavior graphs. These real-time protocols could be complemented by the study of documents like notebooks, articles, patent applications, and interviews.

Technological innovation deserves the same amount of attention by psychologists as scientific discovery and the line between the two is often blurry—Nersessian's researchers referred to themselves as engineering scientists, because they were discovering and applying principles of discovery and innovation in a continuous, iterative fashion. It is hoped that this chapter will inspire future work.

NOTES

1. Bernard Carlson's forthcoming book on Tesla (*Ideal and Illusion: The Life and Inventions of Nikola Tesla*, Princeton University Press) quotes the inventor describing his process: "When I get an idea, I start right away to build it up in my mind. I change the structure, I make improvements, I experiment, I run the device in my mind. It is absolutely the same to me whether I operate my turbine in thought or test it actually in my shop. It makes no difference, the results are the same. In this way, you see, I can rapidly develop and perfect an invention, without touching anything. When I have gone so far that I have put into the device every possible improvement I can think of, that I can see no fault anywhere, I then construct this final product of my brain. Every time my devices works as I conceive it should and my experiment comes out exactly as I plan it." This statement makes it sound like Tesla gave precedence to mental models in his invention process.

2. Actually, Bell's father-in-law, Gardiner Hubbard, submitted it for him, probably because he knew Gray was in Washington and whenever Gray came to Washington, he took his latest ideas to the patent office (Gorman, 1998).

3. The story that Bell spilled acid on his pants and called Watson for help is apocryphal.

REFERENCES

Baker, L. M., & Dunbar, K. (2000). Experimental design heuristics for scientific discovery: The use of "baseline" and "known standard" controls. *International journal of human-computer studies, 53*(3), 335–349.

Bradshaw, G. (1992). The airplane and the logic of invention. In R. N. Giere (Ed.), *Cognitive models of science* (pp. 239–250). Minneapolis, MN: University of Minnesota Press.

Bradshaw, G. (2005). What's so hard about rocket science? Secrets the rocket boys knew. In M. E. Gorman, R. D. Tweney, D. C. Gooding, & A. Kincannon (Eds.), *Scientific and technological thinking* (pp. 259–275). Mahwah, NJ: Lawrence Erlbaum Associates.

Bruce, R. V. (1973). *Bell: Alexander Graham Bell and the conquest of solitude.* Boston, MA: Little, Brown.

Carlson, W. B. (2000). Invention and evolution: The case of Edison's sketches of the telephone. In J. Ziman (Ed.), *Technological innovation as an evolutionary process* (pp. 137–158). Cambridge, MA: Cambridge University Press.

Christensen, B. O. T., & Christensen, B. T. (2007). The relationship of analogical distance to analogical function and preinventive structure: The case of engineering design. *Memory & Cognition, 35,* 29–38.

Christensen, B. T., & Schunn, C. D. (2009).The role and impact of mental simulation in design. *Applied Cognitive Psychology, 23*(3), 327–344.

Crouch, T. (1989). *The bishop's boys: A life of Wilbur and Orville Wright.* New York, NY: W. W. Norton.

Dunbar, K. (1995). How scientists really reason: Scientific reasoning in real-world laboratories. In R. J. Sternberg & J. Davidson (Eds.), *The nature of insight* (pp. 365–396). Cambridge, MA: MIT Press.

Gooding, D. (1990a). *Experiment and the making of meaning: Human agency in scientific observation and experiment.* Dordrecht, Holland: Kluwer Academic Publishers.

Gooding, D. (1990b). Mapping experiment as a learning process: How the first electromagnetic motor was invented. *Science, Technology and Human Values, 15*(2), 165–201.

Gorman, M. E. (1998). *Transforming nature: Ethics, invention and design.* Boston, MA: Kluwer Academic Publishers.

Gorman, M. E. (2002). Types of knowledge and their roles in technology transfer. *The Journal of Technology Transfer, 27*(3), 219–231.

Gorman, M. E. (2006). Scientific and technological thinking. *Review of General Psychology, 10*(2), 113–129.

Gorman, M. E. (2008). Scientific and technological expertise. *Journal of Psychology of Science and Technology, 1*(1), 23–31.

Johnson-Laird, P. N. (2005). Flying bicycles: How the Wright brothers invented the airplane. *Mind & Society, 1,* 1–22.

Klahr, D., & Dunbar, K. (1988). Dual space search during scientific reasoning. *Cognitive Science, 12,* 1–48.

Nersessian, N. J. (2005). Interpreting scientific and engineering practices: Integrating the cognitive, social and cultural dimensions. In M. E. Gorman, R. D. Tweney, D. C. Gooding, & A. Kincannon (Eds.), *Scientific and technological thinking* (pp. 17–56). Mahwah, NJ: Lawrence Erlbaum Associates.

Tweney, R. D. (1985). Faraday's discovery of induction: A cognitive approach. In D. Gooding & F. James (Eds.), *Faraday rediscovered: Essays on the life and work of Michael Faraday: 1791–1867* (pp. 189–209). New York, NY: Stockton Press.

Tweney, R. D. (1989). A framework for the cognitive psychology of science. In B. Gholson, W. R. Shadish, R. A. Neimeyer, & A. C. Houts (Eds.), *Psychology of science* (pp. 342–366). Cambridge, MA: Cambridge University Press.

Weber, R. J., & Dixon, S. (1989). Invention and gain analysis. *Cognitive Psychology, 21,* 1–21.

Weber, R. J., & Perkins, D. N. (1989). How to invent artifacts and ideas. *New Ideas in Psychology, 7,* 49–72.

CHAPTER 16

The Psychology of Research Groups: Creativity and Performance

Sven Hemlin and Lisa Olsson

THE CREATIVE RESEARCH GROUP ENVIRONMENT

Group Climate

A psychosocial climate that research group members[1] perceive as favorable stimulates creativity in research (Amabile & Gryskiewicz, 1989; Anderson & West, 1998; Ekvall & Ryhammar, 1999; Hemlin, 2008, 2009). Such a climate is characterized by openness, respect, and harmony among the group members and by personal work autonomy. However, in analyzing work attitudes and social behaviors, researchers have mixed opinions about the effect of mood on creative outcomes. Isen (1987) concludes from a number of studies that a positive mood promotes creativity, whereas George and Zhou (2002) in one study find that a negative mood may promote creativity. Moreover, it is suggested that a research group may be more creative when there is some intellectual tension and competition among its members (Gulbrandsen, 2004; Hemlin, Allwood, & Martin, 2008). Nevertheless, excess tension and competition in a research group are clearly injurious to creativity; therefore, an important leadership task is to create a work climate where there is the right balance between harmony and disharmony.

Various creativity researchers have addressed the issue of group climate in their attempts to measure creativity and innovation. They conclude that a functional group (and organizational) climate—the perception of the psychosocial micro environment—explains much of the variance in the creativity and innovation in research and other work groups (Agrell & Gustafson, 1994; Amabile, 1996; Amabile, Conti, Coon, Lazenby, & Herron, 1996; Anderson & West, 1998). Using survey instruments, some researchers measure a number of factors; for example, leader support and encouragement, freedom to choose and work with different tasks, access to resources, work pressure (that may exert a negative influence if it exceeds a certain limit), and organizational deficiencies such as the lack of collaboration, the lack of communication, and the presence of fixed routines (Amabile, 1996).

Two survey instruments used to measure group climate are well known: Amabile's (1996) assessing the climate for creativity (KEYS) and Anderson and West's (1998) Team Climate Inventory (TCI). KEYS uses 10 scales to measure environmental factors that either stimulate or stifle creativity and productivity.[2] Anderson and West's TCI, which is frequently used by European researchers (e.g., Agrell & Gustafson, 1994), uses the following four (sometimes, five) dimensions to measure creativity in groups: (a) vision (group members' clear and common goal), (b) participation safety (group members' sense of safety in group discussions), (c) task orientation (group members' focus on excellence), and (d) support for innovation (group members' interest in and aptitude for innovative work). A fifth dimension, interaction frequency, is distinguished from the second dimension.

In a survey study of respondents in excellent and less excellent research groups,[3] Hemlin (2006a) found that the differences in their creativity were generally small, although he highlighted some phenomena that should be studied further. First, the respondents who were satisfied with working in groups appeared more creative and innovative than the respondents who were less content with the group climate. Second, the former group tended to be more supportive of each other, and third, they seemed to enjoy the continuous group contact more. These results suggest the importance of group processes to creativity.

While earlier research reached the same general conclusions (Agrell & Gustafson, 1994; Anderson & West, 1998), the original findings in Hemlin's studies (see also Hemlin 2008, 2009) are that (a) multidisciplinary, biotechnical research groups do not differ from monodisciplinary groups in the relationship between group climate and creativity and (b) the most creative group members exhibit strong group cohesion and group member identity. In his personal interviews with respondents, Hemlin found that group members thought the group climate should be one that promoted members' joy, thrill, relaxation, security, openness, personal freedom, and initiative. In addition, the group members stated that mutual respect, honesty, and common goals were conducive to a creative group climate.

In conclusion, it is clear that the research group's climate plays an important role in the group's creativity. Researchers recognize at least two instruments as valid for measuring research group climate as well as other organizational climates. In general, the findings from empirical studies agree that a creative group climate is one in which there is openness, respect, joy, and security among group members. Such a climate also gives members some work autonomy even as it creates some work tension. Institutional and organizational cultures that limit individuals' and groups' opportunities for developing creative climates typically have a strong and corrective aversion to risk, overly severe time pressures, unsuitable personnel constellations that create conflicts, and/or leadership that does not support harmonious group climates (Amabile, 1996).

Group Size

A research group, often consisting of five to eight members, may vary in size from two to three members to 15 to 20. Currently, the smaller groups are typically in the social sciences and humanities with larger groups typically in the sciences. In most of the social sciences and the humanities, researchers traditionally publish single-author papers and monographs. Thus, research groups in the social sciences and the humanities are not as common as in the so-called "hard" sciences or in two of the social sciences (economics and psychology) where group research, for instance, is reported more commonly in multiauthored papers.

Two literature reviews of empirical studies on the relationship between group/organization size and individual/group productivity (or creative products) in science (Hicks & Skea, 1989; King & Anderson, 1995) indicate that the size of a research group or organization has no significant relationship to productivity. In addition, other studies in the humanities (Hemlin, 2006b) and in science (Martin & Skea, 1992) support this same finding. Moreover, in a study of microbiology research, Seglen and Aksnes (2000) examined a large sample of publications from the years 1992 to 1996. They found no association between the size of the research groups and the number of publications per researcher.

However, in their literature review, Bland and Ruffin (1992) claim that there is a general relationship between group size and productivity in research. A Swedish study in the engineering sciences also found a correlation between group size and individual productivity and literature citations; as the group size increased, so did individual productivity and the number of citations (Wallmark, Eckerstein, Langered, & Holmqvist, 1973).

According to Dundar and Lewis (1998), studies in universities have shown that an association appears to exist between department size and the research productivity of the department (i.e., measured by the number of publications). They account for this association by the correlation between resources and departments—larger departments with more resources publish more. However, Hemlin's (2006b) study in the Humanities Departments in Swedish universities, which focused on research published and doctoral degrees awarded, revealed no correlation between department size and productivity.

Despite these different and opposing findings, there is general support for the idea that the relationship between group size and productivity in work groups is curvilinear. Researchers who hold this opinion believe that the most productive work groups are neither too small nor too large. However, as Cohen and Bailey (1997) note, it is unclear if this relationship measures individual or group productivity. Complicating the analysis is their important point that the composition of groups (i.e., the number of persons in a group) is most strongly influenced by the nature of the group tasks.

In conclusion, researchers disagree on the relationship between research group size and productivity. Their findings are contradictory, possibly because they use different output measures and different aggregate levels and because they study different disciplines. In any case, if we assume, for the sake of argument, that the size of research groups has no significant influence on creative output, one may then ask why larger creative environments—with larger research groups, larger departments, and more researchers—have received greater recognition in the scientific community. For example, Swedish science councils, research foundations, and university boards currently support large "centers of excellence" through the award of Linneaus grants and other long-term research grants.

There may be several reasons for such financial support of large research groups. First, there is the idea of "big science"—certain research efforts require significant resources for the purchase of instruments, the development of technical solutions, and the support of large numbers of researchers. An example of such a research group is European Organization for Nuclear Research (CERN). Second, there is a belief that larger group size results in improved scientific research. The analogy is made to the business concept of "economies of size" where profit results only after a large number of products are produced. Thus, larger research groups and research organizations (e.g., large departments and large universities) produce better and more creative research. It is possible that rationalization cost benefits will be found in larger units, but this result is not linked to creativity. Third, it is argued that large research groups, because of the number of researchers' personal contacts, create an environment that is conducive to creativity. Larger research groups may have more external contacts and more outside stimuli.

In reviewing this mixed evidence and opinion on the relationship in research between group size and creative output, our conclusion is that size is not very important to creativity. Rather, it is important that there is the opportunity for making personal contacts by, for example, information and communication technology (ICT) and by attendance at conferences and meetings.

Knowledge Management

There are both proponents and opponents of the idea that creativity relies on extensive knowledge in a domain. Csikszentmihalyi (1999) argues that knowledge is needed for creativity to flourish. Amabile (1996) claims that knowledge is one of the three main components of creativity (the other two components are intrinsic motivation and creative skills). However, some researchers disagree. For example, Sternberg (2006) argues that knowledge may stifle creativity by locking in new ideas. In other words, extensive established knowledge may be detrimental to creativity. The argument is also made that in problem solving, people who rely solely on established methods and routines (e.g., heuristics) may perceive only one way to use a tool, for instance, a hammer, while the solution to the problem may be that the hammer is used in a new way. Yet Boden (1994) argues that creative problem solving is an act that is always performed within frames and constraints. Without such limitations, she argues, there would be only random processes and probably no creative results.

Among those who argue that extensive knowledge in a domain is necessary for creativity, the discussion often turns to how people make associations among ideas, theories, and other phenomena. For instance, Ziman (1987) believes unusual knowledge may stimulate creative and metaphorical thinking. According to Simonton (2004), flat association hierarchies promote individual creativity more than steep ones. Thus, in making flat associations, the individual may make connections to other fields by combining diverse knowledge; in steep associations, the individual may make a narrower search in seeking links to relevant knowledge (Simonton, 2003).

Some research suggests that ICT supports creativity by providing quick and easy access to new knowledge. Based on a study of ICT and creativity in organizations, Dewett (2003) claims that ICT accelerates creative research efforts. Hemlin's (2008) research supports the view that ICT promotes creativity in biotechnical research.

Other scholars argue that researchers from multiple fields or disciplines—with knowledge from disparate fields—should be involved in the research on certain challenging problems. For example, in a large-scale study of researchers in U.S. biomedicine, Hollingsworth and Hollingsworth (2000) found that one significant explanation for the success in research groups was their multidisciplinarity.

In conclusion, with some exceptions, the consensus in the literature is that extensive domain knowledge is a prerequisite for creative research. This opinion is most strongly held for areas where knowledge is developing rapidly (e.g., biotechnology). Although there is still a discussion on whether creativity benefits from multi- or cross-disciplinary science research, when several researchers work on the same problem, we believe a more useful discussion should be on how problems and ideas are generated and phrased (Hemlin et al., 2008).

Motivation and Incentives

The literature describes a number of motivational incentives for individual researchers working alone and for researchers working in groups. Among these incentives, an important one is recognition by the academic community (Hagstrom, 1965; Merton, 1973). Researchers tend to be more creative in research environments where the opportunity for recognition exists. Typically, such recognition is achieved through publications in prestigious journals and by citations to publications indexed in the web of science and other databases. (A measure of journal prestige, the impact factor, is based on the average number of citations that journal articles per author[s] receive each year.) In addition, other benefits flow from a record of good publications. These include academic appointments, including full professorships and chairs, membership in influential societies (e.g., for scientists, the Royal Society and the National Academy of Sciences), and various awards and prizes, including the most prestigious award, the Swedish Nobel Prize for achievement in the sciences and in literature.

There have been various studies of the motivation and incentives for researchers. For example, in her recent doctoral dissertation, van der Weijden (2007) analyzed performance indicators for Dutch research groups in biomedicine and the caring sciences. She found that these researchers are motivated by praise from their superiors, scientific prestige, salaries, career opportunities, international contacts, and creative climates.

In a study of organizational control systems for university researchers, using peer-reviewed publications as the measure, Omta and de Leeuw (1997) identified four factors that influence research performance: (a) communications with research funding agencies (15% of the explained variance), (b) human resource management (e.g., the salary system, career possibilities, scientific recognition—10%), (c) administrative leadership (e.g., resource control—8%), and (d) international contacts (e.g., the number of conference participations—4%). The results of this study suggest that organizational and control systems influence research performance and, to some extent, creativity. Although they found that the researchers ranked the rewards system highly, Omta and de Leeuw (1997) concluded that research financing was the researchers' most significant incentive.

It is also of interest that in the Omta and de Leeuw study, university researchers ranked international contacts lowest of the four motivational factors. In contrast, in the aforementioned study, van der Weijden found that industry researchers attended numerous international meetings. The explanation of this difference may be that the resources available for travel to international meetings are relatively limited in universities. However, while generally there is more travel money available in business than in universities, this is not always the case in smaller industry groups. In their study of research group members, Hemlin and Olsson (2011) found that university researchers, more than industry researchers, emphasized the importance of travel abroad to their creativity.

In conclusion, individuals and groups in creative research are motivated by the possibility of gaining recognition for their work and by various other rewards and benefits. Researchers who seek recognition by their peers need to work in a creative research environment if they are to achieve their objectives.

Resources

Obviously research, especially scientific research, requires a significant amount of financing and other resources. In general, the major costs of scientific research are the salaries of the researchers/support staff and the outlay for facilities and equipment. However, in recent years, the cost of administrative overhead has grown significantly. Ziman (2000) writes about the bureaucratization of academic research in the post Second World War era and the resulting problem for researchers who are dependent on external grants. Today, researchers who make grant applications are required to apply for administrative overhead costs in addition to researcher and support staff salaries, equipment, and materials.[4]

Although it is often stated that more financial resources lead to research that is of higher quality and more originality, there is an ongoing discussion as to whether this is a valid assumption. Generally, the more researchers and doctoral students at a university or a research organization, the more publications and dissertations will be published. However, the association between these resources and creative research results is not conclusive. In a U.S. study of 1,300 4-year colleges and universities, a strong correlation was found between the amount of research resources and the number of publications by all faculty and doctoral students (0.93) and between the amount of research resources and the number of publications by full-time researchers (0.78) (Toutkoushian, Porter, Danielson, & Hollis, 2003). In similar studies in the humanities in Sweden, the correlations were lower (0.45–0.77) (Hemlin,

2006b). It appears that creative research is not linearly related to how much money is provided (Hemlin, 2006b; Pelz & Andrews, 1966).

The availability and design of physical resources in research environments seem to influence creativity less than the financial resources discussed above. An exception may be how research facilities with high configurational accessibility and shorter walking distances, for example, positively influence the chances for contacts among researchers and thus their innovation results (see Toker & Gray, 2008). We argue that researchers need common areas that encourage contacts as well as private spaces for reflection on ideas and analyses of other researchers' work (Hemlin, Allwood, & Martin, 2004; Hemlin et al., 2008).

Of course, in the natural, medical, and technical sciences—particularly in "big science"—laboratories, technical equipment, and new technologies are required resources for research (Hemlin, 2008). In some instances, the development of technical instruments in itself is the creative research result.

In conclusion, various research resources—especially financing—are necessary for conducting creative research. The importance of other resources (e.g., physical facilities and equipment) varies depending on the research field.

Social Networks

A research group cannot function creatively in the absence of contacts and networks in the scientific or academic communities. Contacts are made with fellow researchers in the internal research environment, and networks are formed with external researchers across the globe. In a review of 13 teams—primarily research and development (R&D) groups in the high tech industry who were engaged in knowledge-intensive research—Cohen and Bailey (1997) found that external communication was positively related to group performance. They also found that managers, who assess project groups' innovative work in creating new products and solving problems, require their members and leaders to establish and maintain frequent contacts with experts and with other groups doing similar work.

Without a good network of researchers and access to outside expertise, researchers are unlikely to produce the most creative ideas and solutions. Kasperson (1978) reached this conclusion in a study that showed that researchers' and engineers' access to information sources was the most important influence on their creativity. In a more recent study at a pharmaceutical company, Sundgren, Dimenäs, Gustafsson, and Selart (2003) found that information exchange with people outside the group influenced group members' intrinsic motivation and organizational learning positively. Moreover, they found that intrinsic motivation and organizational learning were related to the creative climate of the organization.

According to the empirical evidence on social networks and social capital, senior managers rate industry managers with large and substantive networks more highly than industry managers who lack those networks. Industry managers in the first group also have more fast-track careers and higher salaries (Burt, Hogarth, & Michaud, 2000). The research suggests that research leaders also have a positive view of extensive social networks. For example, Crane (1969)—in her test of the "invisible college" hypothesis—and Cole, Rubin & Cole (1977)—in their examination of the "old boys' network" hypothesis—support this idea. The conclusion is that the creative environment of a research group depends on both the leaders' and the members' networks of contacts.

We conclude that a creative research environment for a research group primarily depends on a large and strong researcher network created by the group leader and by the group members. Such a network opens the gate to research contacts in the research groups' area(s) of interest.

Personal Interaction

In his book on the history and sociology of intellectuals, Randall Collins (1998) describes the importance of face-to-face communication. According to Collins, even in early times, philosophers, despite long distances, travelled (by horse and carriage) to meet personally. Collins calls these meetings interaction rituals that must take place in a personal way if scientific contacts and communications are to be most fruitful. Hemlin (2008) confirms this finding in his study of researchers in biotechnology in which he found that many and frequent personal contacts in the field are important to creative research. This study looked at domestic personal contacts with co-researchers and peers as well as foreign contacts with researchers encountered at meetings and conferences.

There is support for the hypothesis that researchers who make contacts with other researchers in their own and related disciplines through scientific meetings and research visits abroad are more creative and productive than researchers who do not. Kyvik and Larsen (1994) tested this hypothesis in a large empirical study in various academic fields (arts and humanities, social sciences, natural sciences, medical sciences, and technology). In their research, they used data from employed researchers who had graduated from one of the four universities in Norway in 1992. For output measures, they used scientific journal articles, book chapters, conference proceedings contributions, scientific monographs, and research reports. The study showed that there is a relatively strong relation between researchers' international, and their publication activity. Several conclusions may be drawn from this review of the research on the relationship between the creative research environment and international contacts, international meetings, and international research visits. First, international publishing productivity is higher among those researchers who attend international conferences and visit foreign colleagues because they are spurred by international research. Second, researchers, irrespective of their disciplines, benefit from international contacts. If the number of international publications (i.e., peer-reviewed articles) is a measure of creativity, then international contacts appear to stimulate creativity. Third, researchers who have many foreign contacts can demonstrate the value of international contacts to other members in their research groups (e.g., junior researchers and doctoral students). In conclusion, personal interaction with other researchers, including travel, conference participation, and research visits abroad, are extremely valuable for promoting a creative research environment. We believe a research community needs to allocate financial resources for such activities.

LEADERSHIP AND CREATIVITY IN RESEARCH GROUPS

In this section, we examine leadership and creativity in research groups. For some years, research into the psychology of research groups has focused on group climates and group leaders. Nearly 50 years ago, Pelz and Andrews (1966) studied leaders and leadership in their book on scientific research groups' productivity. Denti and Hemlin's (2010) review of leadership and innovation research from the last 30 years reveals that empirical studies have established a relationship between leadership and creativity and innovation, mediated and moderated by various components. Research group leaders are, therefore, in a position that allows them to promote group members' creativity. The moderators and mediators that influence this relationship include the seven factors previously reported as relevant to the research group environment in this chapter: group climate, group size, knowledge management, motivation and incentives, resources, social networks, and personal interaction (see "The Creative Research Group Environment" section). In this section, we widen the focus with our claim that group members, as well as leaders, make up leadership in a research group. Leadership, we propose, is a relational phenomenon. Therefore, we also look at the leadership role played by research group members as that role relates to the creativity of the group.

As described above, a creative knowledge environment consists of the social, physical, and cognitive factors that influence individuals, groups, and organizations to excel creatively (Hemlin et al., 2008). Leaders can encourage, motivate, and stimulate creativity in individuals, in groups, and even in organizations. In groups, both the group leader and the group members contribute to the creative environment, but the group leader is formally responsible for exercising a positive influence on it. Thus, group leaders, unlike group members, have responsibility for several factors in the creative knowledge environment, such as the group climate, composition of groups, the distribution of resources, and the allocation of time (Hemlin et al., 2008). Finally, group leaders influence group members' creativity by the manner in which they communicate with colleagues in the organization (Balkundi & Kilduff, 2005) and with external resource providers.

There are numerous studies on both creativity and leadership as separate topics, but leadership is rarely studied as a creativity driver (Elkins & Keller, 2003; Mumford, Scott, Gaddis, & Strange, 2002). At this point, we define more specifically what we mean by leadership and creativity. We define leadership as the influence process exerted by an individual over a group of individuals as they strive toward a common goal (Strannegård & Jönsson, 2009). Consistent with the definition that several other researchers have used, we define creativity as the ability to generate ideas or products that are novel, have good quality, and are useful (Amabile, 1996; Hemlin et al., 2008; Mumford & Gustafson, 1988).

A promising theory to use in leadership research is leader-member exchange (LMX) theory, which is based on social exchanges. In LMX theory, leadership is viewed as the relationship that develops between leaders and group members (e.g., Schriesheim, Castro, & Cogliser, 1999). Leaders are thought to develop unique relationship to each group member within their work group. Each leader-member unit is called a dyad and the quality of the relationship that develops within each dyad is thought predictive of desirable individual and organizational outcomes. We use this theory in one of our studies of research groups that we summarize later in this chapter.

These studies are two empirical studies we recently conducted on leadership and creativity in academic and commercial, biomedical and biotechnical research settings.[5] These studies contribute to psychological research on leadership and creativity in research groups in two ways: (a) by their use of a multilevel analysis that allows simultaneous comparison of the effects of individual and group level LMX on individual creativity; and (b) by their identification of member reported leader-related situations and behaviors that influence group member creativity.

Research Group Leadership

In the large body of leadership literature, only a few studies have treated leadership in research environments. In a review of such studies, Elkins and Keller (2003) offer four propositions on leadership in R&D groups. First, effective research leaders should lead the group internally at the same time as they actively interact with external contacts. Second, transformational leadership[6] is positively related to project effectiveness in R&D organizations. Third, less-than-high-quality leader–member relationships will not contribute to project effectiveness. Fourth, leaders who show consideration toward group members should be more important for group member satisfaction and performance in research projects than in development projects where tasks tend to be more structured and less varied.

LMX theory is based on social exchange theory (e.g., Blau, 1964) with a focus on the dyadic relationship between leaders and group members. A number of positive outcomes are assumed to result from the quality of the dyadic exchanges between a leader and a group member that are at the core of LMX theory. For example, in a meta-analytical study of 164 LMX studies in different settings, Gerstner and Day (1997) associated LMX quality with members' job performance, satisfaction, and commitment. The effects of LMX on leaders' output has rarely been examined (Wilson, Sin, & Conlon, 2010).

Researchers have also emphasized the mutuality of the leader–member relationship. They suggest that *parallel* leader and member ratings of LMX (i.e., ratings that independently target both leader and member exchange activity) rather than *mirrored* ratings (i.e., leader and member ratings that target only the member exchange activity) provide a more trustworthy view of what leader–member relationships are (Greguras & Ford, 2006).

The mutuality in leader-member relationships may be particularly important in the research setting since the decisions of a leader may be dependent on the unique and perhaps nonredundant expertise of each research group member. Therefore, in a sense all members of a research group participate in the construction of leadership. Therefore an approach to leadership that understands leadership as a relational phenomenon (as in LMX research) or explains interpersonal relationships (as in social network research), is suitable for studying leadership in a research setting. In the research environment, high-quality interaction between group members and leaders (as well as the interaction between group members, and between members and researchers external to the group) seems essential if creativity is to flourish and innovative results are to be achieved.

According to Mumford et al. (2002), leadership in creative work, such as in the research environment, differs from leadership in noncreative work because of the special traits of the people engaged in creative work and the exceptional nature of their work. Because creative people at work are said to be especially independent and highly motivated, leaders of creative people (e.g., researchers) need to provide them with intellectual stimulation, grant them the freedom to explore, and protect them from unrelated work. Since creative work is an uncertain venture, creativity-stimulating leaders cannot rely on routine action plans. Mumford et al. (2002) recommend an integrative leadership style that coordinates expertise, people, and relationships during the different phases of creative work.

Leaders of creative people must have expertise relevant to their domain(s) that they can use to solve problems and coordinate the competences of group members (e.g., Amabile, 1996; Mumford et al., 2002). Studies conducted in experimental settings suggest that such leaders need cognitive problem-solving skills (Mumford, Connelly, & Gaddis, 2003) and creative thinking skills (Jaussi & Dionne, 2003) so that they can evaluate and promote group members' creative ideas.

A critical aspect of creative work is idea evaluation. According to Mumford et al. (2003), most models of creative thought assume that leaders evaluate ideas after employees produce them. However, based on their analyses of experimental and case study findings, Mumford and his coauthors argue that creativity-stimulating leaders contribute to and participate in the creative activity, not only as passive supporters but also as active collaborators. Therefore, evaluations by leaders, based on their practical expertise and focused on the organization's mission, can stimulate idea generation and advance creative thought among group members. Thus, leaders and group members who work together creatively should produce more creative results. Such research leaders, who act out a role that is more collaborative than managerial, are likely to be comfortable with a Scandinavian management style in which group members are trusted with high degrees of delegated responsibility (Smith, Andersen, Ekelund, Graversen, & Ropo, 2003).

Creative work also requires leader support (Amabile, Schatzel, Moneta, & Kramer, 2004). A supportive leader monitors progress efficiently and fairly, consults with group members on important decisions, provides emotional support to group members, and recognizes their good performance (Amabile et al., 2004). Additionally, a supportive leader fosters group members' creativity by proposing new ideas and by acting as a role model (Jaussi & Dionne, 2003).

There are limits, however, to the extent to which leaders can influence group members' performance. Contingency theories of leadership highlight the importance of taking these limitations into account when studying leadership effectiveness (Yukl, 2005). Some theories, such as leadership substitutes theory or critical leadership theories, deemphasize or challenge the very necessity of leadership (Yukl, 2005). Such perspectives are worthy of consideration, especially in the context of creative people and creative work.

The consensus is that creativity in the research environment is inherently difficult to manage. This has to do with the uncertainty inherent in discovery and scientific research, but may also have to do with the nature of scientists. Feist (2006) claims that scientists are more autonomous, driven, and ambitious than nonscientists. Amabile and Gryskiewicz (1987) conclude that knowledge workers typically place a very high value on work autonomy. However, Mumford et al. (2002) argue the management problems that arise from leading autonomous people should not be interpreted to mean that leadership is redundant or useless in research groups. Rather, such problems simply mean that the research environment poses specific and challenging managerial demands. In biomedical research, for example, the successful development of a pharmaceutical drug depends on the contributions by many individuals with different competencies. Therefore, research group leaders need to "orchestrate" the various competencies of the group members (Mumford et al., 2002) without restricting their individual freedom and professional autonomy. We agree, as Hackman and Wageman (2007) propose, that it is more constructive to ask when and how leadership works effectively than to ask if leadership is needed.

Creativity in Research Groups

According to Chen and Kaufmann (2008), there are four research approaches to creativity in an R&D context. First, in the evolutionary approach, researchers study creativity as a process of blind idea generation followed by a selection from the ideas generated. Second, in the cross-disciplinary approach, researchers study creativity by combining several disciplines and taking different backgrounds, experiences, skills, knowledge bases, and cultures into account. Third, in the social system approach, researchers study creativity as a systemic rather than an individual phenomenon. Therefore, the study of creativity should address the interactions at individual, team/group, and organizational levels (or of persons, domains, and fields). Fourth, in the social network approach, researchers study creativity by examining interpersonal exchanges and interactions.

Creativity is the creation of something original or novel and at the same time appropriate or useful (Amabile, 1996; Hemlin et al., 2008; Mumford & Gustafson, 1988). When defining creativity in the workplace it is common to view creativity as an outcome, a product (Amabile, 1996).[7] In R&D such products can be patents or scientific publications. Bland and Ruffin's (1992) literature review of studies in the productive research environment reveals that beginning with Pelz and Andrews's (1966) classic study of creativity and productivity in research group environments, followed by similar studies (e.g., Andrews, 1979; Stankiewiecz, 1980), researchers have generally discussed research group leadership in terms of its decisive influence on the productivity of groups.

However, it is sometimes charged that research output (e.g., number of publications) is a measure of productivity, not creativity. A controversy has arisen because of the tendency to allocate resources and award funding based on a researcher's or research group's number of publications or citations. Yet there is a substantial overlap between research productivity and creativity if the output measure is the number of articles published in prestigious, international scientific journals. Since one acceptance criterion for manuscripts submitted to such peer-reviewed journals is originality (i.e., creativity), the measure is not necessarily invalid.

It is often recommended that creative results should be evaluated using both objective and subjective measures (Amabile, 1996). In empirical R&D studies, in addition to the number of publications and citations, several other creativity indicators are used to assess creativity (e.g., patents, self-reported or supervisor-assessed creativity, and creativity tests). In addition to KEYS and TCI (see Group Climate), Ekvall's (1996) Creativity Climate Questionnaire (CCQ) measures creative climate, and the Kirton Adaptation-Innovation inventory (KAI) (see, e.g., Scott & Bruce, 1994) measures the innovative styles of individuals.

Typically, scholars conduct research group studies in industrial R&D departments and in governmental research institutes where the focus is on the development of research findings (Pelz & Andrews, 1966). There have been fewer studies of university research group creativity and even fewer that compare creativity between academic and commercial groups (Hemlin et al., 2008). To fill this research gap, we made two comparative studies in which we studied creativity-stimulating leadership in academic and commercial research groups working in the same research fields (biomedicine and biotechnology). In the first study, using a survey study, we examined the relationship between leadership and creativity; in the second study, using the critical incident technique, we identified creativity-stimulating leader behavior that the research group members reported. Following the description of these studies in the next sections, we draw conclusions about leadership based on our findings.

Study I: The Leadership and Creativity Relationship in Research Groups

In a survey study, we investigated the relationship between LMX and creativity in selected biomedical and biotechnical research groups in Sweden (Olsson, Hemlin, & Pousette, 2012). For the measure of creativity, we used the number of publications by group members under their current leader. Also, for each group member we measured their leaders' number of publications achieved during the same time period. In the study, we examined 137 leader–member dyads from 54 (30 academic and 24 commercial) research groups. (A dyad consists of the group leader and a group member.) We sampled two or three group members from each research group. The leaders used the 12-item multidimensional LMX scale (SLMX-MDM)

(Greguras & Ford, 2006), and the group members used the LMX-MDM scale (Liden & Maslyn, 1998). In responding to the survey questions, both leaders and group members offered their opinions on the LMX sub-dimensions of *affect, loyalty, contribution,* and *professional respect.* We conducted a multilevel analysis since participants were nested in groups (i.e., the data violated the statistical assumption of independent observations). Moreover, a multilevel analysis allowed us to investigate whether the relationship between LMX and group member or leader publications existed on the dyadic level (as suggested by LMX theory), on the group level, or on both levels.

Our results showed that LMX positively predicted creativity in university groups. Leaders' average ratings of LMX within groups (i.e., the leader-rated group level LMX) predicted the number of group member publications and tended to predict leaders' publications. However, our results showed that the leader-rated dyadic level LMX was unrelated to the number of publications. The conclusion is that membership in a group where the group leader, on average, rated LMX highly is conducive to creativity in academia. In commercial research groups, however, the associations between LMX and publications were negative. Leader-rated LMX at the dyadic level negatively predicted group member publications, and member-rated LMX at the group level negatively predicted group member publications. A possible conclusion is that high member-rated LMX is conducive to creativity in university settings but not in commercial research settings since scientific publishing is a more sought after outcome in academic than in commercial settings. Although group members' ratings of LMX at the dyadic level tended to predict leader, publications are a less-than-perfect indicator of creativity in commercial setting.

With respect to the LMX subdimensions, leader-rated *loyalty* predicted leader publications and leader-rated *contribution* predicted group member publications in academia. Therefore, leaders' individual creativity in university research groups flourishes when leaders perceive their follower as loyal. A follower's individual creativity flourishes when the leader reports being willing to contribute to the fulfilment of that follower's work. In commercial research groups, leader publications were negatively predicted by leader ratings of *affect* and *professional respect,* but positively predicted by group member-rated *loyalty.* Group member publications were negatively predicted by member-rated *loyalty* and member-rated *professional respect.* This means that leaders who like working with their followers and have professional respect for them had published less in the commercial setting. Group members in commercial research groups had published less when they perceived their leaders as loyal and had high professional respect for them. The only instance where an LMX subdimension positively predicted publications in the commercial setting was when group members rated leaders as loyal. In this instance, leaders had published more.

The overall tendency, we conclude is that LMX is positively associated with creative performance (measured as publications) in academic research groups but negatively in commercial research groups. Moreover, this study concludes that the dyadic exchanges between a leader and a group member are nested in a higher order structure, namely, the work group.

Study II: Research Group Members' Perception of Their Leaders' Creativity-Stimulating Behavior

Using the method of critical incident technique, we interviewed a subsample from the group members in our first study. We asked each group member to recall a specific incident in which his/her leader had stimulated the group's creativity or his/her own creativity. Of the 93 group members interviewed, 75 reported at least one creative incident involving their group leader. When the member had recalled such an incident, we next asked him/her to describe the incident, the leader's behavior, and the reason for the importance of the incident.

Using the procedure of content analysis procedure, we analyzed and categorized 153 reported critical incidents (Miles & Huberman, 1994) based on their meaning content. The main categories we used were the situation, the leader behavior, and the reasons for incident importance. Then we subdivided the reported leader behaviors into the following six subcategories: Provide expertise, Coordinate group research, Assign tasks, Support group conditions, Enhance external contacts, and Promote independence. Using these codings that were confirmed by an independent judge, we calculated kappa statistics that yielded satisfactory agreement.

In summary form, we next describe the most significant findings from the study. The most frequently reported creativity-stimulating leader behavior was the leader's ability to "Provide expertise"—specifically, the leader's ability to generate ideas and evaluate group/individual progress. For the behavior, "Coordinate group research," the group members' reports indicated that leaders could promote greater creativity by making plans and holding meetings rather than by relying only on spontaneous interactions with group members and with other researchers. The group members described the behavior, "Assign tasks," as creativity stimulating. We conclude that adding more structure provided greater focus for their work. The group members said the behavior, "Support group conditions," had a favorable influence on their creativity. The conditions for creativity appeared to improve when research group leaders provided both professional encouragement and rewards.

In summary, there are three main findings from Study II. First, group members in research can identify leader-related creative incidents. Second, the identified creativity-stimulating leader behaviors support the argument that leader expertise and support, directed at individuals or groups, stimulate creativity (see Amabile et al., 2004; Elkins & Keller, 2003; Mumford et al., 2002). Third, the greater part of the leaders' behaviors was change-oriented in accordance with Yukl, Gordon, and Taber's (2002) taxonomy of general leader behaviors. Change-oriented behaviors indicate that leaders are inclined to support creativity.

CONCLUSIONS

The primary objective of this chapter is to propose a better psychological understanding of the creative process in research groups. To that end, we reviewed the literature that discusses eight factors (group climate, group size, knowledge management, motivation and incentives, resources, social networks, personal interaction, and group leadership) that are assumed to contribute to the creativity of research groups. Of these eight factors, our main focus was on group leadership. In the discussion of these factors, we showed how research group conditions, relationships and exchanges between group members influence creative performance.

Perhaps, the most important feature of a research group vis à vis creativity is its creativity-stimulating climate. To some extent, group leaders—often such leaders are the senior scientists who select the group members—are responsible for creating such climates. Group leaders should choose researchers whose individual skills and education contribute to a good mixture of the knowledge bases needed to achieve overall group goals. This selection task is a critically important responsibility since there must be a fit among the competencies and personalities of the group members. However, once selected, it is the responsibility of the group members to establish interpersonal relationships and to interact with each other as well as with their leader. Ultimately, creativity in the group depends to a great extent upon individual efforts and the climate that arises from such relationships and interactions.

An open, secure, joyful, and communicative group climate that strikes a balance between collaboration and competition appears to be particularly fruitful in research groups—both in industry and academia. An emphasis on a common group goal is also beneficial to creative outcomes. It should be noted, however, to a certain degree, group climate depends on the organizational context such as found in a university department. Certain aspects of the group climate are often outside the influence of the group or its leader. For example, university department research group leaders and members may be unable to influence the hiring and salary policies that the central administration controls.

As our two studies show, leader–member relationships are influential as far as research group creative processes and creative results. We claim that leaders who stimulate creativity in research groups do so by demonstrating their expertise and by exercising their social skills. Moreover, our results point to the usefulness of applying a multilevel approach when studying leadership and creativity that are both individual and group level phenomena. Since there are important differences in the academic and industrial organizational contexts, it is necessary to develop creative output measures that adequately reflect creativity in both settings. However, we conclude that academic and commercial settings display many similarities when it comes to leadership and creativity.

Implications

Research group leaders who understand the creative process as it relates to individuals, groups, and organizations are better prepared to promote and support the discovery of new ideas. Because leaders are key figures in a creative knowledge environment, they should pay attention to the physical, cognitive, and social factors at work that affect their group members. However, research group leaders have no check-the-boxes creativity checklists. Thus, since the conditions that promote creativity are unclear, there are limits to what a group leader can achieve. What is clear, however, is that the research group leader's expertise is the sine qua non of group members' creativity.

Future Research

Although much is known about research group psychology, researchers have still pointed to the need for an integrative study of leadership that examines leaders, group members, and relational (leader–member, member–member) and contextual variables (Avolio, 2007; Graen & Uhl-Bien, 1995; Hackman & Wageman, 2007). We also think that the topic of how relationships in complex networks within and between research groups affect the individual researcher is worthy of future research. Creativity researchers have also called for more integrative studies since multiple factors interact when creativity appears (Amabile, 1996; Mumford & Gustafson, 1988; Sternberg & Lubart, 1999). We also suggest that future researchers use a multilevel analysis that incorporates dispositional and contextual variables at individual and group levels in order to test more advanced models of creativity in research. We call for the identification of boundary conditions under which leadership is effective/ineffective in stimulating creativity. Finally, we recommend conducting longitudinal studies of research groups that examine the relationships between leaders and group members over lengthy time intervals.

NOTES

1. In the literature on leadership, individuals in non-leadership positions are commonly referred to as followers, employees, subordinates, or group members. When referring to research staff, we use the more neutral terms "group member" or "research group member" since the formal positions of leaders and group members do not automatically imply that leaders lead and group members follow. Autonomy is an important part of research work, and the terms "follower" or "subordinate" may imply certain behaviors or characteristics of leaders and group members and their relationships that are not recognized by research staff.
2. KEYS has six stimulation factors (organizational support, leader support, work group support, freedom, resources, and challenges), two harmful factors (organizational deficiencies, and workload and pressure), and two work result factors (creativity and productivity) (Amabile, 1996).
3. This distinction was based on a combination of their leader's publication performance (70 or more/fewer than 70) and on an expert review of the groups.
4. For example, in one department, the university administration may demand an additional 114% of the basic research grant be included to cover overhead costs.

5. About 70% of all academic research in Sweden is conducted in medicine, the natural sciences, or technology (VINNOVA, 2006–2007). Historically, Sweden has had a strong presence in biomedical and biotechnical research (VINNOVA, 2005).

6. Transformational leaders are charismatic, inspirational, intellectually stimulating, and considerate (e.g., Brehm, Kassin, & Fein, 2005).

7. While the view of creativity as a product touches the definition of innovation (and innovation and creativity are often used interchangeably in disciplines other than psychology—see Csikszentmihalyi, 1999); creativity typically refers only to idea generation, whereas innovation also involves the implementation of the generated idea (Mumford & Gustafson, 1988).

REFERENCES

Agrell, A., & Gustafson, R. (1994). The team climate inventory (TCI) and group innovation: A psychometric test on a Swedish sample of work groups. *Journal of Occupational and Organisational Psychology, 67,* 143–151.

Amabile, T. M. (1996). *Creativity in context.* Boulder, CO: Westview Press

Amabile, T. M., Conti, R., Coon, H., Lazenby, J., & Herron, M. (1996). Assessing the work environment for creativity. *Academy of Management Journal, 39*(5), 1154–1184.

Amabile, T. M., & Gryskiewicz, N. D. (1989). The creative environment scales: The work environment inventory. *Creativity Research Journal, 2,* 231–254.

Amabile, T. M., & Gryskiewicz, S. S. (1987). *Creativity in the R&D laboratory* (Technical Report No. 30). Greensboro, NC: Center for Creative Leadership.

Amabile, T. M., Schatzel, E. A., Moneta, G. B., & Kramer, S. J. (2004). Leader behaviors and the work environment for creativity: Perceived leader support. *The Leadership Quarterly, 15,* 5–32.

Anderson, N. R., & West, M. A. (1998). Measuring climate for work group innovation: Development and validation of the team climate inventory. *Journal of Organisational Behavior, 19,* 235–238.

Andrews, F. (Ed.). (1979). *Scientific productivity: The effectiveness of research groups in six countries.* Cambridge, UK: Cambridge University Press.

Avolio, B. J. (2007). Promoting more integrative strategies for leadership theory-building. *American Psychologist, 62*(1), 25–33.

Balkundi, P., & Kilduff, M. (2005). The ties that lead: A social network approach to leadership. *The Leadership Quarterly, 16,* 941–961.

Bland, C. J., & Ruffin, M. T. (1992). Characteristics of a productive research environment: Literature review. *Academic Medicine, 67,* 385–397.

Blau, P. M. (1964). Social exchange. In P. M. Blau (Ed.), *Exchange and power in social life* (pp. 88–115). New York, NY: John Wiley & Sons, Inc.

Boden, M. A. (1994). What is creativity? In M. A. Boden (Ed.), *Dimensions of creativity* (pp. 75–117). Cambridge, MA & London, England: The MIT Press.

Brehm, S. S., Kassin, S. M., & Fein, S. (2005). *Social pyschology* (6th ed.). Boston, MA: Houghton Mifflin.

Burt, R. S., Hogarth, R. M., & Michaud, C. (2000). The social capital of French and American managers. *Organization Science, 11,* 123–147.

Chen, M. H., & Kaufmann, G. (2008). Employee creativity and R&D: A critical review. *Creativity and Innovation Management, 17,* 71–76.

Cogliser, C. C., & Schriesheim, C. A. (2000). Exploring work unit context and leader-member exchange: A multi-level perspective. *Journal of Organizational Behavior, 21,* 487–511.

Cohen, S. G., & Bailey, D. E. (1997). What makes teams work: Group effectiveness research from the shop floor to the executive suite. *Journal of Management, 23*(3), 239–290.

Cole, S., Rubin, L., & Cole, J. R. (1977). Peer review and the support of science. *Scientific American, 237,* 34–41.

Collins, R. (1998). *The sociology of philosophies: A global theory of intellectual change.* Cambridge, MA and London, England: The Belknap Press of Harvard University Press.

Crane, D. (1969). Social structure in a group of scientists: A test of the "Invisible college" hypothesis. *American Sociological Review, 34,* 335–352.

Csikszentmihalyi, M. (1999). Implications of a system perspective for the study of creativity. In R. J. Sternberg (Ed.), *Handbook of creativity* (pp. 313–334). Cambridge, UK: Cambridge University Press.

Denti, L., & Hemlin, S. (2010). *Leadership and innovation in organizations: A literature review and relationship model.* Manuscript. GRI/Dept. of Psychology, University of Gothenburg, Sweden.

Dewett, T. (2003). Understanding the relationship between information technology and creativity in organizations. *Creativity Research Journal, 15,* 167–182.

Dundar, H., & Lewis, D. R. (1998). Determinants of research productivity in higher education. *Research in Higher Education, 39*(6), 607–631.

Ekvall, G. (1996). Organizational climate for creativity and innovation. *European Journal of Work and Organizational Psychology, 5*(1), 105–123.

Ekvall, G., & Ryhammar, L. (1999). The creative climate: Its determinants and effects at a Swedish university. *Creativity Research Journal, 12,* 303–310.

Elkins, T., & Keller, R. T. (2003). Leadership in research and developmental organizations: A literature review and conceptual framework. *The Leadership Quarterly, 14,* 587–606.

Feist, G. J. (2006). *The psychology of science and the origins of the scientific mind.* New Haven, CT and London, UK: Yale University Press.

George, J. M., & Zhou, J. (2002). Understanding when bad moods foster creativity and good ones don't: The role of context and clarity of feelings. *Journal of Applied Psychology, 87,* 687–697.

Gerstner, C. R., & Day, D. V. (1997). Meta-analytic review of leader-member exchange theory: Correlates and construct issues. *Journal of Applied Psychology, 82,* 827–844.

Graen, G. B., & Uhl-Bien, M. (1995). Relationship-based approach to leadership: Development of leader-member exchange (LMX) theory of leadership over 25 years: Applying multi-level multi-domain perspective. *The Leadership Quarterly, 6,* 219–247.

Greguras, G. J., & Ford, J. M. (2006). An examination of the multidimensionality of supervisor and subordinate perceptions of leader-member exchange. *Journal of Occupational and Organizational Psychology, 79,* 433–465.

Gulbrandsen, M. (2004). Accord or discord? Tensions and creativity in research. In S. Hemlin, C. M. Allwood, & B. R. Martin (Eds.), *Creative knowledge environments: The influences on creativity in research and innovation* (pp. 31–57). Cheltenhamn/Northampton, MA: Edward Elgar Publishing Ltd.

Hackman, J. R., & Wageman, R. (2007). Asking the right questions about leadership. *American Psychologist, 62*(1), 43–47.

Hagstrom, W. (1965). *The scientific community.* New York, NY: Basic Books.

Hemlin, S. (2006a). Creative knowledge environments for research groups in biotechnology. The influence of leadership and organizational support in universities and business companies. *Scientometrics, 67*(1), 121–142.

Hemlin, S. (2006b). Research production in the humanities. *VEST Journal for Science and Technology Studies, 19*(1,2), 57–72.

Hemlin, S. (2008). *Kreativa kunskapsmiljöer i bioteknik. En studie av svenska forskargrupper i akademin och industrin.* Lund, Sweden: Nordic Academic Press.

Hemlin, S. (2009). Creative knowledge environments: An interview study with group members and group leaders of university and industry R&D groups in biotechnology. *Creativity and Innovation Management, 18*(4), 278–285.

Hemlin, S., Allwood, C. M., & Martin, B. R. (2004). *Creative knowledge environments: The influences on creativity in research and innovation.* Cheltenham/Northampton, MA: Edward Elgar Publishing Ltd.

Hemlin, S., Allwood, C. M., & Martin, B. R. (2008). Creative knowledge environments. *Creativity Research Journal, 20*(2), 196–210.

Hemlin, S., & Olsson, L. (2011). Creativity stimulating leadership: A critical incident study of leader's influence on creativity in R&D groups. *Creativity and Innovation Management, 20*(1), 49–58.

Hicks, D., & Skea, J. (1989). Is big really better? *Physics World, 2*(12), 31–34.

Hofmann, D. A., Morgeson, F. P., & Gerras, S. J. (2003). Climate as a moderator of the relationship between leader-member exchange and content specific citizenship: Safety climate as an exemplar. *Journal of Applied Psychology, 88*, 170–178.

Hollingsworth, R., & Hollingsworth, E. J. (2000). Major discoveries and biomedical research organizations: Perspectives on interdisciplinarity, nurturing leadership, and integrated structure and cultures. In P. Weingart & N. Stehr (Eds.), *Practicing interdisciplinarity* (pp. 215–244). Toronto, Canada: University of Toronto Press.

Isen, A. M. (1987). Positive affect, cognitive processes and social behavior. In L. Berkowitz (Ed.), *Advances in experimental social psychology* (Vol. 20, pp. 203–253). New York, NY: Academic Press.

Jaussi, K. S., & Dionne, S. D. (2003). Leading for creativity: The role of unconventional leader behavior. *The Leadership Quarterly, 14*, 475–498.

Kasperson, C. J. (1978). An analysis of the relationship between information sources and creativity in scientists and engineers. *Human Communication Research, 4*, 111–119.

King, N., & Anderson, N. R. (1995). *Innovation and change in organizations.* London, UK: Routledge.

Kyvik, S., & Larsen, I. M. (1994). International contact and research performance. *Scientometrics, 29*(1), 161–172.

Liden, R. C., & Maslyn, J. M. (1998). Multidimensionality of leader-member exchange: An empirical assessment through scale development. *Journal of Management, 24*, 43–72.

Martin, B. R., & Skea, J. E. F. (1992). *Performance indicators for academic scientific research.* End-of-award-report to the Advisory Board for the Research Councils and to the Economic and Social Research Council. Science Policy and Research Evaluation Group, Science Policy Research Unit, University of Sussex, UK.

Merton, R. K. (1973). "Recognition" and "excellence": Instructive ambiguities. In N. Storer (Ed.), *The sociology of science: Theoretical and empirical investigations* (pp. 419–459). Chicago, IL: The University of Chicago Press.

Miles, M. B., & Huberman, A. M. (1994). *Qualitative data analysis: An expanded sourcebook.* London, UK: SAGE.

Mumford, M. D., Connelly, S., & Gaddis, B. (2003). How creative leaders think: Experimental findings and cases. *The Leadership Quarterly, 14*(4,5), 411–432.

Mumford, M. D., & Gustafson, S. B. (1988). Creativity syndrome: Integration, application and innovation. *Psychological Bulletin, 103*(1), 27–43.

Mumford, M. D., Scott, G. M., Gaddis, B., & Strange, J. M. (2002). Leading creative people: Orchestrating expertise and relationships. *The Leadership Quarterly, 13*, 705–730.

Olsson, L., Hemlin, S., & Pousette, A. (2010). *A multi-level analysis of leader-member exchange and creativity in research groups.* Working paper. GRI / Dept of Psychology, University of Gothenburg, Sweden.

Omta, S. W. F., & de Leeuw, A. C. J. (1997). Management, control, uncertainty and performance in biomedical research universities, institutes, and companies. *Journal of Engineering Technology Management, 14*(3,4), 223–257.

Pelz, D. C., & Andrews, F. M. (1966). *Scientists in organizations: Productive climates for research and development.* Ann Arbor, MI: Institute for Social Research, University of Michigan.

Schriesheim, C. A., Castro, S. L., & Cogliser, C. C. (1999). Leader-member exchange (LMX) research: A comprehensive review of theory, measurement, and data-analytic practices. *The Leadership Quarterly, 10*(1), 63–113.

Schriesheim, C. A., Castro, S. L., & Yammarino, C. C. (2000). Investigating contingencies: An examination of the impact of span of supervision and upward controlingness on leader-member exchange using traditional and multivariate within- and between-entities analysis. *Journal of Applied Psychology, 85,* 659–677.

Schriesheim, C. A., Castro, S. L., Zhou, X., & Yammarino, C. C. (2001). The folly of theorizing "A" but testing "B". A selective level-of-analysis review of the field and a detailed leader-member exchange illustration. *The Leadership Quarterly, 12,* 515–551.

Schriesheim, C. A., Neider, L. L., & Scandura, T. A. (1998). Delegation and leader-member exchange: Main effects, moderators, and measurement issues. *Academy of Management Journal, 41,* 298–318.

Scott, S. G., & Bruce, R. A. (1994). Determinants of innovative behavior: A path model of individual innovation in the workplace. *Academy of Management Journal, 37*(3), 580–607.

Seglen, P. O., & Aksnes, D. W. (2000). Scientific productivity and group size: A bibliometric study of Norwegian microbiological research. *Scientometrics, 49*(1), 125–143.

Simonton, D. K. (2003). Scientific creativity as constrained stochastic behavior: The integration of product, person, and process perspectives. *Psychological Bulletin, 129,* 475–494.

Simonton, D. K. (2004). *Creativity in science. Chance, logic, genius and zeitgeist.* Cambridge, MA: Cambridge University Press.

Smith, P. B., Andersen, J. A., Ekelund, B., Graversen, G., & Ropo, A. (2003). In search of northern management styles. *Scandinavian Journal of Management, 19,* 491–507.

Stankiewicz, R. (1980). *Leadership and the performance of research groups.* Doctoral dissertation. Research Policy Institute, University of Lund, Lund: Studentlitteratur.

Sternberg, R. J. (2006). Creating a vision of creativity: The first 25 years. *Psychology of Aesthetics, Creativity, and the Arts, 5*(1), 2–12.

Sternberg, R. J., & Lubart, T. I. (1999). The concept of creativity: Prospects and paradigms. In R. J. Sternberg (Ed.), *Handbook of creativity* (pp 3–15). Cambridge, MA: Cambridge University Press.

Strannegård, L., & Jönsson, S. (2009). Ledarskapets lockelse. In S. Jönsson & L. Strannegård (Eds.), *Ledarskapsboken* (pp. 11–27). Malmö, Sweden: Liber AB.

Sundgren, M., Dimenäs, E., Gustafsson, J. E., & Selart, M. (2003). Drivers of organizational creativity: A path model of creative climate in pharmaceutical R&D. *R&D Management, 35*(4), 359–374.

Toker, U., & Gray, D. O. (2008). Innovation spaces: Workplace planning and innovation in U. S. university research centers. *Research Policy, 37,* 309–329.

Toutkoushian, R. K., Porter, S. R., Danielson, C., & Hollis, P. R. (2003). Using publication counts to measure an institution's research productivity. *Research in Higher Education, 44*(2), 121–148.

Van der Weijden, I. (2007). *In search of performance. Research management within the Dutch public medical health sector.* (Doctoral dissertation). Amsterdam, The Netherlands: Vrije Universiteit.

VINNOVA. (2005). *Strategi för tillväxt—Bioteknik, en livsviktig industri i Sverige.* VINNOVA Policy VP 2005:2, Stockholm, Sweden.

VINNOVA. (2006–2007). *Forskning och innovation i Sverige. En lägesbedömning.* Rapport på uppdrag av regeringen. Stockholm, Sweden.

Wallmark, J. T., Eckerstein, S., Langered, B., & Holmqvist, H. E. B. (1973). The increase in efficiency with size of research teams. *IEEE Transactions in Engineering Management, 33,* 218–222.

Wilson, K. S., Sin, H. P., & Conlon, D. E. (2010). What about the leader in leader-member exchange? The impact of resource exchanges and substitutability of the leader. *The Academy of Management Review, 35*(3), 358–372.

Yukl, G. A. (2005). *Leadership in organizations* (6th ed.). Upper Saddle River, NJ: Pearson Education, Inc.

Yukl, G., Gordon, A., & Taber, T. (2002). A hierarchical taxonomy of leadership behavior: Integrating a half century of behavior research. *Journal of Leadership and Organizational Studies, 9*(1), 15–32.

Ziman, J. M. (1987). *Knowing everything about nothing: Specialization and change in scientific careers.* Cambridge, UK: Cambridge University Press.

Ziman, J. M. (2000). *Real science: What it is and what it means.* Cambridge, UK: Cambridge University Press.

CHAPTER 17

The Psychology of Human Research Participation

Anne Moyer

The psychology of science has focused on several aspects of the ways in which scientific investigators, and their work, are influenced (Feist, 2006; Feist & Gorman, 1998; Shadish, Houts, Gholson, & Neimeyer, 1989; Simonton, 1988, 2009). In contrast, there has been less attention to the psychological influences on the people who serve as research participants themselves. Yet, encouraging individuals to enter and participate faithfully in research is a critical and sometimes challenging part of empirical investigation. Understanding research participants' motivations, beliefs, and behavior is especially important because these are directly linked to the integrity of scientific and biomedical research results.

"OH YES, WE LOSE SOME PhD DISSERTATIONS NOW AND THEN BECAUSE OF PROBLEMS LIKE THAT": FROM NUISANCE TO INDEPENDENT VARIABLE

As a doctoral candidate, Robert Rosenthal conducted an unplanned analysis of his dissertation data out of simple curiosity (Rosenthal, 1993). The results revealed that the participants in the three experimental conditions he was comparing in his study on the defense mechanism, projection, differed on pretest measures. This occurred before they had undergone the experimental manipulation, but after they had interacted with the young experimenter, who was aware of the participants' study condition—and, naturally, was invested in the results coming out in favor of his hypotheses. Rosenthal became convinced that he, although subtly and unintentionally, was influencing the ratings of study participants that he was ostensibly treating identically. The quote above (Rosenthal, 1993, p. 6) was the response of faculty mentors when he queried them about the phenomenon, which he first termed *unconscious experimenter bias*. Rather than a lost dissertation, what followed was an influential program of research that demonstrated the effects of experimenter expectancies on their research subjects (both human and animal). This resulted in the routine implementation of control procedures in experimental research and the codification of such procedures in undergraduate methods texts (e.g., Smith & Davis, 2010). Rosenthal's wise realization that what he had initially encountered as a nuisance variable could be deliberately examined (by considering it as an independent variable in its own right; McGuire, 1969) led to valuable discoveries about interpersonal influences (Rosenthal, 1966; Rosenthal & Fode, 1963; Rosenthal & Jacobson, 1968) and improvements in procedures to ensure methodological rigor in research.

As pointed out by other authors, the social psychological aspects of participating in psychological experiments have the potential to contribute to research artifacts, or biases that threaten the validity of causal inferences (Strohmetz, 2008). This is based upon the fact that human research participants are not inanimate objects (or nonhuman animals) but "sentient organisms that do not come to the experimental situation completely naïve and unfettered by previous experiences" (Strohmetz, 2008, p. 861). The likelihood that aspects of the experimental situation will be prone to producing artifacts has been theorized to be related to how receptive research participants are to the experimenter's presumed expectations and participants' motivation and ability to respond along those lines (Strohmetz, 2008). Other psychological aspects of human participation in research have been recognized as potential threats to the validity of research findings; for instance, demoralization resulting from being assigned to an undesired treatment condition is thought to artificially deflate the true effect of a treatment (Cook & Campbell, 1979; Schwartz, Chesney, Irvine, & Keefe, 1997), but little deliberate, systematic research has examined such presumed effects (Moyer, 2009). Similarly, serendipitously documented difficulties, potential artifacts, or confounding effects, may be mentioned parenthetically in research reports, but are seldom written up as the focus of a manuscript (but see Shapiro et al., 2002) or followed up and treated as an independent variable in future studies.

In a large-scale systematic review of psychosocial interventions for cancer patients conducted by the author of this chapter and colleagues (Moyer, Sohl, Knapp-Oliver, & Schneider, 2009), detailed coding of research reports in this large literature (comprising 488 studies conducted over 25 years) brought to light numerous examples of such challenges (Moyer et al., 2009). These intriguing phenomena were most often mentioned as asides in the discussion sections but, taken as a whole, represented a set of issues that could be usefully examined in their own right. This inquiry could produce insights that could be put toward improving the validity of such research by facilitating swifter and smoother recruitment, improving the integrity and distinctness of the treatment that contrasted groups are exposed to, and limiting treatment and study attrition. Such insights would not be expected to be uniquely applicable to one particular literature.

This chapter focuses on the ways in which psychological aspects of serving as a human research participant could be fruitfully considered and addressed in ways that might enhance the methodological rigor of research. These ways include efforts to understand how undergraduate students, a prominent proportion of the participants in psychological research, experience serving as subjects and the potential reluctance of people in general to engage in research. These also include deliberately studying how individuals who have agreed to participate in research may react to practices such as being randomly assigned to treatment conditions and being exposed to novel experimental treatments that are not presupposed to be superior to the standard of care.

"COME ON IN, THE WATER IS NICE": DIVING INTO THE SUBJECT POOL

The extent to which psychological research, in particular, has relied upon college undergraduates as a source of human research participants has been remarked upon for decades (Arnett, 2008; Sears, 1986; Smart, 1966; Wintre, North, & Sugar, 2001). Surveys of North American colleges and universities indicate that a common means of accessing such participants is through subject pools associated with psychology courses (Lindsay & Holden, 1987; Sieber & Saks, 1989). Students in subject pools can participate as volunteers or for extra course credit; if they participate as part of a course requirement, this may or may not involve the option to participate in alternative activities (Landrum & Chastain, 1999; Lindsay & Holden, 1987; Miller, 1981). Critiques of the use of undergraduates have rested on the well-founded concern that such samples, largely Westernized, educated, industrialized, rich, and democratic, are vastly unrepresentative of people in general (Henrich, Heine, & Norenzayan, 2010; King, Bailly, & Moe, 2004; Ward, 1993). However, a less publicly aired worry is the possibility that these individuals, recruited for their convenience, could also be unmotivated, cynical, or even resentful, and how this might interfere with the integrity of research results (Brody, Gluck, & Aragon, 2000; Coulter, 1986; Richert & Ward, 1976).

Although sparse, some research has investigated the experience of being in a subject pool from participants' points of view. These studies often have included evaluations of the educational value of subject pool participation, because providing a learning experience is an ethical imperative if research participation is required as part of college course (Sieber, 1999). In general, undergraduates have evaluated their subject pool research participation experience largely, but not universally, positively (Britton, 1979; Britton, Richardson, Smith, & Hamilton, 1983; Brody et al., 2000; Coren, 1987; Davis & Fernald, 1975; Flagel, Best, & Hunter, 2007; Leak, 1981; Richardson, Pegalis, & Britton, 1992; Waite & Bowman, 1999).

In our work on this topic (Moyer & Franklin, 2011), although we also found that more than half of subject pool participants assessed rated their experience positively, a smaller proportion, approximately 10%, made solely negative comments about the significance of the research they participated in, or what they learned from participating. Negative aspects of their research experience were related to poor debriefing, learning little when they were assigned to a control condition, skepticism about the experimenters' ability to ultimately influence the real-world problems they were conducting research on, and claims that the research was pointless or not applicable to daily life. Other negative experiences involved anticipatory anxiety; finding the tasks tedious, frustrating, or physically tiring; being asked repetitive or overly personal questions; negative reactions to deception or performing poorly on a task; or perceiving the experimenters to be uncaring.

It may be feasible to make subject pool members feel more invested in the research being performed at their institution in order to modulate possible negative reactions to research participation. Just as psychological science's reliance on nonhuman animals has been downplayed in Introductory Psychology textbooks (Domjan & Purdy, 1995), the field's debt to undergraduates is not often emphasized to subject pool participants. Acknowledgments and tokens of appreciation of students' contributions, beyond credits to meet a course requirement or payment, are scarce (but see Kimble, 1987). Students rarely learn about the outcome of the studies that they participated in and are likely unaware that they were part of a sample that appeared in a published article or allowed a graduate student to complete a PhD. Being able to learn about the ways in which they have contributed to science is something participants have noted that they would appreciate (Brody et al., 2000). A simple e-mail could readily accomplish this. Alternatively, if keeping track of students' contact information to inform them of outcomes that may occur years in the future is too onerous, a newsletter summarizing accomplishments by investigators who used the subject pool could be sent with a note of thanks to undergraduates who had been involved in research that semester. In addition, subject pool participants would likely welcome immediate feedback on their responses or performance (Brody et al., 2000). One of the prime motivators that draws people to participate in research on the web is the opportunity to learn something about themselves, since web-based surveys readily can and often do provide such feedback (Fraley, 2007). This is something that routinely is not, but very easily could be, provided to members of subject pools. With simple adjustments such as these, greater vested interest in their research participation may lead to more responsible and enjoyable participation by students. This, in turn, might lead to a decrease in the number of participants whose data are eliminated due to, for instance, a failure to follow experimental instructions.

"I DON'T WANT TO BE A GUINEA PIG": RELUCTANCE TO PARTICIPATE IN RESEARCH

In contrast to research conducted in university departments that have easy access to subject pools, encouraging the public or patients to enter clinical trials to test new biomedical and behavioral treatments can be more challenging (Sung et al., 2003). Despite the importance of such trials to public health, a very small percentage of the population, particularly minorities, participates in research (Wood, Wei, Hampshire, Devine, & Metz, 2006). Even recruiting respondents for relatively low-effort national surveys is becoming increasingly difficult (Curtin, Presser, & Singer, 2005). This lack of willingness to participate may signal a disconnect between the value and priority that researchers and nonresearchers place on scientific inquiry, a lack of investment (particularly in the case of healthy individuals) in the condition being studied (Gilliss et al., 2001), or even wariness or frank mistrust

(Ellis, Butow, Tattersall, Dunn, & Houssami, 2001). One source of reluctance to participate in randomized trials may be the lack of choice inherent in the research design. Participant preferences for one particular treatment condition may affect willingness to enter a trial where there is a chance of being assigned to an undesired condition (Howard & Thornicroft, 2006; King et al., 2005; Llewellyn-Thomas, McGreal, Thiel, Fine, & Erlichman, 1991; Millat, Borie, & Fingerhut, 2005; TenHave, Coyne, Salzer, & Katz, 2003; Welton, Vickers, Cooper, Meade, & Marteau, 1999). Alternatively, for a new treatment that is only available through a clinical trial or for people who might only gain access to treatment through a trial, enrollment can be disproportionately motivated by the desire to receive that treatment (Eng, Taylor, Verhoef, Ernst, & Donnelly, 2005; Minogue, Palmer-Fernandez, Udell, & Waller, 1995; Strohmetz, Alterman, & Walter, 1990). Some authors have noted that those with strong preferences and those who refuse randomization due to preferences differ with respect to demographic variables and pretreatment state from those who do not (Awad, Shapiro, Lund, & Feine, 2000; Feine, Awad, & Lund, 1998), calling into question the representativeness of patients who have sought out a trial mainly to gain access to a particular treatment.

Monetary incentives are one potential solution that has been advocated and implemented to counteract this reluctance to participate in research (Beebe, Davern, McAlpine, Call, & Rockwood, 2005; Singer, Van Hoewyk, Gebler, Raghunathan, & McGonagle, 1999; Willimack, Schuman, Pennell, & Lepkowski, 1995). The use of financial incentives is common in some areas of research (e.g., in genetic studies) and participants report that they are motivating (Kaufman, Murphy, Scott, & Hudson, 2008). Pecuniary compensation has also been shown to be useful in producing good recruitment and retention rates in behavioral trials, which typically require substantial time and effort on the part of participants (Taylor et al., 2003). Even fairly large incentives have been shown to effectively increase the representativeness of follow-up data while not fostering perceptions of coercion in participants, which would be ethically problematic (Festinger, Marlowe, Dugosh, Croft, & Arabia, 2008). Future research should continue to directly investigate the role that incentives might play in recruitment and retention, as we are still learning about how such measures might affect the validity of research (Coogan & Rosenberg, 2004; Grady, 2005; Mack, Huggins, Keathley, & Sundukchi, 1998; Ulrich et al., 2005). For instance, for simple surveys, variations in small monetary incentives ($0, $5, or $15) appear to have only minor influences on the demographic or health characteristics of those who are willing to participate (Moyer & Brown, 2008). Responses to descriptions of hypothetical biomedical research projects with varying levels of risk involving large incentives ($350–$1800), however, indicated that higher payments may make potential participants reluctant to disclose engaging in activities that were restricted for participation in the study (e.g., alcohol consumption), which would have the potential of impacting the integrity of the research findings (Bentley & Thacker, 2004). Whereas some research has indicated that individuals with higher incomes are responsive to higher financial incentives (among hypothetical payments of $100, $1000, and $2000; Halpern, Karlawish, Cassarett, Berlin, & Asch, 2004), other evidence from an actual trial indicates that financial incentives, along with addressing other barriers to participation, resulted in higher proportions of minority and socioeconomically disadvantaged women consenting to participate as opposed to refusing (Webb et al., 2010).

Recently, resources have been directed at initiatives to encourage enrollment in clinical trials and at research addressing potential participants' lack of awareness, distrust, and cultural, linguistic, and logistical barriers. One example is ENACT, the Educational Network to Advance Clinical Trials, a national organization focusing on identifying, implementing, and evaluating innovative community-centered approaches to educating the public about the existence and availability of cancer clinical trials (ENACT, 2010). Numerous psychological theories have been used as frameworks for research to understand, predict, and influence behavioral change particularly with regard to health. Such theories, including the Health Belief Model (Janz & Becker, 1984), the Precaution Adoption Process Model (Weinstein, 1988), and the Transtheoretical Stages of Change Model (Prochaska & Velicer, 1997), may assist in understanding decisions about participating in research. For example, the Transtheoretical Stages of Change Model theorizes that individuals undergo a series of changes in the process of adopting a new behavior. These are: precontemplation, contemplation, preparation, action, and maintenance, and transitions across these stages are driven by 10 cognitive and behavioral processes of change (Prochaska & Velcier, 1997). One process of change, consciousness raising, for example, involves obtaining new information and understandings. Thus, gaining information about the value of participating in research might move potential research participants from the precontemplation stage to the contemplation stage, making them more receptive to an invitation to enter a study. Using this model in the context of a smoking cessation trial, Velicer and colleagues (2005) found that those who consented to enroll were more likely to be in the contemplation or preparation stage of change whereas trial refusers were more likely to be in the precontemplation stage. With this insight, these authors then devised a two-stage recruitment process; in order to prevent alienating smokers in the early stages of change, potential participants were first invited to be in a *smoking* study rather than a *smoking cessation* study. Theoretical frameworks, such as the Diffusion of Innovations Theory (Rogers, 1983), which addresses how new ideas and practices spread through social groups, might be used as a model for understanding how groups as opposed to individuals become more likely to view research participation as a safe, potentially enjoyable, and worthwhile activity.

New insights and tools are being generated and made available to help researchers to more effectively recruit participants. For instance, AccrualNet is a resource that provides links and access to tools and materials (e.g., sample educational brochures, checklists, worksheets, consent scripts), lists of published articles on trial recruitment, training opportunities, and a place for researchers to post questions and share information about what they have learned (National Cancer Institute, 2010). Finally, a novel approach to increasing acceptance of research participation is to borrow principles from the field of marketing. This involves reframing how investigators think of research studies, such that, like businesses, they are recognized as needing a strategy, management, marketing, and sales, and that they have "customers" who must "buy in" to a particular "product." Such marketing strategies include discovering what the people in one's market segment value, and conveying persuasively the relevant benefits of one's product or enterprise to them (Francis et al., 2007). Although marketing principles may be somewhat unfamiliar to psychologists, the underlying theoretical principles are similar to those that social scientists are accustomed to using.

"YOU CAN'T ALWAYS GET WHAT YOU WANT": REACTIVITY TO RANDOM ASSIGNMENT

For participants who enter randomized trials, especially if the trials involve studying a treatment for a medical or behavioral condition, the procedures may feel somewhat artificial, or worse, go against their wishes. Whereas in the real world, patients may have the opportunity to make choices about their treatment in order to best align them with their preferences (Millat et al., 2005), in most trials treatment is assigned randomly. Even more foreign is the ethical principle of equipoise, whereby contrasted treatments examined in a clinical trial should be considered clinically to have equal evidence (or lack of evidence) for their efficacy (Djulbegovic, Cantor, & Clarke, 2003). This can lead to the so-called *therapeutic misconception*, whereby participants come to believe that the goal of treatment in the context of a clinical trial is therapeutic (Miller & Brody, 2003). The artificiality brought about by randomization is accepted by researchers because it is a good method to balance confounding characteristics across groups in order to ensure internal validity (Campbell & Stanley, 1966; Sacks, Chalmers, & Smith, 1982; Shapiro et al., 2002; Stout, Wirtz, Carbonari, & Del Boca, 1994). However, there may be threats to internal validity if participants' beliefs and preferences produce reactivity to treatment procedures and create unintended biases. These biases introduced by trial participants' preferences can influence response expectancies, engagement in the treatments under investigation, and study attrition (Corrigan & Salzer, 2003; Howard, Cox, & Saunders, 1990; Sidani, Miranda, Epstein, & Fox, 2009). These effects may be further moderated by individual participant characteristics. For example, participants who are more motivated to address a problem might be more likely to engage in their assigned condition and remain in a trial if they are assigned to an active treatment as opposed to a control condition. Conversely, participants who are less motivated to address a problem may be more likely to remain in a trial if they are assigned to a control as opposed to an active treatment condition. Because expectancies regarding how effective a treatment will be are related to outcome (Cunningham et al., 2000), motivation resulting from being assigned to a desired treatment condition may inflate its true effect, just as demoralization resulting from being assigned to an undesired treatment condition may deflate its true effect (Cook & Campbell, 1979; Schwartz et al., 1997). It is also possible that compensatory rivalry may cause those assigned to an undesired treatment to strive to do better (Wortman, Hendricks, & Hillis, 1976) or that cognitive dissonance (Festinger, 1957) would lead them to simply report better outcomes.

In contrast to lack of engagement, another type of reactivity to one's group assignment in a randomized trial is *contamination*, whereby participants who are assigned to a control condition try to gain access to the intervention or its elements. This problem is common in trials of behavioral interventions, where "blinding" of participants to their intervention condition is typically not possible. For example, a challenge in physical activity intervention trials involves participants assigned to a control condition subsequently beginning to exercise on their own (Courneya et al., 2003; Mock et al., 2001). Such contamination of control conditions can dilute the observed effects of intervention conditions found in trials. Some instances of reactivity involve a change in participants' psychological state rather than their deliberate behavior. Shapiro et al. (2002) unexpectedly documented what they termed the *premature disclosure effect*, whereby participants' scores on baseline measures were affected by learning their group assignment prior to completing them, and this type of phenomenon has been noted by other authors also (Brooks et al., 1998). However, some evidence suggests that even when participants do not know their treatment group (as when they can be successfully "blinded" to their group assignment), they often guess or suspect, correctly or incorrectly, to which they belong (Brownell & Stunkard, 1982; Fergusson, Glass, Waring, & Shapiro, 2004; Malec, Malec, Gagne, & Dongier, 1996; Strohmetz et al., 1990) and react accordingly.

To date, just a few studies have deliberately examined the effects of reactions to randomization. They found that that randomized participants assigned to a less desirable control condition reported more negative feelings about the investigators and about the study itself (Wortman et al., 1976) and that participants in a study who chose one of two distinct types of enhancement training (vocabulary vs. mathematical) had higher outcome scores than those who were randomized to receive those same types of training (Shadish, Clark, & Steiner, 2008). We (Floyd & Moyer, 2010) found that participants who were not matched to their preferred treatment in an educational intervention trial felt less positive about their experience of being in the trial, although this did not affect their belief in the efficacy of treatment, adherence to or engagement in treatment, or trial attrition.

Concern about artifacts due to participant preferences has spurred the development of alternative research designs that allow them to be taken into consideration (Janevic et al., 2003; Millat et al., 2005; Rucker, 1989; Silverman & Altman, 1996; TenHave et al., 2003; Wennberg, Barry, Fowler, & Mulley, 1993). For instance, in *partially randomized patient preference* designs (Brewin & Bradley, 1989), participants with no treatment preference are randomized to the interventions, whereas participants with a preference are allowed to select their intervention assignment (e.g., de C Williams, Nicholas, Richardson, Pither, & Fernandes, 1999). Other designs involve randomizing participants to a "preference" arm or to a conventional randomized controlled trial arm, or offering potential participants another treatment if they refuse the one to which they are assigned (e.g., Janevic et al., 2003). Some evidence indicates that designs that take preferences into account are useful in achieving high enrollment and low attrition (Henshaw, Naji, Russell, & Templeton, 1993). These types of designs have potential disadvantages, however, including requiring larger samples (Howard & Thornicroft, 2006) and the possibility that the distribution of participants in the preference arm will be unbalanced (Corrigan & Salzer, 2003; Coward, 2002; TenHave et al., 2003). Moreover, a critical drawback of such designs is their vulnerability to bias due to an association of treatment choice and unmeasured confounders, which has been shown to lead to overestimation of treatment effects (Gemmell & Dunn, 2011).

Because research on affective forecasting calls into question people's accuracy in predicting their own feelings about particular outcomes (Kermer, Driver-Linn, Wilson, & Gilbert, 2006; Wilson & Gilbert, 2005), an additional concern is that participants may be drawn to treatments for which they are not suited (Bradley, 1996). Because of the increased resources required for designs that take preferences into account, the extent to which they can truly offset biases resulting from preferences needs to be further investigated, as does the impact of preferences on outcomes (e.g., Renjilian et al., 2001) and factors that predict treatment contamination (e.g., Courneya, Friedenreich, Sela, Quinney, & Rhodes, 2002).

"I AM SO OUT OF HERE!": PREVENTING AND REDUCING ATTRITION

Participant attrition, particularly in longitudinal studies, is a significant problem, compromising the statistical power of analyses and the external validity of results. Dropout rates from drug trials average about one-third (Kemmler, Hummer, Widschwendter, & Fleischhacker, 2005). Although this dropout rate may be related to unpleasant side effects, taking medications is fairly simple compared to the effort, involvement, and sometimes inconvenience of engaging in behavioral interventions, which makes attrition even more likely from these types of trials. Because the informed consent process ensures that research participants are aware of their right to terminate their involvement in a study, the onus is on researchers to prevent negative reactions or apathy in their participants in order to ensure their continued participation.

Although statistical approaches to dropout using "intent-to-treat" analyses are available (Lachin, 2000), they are somewhat limited (Sheiner, 2002) such that preventing dropout is a preferable solution. Some strategies to minimize attrition have been procedural, such as collecting detailed locator information and allowing flexible scheduling (Woolard et al., 2004). Others involve reducing uncomfortable aspects of participation such as managing negative drug side effects (Mathibe, 2007). Additional strategies involve emphasizing the benefits of participation; minimizing respondent burden and giving some level of control to participants; providing incentives and small tokens of appreciation; providing instrumental or tangible support; being persistent but patient; being flexible; enlisting assistance from others and providing social support; maintaining a good tracking system (Coday et al., 2005); and querying participants in advance about potential addressable barriers (Leon et al., 2006). More extensive efforts, such as an orientation using motivational interviewing techniques prior to randomization, have been successful in helping potential participants understand the reasons for alternative trial conditions and reducing attrition (Goldberg & Kiernan, 2005). Importantly, participants report that positive interactions with trial staff are a critical reason for continuing with research (Marmor et al., 1991) and that the relationship established with an interviewer means more to them than incentives (BootsMiller et al., 1998).

Although research investigating demographic and medical factors that are related to attrition from research is available (e.g., Ahluwalia et al., 2002; McCann et al., 1997; Sears et al., 2003), more work investigating more specific personal and individual psychological factors is needed (Goodwin et al., 2000). Despite the fact that decades ago Orne advocated enlisting research subjects as "co-investigators" by asking them to reflect upon their reactions and experiences in studies (Orne, 1969), these types of direct inquiries are not often conducted (Ribisl et al., 1996).

"NOT FEELING TOO COMFORTABLE HERE ...": RITUALS OF RESEARCH AS PECULIAR PRACTICES

Because input from research participants is seldom solicited, investigators may not be aware that certain additional aspects of the research process, taken for granted and well-understood by researchers, can provoke discomfort in those unfamiliar with them. Such unintended discomfort can represent barriers to successful recruitment, engagement, and retention in research. For instance, the wording of consent forms has been noted as a problem in successfully recruiting participants (Buss et al., 2008). As documents that contain specific required language stipulated by investigators' Institutional Review Boards, consent forms' tone and content, such as the outlining of risks and benefits, can be off-putting or disturbing (Angell et al., 2003). Although some organizations have suggested that investigators refer to those who agree to be in research as *participants* rather than *subjects* in order to reflect their active involvement in the scientific process (American Psychological Association, 2001), consent forms may still include the term subject or make reference to contact people whose title involves the term (e.g., Director of Research Compliance, Committee on Research Involving Human Subjects). Research itself may involve procedures, such as asking extremely personal questions, that outside of the research setting would seem intrusive and perhaps impolite (Perez, 2000). Thus, focus groups or research querying patients themselves about, for instance, the types, formats, structure, timing, and ingredients of interventions and trials could be extremely valuable (Sherman et al., 2007). Such feedback will help ensure that researchers are, where possible, conducting research in the most acceptable way. Funding mechanisms that encourage synergy between scientific expertise and community experience, such as the California Breast Cancer Research Program Community Research Collaboration award (California Breast Cancer Research Program, 2009), and the National Center on Minority Health and Health Disparities Community-Based Participatory Research

Initiative (National Institutes of Health, 2009) may help ensure that research practices and procedures will not be inadvertently offensive. Finally, apart from the procedures associated with a particular study or trial, science itself may be something that the public mistrusts or has come to mistrust (Gamble, 1993; Helms, 2002), and investigators themselves may need to be more attuned to, reflective about, and proactive with regard to the potential negative outcomes of their work (Carlson, 2006).

CONCLUSION

Expanding the focus of the psychology of science to more fully and deliberately include the psychology of research participation is likely to be a useful endeavor. Treating research participants and research practices as the subject of inquiry can provide relevant empirical evidence to improve study methodology (Jadad & Rennie, 1998). These types of evidence-based endeavors have been advocated in other areas of research practice, such as using empirical investigation to shape ethical guidelines (Sieber, 2009). Social and behavioral scientists' expertise in perceptions, attitudes, emotions, decision making, and behavior change could very fruitfully contribute to this area of inquiry. In addition, psychological theory is particularly applicable to investigating the processes of deciding whether or not to enter a trial, maintaining the commitment to adhere to and engage in one's assigned treatment condition, and then continuing involvement in a trial through follow-up assessments, regardless of the outcome of treatment. Many questions remain to be more deeply investigated. For example, with respect to recruitment of research participants fruitful questions would include: "Are decisions to avoid research participation based upon accurate beliefs?" and "In what ways do individuals who agree to participate in research differ from those who do not, and how might these differences impact the generalizability of research findings?" With regard to the role of treatment preferences in the context of randomized controlled trials, an important question would be: "To what extent, and under what circumstances, do participants become demoralized when they are assigned to their nonpreferred treatment and how does this impact treatment and trial engagement and outcomes?" A related question is: "How well do designs that take preferences into account resolve potential problems associated with randomization?" With respect to study dropout; "What psychological and motivational factors predict attrition from research?" and "What methods can best be used to enhance retention in research?" are pertinent queries. Answers to these and further questions related to the ways in which psychological variables intersect with the procedures used in research can be used to help ensure and enhance the methodological quality of investigations that use human research participants. Such questions also expand the focus of the psychology of science to include the individuals who serve as the subjects of scientific inquiry whose thoughts and behaviors are, along with those of investigators, elemental to understanding the scientific enterprise.

REFERENCES

Ahluwalia, J. S., Richter, K., Mayo, M. S., Ahluwalia, H. K., Choi, W. S., Schmelzle, K. H., & Resnicow, K. (2002). African American smokers interested and eligible for a smoking cessation clinical trial: predictors of not returning for randomization. *Annals of Epidemiology, 12,* 206–212.

American Psychological Association. (2001). *Publication manual of the American Psychological Association.* Washington, DC: Author.

Angell, K. L., Kreshka, M. A., McCoy, R., Donnelly, P., Turner-Cobb, J. M., Graddy, K.,...Koopman, C. (2003). Psychosocial intervention for rural women with breast cancer: The Sierra-Stanford Partnership. *Journal of General Internal Medicine, 18,* 499–507.

Arnett, J. J. (2008). The neglected 95%: Why American psychology needs to become less American. *American Psychologist, 63,* 602–614.

Awad, M. A., Shapiro, S. H., Lund, J. P., & Feine, J. S. (2000). Determinants of patients' treatment preferences in a clinical trial. *Community Dental and Oral Epidemiology, 28,* 119–125.

Beebe, T., Davern, M., McAlpine, D., Call, K., & Rockwood, T. (2005). Increasing response rates in a survey of medicaid enrollees: The effect of a prepaid monetary incentive and mixed modes (mail and telephone). *Medical Care, 43,* 411–414.

Bentley, J. P., & Thacker, P. G. (2004). The influence of risk and monetary payment on the research participation decision making process. *Journal of Medical Ethics, 30,* 293–298.

BootsMiller, B. J., Ribisl, K. M., Mowbray, C. T., Davidson, W. S., Walton, M. A., & Herman, S. E. (1998). Methods of ensuring high follow-up rates: lessons from a longitudinal study of dual diagnosed participants. *Substance Use and Misuse, 33,* 2665–2685.

Bradley, C. (1996). Patients' preferences and randomised trials. *Lancet, 347,* 1118,1119.

Brewin, C. R., & Bradley, C. (1989). Patient preferences and randomised clinical trials. *British Medical Journal, 299,* 313–315.

Britton, B. K. (1979). Ethical and educational aspects of participating as a subject in psychology experiments. *Teaching of Psychology, 6,* 195–198.

Britton, B. K., Richardson, D., Smith, S. S., & Hamilton, T. (1983). Ethical aspects of participating in psychology experiments: Effects of anonymity on evaluation, and complaints of distressed subjects. *Teaching of Psychology, 10,* 146–149.

Brody, J. L., Gluck, J. P., & Aragon, A. S. (2000). Participants' understanding of the process of psychological research: Debriefing. *Ethics & Behavior, 10,* 13–25.

Brooks, M. M., Jenkins, L. S., Schron, E. B., Steinberg, J. S., Cross, J. A., & Paeth, D. S. (1998). Quality of life at baseline: is assessment after randomization valid? *Medical Care, 36,* 1515–1519.

Brownell, K. D., & Stunkard, A. J. (1982). The double-blind in danger: untoward consequences of informed consent. *American Journal of Psychiatry, 139,* 1487–1489.

Buss, M. K., DuBenske, L. L., Dinauer, S., Gustafson, D. H., McTavish, F., & Cleary, J. F. (2008). Patient/caregiver influences for declining participation in supportive oncology trials. *Journal of Supportive Oncology, 6,* 168–174.

California Breast Cancer Research Program. (2009). *Community collaboration.* Retrieved March 19, 2009, from www.cbcrp.org/community/index.php

Campbell, D. T., & Stanley, J. C. (1966). *Experimental and quasi-experimental designs for research.* Chicago, IL: Rand McNally.

Carlson, E. A. (2006). *Times of triumph, times of doubt: Science and the battle for the public trust.* Cold Spring Harbor, NY: Cold Spring Harbor Laboratory Press.

Coday, M., Boutin-Foster, C., Goldman S. T., Tennant, J., Greaney, M. L., Saunders, S. D., & Somes, G. W. (2005). Strategies for retaining study participants in behavioral intervention trials: Retention experiences of the NIH Behavior Change Consortium. *Annals of Behavioral Medicine, 29,* 55–65.

Coogan, P. F., & Rosenberg, L. (2004). Impact of a financial incentive on case and control participation in a telephone interview. *American Journal of Epidemiology, 160,* 295–298.

Cook, T., & Campbell, D. T. (1979). *Quasi-experimentation: Design and analysis issues for field settings.* Chicago, IL: Rand McNally.

Coren, S. (1987). The psychology student subject pool: Student responses and attitudes. *Canadian Psychology/Psychologie canadienne, 28*, 360–363.

Corrigan, P. W., & Salzer, M. S. (2003). The conflict between random assignment and treatment preference: Implications for internal validity. *Evaluation and Program Planning, 26*, 109–121.

Coulter, X. (1986). Academic value of research participation by undergraduates. *American Psychologist, 41*, 317.

Courneya, K. S., Friedenreich, C. M., Sela, R. A., Quinney, H. A., & Rhodes, R. E. (2002). Correlates of adherence and contamination in a randomized controlled trial of exercise in cancer survivors: an application of the theory of planned behavior and the five factor model of personality. *Annals of Behavioral Medicine, 24*, 257–268.

Courneya, K. S., Friedenreich, C. M., Sela, R. A., Quinney, H. A., Rhodes, R. E., & Handman, M. (2003). The group psychotherapy and home-based physical exercise (group-hope) trial in cancer survivors: physical fitness and quality of life outcomes. *Psychooncology, 12*, 357–374.

Coward, D. D. (2002). Partial randomization design in a support group intervention study. *Western Journal of Nursing Research, 24*, 406–421.

Cunningham, A. J., Edmonds, C. V., Phillips, C., Soots, K. I., Hedley, D., & Lockwood, G. A. (2000). A prospective, longitudinal study of the relationship of psychological work to duration of survival in patients with metastatic cancer. *Psychooncology, 9*, 323–339.

Curtin, R., Presser, S., & Singer, E. (2005). Changes in telephone survey nonresponse over the past quarter century. *Public Opinion Quarterly, 69*, 87–98.

Davis, J. R., & Fernald, P. S. (1975). Laboratory experience versus subject pool. *American Psychologist, 30*, 523–524.

de C Williams, A. C., Nicholas, M. K., Richardson, P. H., Pither, C. E., & Fernandes, J. (1999). Generalizing from a controlled trial: the effects of patient preference versus randomization on the outcome of inpatient versus outpatient chronic pain management. *Pain, 83*, 57–65.

Djulbegovic, B., Cantor, A., & Clarke, M. (2003). The importance of preservation of the ethical principle of equipoise in the design of clinical trials: Relative impact of the methodological quality domains on the treatment effect in randomized controlled trials. *Accountability in Research, 10*, 301–315.

Domjan, M., & Purdy, J. E. (1995). Animal research in psychology: More than meets the eye of the general psychology student. *American Psychologist, 50*, 496–503.

Ellis, P. M., Butow, P. N., Tattersall, M. H., Dunn, S. M., & Houssami, N. (2001). Randomized clinical trials in oncology: understanding and attitudes predict willingness to participate. *Journal of Clinical Oncology, 19*, 3554–3561.

ENACT: Educational Network to Advance Clinical Trials. (2010). *ENACT: Educational Network to Advance Clinical Trials.* Retrieved May 18, 2010, from www.enacct.org

Eng, M., Taylor, L., Verhoef, M., Ernst, S., & Donnelly, B. (2005). Understanding participation in a trial comparing cryotherapy and radiation treatment. *Canadian Journal of Urology, 12*, 2607–2613.

Feine, J. S., Awad, M. A., & Lund, J. P. (1998). The impact of patient preference on the design and interpretation of clinical trials. *Community Dental and Oral Epidemiology, 26*, 70–74.

Feist, G. J. (2006). *The psychology of science and the origins of the scientific mind.* New Haven, CT: Yale University Press.

Feist, G. J., & Gorman, M. E. (1998). Psychology of science: Review and integration of a nascent discipline. *Review of General Psychology, 2*, 3–47.

Fergusson, D., Glass, K. C., Waring, D., & Shapiro, S. (2004). Turning a blind eye: the success of blinding reported in a random sample of randomised, placebo controlled trials. *British Medical Journal, 328*, 432.

Festinger, D. S., Marlowe, D. B., Dugosh, K. L., Croft, J. R., & Arabia, P. L. (2008). Higher magnitude cash payments improve research follow-up rates without increasing drug use or perceived coercion. *Drug and Alcohol Dependence, 96*, 128–135.

Festinger, L. (1957). *A theory of cognitive dissonance.* Oxford, UK: Row, Peterson.

Flagel, D. C., Best, L. A., & Hunter, A. C. (2007). Perceptions of stress among students participating in psychology research: A Canadian survey. *Journal of Empirical Research on Human Research Ethics, 2,* 61–67.

Floyd, A. H. L., & Moyer, A. (2010). The randomized controlled trial: Participant preferences and feelings about participation. *Journal of Empirical Research on Human Research Ethics, 5,* 81–93.

Fraley, R. C. (2007). Using the internet for personality research: What can be done, how to do it, and some concerns. In R. W. Robins, R. C. Fraley, & R. F. Kreuger (Eds.), *Handbook of research methods in personality psychology* (pp. 130–148). New York, NY: Guilford.

Francis, D., Roberts, I., Elbourne, D. R., Shakur, H., Knight, R. C., Garcia, J.,...Campbell, M. K. (2007). Marketing and clinical trials: a case study. *Trials, 8,* 37.

Gamble, V. N. (1993). A legacy of distrust: African Americans and medical research. *American Journal of Preventive Medicine, 9,* 35–58.

Gemmell, I., & Dunn, G. (2011). The statistical pitfalls of the partially randomized preference design in non-blinded trails of psychological interventions. *International Journal of Methods in Psychiatric Research, 20,* 1–9.

Gilliss, C. L., Lee, K. A., Gutierrez, Y., Taylor, D., Beyene, Y., Neuhaus, J.,...Murrell, N. (2001). Recruitment and retention of healthy minority women into community-based longitudinal research. *Journal of Women's Health & Gender-Based Medicine, 10,* 77–85.

Goldberg, J. H., & Kiernan, M. (2005). Innovative techniques to address retention in a behavioral weight-loss trial. *Health Education Research, 20,* 439–447.

Goodwin, P. J., Leszcz, M., Quirt, G., Koopmans, J., Arnold, A., Dohan, E.,...Navarro, M. (2000). Lessons learned from enrollment in the BEST study—a multicenter, randomized trial of group psychosocial support in metastatic breast cancer. *Journal of Clinical Epidemiology, 53,* 47–55.

Grady, C. (2005). Payment of clinical research subjects. *Journal of Clinical Investigation, 115,* 1681–1687.

Halpern, S. D., Karlawish, J. H. T., Cassarett, D., Berlin, J. A., & Asch, D. A. (2004). Empirical assessment of whether moderate payments are undue or unjust inducements for participation in clinical trials. *Archives of Internal Medicine, 164,* 801–803.

Helms, R. (Ed.). (2002). *Guinea pig zero: An anthology of the journal for human research subjects.* New Orleans, LA: Garrett County Press.

Henrich, J., Heine, S. J., & Norenzayan, A. (2010). The weirdest people in the world? *Behavioral and Brain Sciences, 33,* 61–83.

Henshaw, R. C., Naji, S. A., Russell, I. T., & Templeton, A. A. (1993). Comparison of medical abortion with surgical vacuum aspiration: women's preferences and acceptability of treatment. *British Medical Journal, 307,* 714–717.

Howard, K. I., Cox, W. M., & Saunders, S. M. (1990). Attrition in substance abuse comparative treatment research: the illusion of randomization. *NIDA Research Monograph, 104,* 66–79.

Howard, L., & Thornicroft, G. (2006). Patient preference randomized controlled trials in mental health research. *British Journal of Psychiatry, 188,* 303–304.

Jadad, A. R., & Rennie, D. (1998). The randomized controlled trial gets a middle-aged checkup. *JAMA, 279,* 319–320.

Janevic, M. R., Janz, N. K., Dodge, J. A., Lin, X., Pan, W., Sinco, B. R., & Clark, N. M. (2003). The role of choice in health education intervention trials: A review and case study. *Social Science and Medicine, 56,* 1581–1592.

Janz, N. K., & Becker, M. H. (1984). The Health Belief Model: a decade later. *Health Education Quarterly, 11,* 1–47.

Kaufman, D., Murphy, J., Scott, J., & Hudson, K. (2008). Subjects matter: a survey of public opinions about a large genetic cohort study. *Genetic Medicine, 10,* 831–839.

Kemmler, G., Hummer, M., Widschwendter, C., & Fleischhacker, W. W. (2005). Dropout rates in placebo-controlled and active-control clinical trials of antipsychotic drugs: A meta-analysis. *Archives of General Psychiatry, 62,* 1305–1312.

Kermer, D. A., Driver-Linn, E., Wilson, T. D., & Gilbert, D. T. (2006). Loss aversion is an affective forecasting error. *Psychological Science, 17,* 649–653.

Kimble, G. A. (1987). The scientific value of undergraduate research participation. *American Psychologist, 42,* 267,268.

King, A. R., Bailly, M. D., & Moe, B. K. (Eds.). (2004). *External validity considerations regarding college participant samples comprised substantially of psychology majors.* Hauppauge, NY: Nova Science Publishers.

King, M., Nazareth, I., Lampe, F., Bower, P., Chandler, M., Morou, M.,…Lai, R. (2005). Impact of participant and physician intervention preferences on randomized trials: a systematic review. *JAMA, 293,* 1089–1099.

Lachin, J. M. (2000). Statistical considerations in the intent-to-treat principle. *Controlled Clinical Trials, 21,* 167–189.

Landrum, R. E., & Chastain, G. (1999). Subject pool policies in undergraduate-only departments: Results from a nationwide survey. In G. Chastain & R. E. Landrum (Eds.), *Protecting human subjects: Departmental subject pools and institutional review boards* (pp. 25–42). Washington, DC: American Psychological Association.

Leak, G. K. (1981). Student perception of coercion and value from participation in psychological research. *Teaching of Psychology, 8,* 147–149.

Leon, A. C., Mallinckrodt, C. H., Chuang-Stein, C., Archibald, D. G., Archer, G. E., & Chartier, K. (2006). Attrition in randomized controlled clinical trials: methodological issues in psychopharmacology. *Biological Psychiatry, 59,* 1001–1005.

Lindsay, R. C., & Holden, R. R. (1987). The introductory psychology subject pool in Canadian universities. *Canadian Psychology/Psychologie canadienne, 28,* 45–52.

Llewellyn-Thomas, H. A., McGreal, M. J., Thiel, E. C., Fine, S., & Erlichman, C. (1991). Patients' willingness to enter clinical trials: measuring the association with perceived benefit and preference for decision participation. *Social Science and Medicine, 32,* 35–42.

Mack, S., Huggins, V., Keathley, D., & Sundukchi, M. (1998). Do monetary incentives improve response rates in the survey of income and program participation? *Proceedings of the Section on Survey Research Methods, American Statistical Association.*

Malec, E., Malec, T., Gagne, M. A., & Dongier, M. (1996). Buspirone in the treatment of alcohol dependence: a placebo-controlled trial. *Alcoholism: Clinical and Experimental Research, 20,* 307–312.

Marmor, J. K., Oliveria, S. A., Donahue, R. P., Garrahie, E. J., White, M. J., Moore, L. L., & Ellison, R. C. (1991). Factors encouraging cohort maintenance in a longitudinal study. *Journal of Clinical Epidemiology, 44,* 531–535.

Mathibe, L. J. (2007). Drop-out rates of cancer patients participating in longitudinal RCTs. *Contemporary Clinical Trials, 28,* 340–342.

McCann, T. J., Criqui, M. H., Kashani, I. A., Sallis, J. F., Calfas, K. J., Langer, R. D., & Rupp, J. W. (1997). A randomized trial of cardiovascular risk factor reduction: patterns of attrition after randomization and during follow-up. *Journal of Cardiovascular Risk, 4,* 41–46.

McGuire, W. J. (1969). Suspiciousness of experimenter's intent. In R. Rosenthal & R. L. Rosnow (Eds.), *Artifact in behavioral research* (pp. 13–57). New York, NY: Academic Press.

Millat, B., Borie, F., & Fingerhut, A. (2005). Patient's preference and randomization: new paradigm of evidence-based clinical research. *World Journal of Surgery, 29,* 596–600.

Miller, A. (1981). A survey of introductory psychology subject pool practices among leading universities. *Teaching of Psychology, 8,* 211–213.

Miller, F. G., & Brody, H. (2003). A critique of clinical equipoise: Therapeutic misconception in the ethics of clinical trials. *Hastings Center Report, 33,* 19–28.

Minogue, B. P., Palmer-Fernandez, G., Udell, L., & Waller, B. N. (1995). Individual autonomy and the double-blind controlled experiment: the case of desperate volunteers. *Journal of Medical Philosophy, 20,* 43–55.

Mock, V., Pickett, M., Ropka, M. E., Muscari Lin, E., Stewart, K. J., Rhodes, V. A.,…McCorkle, R. (2001). Fatigue and quality of life outcomes of exercise during cancer treatment. *Cancer Practice, 9,* 119–127.

Moyer, A. (2009). Psychomethodology: The psychology of human participation in science. *Journal of Psychology of Science and Technology, 2,* 59–72.

Moyer, A., & Brown, M. (2008). Effects of participation incentives on the composition of health information survey samples. *Journal of Health Psychology, 13,* 870–873.

Moyer, A., & Franklin, N. (2011). Strengthening the educational value of participation in a psychology department subject pool. *Journal of Empirical Research on Human Research Ethics, 6,* 75–82.

Moyer, A., Sohl, S. J., Knapp-Oliver, S. K., & Schneider, S. (2009). Characteristics and methodological quality of 25 years of research investigating psychosocial interventions for cancer patients. *Cancer Treatment Reviews, 35,* 475–484.

National Cancer Institute. (2010). *AccrualNet: Strategies, tools, & resources to support accrual to clinical trials.* Retrieved May 25, 2010, from http://accrualnet.acscreativeclients.com

National Institutes of Health. (2009). *Part I overview.* Retrieved March 19, 2009, from http://grants.nih.gov/grants/guide/rfa-files/RFA-MD-07-003.html

Orne, M. T. (1969). Demand characteristics and the concept of quasi-controls. In R. Rosenthal & R. L. Rosnow (Eds.), *Artifact in Behavioral Research* (pp. 143–179). New York, NY: Academic Press.

Perez, M. A. (2000). Prostate cancer patients and their partners: Effectiveness of a brief communication enhancement intervention prior to undergoing radical prostatectomy. *DAI, 62,* 113.

Prochaska, J. O., & Velicer, W. F. (1997). The transtheoretical model of health behavior change. *American Journal of Health Promotion, 12,* 38–48.

Renjilian, D. A., Perri, M. G., Nezu, A. M., McKelvey, W. F., Shermer, R. L., & Anton, S. D. (2001). Individual versus group therapy for obesity: effects of matching participants to their treatment preferences. *Journal of Consulting and Clinical Psychology, 69,* 717–721.

Ribisl, K. M., Walton, M. A., Mowbray, C. T., Luke, D. A., Davidson II, W.S., & Bootsmiller, B. J. (1996). Minimizing participant attrition in panel studies through the use of effective retention and tracking strategies: Review and recommendations. *Evaluation and Program Planning, 19,* 1–25.

Richardson, D. R., Pegalis, L., & Britton, B. (1992). A technique for enhancing the value of research participation. *Contemporary Social Psychology, 16,* 11–13.

Richert, A. J., & Ward, E. F. (1976). Experimental performance and self-evaluation of subjects sampled early, middle, and late in an academic term. *Psychological Reports, 39,* 135–142.

Rogers, E. M. (1983). *Diffusion of innovations* (3rd ed.). New York, NY: Free Press.

Rosenthal, R. (1966). *Experimenter effects in experimental research.* East Norwalk, CT: Appleton-Century-Crofts.

Rosenthal, R. (1993). Interpersonal expectations: Some antecedents and some consequences. In P. D. Blanck (Ed.), *Interpesonal expectations: Theory, research, and applications* (pp. 3–24). Cambridge, UK: Cambridge University Press.

Rosenthal, R., & Fode, K. L. (1963). The influence of experimenter bias on the performance of the albino rat. *Behavioral Science, 8,* 183–189.

Rosenthal, R., & Jacobson, L. (1968). *Pygmalion in the classroom: Teacher expectation and pupil's intellectual development.* New York, NY: Rinehart and Winston.

Rucker, G. (1989). A two-stage trial design for testing treatment, self-selection and treatment preference effects. *Statistics in Medicine, 8,* 477–485.

Sacks, H., Chalmers, T. C., & Smith, H., Jr. (1982). Randomized versus historical controls for clinical trials. *American Journal of Medicine, 72,* 233–240.

Schwartz, C. E., Chesney, M. A., Irvine, M. J., & Keefe, F. J. (1997). The control group dilemma in clinical research: applications for psychosocial and behavioral medicine trials. *Psychosomatic Medicine, 59,* 362–371.

Sears, D. O. (1986). College sophomores in the laboratory: Influences of a narrow data base on social psychology's view of human nature. *Journal of Personality and Social Psychology, 51,* 515–530.

Sears, S. R., Stanton, A. L., Kwan, L., Krupnick, J. L., Rowland, J. H., Meyerowitz, B. E., & Gantz, P. E. (2003). Recruitment and retention challenges in breast cancer survivorship research: results from a multisite, randomized intervention trial in women with early stage breast cancer. *Cancer Epidemiology Biomarkers & Prevention, 12*, 1087–1090.

Shadish, W. R., Clark, M. H., & Steiner, P. M. (2008). Can randomized experiments yield accurate answers? A randomized experiment comparing random to non-random assignments. *Journal of the American Statistical Association, 103*, 1334–1356.

Shadish, W. R., Houts, A. C., Gholson, B., & Neimeyer, R. A. (1989). The psychology of science: An introduction. In B. Gholson, W. R. Shadish, R. A. Neimeyer, & A. C. Houts (Eds.), *The psychology of science: Contributions to metascience* (pp. 1–16). Cambridge, UK: Cambridge University Press.

Shapiro, S. L., Figueredo, A. J., Caspi, O., Schwartz, G. E., Bootzin, R. R., Lopez, A. M., & Lake, D. (2002). Going quasi: the premature disclosure effect in a randomized clinical trial. *Journal of Behavioral Medicine, 25*, 605–621.

Sheiner, L. B. (2002). Is intent-to-treat analysis always (ever) enough? *British Journal of Clinical Pharmacology, 54*, 203–244.

Sherman, A. C., Pennington, J., Latif, U., Farley, H., Arent, L., & Simonton, S. (2007). Patient preferences regarding cancer group psychotherapy interventions: a view from the inside. *Psychosomatics, 48*, 426–432.

Sidani, S., Miranda, J., Epstein, D., & Fox, M. (2009). Influence of treatment preferences on validity: a review. *Canadian Journal of Nursing Research, 41*, 52–67.

Sieber, J. E. (1999). What makes a subject pool (un)ethical? *Protecting human subjects: Departmental subject pools and institutional review boards* (pp. 43–64). Washington, DC: American Psychological Association.

Sieber, J. E. (2009). Evidence-based ethical problem-solving (EBEPS). *Perspectives in Psychological Science, 4*, 26,27.

Sieber, J. E., & Saks, M. J. (1989). A census of subject pool characteristics and policies. *American Psychologist, 44*, 1053–1061.

Silverman, W. A., & Altman, D. G. (1996). Patients' preferences and randomised trials. *Lancet, 347*, 171–174.

Simonton, D. K. (1988). *Scientific genius: A psychology of science*: Cambridge. Cambridge University Press.

Simonton, D. K. (2009). Varieties of (scientific) creativity: A hierarchical model of disposition, development, and achievement. *Perspectives on Psychological Science, 4*, 441–452.

Singer, E., Van Hoewyk, J., Gebler, N., Raghunathan, T., & McGonagle, K. (1999). The effect of incentives on response rates in interviewer-mediated surveys. *Journal of Official Statistics, 15*, 217–230.

Smart, R. G. (1966). Subject selection bias in psychological research. *Canadian Psychologist/ Psychologie canadienne, 7*, 115–121.

Smith, R. A., & Davis, S. F. (2010). *The psychologist as detective*. Upper Saddle River, NJ: Prentice Hall.

Stout, R. L., Wirtz, P. W., Carbonari, J. P., & Del Boca, F. K. (1994). Ensuring balanced distribution of prognostic factors in treatment outcome research. *Journal of Studies on Alcohol (Suppl), 12*, 70–75.

Strohmetz, D. B. (2008). Research artifacts and the social psychology of psychological experiments. *Social and Personality Psychology Compass, 2*, 861–877.

Strohmetz, D. B., Alterman, A. I., & Walter, D. (1990). Subject selection bias in alcoholics volunteering for a treatment study. *Alcoholism: Clinical and Experimental Research, 14*, 736–738.

Sung, N. S., Crowley, W. F., Jr., Genel, M., Salber, P., Sandy, L., Sherwood, L. M., . . . Rimoin, D. (2003). Central challenges facing the national clinical research enterprise. *JAMA, 289*, 1278–1287.

Taylor, K. L., Lamdan, R. M., Siegel, J. E., Shelby, R., Moran-Klimi, K., & Hrywna, M. (2003). Psychological adjustment among African American breast cancer patients: one-year follow-up results of a randomized psychoeducational group intervention. *Health Psychology, 22*, 316–323.

TenHave, T. R., Coyne, J., Salzer, M., & Katz, I. (2003). Research to improve the quality of care for depression: alternatives to the simple randomized clinical trial. *Genearl Hospital Psychiatry, 25*, 115–123.

Ulrich, C., Danis, M., Koziol, D., Garrett-Mayer, E., Hubbard, R., & Grady, C. (2005). Does it pay to pay?: A Randomized trial of prepaid financial incentives and lottery incentives in surveys of nonphysician healthcare professionals. *Nursing Research, 54,* 178–183.

Velicer, W. F., Keller, S., Friedman, R. H., Fava, J. L., Gulliver, S. B., Ward, R. M., . . . Cottrill, S. D. (2005). Comparing participants and nonparticipants recruited for an effectiveness study of nicotine replacement therapy. *Annals of Behavioral Medicine, 29,* 181–191.

Waite, B. M., & Bowman, L. L. (1999). Research participation among general psychology students at a metropolitan comprehensive public university. In W. M. Bradley, B. L. Laura, & C. D. Garvin (Ed.), *Protecting human subjects: Departmental subject pools and institutional review boards* (pp. 69–85). Washington, DC: American Psychological Association.

Ward, E. A. (1993). Generalizability of psychological research from undergraduates to employed adults. *Journal of Social Psychology, 133,* 513–519.

Webb, D. A., Coyne, J. C., Goldenberg, R. L., Hogan, V. K., Elo, I. T., Bloch, J. R., . . . Culhane, J. F. (2010). Recruitment and retention of women in a large randomized control trial to reduce repeat preterm births: the Philadelphia Collaborative Preterm Prevention Project. *BMC Medical Research Methodology, 10,* 88.

Weinstein, N. D. (1988). The precaution adoption process. *Health Psychology, 7,* 355–386.

Welton, A. J., Vickers, M. R., Cooper, J. A., Meade, T. W., & Marteau, T. M. (1999). Is recruitment more difficult with a placebo arm in randomised controlled trials? A quasirandomised, interview based study. *British Medical Journal, 318,* 1114–1117.

Wennberg, J. E., Barry, M. J., Fowler, F. J., & Mulley, A. (1993). Outcomes research, PORTs, and health care reform. *Annals of the New York Academy of Science, 703,* 52–62.

Willimack, D., Schuman, H., Pennell, B., & Lepkowski, J. (1995). Effects of a prepaid nonmonetary incentive on response rates and response quality in a face-to-face survey. *The Public Opinion Quarterly, 59,* 78–92.

Wilson, T. D., & Gilbert, D. T. (2005). Affective forecasting: Knowing what to want. *Current Directions in Psychological Science, 14,* 131–134.

Wintre, M., North, C., & Sugar, L. A. (2001). Psychologists' response to criticisms about research based on undergraduate participants: A developmental perspective. *Canadian Psychology Psychologie Canadienne, 42,* 216–225.

Wood, C. G., Wei, S. J., Hampshire, M. K., Devine, P. A., & Metz, J. M. (2006). The influence of race on the attitudes of radiation oncology patients towards clinical trial enrollment. *American Journal of Clinical Oncology, 29,* 593–599.

Woolard, R. H., Carty, K., Wirtz, P., Longabaugh, R., Nirenberg, T. D., Minugh, P. A., . . . Clifford, P. R. (2004). Research fundamentals: Follow-up of subjects in clinical trials: addressing subject attrition. *Academic Emergency Medicine, 11,* 859–866.

Wortman, C. B., Hendricks, M., & Hillis, J. W. (1976). Factors affecting participant reactions to random assignment to ameliorative social programs. *Journal of Personality and Social Psychology, 33,* 256–266.

CHAPTER 18

Heuristics and Biases That Help and Hinder Scientists: Toward a Psychology of Scientific Judgment and Decision Making

Joanne E. Kane and Gregory D. Webster

You have probably experienced x-rays numerous times before as a standard part of dental exams or other medical procedures. Although x-rays have become a relatively mundane part of medical treatment, they do carry risks: exposure to radiation through medical imaging increases the probability that an individual will develop cancer during his or her lifetime. Although the health risks from x-rays are real, they are small and generally far outweighed by their benefits. What about N-rays? How concerned should you be about N-ray exposure you have experienced so far? How do the benefits of N-rays compare to their risks?

N-rays were first discovered by a celebrated French physicist named Rene-Prosper Blondlot. Soon after Blondlot's discovery of them in 1903, research on the subject of N-rays exploded: About 300 professional papers written by 100 scientists were dedicated to the study of N-rays. More than 40 scientists reported first-hand direct observation of N-rays. The discovery of N-rays was considered a great achievement for Blondlot himself and for France (Rousseau, 1992).

N-rays seemed to be the discovery of the century until Robert Wood dramatically and conclusively demonstrated that N-rays simply do not exist. (For an entertaining and thorough account and analysis of the debunking of N-rays, see Ashmore, 1993.) Enthusiasm for N-ray research quickly waned among most of the scientific community. Blondlot, however, hung tenaciously to his belief in N-rays and defended their existence for decades.

The now-infamous N-ray episode helps illustrate some of the heuristics and biases that help and hinder scientists in their judgment and decision making. Why were 300 papers written on an imaginary phenomenon? Why did Blondlot maintain his belief in N-rays even in the face of overwhelming evidence that there was no such thing? More broadly, the N-ray example illustrates the importance of understanding how scientists think. In the following pages, we outline some of the key developments in decision science and how they may influence scientific thought. We then highlight several specific heuristics and biases that we believe affect the judgments and decisions made by scientists. We conclude by citing some limitations to this approach, offering some recommendations to our fellow scientists, discussing implications for the psychology of science, and suggesting avenues for future research on heuristics and biases in scientific thinking.

DOMINANT PERSPECTIVES IN DECISION SCIENCE

Psychological research on decision making is generally guided by one of the two major theoretical perspectives: the *heuristics and biases* perspective (e.g., Tversky & Kahneman, 1974), which tends to focus on the limitations of human cognition in ecologically novel situations, and the *bounded rationality* perspective (e.g., Gigerenzer, 2008; Gigerenzer & Selten, 2001; Gigerenzer, Todd, & The ABC Research Group, 1999; Simon, 1972), which tends to focus on the adaptive efficiencies of human cognition despite a complex, dynamic world. We borrow heavily from each of these perspectives.

Heuristics and Biases

The heuristics and biases approach to decision science largely focuses on the mental shortcuts that humans make in everyday life (heuristics), and how these mental shortcuts can result in misleading, erroneous, or otherwise suboptimal judgments and decisions (biases). It would be difficult to overstate the influence the program of research investigating heuristics and biases has had on both decision science and psychology more broadly. This research has been key in demonstrating that standard economic models do not predict real behavior well. Hundreds of experimental studies have been conducted in laboratory and field settings investigating where individuals and groups go wrong in judgment, reasoning, and decisions.

Bounded Rationality

For the last half-century, economists and psychologists have been aware that humans are "boundedly rational" (Simon, 1957, 1983). Due to limited motivation, time, attention, working memory capacity, and cognitive resources, people often rely on cognitive shortcuts in everyday judgment and decision making. These cognitive shortcuts are thought to be adaptive; they help people filter out useless information and attend instead to important information (e.g., Gigerenzer, 2008). Whereas the heuristics and biases approach focuses on how people "get it wrong," the bounded rationality approach tends to focus on how people "get it right" (or as close to right as necessary for everyday functioning).

Examples From Modern Psychology

Small and seemingly insignificant changes in an experimental setting can have large effects on judgments and decisions. For example, subtle differences in elicitation modes can have profound effects on judgment and choice leading to preference reversals (e.g., Shafir, 1993) or choice avoidance (e.g., Iyengar & Lepper, 2000; but see also Scheibehenne, Greifeneder, & Todd, 2009). A 1-hour exercise related to social belonging can impact students' academic achievement and physical health over 3 years or longer (Walton & Cohen, 2011). Even the fonts in which experimental materials are written could influence cognition and decisions (e.g., Oppenheimer & Frank, 2007). The potentially large impact of small manipulations motivates and characterizes much exciting research.

Every research paradigm is effectively comprised of hundreds or thousands of small decisions made by researchers. Just as considering every mundane decision in formal, deliberative ways would be crippling to individuals' functioning in everyday life, considering every mundane decision in a research context would be crippling to the research process and to researchers. The fact that scientists are able to focus on a few important concepts and conduct a few analyses in systematic and reasoned ways is impressive given all of the possibility and potential distraction of choice. But given the necessary narrowing of scientific inquiry, how can scientists protect against "butterfly effects" set in motion by small and seemingly trivial choices (such as font) made in the context of an experiment?

In practice, scientists have at least partially addressed this problem through replication. An effect that has been found in a number of studies conducted by different researchers, in different settings, with different samples, and at different times is less likely to be due to some specific condition of observation than an effect found in any single study. The generalizability of the results is of course stronger if the replications involve variations in research design, materials, and assessment instruments consistent with the proposed interpretation of the effect. Results that can be consistently replicated over a range of conditions are unlikely to be due to experimenter bias or other artifacts. As noted in the introduction, however, replication alone is not enough to guard against experimenter bias; over 40 scientists reported *first-hand* observations of the nonexistent N-rays. Replication is an important first step, but we must devise additional ways of preventing bias from derailing scientific progress.

THEORY: BENEFITS AND COSTS

Theoretical advances are crucial to all scientific disciplines. Theories help guide the development of specific hypotheses and allow researchers to discuss important problems using a common language. Good scientific theories organize a mass of research findings into a coherent set of beliefs, predictions, and explanations that may be shared across investigators and across time. A "top-down" scientific approach—beginning with theoretical predictions and confirming or failing to confirm them—is one of the hallmarks of a mature science.

Theory building is a central task of science and scientists, and scientists regularly attempt to reconcile new data with existing theories. When data and theory are inconsistent, scientists tend to criticize data and methods quickly, and to criticize theory only after ruling out data-based or methodological problems. How much empirically validated evidence does it take to falsify a tenacious theory, or at least to conclusively illustrate that a theory is in need of revision?

Although good theory is crucial in guiding fields of scientific research, adhering too closely to theory can thwart important advances. Taleb (2005, 2010), who has written about the epistemological perils of theory, suggests that scientists should strive to be true skeptical empiricists; scientific knowledge should be more data driven than theory driven. For example, consider that many of what are now considered to be the greatest scientific advances and discoveries in human history were the result of random error, chance, dumb luck, exhaustive trial and error, or unforeseen adaptation. Archimedes's discovery of how to measure volume via displaced liquid was just that—a chance discovery made in the bathtub rather than a federally funded research institute. "Eureka!" indeed, though chance favors the prepared mind.

Perhaps the most famously serendipitous discovery was that of penicillin. Alexander Fleming was not looking for an antibiotic; he was merely curious about the mold that was consuming his bacterial cultures in a Petri dish. No experiments. Not even trial and error. Just chance. In his writings, Taleb (2005, 2010) also notes how Viagra and the Internet were each developed for a specific purpose (curing erectile dysfunction and computer-based file and program sharing), but then by chance, each had a much larger and wider impact on medicine and technology than anyone anticipated. Although these are merely anecdotal examples, many scientific discoveries are made not through grand theoretical thinking, but rather through chance, luck, and trial-and-error learning (Taleb, 2010).

Of course, not just anyone could appreciate an overflowing bathtub or a mold colony as much as the scientists who used them to make important scientific progress. Chance favors the prepared mind, and knowledge of scientific theory is important to recognizing the significance of observations that do not fit with the extant theories or conventional wisdom of the day. It is probably uncontroversial to claim that there is a place for both theory-driven and data-driven research practice in science. Without theory-driven approaches, it seems unlikely that science and knowledge would grow or evolve over time; without widely accepted research paradigms, methodological conventions, and shared vocabulary, communicating with other scientists and the public would be inefficient or impossible. At the same time, adhering too closely to theory can delay or impede scientific progress and can, in some cases, lead researchers astray. As we will explore in the next section, scientists are subject to heuristics and biases at many stages of investigations, despite the safeguards against bias that the scientific method provides.

THE SCIENTIFIC METHOD

Scientific research is a human enterprise, and therefore subject to human bias and error. This bias can potentially affect nearly every aspect of the scientific method from design to data analysis to interpreting results. In this section, we discuss some heuristics and biases that can both aid and impinge upon scientific progress, and we discuss a few ways to prevent bias from hindering research programs.

Design, Methods, and Data Collection

Biases unchecked by research design pose a threat to the internal validity of scientific experiments. Research on Hawthorne effects and response bias, observer-expectancy effects, and selection bias suggests that researchers may inadvertently cause an effect to appear larger than it actually is, or to appear to exist when it does not.

The Hawthorne Effect

The Hawthorne Effect describes the tendency for people to perform differently when they are being observed as a direct result of the observation itself (rather than of an experimental manipulation or intervention). In many cases, scientists are aware of this problem and adjust their research to include appropriate control conditions including placebo control groups. Hawthorne effects and expectancy effects on the part of both the observer and the participants are sufficiently large threats to the internal validity of scientific investigations that warnings against them are worth repeating; the United States Food and Drug Administration—the federal organization charged with clinically testing food and drugs—has mandated procedures for including appropriate control conditions since the 1970s.

Response Bias

Response bias describes the tendency for participants in studies to act in ways that they think are either expected by researchers or considered socially appropriate (e.g., Cronbach, 1950). Self-presentational concerns may lead participants to underreport "undesirable" behaviors and to overreport "desirable" ones. On the surface, response bias appears to be driven by participants' conscious biases. However, it seems likely to us that response bias may also be driven by the strength of the bias exhibited by the experimenter, and may at times operate outside of the awareness of participants and experimenters alike.

Both response bias and Hawthorne effects threaten internal validity by undermining the causal link between a manipulated independent variable and the dependent variable. In other words, what appears to be the expected causal effect may actually be a spurious one if response bias or Hawthorne effects are related to one or more levels of a manipulated variable. To avoid response bias, researchers should take care to avoid leading questions in surveys and should consider including measures of acquiescence and social desirability. Double-blind research designs and participation from researchers unaware of the experimental hypotheses may further defend against these threats.

Experimenter Expectancy Effects

Expectancy effects negatively impact scientific progress in at least two ways. First, experimenter expectancy effects may cause investigators to perceive greater support for their theories than is warranted (e.g., Sheldrake, 1998). Double-blind experimental procedures may help guard against this danger. Second, expectancy effects may cause researchers to discount or entirely miss (fail to perceive) important evidence that might be inconsistent with their theories, but might also lead to new breakthroughs in scientific discovery (Kuhn, 1962). The strength of prior expectations in shaping perceptions and interpretations of data should not be underestimated. Like most humans, scientists tend to see what they expect to see and, conversely, ignore what they expect to be unimportant. Greater tolerance for a data-driven approach to research may help decrease the chance that researchers miss important results due to the blinders sometimes imposed by theory.

Hypothesis Testing: Alternatives to Null Hypothesis Significance Testing

Once researchers have a clear, falsifiable hypothesis in mind and have developed an experimental design with the appropriate control conditions, it is time to create a plan for hypothesis testing. Researchers often test hypotheses through null hypothesis significance testing. Significance testing provides protection against one kind of error—accepting differences in experimental outcomes attributable to sampling variability associated with the selection or assignment of the units of interest (plants, people, etc.) to experimental and control groups. Significance testing does *not* provide any protection against systematic effects that could interfere with the *interpretation* of research results. In particular, a finding of a significant difference in outcomes between two groups does not indicate what *caused* the difference.

Despite an increase in objections to null hypothesis significance testing (Cohen, 1990, 1994; Robinson & Levin, 1997; Schmidt, 1996; Shrout, 1997; Thompson, 1996), many scientists continue to rely on this simple, binary decision approach to data analysis. Indeed, recent research indicates that null hypothesis significance testing is psychology's most widely employed model of statistical inference (Denis, 2003). There are numerous well-known concerns about null hypothesis significance testing, including arguments that the practice violates Bayesian reasoning (e.g., Kirk, 1996); fails to reveal the magnitude of the effects (e.g., Denis, 2003); is extremely sensitive to sample size; adds a sense of certainty to research results when none may be warranted; and is generally misleading for students, researchers, and the general public (e.g., Gliner, Leech, & Morgan, 2002).

The possibility that null hypothesis significance testing might mislead consumers of science is troublesome. It is particularly troublesome when the consumers are policymakers attempting to use scientific knowledge to guide regulatory measures or courses of action. One could imagine using simplifying strategies in statistical analyses so that basic results from experiments could be easily and efficiently communicated to other researchers, policymakers, or to the public. However, to the extent that null hypothesis significance testing is affirmatively misleading to these groups, it is clear to us that different approaches to data analysis should be preferred. It is probably not appropriate to advocate for entirely "banning" the significance test as others have done (Hunter, 1997; Shrout, 1997), but it is also inappropriate to continue relying primarily or exclusively upon significance testing in social science research practice. A focus on effect sizes or practical significance may help reduce some of the bias inherent in interpreting all-or-nothing significance testing (see Kline, 2004, for a clear and accessible discussion of alternatives to null hypothesis significance testing).

An additional source of potentially misleading conclusions based on hypothesis testing is "hypothesizing after the results are known," or "HARKing" (Kerr, 1998; but see Bem, 2003). Many researchers have probably themselves experienced a tension between accurately representing original research hypotheses and writing up results of research investigations in ways that are as clear as possible for readers. Presenting research hypotheses in the introduction of a paper that turn out to be inconsistent with the data might be misleading or confusing for readers. Furthermore, doing so would probably render the work unpublishable.

Awareness of the high costs of a mismatch between hypotheses and results can encourage researchers to present post-hoc hypotheses as though they were a priori. This might not be objectionable if it truly helped readers to understand better the experimental results. However, there is no direct evidence to indicate that this is the case, and to the extent that the practice occurs, it may further undermine the logic behind hypothesis testing; moreover, it is especially problematic if the results are reported without appropriately adjusted p-values.

Some researchers may HARK despite knowing that it is "bad science" to do so. Others may underestimate the consequences of HARKing. We suspect, however, that many researchers are simply unaware of the fact that they are HARKing when they unintentionally adjust their initial hypotheses in subtle ways, or simply misremember their initial hypotheses after seeing the results of their experiments. Like many biases operating outside of conscious awareness, individual researchers may simply lack introspective access to their thought processes and may therefore be unable to accurately assess the extent to which they are "adjusting" their initial hypotheses after seeing their results. In light of scientific falsification (Popper, 1935/2002), it is important to recall that a central tenet of "good science" is the ability to falsify one's theory or hypothesis; sometimes *not* finding what you expected to find can be more important than confirming prior beliefs.

Of course, there are times when scientists deliberately hypothesize after the results of a study are known. Exploratory studies are often exciting for researchers and important for developing new lines of inquiry into a subject.

Our purpose here is not to advise against exploration. On the contrary, we encourage it! The goal of this chapter, rather, is to remind researchers that hypothesizing after the results of an exploratory study is different from hypothesizing after the results of a failed confirmatory study. "Intentionality" factors in most legal decisions, and we think it should be a factor in reporting and evaluating experiments as well.

Interpreting and Reporting Results

Heuristics and biases impact not only data collection and hypothesis testing, but also the entire logic of research programs once the results of experiments are known. In this section, we explore how confirmation bias, omission bias, belief bias, clustering illusions, and the Texas sharpshooter fallacy may negatively affect scientific progress.

Confirmation Bias

Confirmation bias describes the tendency for people to prefer information that confirms their preconceptions over disconfirming evidence. This tendency clearly poses a threat to scientific progress. Popper warned against the confirmation bias (without using the term) after observing psychoanalysts Freud and Adler interpret all evidence offered to them in support of each psychologist's own theory (Popper, 1963). To the extent that individuals craft experimental paradigms designed solely to confirm prior beliefs, or view data through such strong theoretical lenses that disconfirmation of prior theory is impossible, paradigm shifts are unlikely and progress within a given scientific discipline may be stifled.

A classic example of confirmation bias is provided by the Wason (1968) selection task. In the original task, participants were presented with four cards: One card had a vowel, one a consonant, one an even number, and one an odd number (e.g., "A," "K," "4," "7"). Participants were then invited to turn cards over to determine whether the following rule was true: *If a card has a vowel on one side, then it has an even number on the other side.* Participants were instructed to turn over only the cards necessary to test the rule. Results from the original Wason task indicated that college students tended not to perform well on the task. Most students turned the vowel ("A") and the even number ("4"). Some students turned the consonant ("K"). Only a few turned the odd number ("7"). The correct answer, however, is to turn the vowel and the odd number, "A" and "7." The explanation for the students' failure tends to be that they are seeking to confirm the rule rather than to disconfirm it (i.e., in terms of logic, they ignore the contrapositive) (Wason, 1968; Wason & Johnson-Laird, 1970; but see Cosmides & Tooby, 1992).

Confirmation bias is especially strong when a person (researcher) is motivated to believe the hypothesis they are testing (Gilovich, 1991). Researchers may be generally motivated to protect the theories they have developed or to find support for new hypotheses, but emotions run high when the stakes are high and careers or reputations may be on the line. It is thus not surprising that the person whose reputation benefitted most from N-rays when they were first discovered, Rene-Prosper Blondlot, was the person least willing to give up on the possibility that they existed. David Schneider quotes Demosthenes as saying, "Nothing is easier than self-deceit, for what each man wishes, that he also believes to be true" (2007, p. 254).

Confirmation bias has been extensively studied by social psychologists interested in everyday reasoning and naïve hypothesis testing (e.g., Lord, Ross, & Lepper, 1979). Results of empirical investigations of confirmation bias indicate that individuals are more critical of evidence that *opposes* their preexisting theories than they are of evidence that *supports* their preexisting theories. When participants are led to believe that there is error in the data they receive, they tend to believe that the error in the data is inconsistent with their preexisting hypotheses (Gorman, 1989; but see Penner & Klahr, 1996). Scientists might exhibit confirmation bias by testing for outliers only *after* an initial failure to confirm a hypothesis.

Charles Darwin was well aware of the tendency of confirmation bias to impede scientific progress. Darwin acknowledged the necessity of noting disconfirming evidence after he recognized his tendency to easily remember evidence in support of a theory, but to quickly forget evidence against a theory (Clark, 1984). For example, Darwin's observation that the peacock's evolutionarily "costly" tail was inconsistent with his theory of natural selection is often credited with inspiring his theory of sexual selection (Buss, 2012). Without vigilant attendance to the threat of confirmation bias, disconfirming evidence may be passively ignored or actively dismissed in everyday life and in scientific research (for an excellent summary of the perils of confirmation bias for scientific progress, see Gilovich, 1991, pp. 56–60).

Confirmation Heuristic

So far, we have discussed the preference for confirming hypotheses rather than disconfirming them as a threat to scientific progress. However, confirmation can have a constructive impact as well (Hacking, 1983; Kuhn, 1962; Lakatos, 1970). Replication is a critically important part of the scientific process. As Feist (2006a, 2006b) notes, the best and most creative scientists tend to follow the practice of "confirm-early, disconfirm-late." Scientists should seek to replicate their new discoveries and to test the boundaries of new findings through extensions.

It seems important to pause here to distinguish among a few related concepts: confirmation bias, replication and extension, and the confirmation heuristic. Confirmation *bias* is the tendency to ignore or forget evidence that contradicts a favored hypothesis, or the failure to seek out disconfirming evidence after having some demonstration of proof. The confirmation *heuristic* is the tendency to seek out evidence in support of hypotheses soon after first developing them. The major difference between the confirmation bias (a hindrance to progress) and the confirmation heuristic (a help to progress) may be a simple question of timing. Confirmation *bias* describes the tendency not to challenge hypotheses or theories after they have been established; it is problematic from a falsification perspective. The confirmation *heuristic* is the tendency for researchers to develop hypotheses by seeking preliminary support for them; it is useful from an experimental perspective. Straight confirmation is a sign to researchers that a project is worth pursuing or that a hypothesis is worth testing more vigorously, and therefore can be a valuable part of a research investigation (Gorman, 2006).

The Wason selection task example has often been presented as a demonstration of the confirmation *bias*—a failure of participants' logical reasoning. However, the task might actually be an illustration of the confirmation *heuristic* at work—an effective first step toward a hypothesis worth formally attempting to falsify. Falsification is perhaps best left to researchers less engaged with the development of the initial hypothesis. As Cronbach famously noted, "Falsification, obviously, is something we prefer to do unto the constructions of others" (1989, p. 153).

Through *replication and extension,* other researchers test the boundary conditions and generalizability of new phenomena (Gorman, 1989). If either the original researcher or other scientists repeatedly fail to replicate a result, then the original hypothesis should be rejected. If, on the other hand, an effect is replicated across time and experimenters, then confidence in its robustness increases. However, recall the N-ray example: Both Blondlot himself and many other researchers managed to "confirm" the existence of the nonexistent N-ray in over 100 studies. This is a clear failure of the scientific method; why were so many researchers fooled in this case? It is perhaps no surprise that Blondlot sought to confirm his initial findings (guided by the confirmation heuristic) and then continued to believe in N-rays after they were conclusively shown not to exist (confirmation bias). But how is it that other scientists from other labs also were able to find evidence to support his incorrect hypotheses? How is it that they could replicate and extend hypotheses that later proved completely false? We propose that confirmation bias played a strong role in the failure of replication and extension in the case of N-rays, and suggest a renewed focus on falsification in the psychological literature.

Disconfirmation Bias

An interesting exception to the confirmation bias or confirmation heuristic is the tendency for scientists to prematurely reject or ignore evidence that actually supports a hypothesis. Whereas more junior scientists might be more likely to fail to seek out contradictory evidence, ignore contradictory evidence completely, or make only minor modifications to hypotheses in light of new contradictory evidence, more senior scientists might swing too far in the other direction and reject promising hypotheses in the face of puzzling new evidence (Dunbar, 1993). Dunbar attributes this tendency to the negative past experiences more senior scientists are likely to have had with seeing their ideas proved wrong. These competing tendencies among more junior and more senior researchers, to hold on to hypotheses too long or to dismiss them prematurely, might make a group-based approach to scientific inquiry particularly valuable. It might also speak to the need to include demographic differences in new models of scientific thought.

Omission Bias

The omission bias describes the tendency to judge omissions resulting in harm as less morally objectionable than commissions resulting in harm (e.g., Ritov & Baron, 1990). Numerous experiments have demonstrated that committing an action that directly harms a person is more ethically objectionable than failing to commit an action that would prevent harm to a person. We suspect that the same bias may operate when researchers decide what to report in a write-up of their methods and experimental results.

Failures to observe anticipated effects may result in underreporting of elements of research designs or hypotheses tested, and these omissions might seem relatively unproblematic to researchers. However, as we have noted previously, sometimes an experimental failure can be as valuable as an experimental success. Here, again we observe a tension between describing experiments in a simple and short way—so as to not confuse or mislead readers—and fully describing the experimental paradigm or content of an experimental survey.

As journals move online and space for supplemental materials becomes more readily available, we advocate for the inclusion of a description of all experimental materials used in an experiment; not just the elements of the design that "worked." An even better approach would be to require scientists to report the elements included in experimental design before even running the experiment to guard against the motivated forgetting that might occur once the experiment is completed and the results are known.

The Belief Bias

The belief bias describes the tendency to evaluate the logical strength of an argument based on the believability of the conclusion. A major goal of psychology as a discipline is to arm individuals with the tools they need to evaluate the validity of commonly held beliefs (Schneider, 2007). Some research papers include general discussions that include claims that reach far beyond the scope of the data presented in their studies and experiments. These claims may tend to be evaluated in terms of their general plausibility rather than by the strength of their premises (data). The best protection against the belief bias is probably vigilance on the part of both writers and reviewers.

The Clustering Illusion

The clustering illusion generally describes the tendency to perceive patterns in data that truly have no patterns. It is easy to "see" patterns in tea leaves or to "see" patterns in stars as constellations, once someone else has told you what to look for (a bear, a dipper, etc.). Cancer clusters in neighborhoods and "hot hand" streaks in sports or gambling are examples of the clustering illusion in action. Imagine a fair coin being flipped 10,000 times. During those times, people are likely to disproportionally attend to long streaks of heads or tails, so much so that they may even begin to doubt the fairness of the coins or the underlying randomness of the flips (Oskarsson, Van Boven, Hastie, & McClelland, 2009). To some degree, statistical tests protect researchers from the clustering illusion. Rather than relying on an "interocular test," for example, scientists tend to rely on inferential statistics to help determine whether an effect is different from what would be expected by chance alone.

Popper (1935/2002) has warned against what he calls naïve collectivism, which is the tendency to represent objects like groups of people as meaningful and distinguishable groups (e.g., "the middle class"). We suggest measuring the variables of interest underlying such social constructs (e.g., "annual income") rather than relying on socially constructed group membership distinctions in experimental design, data analysis, and reporting of experimental results.

The Texas Sharpshooter Fallacy

We have already discussed the threats to scientific progress and discovery posed to data analysis by hypothesizing after the results are known. HARKing is quite similar to a phenomenon called the "Texas sharpshooter fallacy," whereby researchers select or adjust a hypothesis after the data have already been collected and analyzed, and then present the *new* hypothesis as though it had been the hypothesis all along. The sharpshooter fallacy gets its name from the possibility of shooting at a barn first and drawing a bull's-eye around the shots later. As we have argued before, most researchers would probably say that HARKing and "post-hoc sharpshooting" are "bad science."

There is, however, a place for post-hoc analyses. There is certainly nothing wrong with modifying a theory in response to data that suggest some limitation in the theory. Post-hoc analyses simply need to be treated and described as such. Sometimes such findings are purely chance or error; other times, albeit rarely, they can lead to new avenues worthy of empirical scientific exploration.

Revisions to Theory

We have argued rather strenuously that a key goal of scientific research is to subject theories to empirical checks and thereby to identify theories that approximate reality reasonably well; in this vein, it is clearly preferable to falsify weak theories rather than support them. When theories fail, they should be replaced with stronger theories that accurately reflect or predict empirical observations. Lakatos (1978) points out that in the face of threats to a theory, researchers tend to expand their theories with auxiliary hypotheses rather than replace older theories completely. This tendency to protect fledgling theories from premature rejection serves to allow initially weak theories to be developed more fully. However, there is an important caveat: Revisions to theory must represent improvements in the structure of theory rather than simply being ad-hoc additions designed to deflect negative evidence (Lakatos, 1978).

THE SCIENTIFIC PROFESSION

Just as heuristics and biases can affect judgments and decisions about science, so too can they affect the ways in which scientists view and evaluate their professional activities. In this section, we discuss ways in which biases view may influence the profession of being a scientist, including academic publication game (writing and reviewing manuscripts) and hiring.

Publication Practices

The File-Drawer Effect

The file-drawer effect is a potentially serious problem affecting many scientific disciplines (Rosenthal, 1979). What should we make of failures to replicate previous research, or failures to support a hypothesis directly derived from an established theory? Failures to replicate or support previous work might be taken as evidence that previous work is not as robust as might have been thought, or as evidence that previous theory was simply "wrong." On the other hand, failures to replicate or support previous work might simply indicate that the newer work is underpowered, sloppy, or improperly operationalized. It may be tempting to dismiss failures to replicate established results or to support well-regarded theories as anomalies or as poorly conducted research. Given what we know about confirmation bias, however, scientists should be wary of this biased approach to reasoning about the implications of failures to support theory.

It is not clear what can be realistically done to guard against the threats posed by the file-drawer problem. One idea is to create journals dedicated to the publication of replications and null results, but it is unclear how such a journal would be funded or how useful it would be. Some previous attempts to create such journals have failed (e.g., *Representative Research in Social Psychology*, 1970–2005), suggesting to us that it may not be the best way to share information about null results with other researchers. Another approach might be to create large and searchable online directories of failed experiments. This could save individual researchers a great deal of frustration by allowing them to explore their ideas online before wasting time, effort, and money to essentially "replicate" previous failures to find particular effects. Researchers who advocate for the creation of such an online searchable database note that it would greatly increase the feasibility and efficiency of meta-analyses. In the meantime, meta-analysts have introduced a variety of mathematical formulae that can, to some extent, statistically adjust for the file-drawer effect in meta-analyses (e.g., Hedges & Vevea, 2005; Vevea & Woods, 2005).

Resistant Reviewers

Research findings that do affirmatively support a hypothesis or theory still may not make it into major journals if the report is met with critical resistance from reviewers. Reviewers are far more likely to recommend publication for articles that support, rather than contradict, their own research (Mahoney, 1977) or those that present positive, confirming results rather than negative, disconfirming results (see also Cicchetti, Conn, & Eron, 1991; Gorman, 1992). In many cases, reviewers may not be aware that they are suppressing contradictory work—the tendency to do so may be a natural result of heuristic reasoning processes. However, some researchers have argued that reviewers do deliberately suppress unpopular findings (Resch, Ernst, & Garrow, 2000) and findings that contradict reviewers' own theories (Ioannidis, 2005) through the peer-review process. These tendencies are clearly not in the best interest of scientific progress, and are probably not in the best interest of the reviewers; disagreements in the field seem likely to open the door to further research and to increase interest in particular research findings rather than to diminish them.

Editors and reviewers of peer-reviewed journals also typically value "newsworthy" articles, research that is not only scientifically sound, but will be of wide interest to a scientific—and possibly popular—audience (Gorman, 1992). In many social sciences, journal editors and reviewers may prefer to publish research that is consistent with current social movements or historical narratives (Gergen, 1973). Together, these pressures may further devalue replications in the eyes of the "gatekeepers" of the publication game. That only about 5% of published studies in the social science are replications is indeed sobering (Mahoney, 1987).

Journals, and the editors and reviewers that serve them, sometimes form *invisible colleges*—informal (sometimes latent) groups of scientists who are generally supportive of each other's theories or research paradigms (Gorman, 1992). When the same group or network of people write, peer review, and cite each other's work, it can cause a positive feedback loop, where confirming studies are published and disconfirming ones are not. Some invisible colleges of researchers may, over time, unintentionally "hijack" a journal, whereby it is only interested in publishing its brand of research (be it a theoretical or methodological perspective) and not alternative perspectives.

Being human, editors and reviewers are also susceptible to other social pressures such as perceived prestige. In a study of eminent scholars who served as reviewers for ostensibly new papers—which had actually been published in the past by major journals in their field—they were more likely to recommend the papers for rejection when they were stripped of their authors' names and given a low-ranking institutional affiliation (Ceci & Peters, 1982). This pioneering research suggests that institutional prestige plays a nontrivial role in the publication game; editors and reviewers may be biased toward papers from higher-ranking institution (or perhaps biased against papers from lower-ranking institutions).

What should scientists do to guard against the possibility that biased reasoning may be preventing them from having a clear picture of the accuracy of their theories? One solution would be to follow the prescription of medical researchers and require preregistration of research protocols (Ioannidis, 2005). Having a centralized record of research attempts would make it much easier to track the success rate of attempts to test critical theoretical assumptions. As useful as such a record would be, it is not clear who would administer and maintain such a database or how it would be financed.

Availability Cascades

An availability cascade is a self-reinforcing process in which a collective belief gains increasing plausibility through its repetition in public discourse (or journals). Availability cascades almost certainly lead to bandwagon effects in research or herd behavior. For example, analyses of citations of scientific journal articles have revealed that, for better or worse, citations roughly follow Pareto's 80/20 rule, where roughly the top 20% of most highly cited articles (or researchers) are often responsible for roughly 80% of all citations (e.g., Duffy, Jadidian, Webster, & Sandell, 2011; Taleb, 2010; Webster, Jonason, & Schember, 2009). This effect has also been referred to as the "Matthew effect" (Merton, 1968) or the "rich get richer" effect. To the extent that these additional studies are subject to the confirmation bias and to the file-drawer problem, which tend to limit the opportunity for disconfirmation, the bandwagon can go on for longer than it might if these effects were not operating.

Hiring and Promotion

It is important to consider not only the direct impact of heuristics and biases on the work scientists produce, but on entire careers of individual scientists. The career trajectories and successes of individual scientists in a field shapes the field as a whole; thus, we should be concerned about bias in hiring and promoting individual researchers both for its own sake and in the service of creating a stronger scientific field of researchers.

Stereotyping, Prejudice, and Discrimination

Stereotyping, prejudice, and discrimination are major fields of study within social psychology, sociology, and other social sciences. We have come a long way in understanding the importance of guarding against implicit or explicit prejudice in hiring practices. One well-known example of an intervention designed to eliminate prejudice includes the introduction of screens or barriers in auditions for major orchestras, which dramatically increased the number of female musicians hired after auditions (Goldin & Rouse, 2000). Another may be the double-blind review process adopted by some progressive journals and their editors, where neither the reviewers nor the authors know each other's identities. Here, the question of prejudice is not one of age, class, color, creed, gender, or sexual orientation, but one of theoretical perspective, methodological approach, social networks, academic pedigree, and—dare we say it—fame (or occasionally infamy). We encourage other "judges" and "gatekeepers" of sciences (reviewers, editors, conference program chairs) to adopt double-blind review processes to lessen the possibility of these types of prejudice against other scientists.

The Affect Heuristic

The affect heuristic describes the basic tendency for liking to influence a decision. Rather than deliberating carefully about a particular choice, people simply rely on their feelings to guide them toward the best option (Finucane, Alhakami, Slovic, & Johnson, 2000). Sometimes reliance on feelings is an appropriate decision strategy. For example, it is better to choose a painting based on gut feelings rather than deliberated reasons (Wilson & Schooler, 1991). Some reliance on feelings might be appropriate in making workplace decisions: Interpersonal skills matter, particularly for academics working on committees or in classrooms. Relying solely on feelings, however, is not defensible in hiring or promotion practices, and caution should be taken to ensure that these practices are fair.

Recency and Availability

Other biases that could threaten fairness in hiring practices are recency and availability. *Recency* describes the tendency to give most weight to the most recent occurrence when making a decision. *Availability* describes the tendency to give most weight to the occurrence that most easily comes to mind. Both of these biases could lead hiring committees to favor the candidate whose application they happened to read most recently or who was the latest to interview (Huber, Van Boven, & McGraw, 2011). Simple laboratory studies could be conducted to test this lay theory. Until the impact of serial position on candidate selection is better understood, we propose that application packets should be counterbalanced across readers at least (e.g., simply sorting alphabetically by surname or temporally by application receipt date may not suffice).

Egocentric Overclaiming

Egocentric overclaiming describes the tendency for an individual to claim more credit for tasks than is their due (Ross & Sicoly, 1979). In collaborating on this particular project, for instance, each of the coauthors has contributed at least 65% of the effort (and we suspect the same is true of the coeditors of this handbook). It seems to be common knowledge among new graduate students at least that faculty advisors with tenure are more likely to grant first-author status to graduate students working with them on projects than faculty advisors who have not yet achieved tenure. Indeed, graduate students are often explicitly advised to take this into consideration, along with many other factors, when choosing among potential faculty mentors. The untenured faculty member, the reasoning of this folk knowledge goes, needs the publication and is therefore more likely to take credit as first author. There are no systematic studies, to our knowledge, investigating whether or not this is true. A simple, systematic study could investigate the likelihood of a faculty member's being first author on publications with students as a function of whether or not the faculty member had tenure at the time of the publication.

Even if it were true, however, that faculty members with tenure were more likely to have publications with their students as a first author, it would be unclear whether this could be taken as proof of "unfairness" or a "bias." It could be that well-established faculty members simply attracted better students capable of taking the lead on research projects and manuscripts. It could also be that established faculty members had less time to work actively on projects due to increased obligations elsewhere, and therefore took a backseat on the research projects.

An analysis like the one we have described might be interesting, but it would be far from definitive proof of bias in authorship decisions. Some medical researchers have suggested that to combat this problem, authors should write up a brief summary of what each person contributed to both the project and the manuscript (Bhandari, Einhorn, Swiontkowski, & Heckman, 2003). This summary could be published in the author note of the article, or simply included as supplementary material.

CONCLUDING REMARKS

In this chapter, we have covered several of the heuristics and biases that we believe influence scientific judgments and decision making. Although we have listed several heuristics and biases (Table 18.1), due to the availability heuristic and the bounded rationality of our own minds, our list is neither exclusive nor exhaustive. It is likely that there are other heuristics and biases that affect scientific thought of which we are unaware. To be sure, such "unknown unknowns" may be particularly perilous when it comes to scientific bias. In this concluding section, we acknowledge some limitations of our approach, recount the possible implications of heuristic and biased scientific thinking, offer recommendations on reducing bias to our fellow scientists, and outline some possible new directions for future research.

Limitations, Implications, and Recommendations

Although we have attempted to map some of the heuristics and biases of decision research onto scientific judgment and decision making, our mapping is incomplete and has poor resolution. We wish to reiterate that our list is neither exclusive nor exhaustive; other influential heuristics and biases likely remain to be discovered. Perhaps more importantly, there has been surprisingly little empirical testing on the extent to which scientists from a variety of disciplines are affected by these heuristics and biases. As we have stated earlier, we believe that any rigorous science demands empirical validation, and there has been too little research in this area, which should be of great importance to all scientists. We therefore challenge *you*, the reader, to think about the ways in which the heuristic and biases may affect you, and to go one step further toward empirically testing these potential biases in a

Table 18.1 *Summary of Heuristics and Biases With Recommendations*

Heuristic or Bias	Description	Recommended Action
Affect Heuristic	The tendency for liking to influence a decision	Create conditions that allow for fair practices in hiring and publication, including blind procedures
Availability Cascade	A collective belief gains plausibility through its increasing repetition in public discourse	Limit the threat of the file-drawer problem by creating online repositories; avoid confirmation bias
Belief Bias	The tendency to evaluate the logical strength of an argument based on the believability of the conclusion	Maintain vigilance as writers and reviewers. Examine underlying assumptions
Bias Blind Spot	The tendency to see others' biases more easily than our own bias	Observe the conditions under which scientists fall prey to common biases, then look inward
Clustering Illusion	The tendency to perceive patterns in data that truly have no patterns	Rely on statistical analyses. Avoid naïve collectivism by measuring underlying variables rather than relying on social constructs
Confirmation Bias	The tendency to search for or interpret information in a way that confirms one's well-established preconceptions	Be skeptical of data and of theories when data do not conform to expectations; make special note of disconfirming evidence; avoid overinterpretation of findings through reliance on hypothesis testing
Confirmation Heuristic	The tendency to search for evidence in support of novel hypotheses rather than contradictory evidence	None—this tendency can help to build solid, testable, falsifiable hypotheses
Egocentric Overclaiming	The tendency for an individual to claim more credit for tasks than is their due	Publish a description of each author's contribution to a joint work in the author note
Experimenter Expectancy Effects	The tendency for prior expectations to shape perceptions and interpretations of data	Use double-blind experimental procedures; increase tolerance for a data-driven approach to research
Falsification Bias	The tendency for an individual to discard data that support a particular hypothesis out of concern that the hypothesis might be rejected later	Work in diverse teams to help balance or correct for biases
HARKing—Hypothesizing After the Results are Known	Creating or shifting hypotheses after knowing the results	Keep clear records of original hypotheses; label and statistically test post-hoc hypotheses as such
Hawthorne Effect	The act of observing changes in the thing observed	Make use of placebo control groups

(continued)

Table 18.1   *Summary of Heuristics and Biases With Recommendations*   *(continued)*

Heuristic or Bias	Description	Recommended Action
Null Hypothesis Significance Testing	A relatively simple, binary decision approach to data analysis	Include measures of effect size and descriptions of substantive significance
Omission Bias	The tendency to judge commissions as worse or more harmful than omissions	Mandate inclusion of a thorough description of all materials used in an experiment (might be increasingly possible as journals move online)
Recency and Availability Heuristics	Making decisions based on or in favor of the most recent information or information that most readily springs to mind	Counterbalance order of presentation when and where possible
Response Bias	Participants respond in ways they think are expected or socially desirable	Avoid leading questions; include measures of acquiescence and social desirability
Stereotyping, Prejudice, and Discrimination	Using generalizations about a group to make judgments about an individual; (negative) attitudes about members of a group; decisions and actions that hurt members of a stereotyped group	Make use of blind procedures whenever possible
Texas Sharpshooter Fallacy	Selecting or adjusting a hypothesis after the data have already been collected and analyzed and then presenting the new hypothesis as though it had been the hypothesis all along	Precommit to hypothesis (possibly through online repositories); treat post-hoc hypotheses as such when creating manuscripts and reports

systematic way. Or, in light of research indicating that it is easier to observe bias in others than in ourselves (e.g., Pronin, Lin, & Ross, 2002), it might be more profitable to think about the ways in which heuristics and biases may affect your *colleagues* and then take steps toward empirically testing whether these potential biases might be operating in *their* work.

Researching biased scientific thinking has potentially broad implications for the new field of the psychology of science. Specifically, biased scientific thinking is a diverse topic of research that may provide fertile ground for psychologists who are interested in studying how scientists think and make decisions. The little research that has already been done has consistently shown that scientists are susceptible to heuristical and biased thinking in much the same way everyone else is. Thus, scientists should make a mental note of these heuristics and biases that have this dual effect of making them more efficient thinkers that occasionally reach the wrong conclusion.

SUMMARY

In this chapter, we have highlighted many areas of concern when evaluating social psychological research endeavors. These potential sources of bias are the results of heuristics that are quite useful in many contexts. Like all heuristics, however, they can lead to fallacious reasoning in particular cases. Indeed, they may operate in our research despite that they are precisely the topics we study; confirmation bias, for example, might—ironically—be operating in research investigations of confirmation bias.

We recommend that scientists occasionally glance at Table 18.1 to remind themselves of some biases that might impact their work. However, we note that it is not enough to simply be aware of bias to avoid it. In practice, it is difficult to know whether it is necessary to "correct" for bias in some way, and difficult to know how much to correct for bias if it is suspected. Individuals must carefully avoid "overcorrecting" if an attempt to adjust for a bias is made. We have proposed that safeguards are necessary to minimize opportunities for bias, and include some suggestions for what these safeguards might be in this text and the accompanying table.

Much progress has already been made to reduce or eliminate bias from scientific endeavors both directly—through adherence to the scientific method and research standards—and indirectly—through double-blind review and fair hiring practices. Nevertheless, we suggest that more can be done. For example, we encourage continued discussion of hypothesis-testing approaches, invite individual researchers to personally commit to be aware of the temptation to HARK and to guard against it, and recommend increased efforts toward truly blind review in evaluating scientific research. We raise the possibility of online repositories of hypotheses and results (to help avoid the Texas sharpshooter fallacy and the file-drawer problem), and suggest publication of both author contribution summaries and thorough descriptions of all experimental materials. This seems increasingly possible as journals move online and more space is available for supplemental materials.

We cannot eliminate the potential for harm in the use of heuristics without also giving up the great advantages they confer. However, by being aware of the potentially misleading tendencies in these heuristics, we can mitigate the potential for negative impact while taking advantage of the benefits to be derived from them. Thus, as scientists, we may guard ourselves against research that goes the way of the N-ray.

REFERENCES

Ashmore, M. (1993). The theater of the blind: Starring a Promethean prankster, a phoney phenomenon, a prism, a pocket, and a piece of wood. *Social Studies of Science, 23,* 67–106.

Bem, D. J. (2003). Writing the empirical journal article. In J. M. Darley, M. P. Zanna, & H. L. Roediger, III (Eds.), *The compleat academic: A career guide* (2nd ed., pp. 185–220). Washington, DC: American Psychological Association.

Bhandari, M., Einhorn, T. A., Swiontkowski, M. F., & Heckman, J. D. (2003). Who did what? (Mis)perceptions about authors' contributions to scientific articles based on order of authorship. *The Journal of Bone and Joint Surgery, 85,* 1605–1609.

Buss, D. M. (2012). *Evolutionary psychology: The new science of the mind* (4th ed.). Boston, MA: Pearson/Allyn & Bacon.

Ceci, S. J., & Peters, D. P. (1982). Peer review: A study of reliability. *Change, 14,* 44–48.

Cicchetti, D. V., Conn, H. O., & Eron, L. D. (1991). The reliability of peer review for manuscript and grant submissions: A cross-disciplinary investigation. *Behavior and Brian Sciences, 14,* 119–186.

Clark, R. W. (1984). *The survival of Charles Darwin: A biography of a man and an idea.* New York, NY: Random House.

Cohen, J. (1990). Things I have learned (so far). *American Psychologist, 45,* 1304–1312.

Cohen, J. (1994). The world is round (*p* <.05). *American Psychologist, 49,* 997–1003.

Cosmides, L., & Tooby, J. (1992). Cognitive adaptations for social exchange. In J. H. Barkow, L. Cosmides, & J. Tooby (Eds.), *The adapted mind: Evolutionary psychology and the generation of culture* (pp. 163–228). New York, NY: Oxford University Press.

Cronbach, L. J. (1950). Further evidence on response sets and test design. *Educational and Psychological Measurement, 10,* 3–31.

Cronbach, L. J. (1989). Construct validation after thirty years. In R. E. Linn (Ed.), *Intelligence: Measurement, theory, and public policy* (pp. 147–171). Urbana, IL: University of Illinois Press.

Denis, D. J. (2003). Alternatives to null hypothesis significance testing. *Theory & Science,* 4(1), 02.

Duffy, R. D., Jadidian, A., Webster, G. D., & Sandell, K. J. (2011). The research productivity of academic psychologists: Assessment, trends, and best practice recommendations. *Scientometrics, 89,* 207–227.

Dunbar, K. (1993). How scientists really reason: Scientific reasoning in real-world laboratories. In R. J. Sternberg & J. Davidson (Eds.), *Mechanisms of insight* (pp. 365–395). Cambridge, MA: MIT Press.

Feist, G. J. (2006a). *The psychology of science and the origins of the scientific mind.* New Haven, CT: Yale University Press.

Feist G. J. (2006b). Why the studies of science need a psychology of science. *Review of General Psychology, 10,* 183–187.

Finucane, M. L., Alhakami, A., Slovic, P., & Johnson, S. M. (2000). The affect heuristic in judgments of risks and benefits. *Journal of Behavioral Decision Making, 13,* 1–17.

Gergen, K. J. (1973). Social psychology as history. *Journal of Personality and Social Psychology,* 26, 309–320.

Gigerenzer, G. (2008). Why heuristics work. *Perspectives on Psychological Science, 3,* 20–29.

Gigerenzer, G., & Selten, R. (Eds.). (2001). *Bounded rationality: The adaptive toolbox.* Cambridge, MA: MIT Press.

Gigerenzer, G., Todd, P. M., & the ABC Research Group. (1999). *Simple heuristics that make us smart.* New York, NY: Oxford University Press.

Gilovich, T. (1991). *How we know what isn't so: The fallibility of human reason in everyday life.* New York, NY: The Free Press.

Gliner, J. A., Leech, N. L., & Morgan, G. A. (2002). Problems with null hypothesis significance testing (NHST): What do the textbooks say? *The Journal of Experimental Education, 71,* 83–92.

Goldin, C., & Rouse, C. (2000). Orchestrating impartiality: The impact of "blind" auditions on female musicians. *The American Economic Review, 90,* 715–741.

Gorman, M. E. (1989). Error, falsification, and scientific inference: An experimental investigation. *Quarterly Journal of Experimental Psychology: Human Experimental Psychology, 41A,* 385–412.

Gorman, M. E. (1992). *Stimulating science: Heuristics, mental models, and technoscientific thinking.* Bloomington, IN: Indiana University Press.

Gorman, M. E. (2006). Scientific and technological thinking. *Review of General Psychology, 10,* 113–129.

Hacking, I. (1983). *Representing and intervening: Introductory topics in the philosophy of natural science*. New Rochelle, NY: Cambridge University Press.

Hedges, L. V., & Vevea, J. L. (2005). Selection method approaches. In H. Rothstein, A. Sutton, & M. Borenstein (Eds.), *Publication bias in meta-analysis: Prevention, assessment and adjustments* (pp. 145–174). Chichester, UK: Wiley.

Huber, M., Van Boven, L., & McGraw, A. P. (2011). Donate different: External and internal influences on emotion-based donation decisions. In D. M. Oppenheimer & C. Y. Olivola (Eds.), *The science of giving: Experimental approaches to the study of charitable giving*. New York, NY: Taylor & Francis.

Hunter, J. E. (1997). Needed: A ban on the significance test. *American Psychological Society, 8*, 3–7.

Ioannidis, J. (2005). Why most published research findings are false. *PLoS Medicine, 2*, e124.

Iyengar, S. S,. & Lepper, M. R. (2000). When choice is demotivating: Can one desire too much of a good thing? *Journal of Personality and Social Psychology, 79*, 995–1006.

Kerr, N. L. (1998). HARKing: Hypothesizing after the results are known. *Personality and Social Psychology Review, 2*, 196–217.

Kirk, R. E. (1996). Practical significance: A concept whose time has come. *Educational and Psychological Measurement, 56*, 746–759.

Kline, R. B. (2004). *Beyond significance testing: Reforming data analysis methods in behavioral research*. Washington, DC: American Psychological Association.

Kuhn, T. S. (1962). *The structure of scientific revolutions*. Chicago, IL: University of Chicago Press.

Lakatos, I. (1970). Falsification and the methodology of scientific research programs. In I. Lakatos & A. Musgrave (Eds.), *Criticism and the growth of knowledge* (pp. 91–196). London, UK: Cambridge University Press.

Lakatos, I. (1978). *The methodology of scientific research programmes: Philosophical papers volume 1*. Cambridge, UK: Cambridge University Press.

Lord, C. G., Ross, L., & Lepper, M. R. (1979). Biased assimilation and attitude polarization: The effects of prior theories on subsequently considered evidence. *Journal of Personality and Social Psychology, 37*, 2098–2109.

Mahoney, M. J. (1977). Publication prejudices: An experimental study of confirmatory bias in the peer review system. *Cognitive Therapy and Research, 1*, 161–165.

Mahoney, M. J. (1987). Scientific publication and knowledge politics. *Journal of Social Behavior and Personality, 2*, 165–176.

Merton, R. K. (1968). The Matthew effect in science. *Science, 159*, 56–63.

Oppenheimer, D. M., & Frank, M. F. (2007). A rose in any other font would not smell as sweet: Effects of perceptual fluency on categorization. *Cognition, 106*, 1178–1194.

Oskarsson, A., Van Boven, L., Hastie, R., & McClelland, G. (2009). What's next? Judging sequences of binary events. *Psychological Bulletin, 135*, 262–285.

Penner, D. E., & Klahr, D. (1996). When to trust the data: Further investigations of system error in a scientific reasoning task. *Memory and Cognition, 24*, 655–668.

Popper, K. (1963). *Conjectures and refutations*. London, UK: Routledge Classics.

Popper, K. (2002). *The logic of scientific discovery*. London, UK: Routledge. [Original work published 1935].

Pronin, E., Lin, D. Y., & Ross, L. (2002). The bias blind spot: Perceptions of bias in self versus others. *Personality and Social Psychology Bulletin, 28*, 369–381.

Resch, K. I., Ernst, E., & Garrow, J. (2000). A randomized controlled study of reviewer bias against an unconventional therapy. *Journal of the Royal Society of Medicine, 93*, 164–167.

Ritov, I., & Baron, J. (1990). Reluctance to vaccinate: Omission bias and ambiguity. *Journal of Behavioral Decision Making, 3*, 263–277.

Robinson, D. H., & Levin, J. R. (1997). Reflections on statistical and substantive significance, with a slice of replication. *Educational Researcher, 26*, 21–26.

Rosenthal, R. (1979). The file drawer problem and tolerance for null results. *Psychological Bulletin, 86*, 638–641.

Ross, M., & Sicoly, F. (1979). Egocentric biases in availability and attribution. *Journal of Personality and Social Psychology, 37*, 322–336.

Rousseau, D. L. (1992). Case-studies in pathological science. *American Scientist, 80,* 54–63.

Scheibehenne, B., Greifeneder, R., & Todd, P. M. (2009). What moderates the effect of too much choice? *Psychology and Marketing, 26,* 229–253.

Schmidt, F. L. (1996). Statistical significance testing and cumulative knowledge in psychology: Implications for training of researchers. *Psychological Methods, 1,* 115–129.

Schneider, D. (2007). The belief machine. In R. J. Sternberg, H. Roediger III, & D. Halpern (Eds.), *Critical thinking in psychology* (pp. 110–130). Cambridge, UK: Cambridge University Press.

Shafir, E. (1993). Choosing versus rejecting: Why some options are both better and worse than others. *Memory & Cognition, 21,* 546–556.

Sheldrake, R. (1998). Experimenter effects in scientific research: How widely are they neglected? *Journal of Scientific Exploration, 12,* 73–78.

Shrout, P. E. (1997). Should significance tests be banned? Introduction to a special section exploring the pros and cons. *Psychological Science, 8,* 1–2.

Simon, H. A. (1957). *Models of man, social and rational: Mathematical essays on rational human behavior in a social setting.* New York, NY: Wiley.

Simon, H. A. (1972). Theories of bounded rationality. In C. B. McGuire & R. Radner (Eds.), *Decision and organization* (pp. 161–176). Amsterdam, Holland: North-Holland.

Simon, H. A. (1983). *Reason in human affairs.* Stanford, CA: Stanford University Press.

Taleb, N. N. (2005). *Fooled by randomness: The hidden role of chance in life and in the markets.* New York, NY: Random House.

Taleb, N. N. (2010). *The black swan: The impact of the highly improbable* (2nd ed.). New York, NY: Random House.

Thompson, B. (1996). AERA editorial policies regarding statistical significance testing: Three suggested reforms. *Educational Researcher, 25,* 26–30.

Tversky, A., & Kahneman, D. (1974). Judgment under uncertainty: Heuristics and biases. *Science, 185,* 1124–1131.

Vevea, J. L., & Woods, C. M. (2005). Publication bias in research synthesis: Sensitivity analysis using a priori weight functions. *Psychological Methods, 10,* 428–443.

Walton, G. M., & Cohen, G. L. (2011). A brief social-belonging intervention improves academic and health outcomes of minority students. *Science, 331,* 1447–1451.

Wason, P. C. (1968). Reasoning about a rule. *Quarterly Journal of Experimental Psychology, 20,* 273–281.

Wason, P. C., & Johnson-Laird, P. N. (1970). A conflict between selecting and evaluating information in an inferential task. *British Journal of Psychology, 61,* 509–515.

Webster, G. D., Jonason, P. K., & Schember, T. O. (2009). Hot topics and popular papers in evolutionary psychology: Analyses of title words and citation counts in *Evolution and Human Behavior,* 1979–2008. *Evolutionary Psychology, 7,* 348–362.

Wilson, T. D., & Schooler, J. W. (1991). Thinking too much: Introspection can reduce the quality of preferences and decisions. *Journal of Personality and Social Psychology, 60,* 181–192.

CHAPTER 19

The Psychology of Uncertainty in Scientific Data Analysis

Christian D. Schunn and J. Gregory Trafton

One of the reasons science is so complex is that it involves many layers of uncertainty, as scientists struggle to convert into fact that which is not yet understood at all using instruments and techniques recently developed or newly applied. For this reason, the psychology of science is deeply connected to uncertainty, especially with studies of science in real contexts. As a research strategy, psychologists of science have often chosen to study situations with much uncertainty removed. But, in the end, there must be some consideration of how uncertainty complexifies the psychology of the scientist, for every layer of science.

Studies of behavior in the real world have consistently found that uncertainty has a large influence on behavior. For example, there is a whole subdiscipline of naturalistic decision making focused on judgment under uncertainty (Klein, 1989). Scientific data analysis, as a relatively complex instance of judgment under uncertainty, similarly has a large role for uncertainty as something that needs to be represented (mentally and physically), as something that needs to be diagnosed in particular situations, and as something that needs to be addressed through problem solving. This chapter will address each of these elements beginning with a large contextual description of what uncertainty is.

WHAT IS UNCERTAINTY?

Uncertainty is not an undifferentiated concept. There is the subjective uncertainty a person feels, and there is the objective uncertainty in the information a person has, which we call information uncertainty. This objective uncertainty could be about the past, current, or future state of the world. Uncertainty in scientific data analysis is primarily about informational uncertainty for the past (i.e., the data in hand). However, as data analysis is often an ongoing iterative component and because the goals of some scientific data analyses involve predictions about the future, there can be informational uncertainty about the future component as well. Regardless of the cause, increases in objective uncertainty in the data makes scientific data analysis much more complex, requiring more time and producing lower rates of success (Gorman, 1986; Klayman, 1986; Penner & Klahr, 1996).

Schunn (2010) argued for a distinction not previously emphasized in discussions of uncertainty: the difference between psychological uncertainty and psychological approximation. Uncertainty is the lack of knowledge about possible states (e.g., is the temperature 18°C or 19°C?). Approximation declares a state as falling with a range (e.g., the temperature is between 18°C and 19°C). At first blush, this distinction appears bizarre and without conceptual merit. From a theoretical information or logical perspective, there is no difference between the two. However, a number of elements suggest it is a critical psychological distinction in science problem solving. Sometimes in science, an approximate level of precision is good enough, whereas sometimes it is not, and the uncertainty versus approximation distinction is critical for differentiating those circumstances. Several different datasets suggest that uncertainty and approximation are discriminable constructs in behavior (from both scientist speech and gestures), that they systematically occur in different places and that common problem solving strategies in science serve primarily to convert from uncertainty into approximation (Trickett & Trafton, 2007; Trickett, Trafton, & Schunn, 2009).

In this chapter, to organize the psychology of science literature on uncertainty and to characterize its bounds, we present comprehensive taxonomies of sources of informational uncertainty, strategies for diagnosing informational uncertainty, and strategies for reducing informational uncertainty. These taxonomies were developed from cognitive anthropological work (Hutchins, 1995; Suchman, 1987) we have done over several years, conducting careful observations of many scientists and scientists in training, working in scientific domains with high levels of informational uncertainty (e.g., astronomers, neuroscientists, physicists, and geologists). It came from hundreds of hours watching researchers analyze their data (Schunn, Saner, Kirschenbaum, Trafton, & Littleton, 2007; Trickett, Fu, Schunn, & Trafton, 2000; Trickett, Trafton, Saner, & Schunn, 2007; Trickett et al., 2009), interviews with dozens of experts about the ways in which uncertainty enters into their domain and how they deal with it, attending colloquia and conferences in the domains, and working with developers (e.g., computer scientists, physicists, and mathematicians) of new visualization tools.

Although the taxonomies were developed from observations of many domains, to be efficient here, it will be presented with concrete examples from two domains: cognitive neuroscience using functional magnetic resonance imaging (fMRI) and meteorological forecasting, one more basic science and one more applied science, which are briefly described first to set the context. The goal of fMRI in cognitive neuroscience is to discover both the location in the brain and the time course of processing underlying different cognitive processes. fMRI is a domain of high uncertainty because it has very noisy data with complex data analysis procedures. The procedure of fMRI is as follows: Imaging data are collected in research fMRI scanners hooked to computers that display experimental stimuli to their human subjects. Generally, fMRI uses a subtractive logic technique, in which the magnetic activity observed in the brain during one task is subtracted from the magnetic activity observed in the brain during another task, with the assumption that the resulting difference can be attributed to whatever cognitive processes occur in the one task but not the other. Moreover, neuronal activity levels are not directly measured, but rather one measures the changes in magnetic fields associated with oxygen-rich blood relative to oxygen-depleted blood. The main measured change is not the depletion due to neuronal activity, but rather the delayed overresponse of new oxygen-rich blood moving to active brain areas. The delay is on the order of 5 seconds, with the delay slightly variable by person and brain area. Data is analyzed visually by superimposing color-coded activity regions over a structural image of the brain, looking at graphs of mean activation level by region and/or over time or across conditions, or looking at tables of mean activation levels by region across conditions. Elaborate, multistepped, semiautomated computational procedures are executed to produce these various visualizations, and given the size of the data (gigabytes per subject), many steps can take up to several minutes per subject. Inferential statistical procedures (e.g., t-test, analysis of variance [ANOVA]) are applied to confirm trends seen visually.

The problem in weather forecasting is that there is considerable uncertainty in meteorology data, both in observations and numerical forecast models. Forecasters must make judgments that account for and accommodate that uncertainty, although uncertainty is not typically displayed in any of their tools. Weather forecasters examine observations, summaries of those observations, and predictive forecast models that use those observations as input. While they do explicitly examine actual observations by examining satellite pictures or local wind speed, the majority of their information comes from tools that summarize or use those observations. They see satellite images and loops uploaded onto the Internet. They view predictions made by complex models with varying underlying assumptions. The models are also based on possibly unreliable or sparsely sampled observations. Weather is also extremely chaotic (the "butterfly effect"), and no current numerical forecasting model is able to accurately depict the complete dynamic structure of the atmosphere. The associated uncertainty, unreliability, data sparsity, and underlying assumptions are not explicitly provided to the forecasters. The forecasters must infer the uncertainty, either from experience and training, or because the values are not stable across time or across different instances of the "same" data (e.g., different weather models). The data are transformed in one of many ways (e.g., direct transformation, modeling, combination, and multiple representations) and displayed to the decision maker as stimuli.

WHY DOES SCIENTIFIC DATA ANALYSIS INVOLVE UNCERTAINTY?

A Taxonomy of Sources of Scientific Uncertainty

There are several taxonomies of sources of uncertainty in existence. Some come from psychology and judgment and decision-making research

(Berkeley & Humphreys, 1982; Howell & Burnett, 1978; Kahneman & Tversky, 1982; Krivohlavy, 1970; Lipshitz & Strauss, 1997; Musgrave & Gerritz, 1968; Trope, 1978). Others come from a broad array of particular disciplines, such as geography (Abbaspour, Delavar, & Batouli, 2003), finance (Rowe, 1994), law (Walker, 1991, 1998), medicine (Brashers et al., 2003; Hall, 2002), consumer choice (Sheer & Cline, 1995; Urbany, Dickson, & Wilkie, 1989), negotiation (Bottom, 1998), and military tactics (Cohen, Freeman, & Thompson, 1998). The taxonomies from the disciplines are typically armchair analyses rather than from observations of experts in the field.

We divide sources of information uncertainty into four broad classes: physics uncertainty, computational uncertainty, visualization uncertainty, and cognitive uncertainty. Experimental studies of uncertainty in scientific problem solving have focused on just a small subset of these sources; our taxonomy shows that informational uncertainty stems from many different complex sources and is the norm in science.

Physics Uncertainty

Sometimes uncertainty in scientific data comes from uncertainty in the raw measured information itself. When uncertainty was experimentally introduced in lab studies, it was usually conceptualized as physics uncertainty (Gorman, 1986; O'Connor, Doherty, & Tweney, 1989; Penner & Klahr, 1996; Schunn & Anderson, 1999; Tweney, Doherty, & Mynatt, 1981). This measurement uncertainty subdivides into three subtypes, roughly corresponding to a signal not being measured, signal noise, or noise in the way the signal is being transduced, each of which is described in further detail below (Figure 19.1).

Not measured uncertainty. In many complex domains, the measured signal is inherently ambiguous because key information is not measured either because the information is not recorded or the information is not measurable (with existing technology). Unrecorded information can be a general state of the equipment, or a temporary error producing some missing data points.

In fMRI, the most central unmeasured information is actually the brain activity itself. Instead, fMRI measures the blood oxygenation levels that respond with variable lag by person and brain region to brain activity levels. The variable lag is technically the unmeasured element that causes uncertainty. Because it is not always clear which regions will be important and it is not always feasible to measure the whole brain, some brain regions can be unmeasured in a given study.

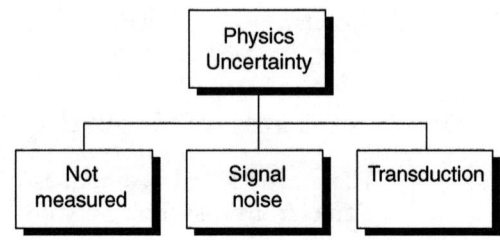

Figure 19.1 *Types of uncertainty coming from physical factors.*

In weather forecasting, weather models make very detailed predictions using previous measurements. These measurements can be very sparse due to weather satellite blind spots and other measurement device placements (especially at sea); point observations may not be very representative of what happens outside the point.

Signal noise uncertainty. The information signal, before it reaches the measurement device, can have stochastic variability, unknown levels of bias, and extraneous signal sources, which introduces additional information uncertainty. In fMRI, electromagnetic noise in the surrounding environment, the participant's body beyond the brain (especially the eyes), and the way the signal moves from the brain through the skull introduce considerable signal noise uncertainty. The bias produced from some of these effects is only partially predictable (e.g., there are weaker signals from deeper in the brain, and there are artifactual signals near the eyes due to eye movements). In weather forecasting, cloud base height (the bottom of the cloud deck) and cloud height are measured by a device called a ceilometer. The ceilometer measures particulates in the atmosphere and, thus, smoke, dust, and even precipitation can lead to an inaccurate reading. The reading may also be erroneous if there is a small break in the clouds above the instrument. Such problems are not the fault of the measurement device, but inherent in a noisy signal.

Transduction uncertainty. The third subtype of physical uncertainty is a function of the measurement device itself. In the process of transduction (converting incoming physical energies such as light, sound, magnetic fields, and heat into data), a measurement device may reduce the quality of the incoming information by simply misreading an incoming signal or compressing dimensionality. Alternatively, the instrument might just fail for a period of time; this form of transduction uncertainty appears to be especially difficult to resolve (O'Connor et al., 1989).

In fMRI, even the latest equipment is still not as spatially precise as some researchers would like and not as temporally precise as other researchers would like. Further, many fMRI researchers share the very expensive device, and this high-use equipment can move out of alignment across participants. In weather forecasting, a radiosonde (weather balloon) may transmit uncertain information about relative humidity under some situations. When the radiosonde gets very dry or very cold, its ability to transmit accurate relative humidity degrades by more than 10%.

Computational Uncertainty

When large amounts of data are measured, many scientists use a number of potentially elaborate computation procedures on that data before they see the data. These computational procedures can add new sources of uncertainty in three general ways (Figure 19.2). Psychology of science has tended to ignore this source of uncertainty.

Future prediction uncertainty. Data is collected at a certain point in time, and the world continues to change beyond that point in time. The computational procedures either make no corrections for these changes or they make only partially accurate corrections, and this introduces a potentially large

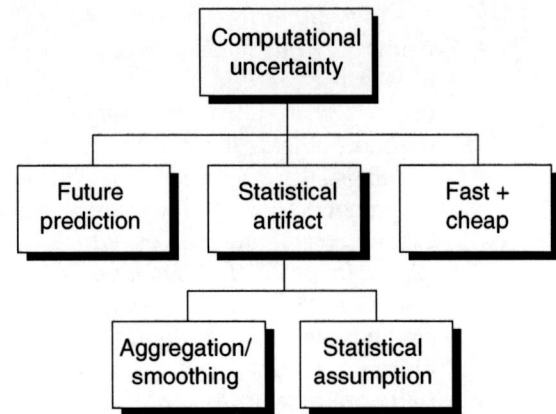

Figure 19.2 *Types of uncertainty coming from computational factors.*

source of uncertainty. Weather forecasting has a large degree of future prediction uncertainty because the forecasters must make predictions for times that are hours, days, weeks, or even months away. In fMRI, brain structural images are collected typically only at the beginning of a scanning session. These structural images are crucial for interpreting the functional location of a given area of activation. Yet, the head often moves slightly during the experiment and the brain itself undergoes minor deformations over time (e.g., the areas near the nasal passages deform during breathing). These changes over time since last measurement introduce uncertainty into the analyses.

Statistical artifact uncertainty. Many statistical algorithms/procedures are applied to scientific data to deal with physics or future prediction uncertainty. These statistical algorithms have the potential of introducing artifacts in the displayed data. We distinguish two main subtypes of statistical artifact uncertainty.

The first subtype is *aggregation/smoothing uncertainty* that removes real features from the data. Statistical algorithms typically try to find a simpler, smoother underlying explanation of the data, filtering out "noise" and possibly aggregating data across larger spatial or time scales than when the data was collected. However, sometimes the features being filtered out are in fact real features of the external world rather than noise. In weather forecasting, microclimates occur in protected valleys, atop high mountains, adjacent to large bodies of water, and so on. These may be more sharply defined than the analysis model predicts. In fMRI, the search for activated regions uses a statistical thresholding procedure that can require sustained activation across time and across adjacent areas. Thus, single, very small areas of activation are removed by this procedure.

The second subtype is *statistical assumption uncertainty*. To infer underlying properties or to further filter out the noise, the computational algorithms/procedures often depend upon certain statistical assumptions. These assumptions can be globally very accurate (i.e., correct most of the time), but locally inaccurate (i.e., untrue for a particular time and place). They can introduce false features into the data, as well as remove true features. The most common statistical assumptions across domains are linearity (changes over time and space are linear) and persistence (things tend to stay the same over small units of time). In weather forecasting, a good predictor of future weather is current weather. Thus, persistence is included into weather models and all weather models make heavy use of the current weather in their predictions. Yet on a local basis, some areas can change weather much more quickly than the models would normally predict. In fMRI, there are algorithms that correct for head motion. The algorithms often assume linear movement over time (e.g., a head gradually sinking into a pillow). Yet, head movements can often be nonlinear (e.g., sudden or back-and-forth).

Fast+cheap uncertainty. In many complex domains, there exist elaborate algorithms that have very high levels of accuracy. Unfortunately, even on modern high-power computers, these elaborate algorithms require considerable time to complete and often longer than the problem solver usually wants to wait. Thus, in these scientific domains, there exist many algorithms that are more approximate in accuracy but much faster to run.

In fMRI, data analysis is very much an iterative process. Before conducting the most detailed and accurate analyses, researchers will typically use more approximate but significantly faster analyses. In weather forecasting, some weather models use exact, micro models of the physics of weather change, but it is not possible to run such models for even medium-sized regions. The time to run the algorithms for even medium-sized regions would introduce additional future prediction uncertainty because it will have been a long time since the last data input to the model.

Visualization Uncertainty

After data are measured and processed through procedures and algorithms, the information must be conveyed somehow to the problem solver, most typically with a visualization (e.g., map, table, graph). Correspondingly, a large spectrum of psychology of science work falls under the use of visual representations (Shah & Hoeffner, 2002). But sometimes the visualizations introduce informational uncertainty about what information can be logically derived from a visualization (this section), or uncertainty regarding human errors or confusions about how to interpret information from the visualization (see the next section on cognitive uncertainty) (Figure 19.3).

Nonrepresented information. The most obvious form of visualization uncertainty occurs when information (that is logically necessary for developing an accurate understanding of a situation) is missing from the visualization entirely. In fMRI, sometimes key regions of activation may be missing from the current visualization because a subset of the brain slices is being displayed. In weather forecasting, some of the displayed measures are not very stable over the time window being displayed, but the visualization provides no information about the certainty of the displayed means.

Composite information uncertainty. Sometimes multiple dimensions that are typically correlated in value are represented with a single composite measure. When the functional combination and the correlation among values are not perfect, this composite measure introduces uncertainty. The most common composite is one that shows a combination of duration and amount so that one does not know whether it was actually a very large amount for a small percentage of the given time interval, or a smaller amount for the

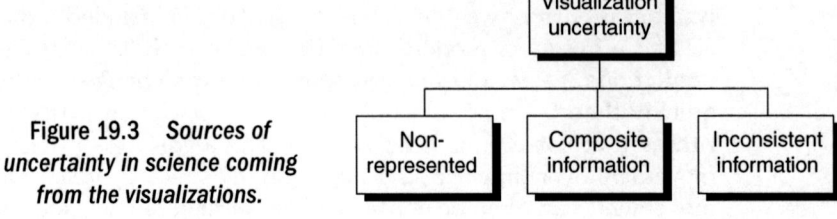

Figure 19.3 *Sources of uncertainty in science coming from the visualizations.*

full time interval. In fMRI, one sometimes has tables that show total signal change of an area (relative to baseline); one does not know whether just a few voxels are very active or whether many voxels are moderately active. In weather forecasting, the display of rain levels does not tell you whether it was 1 cm of rain gradually over 10 minutes, or an onslaught of 1 cm in 1 minute followed by 9 minutes of no rain.

Inconsistent information uncertainty. In many complex domains, there are multiple visualizations shown at the same time. Similar to the joke about the man with two watches never knowing what time it is, multiple displays produce uncertainty when the displays do not match. In weather forecasting, the forecaster frequently examines predictions of different models and these predictions often do not agree in all their details. In fMRI, there are different ways of slicing the data. For example, one can examine which regions activate by time or which regions activate by condition. Sometimes these different visualizations do not agree.

Cognitive Uncertainty

In most psychology of science investigations of uncertainty, the uncertainty sources that were explicitly discussed were external, as in the previous categories. However, humans are a key part of the information system, especially in the complexity of science situations. Humans act as encoding devices, information storage/retrieval devices, and procedure enactors. Correspondingly, they are also a common source of information uncertainty, as possible errors can be introduced in encoding, retrieval, or procedural enactment. In fact, one reaction to external uncertainty is to attribute it to cognitive factors like misencoding or other human error (Chinn & Brewer, 1992). Cognitive uncertainty can reside in the focal problem solver or it can come from the information that team members provide (Hutchins, 1995) (Figure 19.4).

Perceptual error. Early astronomers noted a human perceptual source of uncertainty, referring to a "personal equation" (Schaffer, 1988). In general, information from measurement devices and computers is conveyed to the scientist through perceptual input, typically visually. The scientist must then perceive this information, which includes a transduction process, an attention process and a pattern recognition process. Each step of perception can introduce errors. Particularly important to science, the pattern recognition process is highly influenced by experience (Biederman, 1987; Polk & Farah, 1995) and expectations (Brewer & Treyens, 1981) and can produce errors of omission (i.e., failure to categorize a present complex object) and errors of

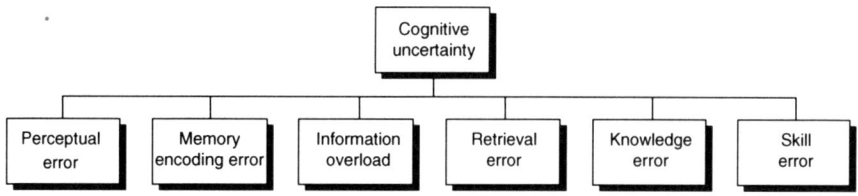

Figure 19.4   *Sources of uncertainty in data analysis coming from cognitive factors.*

commission (i.e., miscategorizations). This expectation-driven misperception may be connected to confirmation bias (Penner & Klahr, 1996).

In fMRI, sometimes the researcher will not notice certain areas of activation on the display in front of them, especially if the activation is in an unexpected area. Sometimes, researchers misperceive what condition is being displayed in an activation map—the font size on the condition labels is often too small. In weather forecasting, many of the weather models show quantitative data by displaying a color legend. Unfortunately, different color schemes can cause different perceptual illusions or cause some colors to be perceived incorrectly.

Memory encoding error. In many sciences, truly impressive amounts of information are often directly displayed in front of a person across large and multiple monitors. However, at any one time, only a very small amount of information is directly perceived. This deep limit on visual input implies that complex inputs must be stored and retrieved from memory. The human memory encoding process can be error prone in two different senses: stochastic selection and biased selection. First, only a small, fairly unpredictable subset of information that a person encounters is encoded correctly in memory. Second and especially relevant to confirmation biases in science, sometimes only particular aspects of the encountered situation are correctly entered into memory. Highly familiar information that can be grouped into familiar, meaningful chunks, for example, is more likely to be encoded (Chase & Simon, 1973). Information that violates expectations or is perceptually distinctive is more likely to be encoded (Bower, Black, & Turner, 1979). Information of emotional relevance is also more likely to be encoded (Cahill & McGaugh, 1995). Human memory encoding errors include both failures to encode information and misencodings of information.

In fMRI, patterns that occur in regions of the brain that a researcher is more familiar with are more likely to be encoded, whether they be expected patterns (and thus decomposable into familiar chunks) or unexpected patterns (and thus violations of expectations). In weather forecasting, the forecaster must compare and contrast not only different weather models with each other, but also must examine different weather models with "truth"from some time in the recent past. Most of these comparisons occur when the forecaster can only see one visualization at a time (i.e., it is difficult to geo- and time-reference different visualizations).

Information overload. A scientist can consider or be aware of only so much information at once, and many sciences involve large amounts of potentially relevant information. Although it is thought that there are no hard limits on how much information can be kept in working memory, the more information that is kept in working memory, the harder it is to rehearse all the elements to keep them there. By organizing information into templates for a particular situation (e.g., like the method of loci), experts are able to include keep truly impressive amounts of information in working memory. However, even expert working memory has clear limits that complex domains often exceed, and disconfirming evidence that does not "fit" into an overall model might be more likely to be dropped.

In fMRI, there are many regions of the brain that can become active, and most thought processes activate a number of areas in the brain. A given participant's brain activation is displayed across 10 to 30 brain slice images, and a given study often has 10 or more participants. In weather forecasting, the forecaster has access to thousands of weather visualizations showing different weather models, blends, satellite images, time courses, and so on. Because many of these variables interact with other variables at different geographical or atmospheric levels, the forecaster needs to keep track not only of different weather models, but also different variables within each weather model.

Retrieval error. Once information is actually encoded in memory there is no guarantee the information will be retrieved at a later point in time due to either inference and/or decay processes (Baddeley & Scott, 1971; Gillund & Shiffrin, 1984). Moreover, not only can information fail to be retrieved at all but also erroneous information could be retrieved instead. In either case, the retrieval errors might reflect a stochastic factor reflecting neural firing variability, or a bias toward remembering particular information (e.g., consistent with current expectations or the current theory under test, or more recently or frequently encountered information).

In fMRI, data from participants are often examined from one participant at a time because each participant can generate gigabytes of data and the computers can often process only one participant at a time. These memory retrievals for across-participant comparisons are prone to error, and researchers will bring back on the screen old images (or recompute images) to double check for such errors. In weather forecasting, forecasters will frequently attempt to retrieve a similar situation to the current forecasting problem. For example, a hurricane forecaster may attempt to retrieve a similar case but because there are very few tools for this kind of case-based reasoning, the forecaster's memory will likely be quite faulty.

Background knowledge error. Uncertainty can also be caused by failing to bring in appropriate background knowledge that changes the interpretation or prediction for a particular situation. These errors contrast with retrieval errors in that retrieval errors are information about a particular state of the world, whereas background knowledge errors are either a lack of knowledge or failed retrievals of knowledge about the general state of the world. In general, the same factors that produce episodic memory retrieval errors also produce semantic memory retrieval errors, and similarly confirmation biases can influence data analysis through selective retrieval of relevant semantic memories.

In fMRI, even experts will forget what particular regions on a brain map are. There are several different taxonomies for labeling brain areas, and particular researchers specialize in particular brain circuits. But activation that occurs in a new experiment could involve a region not previously examined by the particular researcher. In weather forecasting, background knowledge is required to adjust weather predictions, and experts sometimes forget (or not have) the relevant background knowledge for a particular situation. For example, one might not know that a given power plant is off on Sundays and this changes local weather patterns.

Skill error. Uncertainty can also arise from procedural errors in the data transformation or interpretation processes. For example, the scientist may fail to do a mental or external transformation step. Here the informational uncertainty is whether the transformation was done, not the error itself. Skill errors are more likely to occur in steps that are less practiced (Singley & Anderson, 1989; Woodrow & Stott, 1936), less recently practiced (Kyllonen & Alluisi, 1987), or when hurried (Grice & Spiker, 1979), fatigued (Krueger, 1994), or stressed (Alkov, Gaynor, & Borowsky, 1985; Beilcock & Carr, 2001).

In fMRI, the analysis process has many steps, is quite complex, has procedures that can vary depending on the situation, and is continually changing as new procedures are being introduced. Many researchers immediately want to use the latest innovations in analysis techniques. As a result, they will put up with having to do many steps by hand and suffer poorly designed and poorly documented software. In weather forecasting, coordination of weather models can be complex and errors occur in this coordination process. For example, predictions from the global weather model are provided as inputs to the mesoscale model. Sometimes forecasters, in their haste to develop a prediction, fail to verify accuracy of the global inputs before running the mesoscale model, or they (mistakenly) assume that their partner already validated the global inputs.

INDICATORS OF UNCERTAINTY IN SCIENTIFIC DATA ANALYSIS

Although scientific data always have some level of informational uncertainty in them, the level of uncertainty does fluctuate over time. Indeed, one dimension of expertise is the ability to detect moments of relatively high uncertainty in data analysis (Schunn, 2010; Schunn & Anderson, 1999; Schunn et al., 2007). What kind of indicators do scientists use to track the level of informational uncertainty? Our cognitive anthropology suggests there are four general kinds of indicators that scientists can use, which can be cast in terms of a general process of trying to find meaningful patterns in the data (see Figure 19.5), although they also allow for systematic biases to be introduced by the scientists.

See No or Unusually Weak Pattern

Not seeing any pattern or highly noisy patterns in data are good indicators of informational uncertainty. The world around us contains innumerable patterns at various levels, and when we do not see patterns in the data, we have good grounds to be suspicious of what we are seeing—perhaps the sensory equipment is broken or disconnected, perhaps an analysis transformation step was forgotten, and so on.

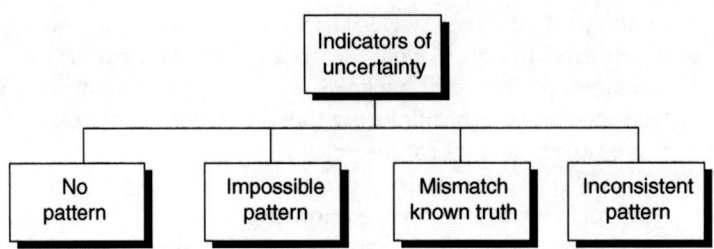

Figure 19.5   *Indicators used to detect relatively high levels of uncertainty.*

There are two important notes about this indicator. First, different sciences can have very different kinds of patterns to be seen, and being able to see patterns in a given scientific domain is a part of expertise in that domain (Kaplan & Simon, 1990; Kulkarni & Simon, 1988). That is, what appears as just noise to the novice may be highly ordered and informative to the expert. Second, most sciences have some level of noise, and another component of expertise is knowing what level of noise is acceptable or normal (Schunn & Anderson, 1999).

In fMRI, the activation profile (change in activation levels of a region over time) can follow a nice smooth curve or can be very jagged, indicating high uncertainty. Alternatively, an activation map (topographic map of areas of activation in a condition) can have clear clumps of activity or there can be many small points of activity with no clear organization, again indicating high uncertainty.

See an Impossible Pattern

A second important indicator of high levels of uncertainty is when an observed pattern clearly violates domain expectations. The most common form is an observation with values that are out of an acceptable range. However, sometimes the values are out of range for the particular type of situation being examined. In fMRI, it is not good to see data displays with extra coronal areas of activation (i.e., apparently activated brain areas outside the skull), and it makes one uncertain about all of the data one is seeing. In weather forecasting, the detectors of impossible patterns are sometimes automated. There is a program that notices anomalies in the data. Also, weather satellites sometimes have out of range tests in their instruments and will flag a probable error state when those ranges are exceeded (e.g., ground temperature readings that are much too high or much too low). The detection of impossible patterns also occurs in the meteorologist (e.g., noting predictions of snow over Georgia in the summer).

See a Pattern That Mismatches the Known State of the World

Not all data are considered equal; some data are considered to be direct measures of truth, whereas other data are considered as much more inferential and subject to issues of uncertainty. Thus, an important indicator of uncertainty is when a pattern from one of the less direct measures mismatches the pattern considered a known state of the world—it makes the problem solver doubt all of the patterns being currently displayed in the mismatching display. The difference between mismatching the known state of the world and being an impossible pattern is that the impossible pattern requires no reference to other data collection or manipulation of the world, whereas the mismatch to the known state of the world does require reference to some other data about the current state that is considered truth. One of the strongest forms of known truth comes from a manipulation of the environment—if the problem solver manipulates the environment, the problem solver may have very clear expectations for some of the impacts of this manipulation and these expectations are considered a known truth about the state of the world (cf. Baker and Dunbar [2000] on known control trials). But sometimes the current facts and laws in science that are taken as known truths are not completely accurate, and requiring all new data to fit existing facts and laws can be problematic. Even more problematic is the case when a scientist considers his or her own theory as a "fact," ignoring disconfirming evidence (Chinn & Brewer, 1992; Mitroff, 1974).

In fMRI, good researchers typically include validity tests in their designs—conditions for which the impact on brain activation is so well established that it is treated as a known. For example, one might have a simple visual stimulus and expect to see activation in early visual areas of the brain. In weather forecasting, there is less opportunity for manipulation but some sources of data are considered truth. For example, forecasters will sometimes overlay weather model predictions for a current or past time on top of a satellite image for that same time to see whether the weather model predictions match the current "truth."

See a Pattern That Is Inconsistent Across Data Sources

Finally, scientists can also use mismatches across multiple (equal status) data sources or (equal status) analysis methods. If a mismatch across data sources or analysis methods is detected, then data from all sources/methods are considered uncertain until further problem solving resolves which source is more likely to be correct. In fMRI, a scientist might use different analysis procedures and see whether the different analysis procedures produce similar results. In weather forecasting, forecasters will compare different data sources (e.g., National Oceanic and Atmospheric Administration [NOAA] vs. Navy sources) or different weather model predictions.

TAXONOMY OF STRATEGIES FOR DEALING WITH UNCERTAINTY

Having diagnosed the situation as being uncertain, what kind of strategies do scientists use to deal with uncertainty? In scientific data with high levels of uncertainty, much of problem solving can be focused on reducing uncertainty. In general, actively manipulating one's environment allows for greater success in finding patterns in uncertainty data (Klayman, 1986). Strategies can be used to reduce uncertainty but they are heuristic and can introduce other forms of information uncertainty (e.g., statistical uncertainty or cognitive uncertainty). Thus, it is not surprising that there are at least seven different kinds of strategies that we have observed. There is no strict chronological ordering or preference of use among these strategies. The order listed below, however, is roughly descriptive of the order in which strategies are likely to be applied (Figure 19.6).

We use the term strategies to refer to the kinds of responses experts use to deal with or reduce uncertainty. We prefer the term strategies to heuristics, algorithms, procedures, or methods because the term strategies highlights the facts that (a) scientists typically have multiple behaviors they can and do use to deal with uncertainty in data, (b) the choice of behavior is influenced by situation features, and (c) the behaviors are generally adaptive but not necessarily optimal.

Check for Likely Errors

When results are particularly surprising and especially toward the beginning of a problem-solving episode, the common first step is to look for likely sources of error. The particular sources to look for can depend upon the particular tools one is using—not all tools have the exact same set of common errors. However, we list some typical kinds of errors to look for in each of the three domains.

In fMRI, problem solvers typically look for common skill errors. For example, problem solvers often double check the splitfile—the way the data is parsed by condition. Conditions can be mislabeled, copied to incorrect directories, or be out of temporal phase with the brain imaging data. Here, the problem solver will reexamine the process by which the data were generated, and perhaps redo the splitfile from scratch. Weather forecasters also will look for skill errors. For example, sometimes a satellite picture or weather model may be mislabeled either on a weather portal or on the title of the visualization.

Focus on More Reliable Sources

Some data sources are more reliable than others, and a common response to high levels of uncertainty is to focus on the sources or kinds of data considered more reliable. It is worth noting, however, that reducing the kinds

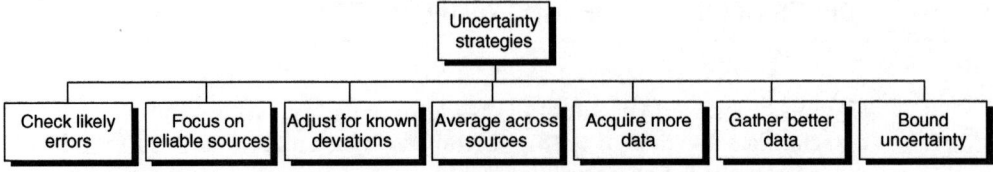

Figure 19.6   *Strategies commonly used to reduce uncertainty in data analysis.*

of information being examined can introduce other kinds of informational uncertainty or perhaps introduce confirmation biases (Gorman, 1992). In fMRI, scientists often focus on more reliable aspects of the data. For example, the scientist will typically adjust statistical thresholds in cases of high uncertainty so that only the most reliable results are displayed. Another common approach is to throw out high noise or low signal participants—in fact, it is not uncommon to throw out 10% to 40% of participants because of high noise or low signal problems. In weather forecasting, a forecaster will learn from experience that some weather models are particularly good at certain situations. For example, the forecaster may use weather model A when there is a low off the coast of Florida and a large Nor'easter (a storm blowing from the northeast) coming in December, but rely on weather model B when there is not a Nor'easter. Because of the relatively tight coupling between the forecast and truth (i.e., it is usually easy to determine if a forecast was correct), a forecaster quickly gains faith in certain models in certain conditions.

Adjust for Known Deviations From Truth

When some sources of data can be considered the true state of the world, then the expert problem solver often uses this information to adjust clear deviations from truth in more inferential, indirect sources of data. This adjustment can be done externally in software or it can be done mentally by the problem solver (Trickett & Trafton, 2007; Trickett et al., 2007). In weather forecasting, there are cases of both external and mental adjustments. Some software allows the forecaster to move features on the display by hand (e.g., move the locations of predicted lows). Another key activity is adjusting past state descriptions that weather models take as input. But verbal protocols of forecasters also show that forecasters will mentally adjust features in a predicted model that they consider in need of adjustment. In fMRI, the manipulations are primarily mental because it is considered unethical to manipulate data subjectively by hand (rather than clearly described and objective algorithms).

Average Across Sources/Analyses

When there are multiple sources of data and they are equally suspect, problem solvers can average (weighted or unweighted) across the sources of data to produce what they consider a more certain outcome. This averaging process, like the adjustment process, can be done in software or mentally (Trickett et al., 2007; Trickett et al., 2009). In fMRI, a common approach with high uncertainty is to move from individual participant data displays to group level average data displays, which is done in software. Weather forecasters will frequently average mentally across different weather models, producing not just a simple numerical average, but rather a complex combination mediated by a qualitative mental model of the overall situation that is developed (Trafton et al., 2000). Recently, forecasters have begun using statistical combinations of different models, a practice that is called ensemble forecasting.

Acquire Additional Data of the Same Kind

To deal with uncertainty, often patience is required; the problem solver must merely wait for more data of the same kind to resolve the uncertainty. A common effect of introducing noise into data is that many more trials are required to discover a pattern (Gorman, 1986; Penner & Klahr, 1996). It is important to note that waiting for more data can introduce additional uncertainty in the form of possible retrieval errors (forgetting what happened before) and future prediction uncertainty (increasing temporal lag since last collection of key data, like just after a maneuver in submarine operations). In fMRI, the researcher will often decide to collect data from additional participants, in order to see whether unexpected trends will continue or noisy data will average out more clearly. A weather forecaster can examine more time slices of recent data, or even reexamine the same data again to refresh his/her memory of the data.

Gather More Reliable Data

Some sources of data are less reliable than others, but may be used at a particular point in time because of cost/accuracy tradeoffs or may just happen to be the data that is currently available. When examining highly uncertain data that cannot be resolved through the simpler strategies described above, scientists often choose to gather more reliable data, which they may not have done prior to this point because of resource management issues. There have been some arguments about what kind of data is selected in response to the possibility of error, but in general there is not an increase in confirmation bias (or H+ tests), but there may be an increase in replication trials (Baker & Dunbar, 2000; Gorman, 1986; Penner & Klahr, 1996).

In fMRI, the researcher can use more detailed/accurate analysis procedures that are slower to run and thus not used first. Also, the researcher can decide to change the experiment to a more conservative structure and then collect more data (Schunn & Klahr, 2000). The more conservative experiment may have more trials per condition, collect baseline data more often (to deal with motion correction issues), or somehow make the manipulation stronger. A weather forecaster can request a weather balloon to be launched, which provides highly accurate and more recent information for a particular small region. Some more detailed models produce more accurate predictions, but are much slower to run, and thus are not typically used first.

Bound Uncertainty in Final Solution

Uncertainty does not always get resolved by the end of data analysis. Papers must be submitted to conferences and journals, and complete resolution of uncertainty is not necessary. The final common strategy that we have observed scientists use is to give explicit bounds on the uncertainty of their final solution. In fMRI, observed results are categorized as clear or marginal. For example, the researcher might state that the location of a particular activated region may actually be in this or another adjacent brain region.

In weather forecasting, ranges of values are often given (e.g., winds between 4 and 7 knots).

MEASURING PSYCHOLOGICAL UNCERTAINTY IN SCIENTISTS

To measure psychological uncertainty during science, one can use surveys, speech cues, or gestures. Using surveys is relatively straightforward. Speech and gesture are more complex and this is described in more detail.

Speech Coding of Uncertainty

One approach to coding uncertainty and approximation is syntactical with verification. Uncertainty hedge words include "probably," "sort of," "guess," "maybe," "possibly," "don't know," "(don't) think," "(not) certain," and "believe." Each instance of the hedge words should be examined to make sure it was being used in an uncertainty sense.

For example, we have been involved in coding of speech uncertainty in two different domains of science (Schunn et al., 2007; Trickett et al., 2009). The first domain involved conversations of earth scientists working at the Jet Propulsion Lab analyzing data as it came down from Mars from two robotic rovers—the Mars Exploration Rovers. The coded conversations were of impromptu meetings held throughout the day between groups of 2 to 10 scientists from several different disciplines (soil and rock scientists, geochemists, geologists, and atmosphere scientists). There were a number of video cameras off to the sides of the large data analysis rooms. The scientists had given informed consent for this video collection but the cameras were relatively small, discretely located, and constantly present. Thus, the scientists generally forgot about the existence of the cameras and the transcripts likely capture very typical problem-solving behaviors in this context.

The second domain involved cognitive neuroscientists analyzing fMRI data. After 30 to 45 minutes of data analysis, they were then shown three or four different minute-long snippets of the videotape that corresponded to critical decision-making moments during data analysis. The scientists were asked to explain what they knew and didn't know at that moment in time. Sometimes problem solvers given think-aloud instructions fall silent exactly at the interesting moments in time, especially when the task is long and complex. This cued recall method was designed to capture additional information about these more interesting moments. Across these two domains, we used the same hedge word technique for coding uncertainty from the transcribed speech, with interrater reliability Kappas greater than 0.8.

One source of validation involves the temporal reoccurrence of uncertainty statements: If one line had an uncertainty word, the next line was more likely to have an uncertainty word—here approximation terms were also coded and these terms were not temporally associated with uncertainty terms. The know/don't know probes in the study of fMRI provide another validation of the uncertainty codes (and coding was done blind to question context). One would expect that there would be more uncertainty speech cues in response to the "what did you not know?" question than in response to the "what did you know?" question. This is exactly what was found.

Gesture Coding of Uncertainty

In science and engineering, much of the data is inherently visual-spatial or is displayed in spatial format (e.g., graphs of temperature varying with time). Thus, much of the uncertainty is expressed about visual-spatial quantities. Because sciences have usually formalized much if not all of the quantities and relationships in symbolic formats (e.g., terms for particular quantitative data patterns, equations to represent quantitative data patterns), much can be studied from coding speech from conversations and think-alouds. However, it is likely that considerable representing, reasoning, and problem solving in science are also happening in a visual-spatial, nonverbal format.

How does one measure internal problem solving on visual-spatial content? All measures of mental representations and problem solving are necessarily indirect. Verbal report is one general source of data regarding mental representation and problem solving. However, for visual-spatial content, it is a suspect source, as verbal data are generally thought to capture the contents of verbal working memory, not spatial working memory (Ericsson & Simon, 1993). Another approach is to use spontaneous gestures. In addition to serving a communicative act between speaker and listener, spontaneous gestures are thought to be an online measure of mental representations much like verbal protocols (Alibali, Bassok, Solomon, Syc, & Goldin-Meadow, 1999; Alibali & Goldin-Meadow, 1993; McNeill, 1992). In spatial tasks, in fact it is disruptive to the problem solver to prevent gesturing from occurring.

In addition to coding gestures for complex representational content, gestures can also be used as a measure of uncertainty or approximation. There are a number of taxonomies of gesture. One common distinction (McNeill, 1992) is between beat gestures (rhythmic, repetitive gestures often cotimed with speech), deictic gestures (pointing to things in the world around the speaker such as the clock on the wall over *there*), iconic gestures (gestures that are literal physical presentations of things absent, such as hand shape holding an implied glass), and metaphoric gestures (a spatial representation of a nonspatial object, such as pointing behind oneself to represent back in time). All of these gestures can have many phases (McNeill, 2005): preparation (optional), prestroke hold (optional), stroke (obligatory), stroke hold (obligatory if the stroke is static), poststroke hold (optional), and retraction (optional). Uncertainty gestures are typically wiggling movements in the stroke of an iconic or metaphoric gesture that represents some quantity (i.e., normally would be static). For example, a pinch indicating a size together with wavering the size or wiggling the hand. In this way, the uncertainty gesture is discriminable from a beat gesture in that there is content to the gesture beyond the movement in an uncertainty gesture of this type but the beat gesture does not have content beyond the movement (i.e., the hand does not indicate a size or distance or volume). However, another common form of an uncertainty gesture involves a shoulder shrug. In this case, one must rely on speech or perhaps another gesture to determine which quantity is producing uncertainty.

How do we know such gestures correspond to psychological uncertainty? We have examined the overlap between uncertainty gesture and speech in the four science/applied science domains mentioned earlier. In every case, uncertainty gestures were statistically more common when there was uncertainty in the speech than when there was not uncertainty in the speech.

To further validate that there is indeed something called an uncertainty gesture that signals an internal state of uncertainty, we can examine gesture data from the fMRI study, focusing on the relative frequency of uncertainty gestures in response to the Know and Not-Know questions. Two percent of segments co-occurred with an uncertainty gesture during the response to the Know question. In response to the Not-Know question, rate of uncertainty gestures more than doubled.

SUMMARY

Uncertainty plays a very complex role in basic and applied science. From the in-depth analysis of the sources of informational uncertainty, it is clear that uncertainty can come from many sources and is likely to be high in most basic and applied science domains—at least in the real world cases, as opposed to the more simplified cases studied in the lab. Importantly, the scientist himself/herself can act as a source of information uncertainty, and this source may be particularly relevant to cases of confirmation biases, in which simple misencodings or misrememberings may occur in favor of a tested hypothesis rather than unethical behaviors.

To influence behavior, the scientist must diagnose states of informational uncertainty, and there are number of strategies that scientists have for making this diagnosis. Here expertise in science overall and in a particular domain of science can play a large role. The second layer of behavior involves mitigation: actions taken to reduce the informational uncertainty. Studies that have manipulated noise in the data even in simple tasks have found large effects on number of trials or probability of success. In complex science domains, there are many different strategies scientists can follow to reduce the informational uncertainty.

Finally, to measure psychological uncertainty, psychologists of science have many measurement tools at their disposal, most prominently speech and gesture. It is likely that the future of the psychology of science of uncertainty will include neuroscientific techniques, such as Evoke Response Potentials and fMRI, which have been used to study uncertainty in other domains such as language processing and simple task learning. And more cognitive anthropology, like the type that produced these presented taxonomies, will be useful to better understand how scientific tasks are constructed to deal with the ever-present informational uncertainty in science.

THE FUTURE OF UNCERTAINTY IN SCIENCE

The taxonomies of uncertainty, especially the indicators and strategies taxonomies, can be used to guide the development of additional automation to support problem solving in science domains with high uncertainty. Consider the case of indicators. Each of the four categories of indicators in our taxonomy can be automated to some degree, and thus one now has four different kinds of indicators that one can try to automate in any domain of high uncertainty. There are already some measures of noise in many domains (e.g., error bars on graphs), but other measures of noise in more complex visual displays can be developed. For example, one can make use of information theoretic measures of the degree of patterns in data. One can develop measures of mismatch across data sources. One can develop measures of mismatch from a particular source that is considered the known current state of the world. Finally, one can develop measures of deviations from theoretically possible states.

Similarly, one can develop automation for the strategies for dealing with informational uncertainty, and our taxonomy provides a set of general strategies that one might wish to automate. The value of automating the strategies is not as clear as in the case of indicators because doing automation (like automatic error correction in typing) is always more controversial than informing automation (like automatic underlining of errors in typing). Whether a given strategy should be automated in a given domain will depend upon two factors: (a) how simple the strategy is to automate relative to how effective people are already at implementing the strategy on their own (e.g., people are good at spatial transformations) and (b) how accurate the automated transformations are.

Another use of the uncertainty taxonomies is for psychology science research. Many researchers are interested in behavior in science domains of high uncertainty but tend to focus on particular sources, particular indicators, or particular strategies. There is certainly nothing wrong with focusing on particular elements in a complex situation. But these taxonomies help specify the contrast set, the set of alternative elements that might be considered as also contributing to performance and perhaps possible confounds in the research on the given element. For researchers beginning work in a previously unstudied domain with high uncertainty, these taxonomies provide a starting place for understanding the ecology of the presumably important uncertainty aspect of the domain.

REFERENCES

Abbaspour, R. A., Delavar, M. R., & Batouli, R. (2003). The issue of uncertainty propagation in spatial decision making. In K. Virrantaus and H. Tveite, (Eds.), *ScanGIS'2003: Proceedings of the 9th Scandinavian Research Conference on Geographical Information Science* (pp. 57–65). Espoo, Finland. Helsinki, Helsinki University of Technology.

Alibali, M. W., Bassok, M., Solomon, K. O., Syc, S. E., & Goldin-Meadow, S. (1999). Illuminating mental representations through speech and gesture. *Psychological Science, 10*(4), 327–333.

Alibali, M. W., & Goldin-Meadow, S. (1993). Gesture-speech mismatch and mechanisms of learning: What the hands reveal about a child's state of mind. *Cognitive Psychology, 25*(4), 468–523.

Alkov, R. A., Gaynor, J. A., & Borowsky, M. S. (1985). Pilot error as a symptom of inadequate stress coping. *Aviation, Space, and Environmental Medicine, 56*(3), 244–247.

Baddeley, A. D., & Scott, D. (1971). Short term forgetting in the absence of proactive interference. *Quarterly Journal of Experimental Psychology, 23*(3), 275–283.

Baker, L. M., & Dunbar, K. (2000). Experimental design heuristics for scientific discovery: The use of "baseline" and "known standard" controls. *International Journal of Human Computer Studies, 53*(3), 335–349.

Beilcock, S. L., & Carr, T. H. (2001). On the fragility of skilled performance: What governs choking under pressure? *Journal of Experimental Psychology: General, 130*(4), 701–725.

Berkeley, D., & Humphreys, P. (1982). Structuring decision problems and the "bias heuristic". *Acta Psychologica, 50*(3), 201–252.

Biederman, I. (1987). Recognition-by-components: A theory of human image understanding. *Psychological Review, 94*(2), 115–117.

Bottom, W. P. (1998). Negotiator risk: Sources of uncertainty and the impact of reference points on negotiated agreements. *Organizational Behavior and Human Decision Processes, 76*(2), 89–112.

Bower, G. H., Black, J. B., & Turner, T. J. (1979). Scripts in memory for text. *Cognitive Psychology, 11*(2), 177–220.

Brashers, D. E., Neidig, J. L., Russell, J. A., Cardillo, L. W., Haas, S. M., Dobbs, L., . . . Nemeth, S. (2003). The medical, personal, and social causes of uncertainty in HIV illness. *Issues in Mental Health Nursing, 24*, 497–522.

Brewer, W. F., & Treyens, J. C. (1981). Role of schemata in memory for places. *Cognitive Psychology, 13*(2), 207–230.

Cahill, L., & McGaugh, J. L. (1995). A novel demonstration of enhanced memory associated with emotional arousal. *Consciousness and Cognition: An International Journal, 4*(4), 410–421.

Chase, W. G., & Simon, H. A. (1973). The mind's eye in chess. In W. G. Chase (Ed.), *Visual information processing* (pp. 215–281). New York, NY: Academic Press.

Chinn, C. A., & Brewer, W. F. (1992). *Psychological responses to anomalous data.* Paper presented at the 14th Annual Meeting of the Cognitive Science Society, Bloomington, IN.

Cohen, M. S., Freeman, J. T., & Thompson, B. (1998). Critical thinking skills in tactical decision making: A model and a training strategy. In J. A. Cannon-Bowers & E. Salas (Eds.), *Making decisions under stress: Implications for individual and team training* (pp. 155–189). Washington, DC: American Psychological Association.

Ericsson, K. A., & Simon, H. A. (1993). *Protocol analysis: Verbal reports as data* (2nd ed.). Cambridge, MA: MIT Press.

Gillund, G., & Shiffrin, R. M. (1984). A retrieval model for both recognition and recall. *Psychological Review, 91*, 1–67.

Gorman, M. E. (1986). How the possibility of error affects falsification on a task that models scientific problem solving. *British Journal of Psychology, 77*, 85–96.

Gorman, M. E. (1992). *Simulating science: Heuristics, mental models and technoscientific thinking.* Bloomington, IN: Indiana University Press.

Grice, G. R., & Spiker, V. A. (1979). Speed-accuracy tradeoff in choice reaction time: Within conditions, between conditions, and between subjects. *Perception and Psychophysics, 26*(2), 118–126.

Hall, K. H. (2002). Reviewing intuitive decision-making and uncertainty: The implications for medical education. *Medical Education, 36*, 216–224.

Howell, W. C., & Burnett, S. A. (1978). Uncertainty measurement: A cognitive taxonomy. *Organizational Behavior and Human Decision Processes, 22*(1), 45–68.

Hutchins, E. (1995). *Cognition in the wild.* Cambridge, MA: MIT Press.

Kahneman, D., & Tversky, A. (1982). Variants of uncertainty. *Cognition, 11*(2), 143–157.

Kaplan, C. A., & Simon, H. A. (1990). In search of insight. *Cognitive Psychology, 22*, 374–419.

Klayman, J. (1986). Cue discovery in probabilistic environments: Uncertainty and experimentation. *Journal of Experimental Psychology: Learning, Memory and Cognition, 14*(2), 317–330.

Klein, G. A. (1989). Strategies of decision making. *Military Review, May,* 56–64.

Krivohlavy, J. (1970). Subjective probability in experimental games. *Acta Psychologica, 34*(2,3), 229–240.

Krueger, G. P. (1994). Fatigue, performance, and medical error. In M. S. Bogner (Ed.), *Human error in medicine* (pp. 311–326). Hillsdale, NJ, England: Lawrence Erlbaum Associates, Inc.

Kulkarni, D., & Simon, H. A. (1988). The process of scientific discovery: The strategy of experimentation. *Cognitive Science, 12,* 139–176.

Kyllonen, P. C., & Alluisi, E. A. (1987). Learning and forgetting facts and skills. In G. Salvendy (Ed.), *Handbook of human factors* (pp. 124–153). Oxford, England: John Wiley & Sons.

Lipshitz, R., & Strauss, O. (1997). Coping with uncertainty: A naturalistic decision-making analysis. *Organizational Behavior and Human Decision Processes, 69*(2), 149–163.

McNeill, D. (1992). *Hand and mind: What gestures reveal about thought.* Chicago, IL: University of Chicago Press.

McNeill, D. (2005). *Gesture and thought.* Chicago, IL: University of Chicago Press.

Mitroff, I. I. (1974). *The subjective side of science.* New York, NY: Elsevier.

Musgrave, B. S., & Gerritz, K. (1968). Effects of form of internal structure on recall and matching with prose passages. *Journal of Verbal Learning and Verbal Behavior, 7*(6), 1088–1094.

O'Connor, R. M., Doherty, M. E., & Tweney, R. D. (1989). The effects of system failure error on predictions. *Organizational Behavior and Human Decision Processes, 44*(1), 1–11.

Penner, D. E., & Klahr, D. (1996). When to trust the data: Further investigations of system error in a scientific reasoning task. *Memory & Cognition, 24*(5), 655–668.

Polk, T. A., & Farah, M. J. (1995). Late experience alters vision. *Nature, 376*(6542), 648,649.

Rowe, W. D. (1994). Understanding uncertainty. *Risk Analysis, 14,* 743–750.

Schaffer, S. (1988). Astronomers mark time: Discipline and the personal equation. *Science in Context, 2*(1), 115–145.

Schunn, C. D. (2010). From uncertainly exact to certainly vague: Epistemic uncertainty and approximation in science and engineering problem solving. In B. Ross (Ed.), *Psychology of learning and motivation* (Vol. 53, pp. 227–252).Burlington, IN: Academic Press.

Schunn, C. D., & Anderson, J. R. (1999). The generality/specificity of expertise in scientific reasoning. *Cognitive Science, 23*(3), 337–370.

Schunn, C. D., & Klahr, D. (2000). Discovery processes in a more complex task. In D. Klahr (Ed.), *Exploring science: The cognition and development of discovery processes* (pp. 161–199). Cambridge, MA: MIT Press.

Schunn, C. D., Saner, L. D., Kirschenbaum, S. K., Trafton, J. G., & Littleton, E. B. (2007). Complex visual data analysis, uncertainty, and representation. In M. C. Lovett & P. Shah (Eds.), *Thinking with data* (pp. 27–64). Mahwah, NJ: Erlbaum.

Shah, P., & Hoeffner, J. (2002). Review of graph comprehension research: Implications for instruction. *Educational Psychology Review, 14*(1), 47–69.

Sheer, V. C., & Cline, R. J. (1995). Testing a model of perceived information adequacy and uncertainty reduction in physician/patient interactions. *Journal of Applied Communication Research, 23,* 44–59.

Singley, M. K., & Anderson, J. R. (1989). *The transfer of cognitive skill.* Cambridge, MA: Harvard Press.

Suchman, L. A. (1987). *Plans and situated action: The problem of human-machine communication.* New York, NY: Cambridge University Press.

Trafton, J. G., Kirschenbaum, S. S., Tsui, T. L., Miyamoto, R. T., Ballas, J. A., & Raymond, P. D. (2000). Turning pictures into numbers: extracting and generating information from complex visualizations. *International Journal of Human Computer Studies, 53*(5), 827–850.

Trickett, S. B., Fu, W. T., Schunn, C. D., & Trafton, J. G. (2000). From dipsy-doodles to streaming motions: Changes in representation in the analysis of visual scientific data. *Proceedings of the 22nd Annual Conference of the Cognitive Science Society*. Mahwah, NJ: Erlbaum.

Trickett, S. B., & Trafton, J. G. (2007). "What if …": The use of conceptual simulations in scientific reasoning. *Cognitive Science, 31*(5), 843–875.

Trickett, S. B., Trafton, J. G., Saner, L., & Schunn, C. D. (2007). *"I don't know what's going on there"*: The use of spatial transformations to deal with and resolve uncertainty in complex visualizations. In M. C. Lovett & P. Shah (Eds.), *Thinking with data* (pp. 65–85). Mahwah, NJ: Erlbaum.

Trickett, S. B., Trafton, J. G., & Schunn, C. D. (2009). How do scientists respond to anomalies? Different strategies used in basic and applied science. *Topics in Cognitive Science, 1*(4), 711–729.

Trope, Y. (1978). Inferences of personal characteristics on the basis of information retrieved from one's memory. *Journal of Personality and Social Psychology, 36*(2), 93–106.

Tweney, R. D., Doherty, M. E., & Mynatt, C. R. (Ed.). (1981). *On scientific thinking*. New York, NY: Columbia University Press.

Urbany, J. E., Dickson, P. R., & Wilkie, W. L. (1989). Buyer uncertainty and information search. *Journal of Consumer Research, 16*(2), 208–215.

Walker, V. R. (1991). The siren songs of science: Toward a taxonomy of scientific uncertainty for decision makers. *Connecticut Law Review, 23*, 567.

Walker, V. R. (1998). Risk regulation and the "faces" of uncertainty. *Risk: Health, Safety, & Environment, 27*, 27–38.

Woodrow, H., & Stott, L. H. (1936). The effect of practice on positive time-order errors. *Journal of Experimental Psychology, 19*, 694–705.

Past and Future of Psychology of Science

Quantitative Trends in Establishing a Psychology of Science: A Review of the Metasciences

Gregory D. Webster

*H*ow does science evolve and how does a new science emerge? What factors contribute to the development of a new scientific field? Is the establishment of a new journal or a regular conference more important to the growth of a new branch of science? These are just some of the questions I will explore in this chapter. I will also present some philosophical perspectives on the nature of knowledge and the evolution of science. I will then summarize some of my prior research (Webster, 2008) to show how establishing a journal and a regular meeting or conference affects a metascience's growth in terms of scholarly work and citations. Finally, I will discuss some new directions for the psychology of science, including why it may be vital to understanding a world of increasingly too much information, rather than too little.

PERSPECTIVES ON THE EVOLUTION OF SCIENCE AND KNOWLEDGE

The Metasciences

There are at least four metasciences that have presented their perspectives on the evolution of science. These are the philosophy, history, sociology, and psychology of science. The oldest of the four metasciences, the philosophy of science, concerns itself with the nature of knowledge, and how it is obtained, processed, and interpreted. The philosophy of science has historically been the most influential of the four metasciences, and has produced some of the most influential work on how science and knowledge evolve over time (e.g., Kuhn, 1962; Popper, 1935/2002). The second oldest metascience is the history of science, which documents the purported antecedents and consequences of key scientific events and achievements over time. The history of science also concerns itself with the lives of individual scientists and the historical contexts that may have contributed to their scientific achievements. The third

488 SECTION VI. PAST AND FUTURE OF PSYCHOLOGY OF SCIENCE

oldest metascience is the sociology of science, which focuses on the broader social, economic, and political contexts of science. For example, some sociologists of science argue that because science does not take place in a social vacuum, it may be biased by many of the same social and economic forces that affect any intellectual enterprise. The newest metascience to emerge is the psychology of science, which, among other things, studies the behavior and development of scientists and how people understand and perceive science (Feist, 2006a, 2006b; Simonton, 1988).

Philosophical and Epistemological Perspectives

There are several philosophical perspectives on how and why scientific knowledge evolves over time. Karl Popper (1935/2002, 1999) proposed that scientific theories must be empirically falsifiable—they must be capable of being shown to be false (e.g., via experimentation). Theories that are not falsifiable (e.g., psychoanalysis), he argued, are simply not scientific. Thus, though many studies may appear to support or verify a given theory, a single study or observation can effectively falsify a theory. Popper believed that scientific progress was an evolutionary process, whereby multiple tentative theories relating to a given problem situation are systematically subjected to rigorous testing that must allow for falsifiability. During this process of selection or error elimination, theories that "survive" empirical scrutiny are, in an evolutionary sense, more "fit" than theories that do not stand up to scrutiny. These relatively robust theories go on to inform or inspire new problem sets or situations, and the cycle repeats itself.

Along similar lines, Thomas Kuhn (1962) proposed a view of scientific progression that advances in fits and starts through scientific revolutions and theoretical paradigm shifts. Kuhn believed that science progresses in three stages. First, in the prescience stage, accumulated knowledge in a given area does not have a central paradigm or formal organizational structure. Second, in the normal science stage, scientists apply a central paradigm to an increasing scope of problems (Kuhn, 1970). When inconsistencies with a central paradigm are observed, they are typically attributed to the researcher(s) rather than to an error in the paradigm (i.e., Popperian falsification can be, for a time, violated). Third, when enough inconsistencies with a central paradigm have accumulated over time, the crisis stage begins, and a new paradigm subverts or subsumes the old one.

In contrast to these views, Feyerabend (1975) proposed that inconsistencies in the ways the scientific method is used both within and between scientific fields—and across time—make it nearly impossible to objectively define—let alone track—scientific progress over time. More specifically, shifts may occur over time in what people consider to be "science" or "scientific." For example, the pseudosciences of astrology, phrenology, and psychoanalysis were once considered by many to be at least marginally legitimate and empirical. In addition, imagine contrasting the medical science 200 years ago with what we would consider medial science today. (Or, if you have a time machine handy, consider comparing the medical science of today with what will be considered medical science 100 years from now.)

Recently, empirical skeptic Nassim Nicholas Taleb (2005, 2010) has offered some alternative epistemological views. In his books *Fooled by Randomness* and *The Black Swan*, Taleb discusses the role chance plays in markets, life, and even scientific discoveries. He stresses that many of the greatest scientific achievements in history were not planned or predicted by theory, but were instead the result of chance, serendipity, dumb luck, or exhaustive trial-and-error learning. For example, Alexander Flemming's discovery of the revolutionary antibiotic panacea penicillin came about purely by accident when he became curious about the mold that was killing the bacteria he was cultivating. Thus, as Taleb points out, the roles of chance and randomness in the evolution of science and knowledge are often unrecognized, underestimated, and woefully underappreciated. Nevertheless, as Louis Pasteur once cautioned, "In the fields of observation, chance favors the prepared mind." Thus, although chance and randomness play key roles in scientific discovery, they may favor scientists who are particularly hard working (iterative trial-and-error learning) or creative (seeing things in new ways).

Given these differing views on epistemological evolution, it may nevertheless be possible to reconcile some of Kuhn's (1962) and Popper's (1935/2002) perspectives. Because of the power that scientific falsification affords, it can result in the type of instant paradigm shifts Kuhn discussed. That is, falsification is not typically a gradualist enterprise. Falsification can be analogous to removing a keystone or cornerstone in a brick structure; when it is removed, it causes several of the bricks above it to collapse together. In this sense, Popperian falsification can result in Kuhnian paradigm shifts. When this occurs on a large scale, new scientific disciplines can emerge from the resulting rubble; the same bricks can be rearranged to build newer, more robust structures.

Metascientific Perspectives

A more specific question within epistemological evolution is this: What factors contribute to the evolution of scientific knowledge and new scientific breakthroughs? The metasciences have identified several possible factors. For example, findings from the history and sociology of science have suggested that massive increases in funding over the last two centuries—from both the public and private sectors—have dramatically contributed to the growth and diversification of scientific knowledge. Throughout the 20th century, many would agree that major political and economic events—such as two World Wars and a Cold War—contributed to scientific progress, for better or worse. Nevertheless, because there are no controlled experiments in history, it is impossible to say for certain how the state of scientific knowledge would have been different without such events. One can certainly imagine counterfactual outcomes (Roese, 2005) or alternative histories in which the same—or different—events cause wildly different outcomes (e.g., "What if the atom had never been split?").

To be sure, technological innovations also affect scientific progress. Inventions such as the Internet and the printing press have revolutionized how information and knowledge are gathered, shared, and communicated.

Sometimes technological innovations have unforeseen social consequences that affect science. Witness the increase in scientific collaboration and the rise of scientific teams over the last 15 years that may be a result of increased communication through e-mail and the Internet (Wuchty, Jones, & Uzzi, 2007).

The recent rise in "team science" also highlights how both individual differences and interpersonal processes can affect the evolution of science. Differences in individual creativity, intellect, and ambition can dramatically affect scientific progress. Indeed, individual differences in ambition, creativity, and work ethic may be partly responsible for the fact that a small number of scientists are responsible for a disproportionate amount of research productivity. In many scientific fields, it is common that about 80% of the research output is produced by only about 20% of all researchers. Such 80% to 20% distributions are often referred to as following Pareto's 80–20 Rule (Barabási, 2003; Taleb, 2010). This concept is related to the Matthew Effect (Merton, 1968) or "the rich get richer" effect, where authors who are highly cited earlier in their careers often receive a stellar reputation, and hence are also cited disproportionately more throughout the rest of their careers.

Supporting the possibility of a Matthew Effect in psychology, a recent study of over 600 counseling and industrial-organizational professors at American universities showed that a minority of researchers were responsible for the majority of research productivity (articles) and impact (Duffy, Jadidian, Webster, & Sandell, 2011). Although the skew was not as severe as a 20:80 ratio (Barabási, 2003; Taleb, 2010), it was nevertheless strong; about one-third of researchers were responsible for about two-thirds of all peer-reviewed journal publications; about one-fourth of researchers were responsible for about three-fourths of all citations. Indeed, the top tenth of cited researchers in these areas of psychology were responsible for the majority of all citations. Given these facts, we cannot easily dismiss the role that individual differences in personality play in the evolution of science. An especially creative or driven scientist is more likely to have ideas that "stick" than is an equally intelligent and capable scientist who lacks these traits.

In addition to *intra* personal processes, *inter* personal processes may also shape the evolution of scientific knowledge. Many scientists and their teams are embedded in social and professional networks that allow for the rapid sharing and dissemination of ideas and information. For example, research has shown that social and personality psychologists are often embedded in "invisible colleges"—networks of researchers that are revealed through authorship and citation networks (i.e., who writes with whom and who cites whom; Quiñones-Vidal, López-García, Peñaranda-Ortega, & Tortosa-Gil, 2004). These invisible colleges are the mass-production machinery of big science. Being part of a large, well-connected network of scientific researchers, allows one to do large-scale science in a relatively efficient way, where a big problem or question is broken down into parts and each researcher or team works on their own part. Although many great scientific discoveries continue to be made by individuals, many more will likely be made through collective scientific endeavors in the future.

THE EVOLUTION OF NEW SCIENTIFIC FIELDS

Although there are many perspectives on epistemological evolution, comparatively little research has focused on the more specific questions of how a new branch of science (or metascience) develops and how to evaluate its growth. Feist (2006a) shed some light on this topic by proposing that metasciences have evolved in three overlapping stages: (a) isolation, (b) identification, and (c) institutionalization. First, in the isolation stage, scholars pursue their work on a given problem separately, often without much collaboration, and often without realizing their nascent discipline. For example, a researcher studying scientific creativity some 60 years ago could not have known that he or she was setting the stage for what would later fall under the emerging field of psychology of science. Second, in the identification stage, researchers give the nascent field a name (e.g., "psychology of science"), and they accept it and identify with it by holding sporadic meetings. Third, in the institutionalization stages, the new scientific field undergoes rapid growth; interested researchers multiply, societies form and meet regularly, dedicated scholarly journals are published, and graduate programs are established (Table 20.1). Regarding the metasciences, the philosophy, history, and sociology of science have progressed to the third stage (institutionalization), whereas the psychology of science, which is comparatively new, is arguably somewhere in between the second and third stages (identification → institutionalization).

A Recent Example: The Evolution of Evolutionary Psychology

The growth of the new science of evolutionary psychology provides a good example of Feist's (2006a) stage model from outside of the metasciences. Evolutionary psychology is the scientific application of biological principles to help explain behaviors (e.g., adaptations). The growth of evolutionary psychology can be measured in its proliferation of journals, meetings, and graduate programs.

Evolutionary psychology published its first dedicated journal, *Evolution and Human Behavior* (originally titled *Ethology and Sociobiology*) in 1979. A review of publication trends in this journal showed a significant increase in empirical articles over time (Webster, 2007c). Moreover, its impact factor—an index of how often the average article is cited in the first 2 years of its publication often tied to a journal's prestige—increased between 1998 and 2009 from 0.957 to 3.594, an increase of 276% (see also Webster, 2007c). With evolutionary psychology's growth came additional journals such as *Human Nature* (1990), *Evolutionary Psychology* (2003), *Journal of Social, Evolutionary, and Cultural Psychology* (2007), and *Frontiers in Evolutionary Psychology* (2010).

In addition to journals, scientific meetings that focus on evolutionary psychology have also proliferated, from the International Society for Human Ethology (1972–present) to the Human Behavior and Evolution Society (1989–present) to the to the NorthEastern (United States) Evolutionary Psychology Society (2007–present). Moreover, in psychology, doctoral programs in evolutionary psychology have been established in multiple universities. Finally, publication and citations that reference "evolutionary psychology" have increased rapidly over time, outpacing and eclipsing the growth of its sister field, sociobiology (Webster, 2007d), and showing substantial growth in influencing related fields like cognitive neuroscience (Webster, 2007a) and social–personality psychology (Webster, 2007b). This brief history of evolutionary psychology suggests that it has reached the institutionalization stage of Feist's (2006a) model.

A Detailed Analysis: The Evolution of the Metasciences

Background and Predictions

Another way to examine the evolution of new scientific fields is to do so empirically. One can gauge a science's growth and influence by counting the works its researchers produce and citations to those works. In this part of the chapter, I will summarize my prior work (Webster, 2008)—a quantitative analysis of publication trends in the metasciences. In this research, I made two central predictions:

1. Establishing a journal will lead to immediate increases in academic search engine hit counts (i.e., number of scholarly works and citations to them). Specifically, positive shifts in the mean (intercept discontinuity), change-over-time slope (slope discontinuity), or both, in the number of hit counts will immediately follow the publication of an inaugural journal issue in each of the metasciences, except the psychology of science, which is comparatively too new
2. Establishing a dedicated conference will show the same effects as those predicted for establishing a dedicated journal

Method

To gather data, I used the Google Scholar Internet search engine. This allowed me to record the hit counts (i.e., number of scholarly publications and citations) for all four metasciences over time. Specifically, I made searches for "philosophy of science," "history of science," "sociology of science," and "psychology of science" at 1-year intervals from 1900 through 2005. I recorded the number of hits per year for each metascience. I chose the year 1900 because it allowed me to establish a credible baseline prior to the establishment of a journal or scientific meeting for each metascience. I chose the year 2005 because there is a time lag for articles and citations to be indexed on Google Scholar, which catalogs thousands of books and journals. Having a time span of this length allowed me to conduct interrupted time-series analyses for each metascience (see Shadish, Cook, & Campbell, 2002).

I chose to log-transform the hit counts because they were count data and, consequently, severely positively skewed (see McClelland, 2000). This log transformation sufficiently normalized the data. I ran a series of interrupted time-series analyses using Feist's (2006a) review of key historical events in

Table 20.1   *Journal and Conference Establishment in the Metasciences:*
Dates and Descriptions

First Dedicated Journal			First Regular Conference	
Metascience	Year	Title	Year	Name
Philosophy of science	1934	*Philosophy of Science*	1949	First International Congress for the Philosophy of Science
History of science	1912	*Isis*	1924	History of Science Society (United States) starts holding annual conferences
Sociology of science	1971	*Science Studies* (renamed *Social Studies of Science*)	1988	National Association for Science, Technology, and Society (NASTS) forms
Psychology of science	2008	*Journal of Psychology of Science and Technology*	1986	The Psychology of Science conference at Memphis State University

the metasciences as guideposts. Thus, an interrupted time-series analysis was performed at two time points for each metascience: (a) establishment of a dedicated journal and (b) establishment of a dedicated, recurring conference. I followed quasi-experimental procedures outlined by Shadish et al. (2002; see also Webster, 2007d). Interrupted time series analyses allowed for the assessment of discontinuities in intercepts (mean shifts) and change-over-time slopes (trend or slope shifts) at a specific point in time (e.g., immediately following the publication of a new journal or the establishment of a new conference).

Interrupted time-series designs are a family of quasi-experimental methods, whereby the dependent variable is measured repeatedly over several (preferably equal) intervals of time, and the independent variable often represents a single important (often historical) event or the presence or absence of something important over the course of the time series (e.g., an influential new work is published, a terrorist attack occurs, a new law is passed, a drug is administered). The independent variable is typically coded dichotomously with dummy codes (0, 1) or effects codes (−0.5, 0.5) representing the repeated observations before and after the event of interest.

The first step tests for an intercept shift by regressing the dependent variable onto time and the independent variable simultaneously. The effect of interest is the independent variable, which tests the intercept shift—the extent to which the independent variable's introduction is associated with a discontinuity in the regression line over time. The second step tests for a slope shift by regressing the dependent variable onto time, the independent variable, and the time × independent variable interaction simultaneously. The effect of interest is the interaction, which tests the slope shift—the extent to which the change-over-time slope differs before and after the introduction of the independent variable.

Any combination of null and significant effects for intercept and slope shifts is possible. Mean-centering the time variable (by subtracting its mean from each observation) and the independent variable (by using effects codes [−0.5, 0.5] instead of dummy codes [0, 1]) can often aid in the interpretation of the regression coefficients; this was done in the present analyses.

Results

Results were mixed. As predicted, each of the established metasciences (i.e., all except the psychology of science) saw a significant increase in either intercept (mean hit counts) or slope (change over time in hit counts) immediately after the establishment of a dedicated journal (Prediction 1). Contrary to prediction, none of the metasciences' hit counts showed a significant shift in intercept or slope immediately following the establishment of a regular conferences (Prediction 2). I describe detailed results for each metascience below.

Philosophy of science. I found significant intercept and change-over-time slope shifts in hits for philosophy of science following the establishments of its eponymous journal, *Philosophy of Science*, first published in 1934 (Table 20.2, Figure 20.1). That is, establishing a journal was related not only to an immediate increase in scholarly works and citations attributed to the philosophy of science, but also a nontrivial increase in its trajectory over time.

History of science. I initially ran an interrupted time-series analysis on all the data from 1900 to 2005. This preliminary analysis revealed no significant effects (Table 20.3, left). Indeed, the intercept shift was actually negative immediately following the publication of the journal *Isis* in 1912. The range of years (1900–1912), however, was insufficient for establishing a stable baseline (only 13 years), and thus underpowered for detecting an effect, should one exist. To establish longer, more reliable and stable baseline, I chose to expand the sample back to 1880, thus allowing for a prejournal baseline of about 33 years. When I reran the analysis, I found a negative shift in the intercept, but more importantly, I also found a significant positive shift in the change-over-time slope (Table 20.3, Figure 20.2). Thus, establishing a journal for the history of science did not have an immediate positive impact on publications and citations, but it did have a substantial long-term impact on the same. In other words, establishing a journal in this case appeared to have a delayed, but meaningful, effect.

Sociology of science. I first conducted analyses on log hit counts from 1900 to 2005. These analyses showed significant shifts in both the intercept and change-over-time slope for the sociology of science. These analyses, however, were biased because hit counts for sociology of science did not begin to regularly deviate from zero until the middle 1930s. Including all of these years with hit counts near zero made the baseline artificially low, which made the prejournal slope flatter than expected, which in turn may have biased the slope-shift test in a liberal direction (Type I error). To remedy this possible bias, I conducted more conservative analyses using the years 1936 to 2005. The new analyses showed a significantly positive intercept shift following the establishment of a dedicated journal; however, the change-over-time slope shift was no longer significantly positive (Table 20.4, Figure 20.3). These

Table 20.2   *Interrupted Time-Series Results for the Effect of Journal Establishment on Log Google Scholar Hits for Philosophy of Science, 1900–2005*

Variable	*b*	*t*	*pr*
Step 1: Mean shift			
Intercept	3.060	78.86[b]	0.99
Time (publication year)	0.061	33.96[b]	0.96
Journal effect (1934)	**1.116**	**9.43[b]**	**0.68**
Step 2: Slope shift			
Intercept	2.940	42.92[b]	0.98
Time (publication year)	0.056	18.56[b]	0.88
Journal effect (1934)	1.269	9.26[b]	0.68
Time × journal effect	**0.013**	**2.11[a]**	**0.20**

Note. Effects of interest appear in boldface.
[a] $p < 0.05$; [b] $p < 0.001$.

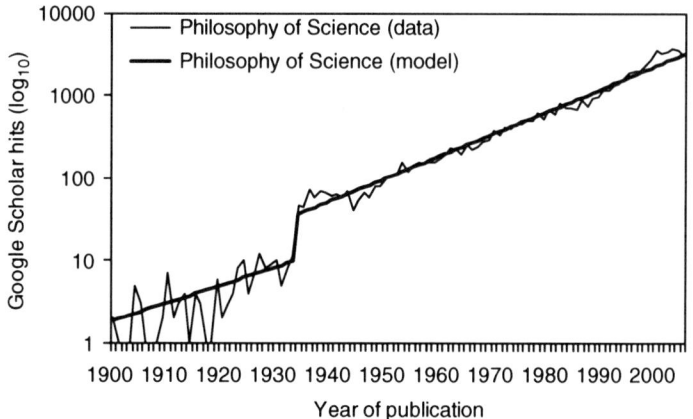

Figure 20.1   *Interrupted time-series analysis results for Google Scholar hits (Log$_{10}$ Scale) for "Philosophy of Science" as a function of publication year (1900–2005) and journal establishment (1934), 1900–2005.*

results suggest that establishing a journal for the sociology of science had an immediate impact on the field's articles and citations, but it did not have a significant impact of the field's long-term growth trajectory.

Psychology of science. The psychology of science's only stand-alone journal, *Journal of the Psychology of Science and Technology,* was first established in 2008. Thus, it was too recent to allow for an empirical test of its affect on publication and citations in the field. Instead, I chose to conduct a simple change-over-time regression on log publication and citation hits for psychology of science (Table 20.5, Figure 20.4). As with the sociology of science, there were too many years during the first half of the 20th century with

Table 20.3   *Interrupted Time-Series Results for the Effect of Journal Establishment on Log Google Scholar Hits for History of Science*

Variable	1900–2005			1880–2005		
	b	*t*	*pr*	*b*	*t*	*pr*
Step 1: Mean shift						
Intercept	2.425	49.75[a]	0.98	2.548	63.96[a]	0.99
Time (publication year)	0.055	50.76[a]	0.98	0.054	42.79[a]	0.97
Journal effect (1912)	**−0.262**	**−2.52[b]**	**−0.24**	**−0.456**	**−4.32[b]**	**−0.36**
Step 2: Slope shift						
Intercept	2.413	27.57[a]	0.94	2.418	35.47[a]	0.95
Time (publication year)	0.053	4.42[a]	0.40	0.047	14.48[a]	0.80
Journal effect (1912)	−0.239	−1.37	−0.13	−0.249	−1.83[c]	−0.16
Time × journal effect	**0.004**	**0.16**	**0.02**	**0.015**	**2.34***	**0.21**

Note. Effects of interest appear in boldface.
[a] $p < 0.001$; [b] $p < 0.05$; [c] $p < 0.10$.

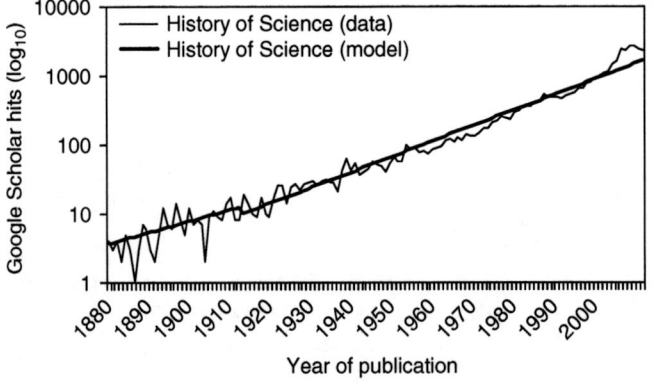

Figure 20.2   *Interrupted time-series analysis results for Google Scholar hits (Log$_{10}$ Scale) for "History of Science" as a function of publication year (1880–2005) and journal establishment (1912).*

zero hit counts; any analyses would be biased by this fact. Thus, I ran analyses using the years 1950 to 2005, because 1950 was the year in which hit counts for psychology of science first began to deviate regularly from zero. As expected, the change-over-time slope was significantly positive for psychology of science, suggesting that it was a field with growing influence in the scientific literature.

Comparing the four metasciences. To compare the growth trajectories of the four metasciences over time, I conducted four simple regressions on log hit counts—one for each metascience—from 1950 to 2005 (Table 20.5, Figure 20.5). These results showed that, despite starting at different times, the philosophy,

Table 20.4   *Interrupted Time-Series Results for the Effect of Journal Establishment on Log Google Scholar Hits for Sociology of Science*

	1900–2005			1936–2005		
Variable	*b*	*t*	*pr*	*b*	*t*	*pr*
Step 1: Mean shift						
Intercept	3.344	57.05[a]	0.98	3.308	81.24[a]	0.99
Time (publication year)	0.045	16.19[a]	0.85	0.083	20.58[a]	0.93
Journal effect (1971)	**1.977**	**10.84[aa]**	**0.73**	**0.732**	**4.49[a]**	**0.48**
Step 2: Slope shift						
Intercept	2.913	31.26[a]	0.95	3.272	39.94[a]	0.98
Time (publication year)	0.063	15.73[a]	0.84	0.083	20.47[a]	0.93
Journal effect (1971)	1.452	7.79[a]	0.61	0.732	4.47[a]	0.48
Time × journal effect	**0.044**	**5.56[a]**	**0.48**	**0.004**	**0.51**	**0.06**

Note. Effects of interest appear in boldface.
[a] $p < 0.001$.

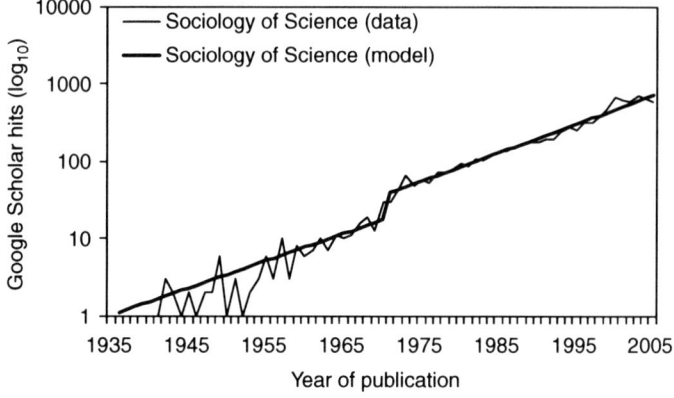

Figure 20.3   *Interrupted time-series analysis results for Google Scholar hits (Log$_{10}$ Scale) for "Sociology of Science" as a function of publication year (1936–2005) and journal establishment (1971).*

history, and psychology of science were all growing at roughly the same rate, at least in the log metric. The sociology of science had a notably steeper (more positive) change-over-time slope than the other three metasciences. It could be that, comparatively speaking, the sociology of science has showed more growth over the last 60 years that the other three metasciences.

Discussion

Overall, Prediction 1 was supported by the data. For the philosophy, history, and sociology of science, establishing a dedicated journal was related to either an immediate mean increase in hit counts of publications and

Table 20.5   *Log Google Scholar Hits as a Function of Publication Year (1950–2005) for Each of the Four Metasciences*

Variable	b	t	R^2
Philosophy of science	0.065	54.79[a]	0.98
History of science	0.065	41.20[a]	0.97
Sociology of science	0.106	40.01[a]	0.97
Psychology of science	0.076	22.63[a]	0.90

[a] $p < 0.001$.

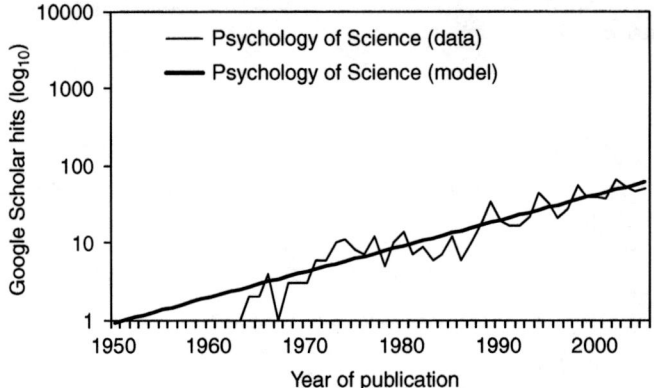

Figure 20.4   *Linear regression analysis results for Google Scholar hits (Log$_{10}$ Scale) for "Psychology of Science" as a function of publication year (1950–2005).*

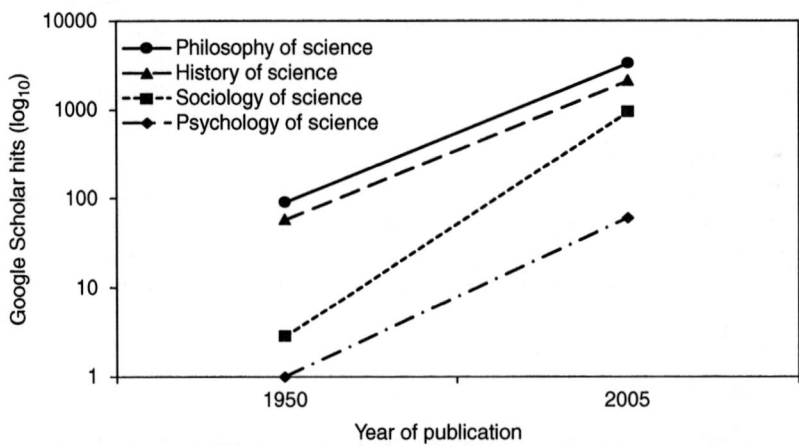

Figure 20.5   *Comparative linear regression analyses for Google Scholar hits (Log$_{10}$ Scale) for all four metasciences as a function of publication year (1950–2005).*

citations (intercept shift), or an immediate change-over-time increase in the same (slope shift). In contrast to establishing a journal, establishing a regular scholarly meeting or conference was not reliably related to hit counts in any consistent way. In other words, holding a dedicated, regular meeting in a metascience appeared to do little with respect to its research output.

Implications. What are the implications of these findings for the metasciences and the evolution of science more generally? Establishing a journal appeared to be important in accelerating the development of scholarly works in each of the metasciences. Although this research suggested that establishing a journal may be more effective than establishing a regular conference, a dedicated journal may not be the only way to stimulate growth in a new science. For example, using a series of interrupted time-series analyses, Webster (2007d) showed that the scientific fields of sociobiology and evolutionary psychology had significant intercept shifts following the publication of a seminal book (E. O. Wilson's [1975] *Sociobiology*) or book chapter (Tooby & Cosmides, 1992) in their respective fields.

Overall, these results are generally consistent with Kuhn's (1962) perspective on scientific revolutions. That is, although scientific progress may appear to be gradual, it can make rapid advances on the basis of a single, influential event (a new journal) or work (a widely cited book). In other words, the seemingly linear growth of scientific knowledge may actually be dynamic and punctuated with disproportionately influential paradigm shifts. The works and events that are true game-changers appear to be those that synthesize disparate perspectives, reinterpret old perspectives in a novel way, or approach a problem from a completely new perspective.

The implications of this research for the psychology of science are generally encouraging, but they should nevertheless be interpreted with caution, primarily because it is such a new field. While it appears true that scholarly works referring to the psychology of science are increasing over time, it remains unclear how recently developments such as a dedicated journal and a regular conference will affect this. According to Feist's (2006a) classification, the psychology of science appears to be straddling the identification and institutionalization stages. Assuming the psychology of science is similar to its three older "sibling" metasciences, we can expect its influence to grow for at least the immediate future. Works such as this handbook, influential books by Feist (2006a) and Simonton (1988), and an edited special issue of the *Review of General Psychology* (Feist, 2006b) have helped lay the necessary groundwork for the psychology of science to emerge as the fourth metascience.

Limitations. Can we be certain that establishing a dedicated journal caused a boost in increased growth in the three metasciences? No. Interrupted time-series designs cannot rule out possible "third variable" effects. That is, there could be an unknown, unmeasured variable or event that is confounded with time, hit counts, or the events of interest (i.e., establishing a journal or meeting). For example, a "bottom-up" groundswell of interest in a given area of science may contribute to both (a) the appearance of an influential book or new journal and (b) a sudden spike in publications and citations to that particular topic or scientific field. In other words, a third variable—in this case increased interest—could be at least partially responsible for the observed relationship between founding a journal and increased notoriety in the metasciences. If such an event were to occur, then the new journal or work that is fortunate to appear at just the right time by chance will likely reap the reward of being dubbed a groundbreaking achievement, regardless of whether or not it deserves such accolades.

Nevertheless, interrupted time-series designs can show both correlation and temporal precedence, which are both necessary—but not sufficient—for establishing causation (see Kenny, 2004). To make a stronger causal argument, one would have to have two equal worlds and manipulate history in one, but not the other. Despite these limitations, it is fairly remarkable that establishing a journal was related to an increase in a metascience's average number of hit counts—or changes in hit counts over time. In a broader sense, these findings suggest that establishing a dedicated journal in a new field—at least in the metasciences—may be the key tipping point that moves an emerging scientific discipline across the threshold from the identification stage to the institutionalization stage.

A possible shortcoming of this research is that Google Scholar may not be an exact tool for measuring the expansion of scientific knowledge in a given field through its publication and citation counts. When I originally conducted my study (Webster, 2008), Google Scholar—the search engine I used—was still using its preliminary "beta" version. For example, more recent articles (e.g., 1990s) are more likely to be indexed than much older ones (e.g., 1930s). Moreover, the most recent articles (those that have been published in the last few months) may not have yet been indexed due to a lag in indexing. For example, a search for works published in 2010 would likely yield more hits in January of 2012 than it would in January of 2011. In addition, although the Internet is a great equalizer of information availability, high-impact journals and other popular scholarly works are still more likely to be indexed than low-impact journals and less-popular scholarly works. Nevertheless, these indexing biases are not a grave concern to the present findings because they are more focused on year-to-year change. Any systematic bias that appears in one year is likely to be similar to any systematic bias that appears in another year.

To be sure, the establishment of a journal in each of the metascience does not constitute a scientific revolution in the Kuhnian sense, but a first journal in a new field is likely to have a tremendous impact, whereas the 101st journal in an established field is less likely to have a dramatic effect. Perhaps a better model for the creation and institutionalization of new scientific fields might be one based on emerging ties and shared language among small groups of researchers (i.e., "trading zones"; Collins, Evans, & Gorman, 2007). These trading zones are often dynamic, forming at the borders or intersections of two or more related disciplines, and, over time, can form their own discipline with their own language or scientific jargon (e.g., biochemistry's emergence from the intersection of biology and chemistry; Collins et al., p. 658). New scientific fields may also emerge from the needs of a growing community with similar research interests that feels its work is not being published or recognized by existing scientific journals or societies. Consequently, this nascent community or trading zone begins to understand it must identify itself and form its own, journals, societies, and programs of study.

Note, however, that scientific progress may be scale invariant. That is, it may have a fractal structure that reflects symmetry of scale (see Collins et al., 2007) in much the same way that market prices do over time. For example, when viewing graphs of time-series data of the S&P 500 stock market index without information on the x or y axes, it is often impossible to tell whether the time series reflects an hour, a day, a week, a month, a year, or a decade of data. Thus, the definition of which scientific works or events represent truly revolutionary ones or the emergence of new scientific fields will likely have to be defined in a relative temporal sense and will also likely have to be retrospectively with the benefit of hindsight.

Recent Publication Trends in the Psychology of Science

For this chapter, I wanted to give a brief update on the recent changes in Google Scholar hit counts for the psychology of science. To do this, I again performed searches for "psychology of science" each year from 2000 to 2009. Figure 20.6 shows the most recent changes over the last decade in raw counts (rather than log-transformed counts). In this analysis, publications and citation relating to psychology of science grew up from 66 in 2000 to 189 in 2009, a nearly three-fold increase (286%). Although the change over the last few *years* has been largely linear ($R^2 = 0.75$), on a longer time scale, change over the last few *decades* has been largely log-linear, which is consistent with the other metasciences.

NEW DIRECTIONS FOR THE PSYCHOLOGY OF SCIENCE

Although my list is neither exclusive nor exhaustive, below I present four potentially promising avenues for future research for the psychology of science. I briefly summarize each of these possible new directions and evaluate their implications.

Psychology of Science Through the Lens of Scientometrics

How can we use scientometrics to glean insights into scientific trends? In much the same way that people's personalities can be gleaned indirectly from viewing the residual behavioral cues from the physical spaces they inhabit (e.g., office cubicles, dorm rooms; see Gosling 2008; Gosling, Ko, Mannarelli, & Morris, 2002), we can make informed inferences about the psychology of science—and the psychology of scientists—by examining scientometrics. For example, as shown above, I have used scientometrics to show the growth of the metasciences and the impact of establishing a dedicated journal (Webster, 2008). Scientometrics can also be harnessed to make comparisons between scientists or even within scientists over time. For example, all else being equal, a scientist with substantially more publications than another might be inferred to be more ambitious than another. Similarly, we may be able to distinguish between extrinsically and intrinsically motivated scientists by examining their research productivity rates before and after tenure (e.g., Duffy et al., 2011). These are merely some of the many possible avenues that scientometrics can offer to a growing psychology of science.

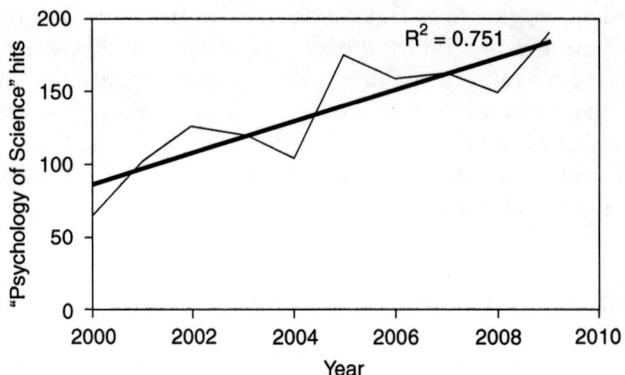

Figure 20.6   *Linear regression analysis results for Google Scholar hits for "Psychology of Science" as a function of publication year (2000–2009).*

Do the Rich Get Richer?

Why do the majority of scientific publications and citations tend to be attributed to only a handful of influential individuals (i.e., why is there a Matthew Effect in science; see Merton, 1968)? Why do scientific productivity (publications) and popularity (citations) tend to follow the same underlying frequency distribution as artists and writers and the work they produce, where a few noteworthy works by a few noteworthy individuals dominate the landscape, while others, and their work, remain relatively obscure? A fair amount of research has examined this phenomenon know as the Matthew Effect or the "rich get richer" effect (Merton, 1968). One possible reason for this variation is that scientists—like musicians and other artists whose popularity also roughly follows an 80–20 Pareto distribution—write scientific papers and books as a costly signal of intelligence or creativity to attract and keep mates while out-publishing same-sex competitors (Kanazawa, 2000; Miller, 2000). Such explanations can be empirically explored by using scientometrics.

Out With the Old, in With the New?

Why do researchers tend to value newer science over older science (i.e., why are newer works typically cited in lieu of older works)? Another question that the psychology of science may seek to answer is why researchers tend to value newer research over older research. To be sure, newer research may be more accurate and may use newer methods, but sometimes scientists "reinvent the wheel" and rediscover findings that had been discovered previously. Thus, there appears to be a strong recency bias in the works that scientists choose to cite. Scientometric analyses bare this out. Nearly all scientific articles show the same pattern in the references they cite; they tend to be extremely positively skewed regarding the age of the works they cite. That is, scientists are far more likely to cite works published during the prior decade than they are they decade before that, and so on. This pattern appears to remain constant over time, and suggests that, for better or worse, people have a tendency to favor newer knowledge over older knowledge. Admittedly, some research becomes so well established that it becomes part of common knowledge, and is thus no longer cited; this may account for some of recency bias in citations. Nevertheless, this pattern poses a temporal paradox: If we agree that current knowledge is in some way more valuable than prior knowledge, must we not also discount our present knowledge knowing that it will likely be replaced by even "better" knowledge in the future? This is just one of many questions facing the psychology of science that can also be partially addressed with scientometrics.

One possible explanation of this pattern of recency bias in citations is that the authors of newer works are more likely to be alive than the authors of older works, and scientists—being human—are more likely to cite their friends, colleagues, and collaborators, as well as other people who have cited them in the past. Thus, there is a distinct social psychological component that is related to who cites whom. Indeed, "invisible colleges"—again, latent networks of scientists who work on the same problem(s)—are often revealed by citation networks, which tend to resemble social networks (Quiñoes-Vidal et al., 2004).

Interestingly, in many scientific fields, the number of references an article contains is often an excellent predictor of the number of citations it will later receive (Price, 1965; Webster, 2010; Webster, Jonason, & Schember, 2009). Specifically, positive relationships between number of references and number of later citations have been observed in the journal *Science* (1901–2000), the *Journal of Clinical and Counseling Psychology* (1968–2000), and *Evolution and Human Behavior* (1979–2000; Webster, 2010; Webster, Jonason, & Schember, 2009). Future research should consider exploring whether the references-citations relationship will change over time as the number of references increases in psychology (Adair & Vohra, 2003) and articles grow longer (Webster, 2007e). Clearly, this is an area of interest that appears ripe for psychologists of science to study.

Information Overload?

How will scientists cope with having too much information after having centuries of too little information? Another pressing issue for the psychology of science is how scientific minds are adapting to a world with too much information after a human history of too little information. Given advances in data collection, storage, distribution, and access, scientists are increasingly shifting from questions like "How do I find and collect some data?" to "How do I manage and analyze all these data?" In addition, advances in computational computing have led to relative ease of analyses. The same complex analyses that may have taken a few hundred punch cards and a reserved hour on a room-sized university mainframe computer 40 years ago can now be performed on a laptop computer in milliseconds. Over the last few decades, this may have led scientists to be more liberal in their hypothesis testing because the cost of running an analysis in terms of time and effort has dramatically diminished. Thus, scientists can now go on data analytic "fishing expeditions," where they search, and search, and search again until they find a potentially publishable result. For example, my own research on publication trends in the metasciences (Webster, 2008) that is summarized above clearly benefited from this tactic when I noticed that testing the time window of 1900 to 2005 was inappropriate for all metasciences, given that they emerged at different times.

SUMMARY

What does this mean for the future of the psychology of science, as well as the evolution of science and knowledge more generally? First, although the psychology of science has been on the rise, it faces a critical juncture where having a new, dedicated, and stable journal will likely be necessary for continued to growth, especially in terms of publications and citations. Without such a journal, the psychology of science's growth will likely be stymied. Second, we have learned that the metasciences appear to benefit more from the establishment of dedicated journals than they do from the establishment of regular meetings or conferences. Whether or not the same could be true for other scientific disciplines, however, remains an open question. Third, scientometrics—many of which are available over the Web, especially via university library subscriptions—can allow researchers to explore some fundamental questions pertaining to the psychology of science. It is hoped that this chapter will provide the reader with not only some insight into the recent growth of the psychology of science, but also the methods necessary to conduct further research in this exciting emerging metascience.

AUTHOR NOTE

Much of this research was originally published as an article (Webster, 2008) and presented at the 19th annual convention of the Association for Psychological Science, Washington, DC, May 2007. This material was based upon work supported by the National Institute of Mental Health under Training Grant Award PHS 2 T32 MH014257 entitled "Quantitative Methods for Behavioral Research" (to M. Regenwetter, principal investigator). Some of this research was performed while the author was a postdoctoral trainee in the Quantitative Methods Program of the Department of Psychology, University of Illinois at Urbana-Champaign. Any opinions, findings, and conclusions or recommendations expressed in this chapter are those of the author and do not necessarily reflect the views of the National Institute of Mental Health.

REFERENCES

Adair, J. G., & Vohra, N. (2003). The explosion of knowledge, references, and citations: Psychology's unique response to a crisis. *American Psychologist, 58*, 15–23.

Barabási, A.-L. (2003). *Linked: How everything is connected to everything else and what it means for business, sciences, and everyday life.* New York, NY: Plume.

Collins, H., Evans, R., & Gorman, M. (2007). Trading zones and interactional expertise. *Studies in the History and Philosophy of Science, 38,* 657–666.

Duffy, R. D., Jadidian, A., Webster, G. D., & Sandell, K. J. (2011). The research productivity of academic psychologists: Assessment, trends, and best practice recommendations. *Scientometrics, 89,* 207–227.

Feist, G. J. (2006a). *The psychology of science and the origins of the scientific mind.* New Haven, CT: Yale University Press.

Feist, G. J. (Ed.). (2006b). The psychology of science [Special issue]. *Review of General Psychology, 10*(2).

Feyerabend, P. (1975). *Against method: Outline of an anarchistic theory of knowledge.* London, UK: New Left Books.

Gosling, S. D. (2008). *Snoop: What your stuff says about you.* New York, NY: Basic Books.

Gosling, S. D., Ko, S. J., Mannarelli, T., & Morris, M. E. (2002). A room with a cue: Judgments of personality based on offices and bedrooms. *Journal of Personality and Social Psychology, 82,* 379–398.

Kanazawa, S. (2000). Scientific discoveries as cultural displays: A further test of Miller's courtship model. *Evolution and Human Behavior, 21,* 317–321.

Kenny, D. A. (2004). *Correlation and causality* (Rev. ed.). Retrieved August 9, 2005, from http://davidakenny.net/doc/cc_v1.pdf

Kuhn, T. S. (1962). *The structure of scientific revolutions.* Chicago, IL: University of Chicago Press.

Kuhn, T. S. (1970). Logic of discovery or psychology of research? Reflections on my critics. In I. Lakotos & A. Musgrave (Eds.), *Criticism and the growth of knowledge* (pp. 1–23, 231–278). London, UK: Cambridge University Press.

McClelland, G. H. (2000). Nasty data: Unruly, ill-mannered observations can ruin your analysis. In H. T. Reis & C. M. Judd (Eds.), *Handbook of research methods in social and personality psychology* (pp. 393–411). New York, NY: Cambridge University Press.

Merton, R. K. (1968). The Matthew effect in science. *Science, 159,* 56–63.

Miller, G. (2000). *The mating mind: How sexual choice shaped the evolution of human nature.* New York, NY: Random House.

Popper, K. (1999). *All life is problem solving.* New York, NY: Routledge.

Popper, K. (2002). *The logic of scientific discovery.* London, UK: Routledge. [Original work published 1935].

Price, D. (1965). Network of scientific papers. *Science, 149,* 510–515.

Quiñones-Vidal, E., López-García, J., Peñaranda-Ortega, M., & Tortosa-Gil, F. (2004). The nature of social and personality psychology as reflected in *JPSP,* 1965–2000. *Journal of Personality and Social Psychology, 86*(3), 435–452.

Roese, N. (2005). *If only: How to turn regret into opportunity.* New York, NY: Broadway Books.

Shadish, W. R., Cook, T. D., & Campbell, D. T. (2002). *Experimental and quasi-experimental designs for generalized causal inference.* Boston, MA: Houghton-Mifflin.

Simonton, D. K. (1988). *Scientific genius: A psychology of science.* Cambridge, UK: Cambridge University Press.

Taleb, N. N. (2005). *Fooled by randomness: The hidden role of chance in life and in the markets.* New York, NY: Random House.

Taleb, N. N. (2010). *The black swan: The impact of the highly improbable* (2nd ed.). New York, NY: Random House.

Tooby, J., & Cosmides, L. (1992). The psychological foundations of culture. In J. H. Barkow, L. Cosmides, & J. Tooby (Eds.), *The adapted mind: Evolutionary psychology and the generation of culture* (pp. 19–136). New York, NY: Oxford University Press.

Webster, G. D. (2007a). Evolutionary theory in cognitive neuroscience: A 20-year quantitative review of publication trends. *Evolutionary Psychology, 5,* 520–530.

Webster, G. D. (2007b). Evolutionary theory's increasing role in personality and social psychology. *Evolutionary Psychology, 5,* 84–91.

Webster, G. D. (2007c, Winter). Increasing impact, diversity, and empiricism: The evolution of *Evolution and Human Behavior,* 1980–2004. *HBES Winter 2007 Newsletter,* pp. 16–19.

Webster, G. D. (2007d). What's in a name: Is "evolutionary psychology" eclipsing "sociobiology" in the scientific literature? *Evolutionary Psychology, 5,* 683–695.

Webster, G. D. (2007e). The demise of the increasingly protracted APA journal article? *American Psychologist, 62,* 255–257.

Webster, G. D. (2008). An emerging psychology of science: A quantitative review of publication trends in the metasciences. *Journal of Psychology of Science and Technology, 1,* 6–14.

Webster, G. D. (2010, August). *Scientists who cite more are cited more: Evidence from over 50,000 Science articles.* Talk given at the 3rd biennial conference of the International Society for the Psychology of Science and Technology, Berkeley, CA.

Webster, G. D., Jonason, P. K., & Schember, T. O. (2009). Hot topics and popular papers in evolutionary psychology: Analyses of title words and citation counts in *Evolution and Human Behavior,* 1979–2008. *Evolutionary Psychology, 7,* 348–362.

Wilson, E. O. (1975). *Sociobiology: The new synthesis.* Cambridge, MA: Belknap Press of Harvard University Press.

Wuchty, S., Jones, B. F., & Uzzi, B. (2007). The increasing dominance of teams in production of knowledge. *Science, 316,* 1036–1039.

CHAPTER 21

Conclusions and the Future of the Psychology of Science

Michael E. Gorman and Gregory J. Feist

The psychology of science has made some important first steps toward becoming an established discipline. This handbook is an affirmation of the movement toward an established discipline. Indeed, handbooks are an important sign that a field has matured enough to warrant a definitive summary and statement of the current state of the art for a discipline. In addition to explicating the field's historical trends, the authors of this handbook have made the case that many of the main domains of psychology have turned an empirical and conceptual eye toward scientific thought and behavior and have important things to say about the development, cognitive process, personality, and social forces that shape scientific thinking and behavior. Indeed, the developments in scientific thinking and theory change are large enough to require their own section. Special topics of the psychology of science include the following:

- Creativity and genius in science
- Gender and science
- Conflict and cooperation among research groups
- Critique of the postmodern denial of scientific epistemology
- Psychobiographical account of scientific theorizing

As stimulating as the foundational and special topics are, in the end the psychology of science also has to offer a practical guide to action for students, scientists, educators, entrepreneurs, and policy makers. The section on the applications of psychology of science offers such guidance. In the final section of the book, we provide a quantitative look backward and a speculative look forward to see where the field has been and where it might be moving.

We have only just begun to take our first steps toward becoming a full-fledged discipline. It is clear that there is as much latent as actual interest in the psychology of science. In other words, there are many psychologists and psychologically oriented scholars who are doing psychology of science but only implicitly. Their identification with the field is not yet explicit. They may not even know the field and society exists.

507

The question we now need to be asking ourselves is "What can we do to ensure that the next generation wants to carry the tradition forward and that scholars start to identify themselves as psychologists of science?" (Feist & Gorman, 2009). If a discipline is to survive and move forward, it needs to excite the interest of young scholars and in this sense the psychology of science is no different from any other discipline and, indeed, it stands to gain from learning how the other studies of science—philosophy, history, and psychology—made this transition. As we recently editorialized, there are at least three major activities that the current generation of psychologists of science need to do to ensure the next generation inherits an active and healthy discipline: Teach exciting courses, carry out and publish exciting research, and reward student excellence (Feist & Gorman, 2009). The first two activities concern getting the next generation into the field, whereas the last concerns keeping them in the field.

TEACH EXCITING COURSES IN THE PSYCHOLOGY OF SCIENCE

As academics, we teach "old" knowledge in the classroom and create "new" knowledge in the laboratory. Ideally, good teachers are also good researchers. Granted some of us tend to be more research oriented and others more teaching oriented. Sometimes this happens by choice and other times because the institutions we are in dictate it. But an ideal thing in higher education is that those who create knowledge should also pass on that knowledge to the next generation by teaching, as well as publishing.

One principle of teaching is that great teachers excite. Typically, they excite students by being excited themselves. As Richard Feynman apparently once said: "Teaching is like sex: If you're not enjoying it, you're not doing it right!" As instructors, we may sometimes forget to focus on enjoyment. Good teaching is exciting and enjoyable for both teacher and student. A second principle is "steal good ideas from others." In contrast to research and business, education is one domain in life where stealing from (really simply being inspired by) others is both acceptable and even encouraged. Teachers should be constantly open to new ideas in the classroom and talk to their colleagues and other instructors. When they hear about something other teachers do that sounds interesting they should not be shy about trying it. No one owns patents or copyrights on good teaching techniques. The more widely disseminated good teaching is, the better.

In the spirit of sharing good teaching ideas, we recently culled from International Society for the Psychology of Science and Technology (ISPST) members examples of syllabi for courses on the psychology of science or some related course. We received a number of syllabi, many of which we have posted on the ISPST website (http://www.psychologyofscience.org). These are just a few of the interesting and creative topics and techniques being used to excite a student's interest in the psychology of science. We can only hope that others will be inspired to incorporate some of these ideas and techniques. If you are not currently teaching a course in the psychology of science, perhaps these course syllabi and ideas will inspire you to propose such a course to your department.

CARRY OUT AND PUBLISH EXCITING RESEARCH IN THE PSYCHOLOGY OF SCIENCE

In the end, teaching depends on research. In science, if a field does not advance, there won't be anything to teach. The purpose behind publishing after all is to not only publish the best and most interesting research in the psychology of science but to encourage it to happen. Research needs not only an outlet for publication once it is completed, it also needs support and facilities to be carried out.

If we succeed as teachers and researchers in inspiring young people to join us and commit themselves to a career as psychologists of science, then they need to be trained and supported. At the Berlin conference in July 2008, Mike Gorman gave voice to the need to develop a collaborative training center so that aspiring PhDs can be trained. Instead of a single graduate program, his idea was to create a graduate program distributed among multiple institutions. Only a few programs might be able to generate an actual degree; distance learning technologies have evolved to the point where psychology of science students at one institution could have learning, research, and mentoring experience with faculty at other institutions. In that way, we could leverage the depth and breadth of the field—and one or two institutions could step forward to offer a degree without having to hire a half-dozen faculty, or provide all the research space. A distributed institutional network also increases the base of contacts a student can take advantage of when looking for a job.

Relatedly, psychologists of science need to obtain funding for their research. Currently, the atmosphere at the federal level is increasingly cognizant of and receptive to psychology of science research. This is particularly true at the National Science Foundation (NSF). For example, the NSF has a "Science of Science and Innovation Policy" (SciSIP) program that funds research geared toward developing knowledge, theories, and human capital in the science of science studies. There is no reason why psychology cannot be one of the major players in these grant proposals.

In addition, one of the core leaders of the psychology of science movement, Michael Gorman, served 2 years as one of the three Program Directors in Science, Technology, and Society (STS) at the NSF. This program "considers proposals that examine historical, philosophical, and sociological questions that arise in connection with science, engineering, and technology, and their respective interactions with society" (STS, n.d.). Although the psychology of science is not explicitly mentioned on the STS NSF website, Gorman was the first NSF program director who identified explicitly as a psychologist of science and, therefore, actively encouraged proposals from psychologists.

REWARD STUDENT EXCELLENCE IN THE PSYCHOLOGY OF SCIENCE

The other incentive that came out of the Berlin conference was that young aspiring psychologists of science not only need support for their research, but they also need to be reinforced for it. Prizes, especially those that involve money and a guaranteed publication, are rewards and reinforcers for scholars, young and old. The idea, therefore, was floated to have the society support a "Best Student Paper Award." The first such award, named in honor of the late David C. Gooding, was bestowed to Jordan Lippman at the 2010 ISPST conference in Berkeley and the second one was awarded to Matthew Lira at the 2012 ISPST conference in Pittsburgh. Of course, reward and reinforcement need not be in the form of a prize. Praise and encouragement by senior faculty of high-quality research and thinking can go a long way.

Finally, the psychology of science is a unique discipline within psychological science. It is one of the few subdisciplines that actually integrates all of the major subdisciplines in psychology, namely cognitive, developmental, social, biological-neuroscience, personality, and even clinical psychologies. The need for psychological science to develop a unifying perspective is as urgent now as it has ever been (Staats, 1999). Not that psychology of science will unify all of psychology, but in fact it offers the possibility of demonstrating how one topic—scientific thought and behavior—requires each and every major subdiscipline within the field of psychology. Indeed, Division 1 of the American Psychological Association (APA), which is devoted to general psychology, has been quite receptive to the psychology of science. In 2006, an entire special issue of Division 1's flagship journal, *Review of General Psychology*, was devoted to the psychology of science. The following year, the division awarded the William James Book Award to Feist's book, *The Psychology of Science and the Origins of the Scientific Mind*. More recently, president-elect of the division, Dean Simonton, has argued he would uphold psychology of science as an exemplar of general psychology during his upcoming presidency.

The psychology of science has begun its journey toward being an established and contributing member to the studies of science and to the rest of psychology (see Chapter 20, this volume).

What would it take to establish psychology as a recognized academic field?

1. Psychology of science must distinguish itself from existing fields. To do psychology of science, one need not be a psychologist. For example, members of ISPST and those who do studies of experience and expertise (SEE) have held a joint conference and plan another. The SEE group includes at least one psychologist (Gorman) along with sociologists, philosophers, historians, and cognitive scientists who want to understand the nature of scientific expertise and its role in policy. The late David Gooding, a distinguished philosopher of science, was the Secretary of ISPST. So psychology of science will differentiate itself by its focus on the psychological aspects of science and technology, including all of the topics in this volume

2. To be successful, psychology of science must also complement existing fields. The papers in this volume demonstrate that psychology of science occupies a unique space at the intersection of the other metascientific disciplines. This volume cites work from philosophy, history, sociology, and anthropology of science, and technology, adapting methods and concepts

from these fields to issues like the development of scientific thinking. For psychologists, this new field provides an opportunity to apply the best work in psychology to a transformative set of human activities

3. Psychology of science will not only adapt the methods of existing fields but it must develop new ones, or at least novel combinations, in order to establish the need for this new research community. So, for example, psychologists of science have developed unique ways of mining and making sense of data from historical records like scientific notebooks, letters, publications, and so on. At the end of this conclusion, we will make suggestions for future research that involve methodological and conceptual innovations

4. Mobilization of the latent psychology of science community is critical, involving activities like creating a society, founding a journal, and finding sources of funding for research. The ISPST is a good start, with over a hundred members and three conferences under its belt. The NSF funded a workshop that linked ISPST and SEE and also provides funding for students to go to ISPST meetings, showing that it is possible to seek NSF funding for new efforts in psychology of science. The *Journal of Psychology of Science and Technology* ran for 2 years, 2008 to 2010 published by Springer Publishing. The journal will have to be revived as a multidisciplinary refereed outlet for scholars doing psychology of science research and for articles on new courses, graduate initiatives, and other matters related to the growth of the field

5. Psychologists of science should produce exciting undergraduate courses that inspire student interest in psychology of science. As academics, we teach "old" knowledge in the classroom and create "new" knowledge in the lab. But courses on the frontier of a new field can allow exploration of issues that cut across multiple research programs and fields. So innovative undergraduate courses can catalyze the kind of thinking that leads to new research—and provide the generation of scholars that will do it. ISPST members continue to provide examples of syllabi for courses on the psychology of science that are posted on the ISPST website (www. psychologyofscience.org)

6. Psychology of science must also have graduate programs or at least degrees in order to create a new generation of scholars. At the Berlin conference in July 2008, Mike Gorman gave voice to the need to develop a collaborative training center so that aspiring PhDs can be trained. Instead of a single graduate program, his idea was to create a graduate program distributed among multiple institutions. Only a few programs might be able to generate an actual degree, but distance learning technologies have evolved to the point where psychology of science students at one institution could have learning, research, and mentoring experience with faculty at other institutions. In that way, we could leverage the depth and breadth of the field—and one or two institutions could step forward to offer a degree without having to hire a half-dozen faculty, or provide all the research space. A distributed institutional network also increases the base of contacts a student can take advantage of when looking for a job

7. Psychology of science must show it can contribute to the solution of problems that existing fields cannot sufficiently address. Psychology is in a

position to provide unique perspectives on discovery, invention, and innovation, and new perspectives to study these issues. One end result will be ways of improving education, policy, and practice

SUGGESTIONS FOR FUTURE RESEARCH

The *Handbook of the Psychology of Science* has made a cogent case for where the field has been and where it is. The future, however, has mostly been left out. Various authors have hinted at gaps and where the field might go in the future, but we want to end with our own thoughts and suggestions for the future of the field. First, we suggest missing content and then offer our ideas about missing and underrepresented techniques and methods from the field.

There are topics and content that are currently missing or nearly missing in a fully developed psychology of science, namely a clinical psychology of science, a psychology of science that focuses on scientific misconduct and ethics in science, and a psychology of scientific interdisciplinarity. Few studies have explicitly looked at mental health issues among the scientific and technological professions. Moreover, the whole topic of what happens *after* scientific research has been carried out but before, during, and after its publication is ripe with ethical and moral issues. Scientific misconduct and fraud are more common than we might think. One sign of this is the significant increase recently in the number of retractions that scientific journals have had to publish (Zimmer, 2012). Relatedly, what can psychologists offer to explain disputes among scientists over the primacy of discovery or any particular person's contribution to the discovery? The history of science is replete with such ego-based battles, and psychology more than any other discipline should be able to offer insight into the motivation behind this less-than-ideal scientific thought and behavior.

Another topic that is ripe for exploration is collaborations among scientists and engineers at various scales, from small groups and teams to distributed multinational collaborations like European Organization for Nuclear Research (CERN). An NSF supported workshop on Interdisciplinary Collaboration in Innovative Science and Engineering Fields resulted in a report that includes useful suggestions for future work (http://csid.unt.edu/nsf/nsf-workshop-report.pdf). Understanding the psychological factors that predict successful interdisciplinary collaboration would be useful; for example, the relative roles of leadership, team composition, individual personalities, group dynamics, and organizational constraints. These factors could be related to measures of productivity and innovation.

It would be useful to have a taxonomy of different kinds of collaborations, in part because different factors might promote success. A first step in this direction is work by SEE scholars on collaborations that stretch across apparently incommensurable barriers. These collaborations can begin with relatively simple trading zones, where participants exchange "goods" like resources and expertise, to ones where the participants are codiscovering or coinventing (Gorman, 2010). This coevolution process requires the development of interactional expertise among the participants, which is the ability to understand each other's languages and concepts at a level sufficient for joint decisions about research strategy and the implications of results. Psychologists of science cannot only study the process by which discoveries, inventions, and innovations occur, they can also facilitate processes like more effective collaboration and the recruitment of underrepresented groups into science, technology, engineering, and mathematics (STEM) disciplines.

Among the techniques that are missing or nearly missing in a fully developed psychology of science are neuroscience, behavioral genetics, and longitudinal growth modeling. These research methods need to become more common if we are to really understand the psychology of the scientist and scientific thinking and behavior. Brain research has brought a revolution to the study of human thought and behavior in general, and yet scientific thought and behavior has mostly been left out of the revolution. The recent work by Jonathan Fugelsang and Kevin Dunbar has been a notable exception to this trend (Dunbar, Fugelsang, & Stein, 2007; Fugelsang & Dunbar, 2004).

In addition, as Simonton's (2008) recent article on the genetics of scientific talent makes clear, there is also a lack of behavioral genetic evidence on scientific thought and talent. With more and more research using twin-adoption paradigm to tease apart heritable and nonheritable contributions to thought and behavior, one can hope and assume that this method will be applied more frequently in the future to scientific thought and behavior to tease apart the relative contribution of the nature and nurture of scientific talent, creativity, and achievement.

Work begun by Julian Stanley in the 1970s and continued by David Lubinski and Camilla Benbow on a 35-year longitudinal investigation into mathematical and scientific talent is a benchmark for the kind of research that is needed to fully understand how scientific interest, aptitude, talent, and achievement evolve from childhood to adolescence and into young, middle, and late adulthood (Park, Lubinski, & Benbow, 2008; Wai, Lubinski, & Benbow, 2009). This investigation notwithstanding, there is an obvious lack of longitudinal studies focusing on the development of scientific interest, reasoning, talent, and achievement. Relatively new and sophisticated statistical techniques such as structural modeling and growth curve modeling can answer longitudinal and causal questions in ways that older statistical techniques could not. Therefore, the field would benefit from more researchers in the future focusing on how the nature and nurture of scientific thought and behavior develop and grow, from childhood to adulthood. Developmental psychologists of science are in a unique position to provide such important programs of research.

Longitudinal and cross-sectional methods could also be applied to detailed studies of the career trajectories of scientists and engineers in different fields. Their personalities, aptitudes, and even genetic predispositions could be initially assessed using measures described in this volume. Then a variety of measures could be used to track volunteers over time, including interviews, occasional observations and self-observations recorded in diaries (Shrager, 2005). In the cross-sectional case, these measures could be used on participants at several stages of their careers; for example, post-doctorate, assistant professor, and full professor in the case of academic scientists and engineers. This kind of monumental study should be piloted first on smaller samples, and would involve multiple investigators with complementary backgrounds in psychology of science. It would also help us understand how scientific and engineering expertise is acquired and deployed in various domains, whether and how motivations change over time, what strategies participants use to advance their careers, when and how they serve as gatekeepers, and so on.

In summary, the psychology of science is a nascent but exciting new addition to the studies of science. Psychologists have a lot to say about the nature and nurture of scientific thought and behavior, and if this handbook contributes to the continued growth and health of the discipline, then all the hard work and effort will have been well worth it.

REFERENCES

Dunbar, K., Fugelsang, J., & Stein, C. (2007). Do naïve theories ever go away? Using brain and behavior to understand changes in concepts. In M.C. Lovett & P. Shah (Eds.), *Thinking with data* (pp. 193–205). Mahwah, NJ: Lawrence Erlbaum Associates Publishers.

Feist, G. J., & Gorman, M. E. (2009). Ensuring a next generation of psychologists of science. *Journal of Psychology of Science and Technology, 2,* 2–4.

Fugelsang, J. A., & Dunbar, K. N. (2004). Brain-based mechanisms underlying complex causal thinking. *Neuropsychologia, 43,* 1204–1213.

Gorman, M. E. (Ed.). (2010). *Trading zones and interactional expertise: Creating new kinds of collaboration.* Cambridge, MA: MIT Press.

Park, G., Lubinski, D., & Benbow, C. P. (2008). Ability differences among people who have commensurate degrees matter for scientific creativity. *Psychological Science, 19,* 957–961.

Shrager, J. (2005). Diary of an insane cell mechanic. In M. E. Gorman, R. D. Tweney, D. C. Gooding, & A. Kincannon (Eds.), *Scientific and technological thinking* (pp. 119–136). Mahwah, NJ: Lawrence Erlbaum Associates.

Simonton, D. (2008). Scientific talent, training, and performance: Intellect, personality, and genetic endowment. *Review of General Psychology, 12*(1), 28–46.

Staats, A. (1999). Unifying psychology requires a new infrastructure, theory, method, and research agenda. *Review of General Psychology, 3,* 3–13.

Wai, J., Lubinski, D., & Benbow, C. P. (2009). Spatial ability for STEM domains: Aligning over 50 years of cumulative psychological knowledge solidifies its importance. *Journal of Educational Psychology, 101,* 817–835.

Zimmer, C. (2012, April 16). A sharp rise in retractions prompts calls for reform. *New York Times, Science.* Retrieved June 12, 2012, from http://www.nytimes.com/2012/04/17/science/rise-in-scientific-journal-retractions-prompts-calls-for-reform.html?_r=1&adxnnl=1&adxnnlx=1334958458-PxivQM3BpvGZR636Xup/Qw&pagewanted=all

Index

Note: Page numbers followed by "*f*" and "*t*" denote figures and tables, respectively.